SPRINGER SERIES ON ENVIRONMENTAL MANAGEMENT

DAVID E. ALEXANDER
Series Editor

Springer

New York
Berlin
Heidelberg
Barcelona
Budapest
Hong Kong
London
Milan
Paris
Santa Clara
Singapore
Tokyo

Springer Series on Environmental Management
David E. Alexander, Series Editor

James M. Coe, Editor
National Marine Fisheries Service
Marine Entanglement Research
Seattle, Washington

Donald B. Rogers, Editor
Alaska Fisheries Science Center
Seattle, Washington

Marine Debris

Sources, Impacts, and Solutions

With 108 illustrations, 4 in full color

 Springer

James M. Coe
National Marine Fisheries Service
Marine Entanglement Research
Seattle, WA 98115-0070
USA

Donald B. Rogers
Alaska Fisheries Science Center
7600 Sand Point Way, NE
Seattle, WA 98115-0070
USA

Series Editor:
David E. Alexander
University of Massachusetts
Department of Geology and Geography
Amherst, MA 01003
USA

Cover photo: A trash-clogged urban tidal creek in the Washington, D.C. area. Photo by E. Durrum.

Library of Congress Cataloging-in-Publication Data

Coe, James M.
    Marine debris : sources, impacts, and solutions / James M. Coe and
Donald B. Rogers.
        p.   cm.—(Springer series in environmental management)
    Includes bibliographical references and index.
    ISBN-13: 978-1-4613-8488-5   e-ISBN-13: 978-1-4613-8486-1
    DOI: 10.1007/978-1-4613-8486-1
        1. Marine pollution—Congresses.   2. Marine debris—Congresses.
    I. Rogers, Donald B.   II. Title.   III. Series.
    GC1081.C64   1996   96-18351
363.73′94′09162—dc20   CIP

Printed on acid-free paper.

Production supervised by Lesley Poliner; manufacturing supervised by Joe Quatela.
Typeset by TechType, Inc., Ramsey, NJ.

9 8 7 6 5 4 3 2 1

ISBN-13: 978-1-4613-8488-5   e-ISBN-13: 978-1-4613-8486-1
DOI: 10.1007/978-1-4613-8486-1

For
Jenny and Diana C.
and
Amanda R.

Series Preface

This series is concerned with humanity's stewardship of the environment, our use of natural resources, and the ways in which we can mitigate environmental hazards and reduce risks. Thus it is concerned with applied ecology in the widest sense of the term, in theory and in practice, and above all in the marriage of sound principles with pragmatic innovation. It focuses on the definition and monitoring of environmental problems and the search for solutions to them at scales that vary from the global to the local according to the scope of analysis. No particular academic discipline dominates the series, for environmental problems are interdisciplinary almost by definition. Hence a wide variety of specialties are represented, from oceanography to economics, sociology to silviculture, toxicology to policy studies.

In the modern world, increasing rates of resource use, population growth and armed conflict have tended to magnify and complicate environmental problems that were already difficult to solve a century ago. Moreover, attempts to modify nature for the benefit of humankind have often had unintended consequences, especially in the disruption of natural equilibria. Yet at the same time human ingenuity has been brought to bear in developing a new range of sophisticated and powerful techniques for solving environmental problems—for example, pollution monitoring, restoration ecology, landscape planning, risk management, and impact assessment. Books in this series will shed light on the problems of the modern environment and contribute to the further development of the solutions. They will contribute to the immense effort by scholars and professionals of all persuasions and in all countries to nurture an environment that is both stable and productive.

David E. Alexander
Amherst, Massachusetts

Preface

This volume was conceived as the definitive treatment of the global marine debris problem. Our desire is to lay the foundation for much wider recognition of this pollution, and to address its management at all social, economic, and political levels.

The papers in this volume were based on invited contributions to the Third International Conference on Marine Debris (May 1994, Miami.). The theme of the conference, "Seeking Global Solutions," was the logical extension of the themes of the two prior international conferences on marine debris. The first, in 1984, considered whether marine debris was a pollutant of consequence. The second, in 1989, was convened to fully analyze all facets of the marine debris problem.

The selected, edited contributions to this book include reports of complex research results, reviews of data and information, legal analyses, and editorial perspectives. Each paper provides definitive information on, or relates key experiences relevant to, an important aspect of the marine debris problem or its solution. We as editors prepared brief introductions for each section highlighting the conceptual foundation and relevance of the subject matter and pointing out challenges important to the topic and to the global-solutions theme. While the papers in this volume offer the first comprehensive treatment of the global marine debris problem, we recognize that much of their content arises from a developed-world perspective, a bias difficult to avoid. We hope that readers with different perspectives will consider the information offered and reach conclusions in light of their own concerns and needs.

The wide range of disciplines involved in the marine debris problem is reflected in the topics in this book, in the diversity of authors (51, from 13 countries), and in the audience we hope to reach. Our intended readership includes marine resource managers and pollution control experts, marine and civil engineers, ship owners and operators, port operators, solid waste managers, coastal zone and urban planners, international bankers, politicians and bureaucrats, sociologists, economists, environmental lawyers, conservationists, and concerned citizens.

James M. Coe
Donald B. Rogers
April 1996

Acknowledgments

This book is built on a foundation of work and dedication by a very large number of people and institutions. The Third International Conference was made possible by the generous sponsorship of the U.S. National Oceanic and Atmospheric Administration's National Marine Fisheries Service, the U.S. Environmental Protection Agency, the U.S. Coast Guard, the U.S. Navy, the U.S. Marine Mammal Commission, the Intergovernmental Oceanographic Commission of UNESCO, the Society of the Plastics Industry, Inc., and the Center for Marine Conservation. The Keynote address was given by Admiral J. William Kime, Commandant of the U.S. Coast Guard. The banquet address was given by Professor Gunnar Kullenberg, Secretary of the Intergovernmental Oceanographic Commission of UNESCO. Mona Bregman of Bregman Associates, Bethesda, Maryland, did a masterful job of coordinating the conference. The Conference Steering Committee included Anders Alm, Dr. George Boehlert, Dr. Bradford Brown, Ronald Bruner, CDR John Clary, Dr. John Davis, Professor Trevor Dixon, LCDR James Farley, Dr. Bernard Griswold, Lawrence Koss, David Laist, Kathryn O'Hara, and David Redford.

The Working Group Chairpersons were responsible for the substance of the Conference. They organized the plenary sessions, identified and invited the authors, and prepared the working group reports. Without their generous assistance, this book would not have been possible. They are Dr. Christine Ribic and Ivan Vining (Amounts, Types and Distribution), Drs. Peter Ryan and Kenneth E. (Ted) McConnell (Impacts), Barbara Wallace (Vessel Sources), Kathryn O'Hara (Recreational Sources), Robert Howard and Christopher McArthur (Urban Sources), and Michael Liffmann (Rural Coastal and Upland Sources). The chairpersons for the plenary sessions were Anders Alm (Amounts, Types and Distribution), Dr. Charles Fowler (Impacts), Captain Dimitros Mitsatsos (Ship Sources), John Brownlee (Recreational Sources), Paul Molinari (Urban Sources), and Usamah Dabbagh (Coastal Rural and Upland Sources).

We wish to acknowledge the Conference authors whose papers could not be included in this volume; their work contributed significantly to the success of the Conference and the quality of the book. Able assistance with the hundreds of details involved in producing this volume was freely given by Pam Moen, Pat Kerlik, Wendy Carlsen (graphics), John Lowell, Kathy Zecca, Susan Calderon, and Gary Duker.

Finally, we wish to thank Dr. William Aron for making it possible for us to undertake this project and Commander John C. Clary III for close support throughout its development.

Contents

PART I
The Status of Marine Debris

SECTION I. Amounts, Types, and Distribution

SECTION II. Biological Impacts

PART II
The Sources and Solutions to the
Marine Debris Dilemma

SECTION III. The Socioeconomics of Marine Debris

SECTION IV. Considering the Maritime
Sources of Debris

Section V. Considering the Land-Based Sources of Debris

Contributors

F.G. (Jerry) Barnett, Regional Consultant, Wider Caribbean International Maritime Organization, Ferguson, MO 63135-2636

H. Arnold Carr, Marine Fisheries Biologist, Massachusetts Division of Marine Fisheries, Sandwich, MA 02563

Ruperto Chaparro, Marine Agent, University of Puerto Rico, Sea Grant Program, Mayguez, PR 00081

James M. Coe, Program Manager, Marine Entanglement Research, National Marine Fisheries Service, Seattle, WA 98115-0070

Emmett Durrum, Environmental Policy Analyst, Department of Public Works, Washington, DC 20009

Jim Ellis, Boat US, Alexandria, VA 22304

Murray R. Gregory, Professor, Department of Geology, University of Auckland, Auckland, New Zealand

Abraham Golik, Professor, Israel Oceanographic & Limnological Research Ltd., Haifa, Israel

Dewayne Hollin, Marine Business Management Specialist, Texas A&M University, Sea Grant, Bryan, TX 77802

James Kirkley, Professor, Virginia Institute of Marine Science, Gloucester Point, VA 23062

Lawrence J. Koss, Head, Ship & Air Systems Branch, Office of the Chief of Naval Operations, Department of the Navy, Washington, DC 20350

David W. Laist, Senior Policy and Program Analyst, U.S. Marine Mammal Commission, Washington, DC 20009

Shirley Laska, Vice Chancellor for Research, University of New Orleans, New Orleans, LA 70148

Michael Liffmann, Assistant Director, Louisiana Sea Grant College Program, Louisiana State University, Baton Rouge, LA 70808

Satsuki Matsumura, Director, Oceanography and Southern Ocean Resources Division, National Research Institute of Far Seas Fisheries, Shimizu, Shizuoka, Japan

Paul J. Molinari, Deputy Director, Water Management Division, U.S. Environmental Protection Agency, New York, NY 10278

Ellen Ninaber, Director, Workgroup North Sea Foundaton, Amsterdam, The Netherlands

André Nollkaemper, Professor of Law, Erasmus University Rotterdam, Rotterdam, The Netherlands

David P. Redford, Chief, Ocean Dumping and Marine Debris Section, U.S. Environmental Protection Agency, Washington, DC 20460

Christine A. Ribic, Assistant Unit Leader, NBS Wisconsin Cooperative Research Unit, Department of Wildlife Ecology, University of Wisconsin, Madison, WI 53706-1598

Martin D. Robards, Environmental Specialist, ENSR Consulting and Engineering, Anchorage, AK 99503

Joth G. Singh, Scientist, Caribbean Environmental Health Institute, Castries, St. Lucia

V. Kerry Smith, Director, Center for Environmental and Resource Economics, Duke University, Durham, NC 27708

Jon G. Sutinen, Professor, Department of Resource Economics, University of Rhode Island, Kingston, RI 02881

R. Lawrence Swanson, Director, Waste Management Institute and Marine Sciences Research Center, The University of New York at Stony Brook, Stony Brook NY 11794-5000

Paul Topping, Program Officer, Environment Canada, Ottawa, Ontario Canada

Wayne R. Trulli, Principal Research Scientist, Battelle Ocean Sciences, Duxbury, MA 02332

Barry A. Wade, Chairman, Environmental Solutions, Ltd., Kingston, Jamaica

Barbara Wallace, Project Director, A.T. Kearney, Inc., Alexandria, VA 22314

Judith E. Winston, Director of Research & Collections, Virginia Museum of Natural History, Martinsville, VA 24112

List of Illustrations

List of Tables

Introduction

For centuries the ocean was viewed as a boundless reservoir of productivity with unlimited capacity to assimilate wastes. This view persisted through the 1970s even as the finite nature of earth's terrestrial and atmospheric systems was being widely recognized. Now, the limits of ocean systems are also becoming evident. In the 1990s, everyday wastes constitute a marine pollution problem of global proportions, removing any doubt of man's capacity to affect the oceans. For many people this previously philosophical issue is made all too tangible by marine debris. Whatever vague security man may have drawn from the notion of the limitless oceans has been utterly shattered by marine debris. While some still consider marine debris an incidental pollutant of arguable impact, this is largely a problem of measurement rather than direction. Perhaps the most significant impact of marine debris is that it forces all of us to recognize our personal role in environmental degradation. No other form of pollution is so universally familiar. This impact may help precipitate a "midlife crisis" in man's tenure on the planet, resulting in a society-wide reevaluation of how environmental resources are managed and consumed. Clearly, dealing effectively with wastes of all types is essential to sustainable occupation of the earth.

Scientists and others who study the amounts, types, and distribution of marine debris generally define it as any manufactured or processed solid waste material (typically inert) that enters the marine environment from any source. The terms "marine litter" and "floatables" are occasionally used in place of the term "marine debris," connoting deliberate release (littering) or debris of a sewage or stormwater origin (floating). Unlike oil spills, this pollution does not come into the marine environment by accident, but rather by default. Its generators, often lacking adequate waste control or disposal options, simply lose, release, abandon, discharge, or dump it as convenience or necessity dictates. Debris released on land may blow, float, or be carried into the aquatic environment, to the sea and, then possibly back to land. Debris released at sea may circulate for weeks or years or be washed ashore immediately. These cycles of mobility ensure that its presence and impacts are felt locally, nationally, and internationally, continually reminding us of our immense capacity to pollute.

The problem is not that a few persons are littering but rather that a large part of the world population is using the open environment as a low-cost disposal alternative for solid wastes. Persistent synthetic materials such as plastics have filled a wide variety of human needs and wants in the post-World War II era, replacing less functional natural products in most applications. Plastic resin production in the U.S. was 37.6 million tons in 1994. The compound annual growth rate for this production was 5% between 1989 and 1994. United States plastic resin production is about 20% of world production, suggesting an annual world production of about 190 million tons. The British Plastics Federation estimated in 1990 that packaging accounts for 30% of worldwide annual plastics consumption (Economist 1990). Assuming all packaging becomes waste and all annual production is consumed, and further ignoring other sources of plastic wastes and amounts recycled, we arrive at a very general estimate of worldwide plastic waste generation of about 57 million tons per year (30% of 190 million tons). Even on a global scale, this is a substantial annual accumulation.

The proportion of this waste that reaches the marine environment is unknown. However, Agenda 21 of the 1992 United Nations Conference on Environment and Development (UNCED) projected that by the year 2000 fully 50% of the urban populations of the developing countries of the world will be without adequate solid waste services. Rural populations are even less likely to have adequate solid waste services. The demand for, and production of, persistent materials has outstripped the growth of institutions, infrastructures, and behaviors necessary to manage the resultant wastes. Public and industrial acceptance of littering and dumping, the absence or weakness of appropriate institutions and legal authorities to control these wastes, and incomplete waste collection and disposal infrastructures each contribute to this problem. From this, we can surmise that a very significant proportion of the plastic waste generated in the world is extant in the environment and that heavily populated coastal watersheds are major sources of marine debris.

The logic of the marine debris problem is clear. Marine debris is a fundamental manifestation of population growth and the industrial-technological revolution. Economic development increases consumption that increases persistent solid wastes, which are released into the environment from many sources. Because most plastics float and are quite durable, over time scales from days to decades (or perhaps centuries) this debris is transported by wind, water, gravity, and human and animal activity, to temporary or permanent "sinks" in the environment. Debris is constantly accumulating in these sinks: shorelines, estuaries, lakes, and the sea floor. In the absence of large-scale efforts to control it, *marine debris will get progressively worse.*

The primary marine debris challenge, then, is to achieve a functional balance between the production of wastes, especially persistent wastes, and their effective disposal. As the full scope of the marine debris problem is revealed and its lesson is digested, the priority of its solution will continue to rise in both developed countries and developing countries. This book assembles selected papers from the 1994 Third International Conference on Marine Debris into both the justification and the basis for working toward this balance at all socioeconomic and political levels. We present a summary review of the amounts, types, and distribution of marine debris around the world as well as its biological, social, and economic impacts. We also address the maritime and land-based sources of marine debris, providing legal reviews, case studies, and strategies for controlling both sources.

At all stages in the generalized "life history" of marine debris, there are impacts. Most are very poorly measured, but some are well documented, such as the entanglement of northern fur seals, ingestion by sea turtles, and the trashing of beaches. Others are a matter of logical extension of limited observations such as the potential effects of "microlitter" on zooplankton and larval fish, the smothering of seabed communities, or the transport of exotic species. The records of the biological impacts of marine debris, although extensive, are largely anecdotal and nonsystematic.

The "costs" of marine debris are varied and, for the most part, unmeasured. They include primary costs associated with its direct impacts such as the death and injury of commercial and protected marine life through entanglement, ingestion, ghost-fishing, smothering, and habitat displacement; the interference with maritime traffic by damage to ships' propulsion and steering systems, and the costs of cleaning, collection, and disposal. The indirect costs associated with marine debris include the aesthetic degradation of natural resources, the cost of unrecovered materials, the costs of efforts to prevent or protect against direct damages, and the opportunity cost of resources committed to these activities.

The marketplace shows us that the benefits society derives from the use of modern materials exceed the cost incurred in their manufacture, use, and disposal. However, the full environmental costs of disposal are not known, and the continuing accumulation of persistent debris and its impacts are steadily adding to those costs. Also, rising public (or consumer) perception of the environmental costs of these materials may be adding significantly to disposal costs. Fortunately, the erosion of these benefits (by growing costs) can be managed through identification, measurement, and minimization of both the direct and indirect costs. It is in this large-scale cost–benefit context that solution strategies involving public perception and behavior, institutional development, industrial policy changes, and waste management infrastructure investments should be considered.

Currently, in much of the world the immediate benefits to individuals from direct discard of persistent wastes, whether at sea or on land, outweigh the immediate costs of improper disposal. As long as society at the local level will tolerate the costs of improper disposal, investments in waste management strategies and systems will be inadequate. A reversal of this perception is needed before people become interested in taking action to manage the sources of marine debris. The inevitable worsening of marine debris impacts worldwide and the apparent broad public valuation of various features of the marine environment may make this perception relatively easy to change. To date, experience suggests that the general public in many countries is genuinely concerned when presented with the compelling logic of the marine debris problem. It is interesting to note that, at least in developed countries, public response to the marine debris issue—a general willingness to act beyond one's immediate self-interest—is somewhat at odds with economists' theoretical expectations. This inconsistency may be signaling a paradigm shift in the socioeconomics of environmental awareness. If a similar response arises in developing countries, the implications for solving the marine debris problem will be enormous. Development of the local and national political will to face the legal, financial, and cultural challenges to controlling the land-based sources of marine debris will be greatly enhanced by an active, dedicated public constituency.

Persistent debris is accumulating in the marine environment from both land

and sea sources, and institutional responses reflect these sources. Separate legal systems and agencies typically exist for the maritime community. As such, the maritime sources of debris have been relatively easy to address. At the international level, Annex V, Regulations for the Prevention of Pollution by Garbage from Ships of the 1973 International Convention of the Prevention of Pollution from Ships as modified by the Protocol of 1978, better known as MARPOL 73/78, provides the framework for control of wastes generated aboard ships of all types and sizes, which includes recreational vessels as well as fixed and floating platforms. Annex V prohibits the discharge of plastics from ships of nations party to the Annex and from all ships operating in waters under the jurisdiction of nations party to the Annex. It also sets restrictions on the discharge of other ship-generated wastes and allows for more stringent control within designated "Special Areas." As with most international treaties, nations electing to be bound by its terms must ratify the treaty and establish their own domestic laws to give force to its provisions. MARPOL Annex V entered into force for its signatories on December 31, 1988. As of July 21, 1995, 77 nations representing nearly 80% of the world's registered shipping tonnage had ratified MARPOL Annex V.

Under international law, waste materials carried to sea for the purposes of disposal are distinguished from those generated during ship operations. The former are the subject of the 1972 Convention for the Prevention of Marine Pollution by the Dumping of Wastes and Other Matter, also known as the London Dumping Convention or LDC. The LDC entered into force in 1975, establishing permit requirements for the disposal of wastes into the sea. These requirements cover all manner of wastes including dangerous substances, radioactive wastes, sewage, and garbage. As of July 21, 1995, 74 nations representing about 70% of the world's registered shipping tonnage had ratified the LDC.

The effectiveness of these maritime waste management treaties is largely a matter of national adoption, interpretation, and enforcement. At the international level, both MARPOL 73/78 and the LDC are administered under the United Nations by the International Maritime Organization (IMO). The successful implementation of MARPOL 73/78, Annex V and its national implementing legislation are key objectives for controlling maritime source debris. To further these objectives, we present and evaluate the challenges to their implementation as they affect shipping, ports, public vessels (military), and island nations.

Legal systems and agencies for controlling marine debris originating from land-based activities are another matter. While the United Nations Environment Programme (UNEP) adopted a Global Programme of Action for the Prevention of Marine Pollution from Land-Based Activities (GPA) in November 1995, there currently is no broadly functional international legal regime addressing the many land-based sources of marine debris. The 1985 Montreal Guidelines for the Protection of the Marine Environment Against Pollution from Land-Based Sources were developed to assist governments in establishing national laws and regional agreements to protect the marine environment from pollution by land-based activities. While these Guidelines are advisory in nature, they form potential bases for future conventions and were considered in the formulation of UNEP's Global Programme of Action. The recent stimulus of the 1992 United Nations Conference on Environment and Development engaged a number of nations in reviews of legal and policy options for addressing land-based sources of marine pollution (including marine debris) at both international and domestic levels, aided in the preparations for the GPA. Unfortunately, the GPA addresses marine debris in only the most general and cursory fashion, leaving a great deal of international and national legal development for the future.

We treat the challenges posed by land-based marine debris in hierarchical fashion, beginning with the overall international policy regime needed to induce action at lower levels, and following with case studies relating to both urban and rural and coastal sources. Drawing from the array of information available, an original framework of strategies for controlling the land-based sources of marine debris is proposed. The overwhelming impressions arising from this collection are that the land-based sources of persistent solid wastes (1) are large and diverse, providing as much as 80% of the marine debris; (2) are ingrained in social and industrial practices; and (3) differ markedly between developed and developing nations. Clearly, the global challenge for the future is to identify and develop the means to minimize the land-based sources of marine debris. The information and insight provided in this volume, we hope, will empower the initial steps toward broad solutions to the land-based sources while stimulating more efficient control of the maritime sources.

James M. Coe
December, 1995

PART I
The Status of Marine Debris

SECTION I
Amounts, Types, and Distribution

SECTION I
Amounts, Types, and Distribution

Introduction

The scientists who were first to recognize the negative effects of
marine debris developed a wide variety of methods to characterize its amounts, types, and distribution. Using these methods,
amounts and concentrations as well as types and compositions
have been measured on beaches, in harbors, at the sea surface,
and on the sea floor at locations all over the globe. Wherever
people have looked for marine debris, they have found it.

Only in the last several decades has the widespread accumulation of trash in the marine environment begun to be seen as
pollution [by those outside the marine ecology community].
Along with this rising perception has come the need to measure
this pollution so as to scope (as well as gauge) societies' behavioral and economic responses. In this section, the presentation
of reliable, international data on the amounts, types, and distribution of marine debris further establishes the foundation for
realistic assessments of the hazards to the marine environment
posed by the global solid waste dilemma. Consistent monitoring
of amounts, types, and distribution of marine debris provides
biologists and other earth scientists, as well as economists, with
information essential for (1) identifying the sources of debris by
types; (2) elucidating the "life history" and the pathways traveled by marine debris; and (3) evaluating benefits to society of
controlling debris. The answers to the questions implied in these
studies would permit the efficient allocation of public and
private resources to solving the many facets of the marine debris
problem. Typically, speculation as to these answers has not been
enough to stimulate significant investment in data collection. As
is seen here and in the following chapters, the study of the
characteristics of marine debris, while advancing steadily over
the past two decades, is far from complete.

Our opening section on the amounts, types, and distribution
of marine debris presents summary reviews of a variety of recent
surveys from a number of regions. Most are based on beach

surveys, by far the most widely employed data-gathering technique. The surveys have been one-time events or irregular, systematic, and even daily series. They have been carried out by scientists and technicians, industry groups, civic groups, and volunteers of all ages. While few of these surveys have been designed to meet rigorous statistical requirements, in most cases the results are assumed to be representative of the level and composition of marine debris at the time of the survey. These results are often used to compare changes from one time or location to another, but the comparisons are usually inappropriate because of gross violations of basic statistical assumptions (e.g., random sampling and normality). Often they have little statistical power because the sample sizes are small relative to the variances of the measured variables (usually debris densities). Throughout this chapter, and indeed this volume, representations about the amounts of marine debris should be considered qualitative and preliminary. Such is the nature of the current global database on the amounts, types, and distribution of marine debris. In her contribution to this section, Dr. Ribic offers some key observations and recommendations to improve this situation.

While global coverage has not been achieved, the reports included in this section from North America (Ribic et al.), the Wider Caribbean (Coe et al.), the Mediterranean (Golik), the North Pacific (Matsumura et al.), and the Southern Hemisphere (Gregory and Ryan) offer a strong indication that persistent solid wastes pollute not only the margins but also the surfaces and floors of all the world's oceans and seas—no nation, however developed, is immune to this plague. And while emphasizing the difficulty of identifying the specific sources of much of the debris surveyed, these reviews strongly support the growing recognition that a large proportion of the persistent wastes found in the marine environment originates from land-based activities.

1.
Debris in The Mediterranean Sea: Types, Quantities, and Behavior

Abraham Golik

Introduction

The Mediterranean Sea, which occupies some 2.5 million km^2, is an enclosed sea with only one opening for water exchange, the 14-km-wide Strait of Gibraltar. In the strait, surface water flows into the Mediterranean Sea and deeper water flows out. The water exchange rate of the Mediterranean is estimated to be 80 years. The sea is bordered by 18 countries, where more than 135 million people inhabit its coastal regions (Blue Plan 1987). The northwestern shores of the sea are heavily populated and highly urbanized, although its southern coast is sparsely populated. Major shipping lanes are found in the Mediterranean, with oil as probably the most important cargo. These physical and demographic conditions of the Mediterranean Sea make it a trap for marine- and land-derived litter.

The deleterious effects of litter may be divided into three categories: the harmful effect debris has on organisms, mostly through entanglement and ingestion; the disruption of navigation by clogging of cooling intake pipes and entanglement of debris in vessel propellers; and the aesthetic aspect, which is the most noticeable on the beach and in coastal waters. In the Mediterranean, the aesthetic effect might be the most important consideration. During the past two to three decades, there has been a significant development of coastal tourism in the Mediterranean because of its benign climatic conditions, warm water and abundant sunshine. Every year millions of people come to bathe on the beaches of the Mediterranean Sea. However, pollution of the sea and its beaches by debris is a severe deterrent to tourists and the tourist industry, which is of great economic importance to Mediterranean countries.

With this background, it is surprising how little has been done to investigate the marine debris problem in the Mediterranean. Until the mid-1980s there were few attempts to survey Mediterranean marine debris (Shiber 1979, 1982, 1987; Morris 1980; Saydam et al. 1985; Gramentz 1988; McCoy 1988) or to estimate the input of litter from vessels (National Academy of Sciences 1974; Horsman 1982; Bingel 1989). Field measurements were initiated in the late 1980s (Golik and Gertner 1990, 1992; Gabrielides et al. 1991; Shiber and Barrales-Rienda 1991; UNEP/IOC/FAO 1991; Galil et al., 1995; Bowman et al. unpublished data). This paper reviews studies of the amounts, types, and distribution of marine debris in the Mediterranean and proposes new approaches to investigate this problem.

Quantities of Litter in the Mediterranean Sea

It is difficult to obtain quantitative information concerning the distribution of marine litter because of the great areal variability in the size and nature of its constituents. A large number of samples must be collected to alleviate this difficulty. In addition, there is the problem of what to measure: one can measure the number of litter pieces, item weight, or the area the item occupies. Distortion will take place using any of these measurements. For example, the importance of Styrofoam and plastic pieces will be greatly reduced if weight of litter is measured, and at the same time the importance of bottle covers might be exaggerated if the number of pieces is considered. Because all these methods have been employed by various researchers, comparison of results of these studies is difficult and may require mathematical manipulations, thereby incorporating assumptions that may cause even more error in the resultant estimates.

Input of Debris from Vessels

Matthews (1975) used statistics of various maritime activities in the Mediterranean Sea to estimate the number of person-days per year that such activities required. He then estimated the trash production, per person per day, for each of these activities. For example, the trash population for a passenger ship was estimated at 1.6 kg/person/day while for a merchant ship it was only 0.8 kg/person/day. In a similar way he also estimated the amount of nonpersonal material, such as wires, plastic covers, and dunnage, that is disposed from a ship. Table 1.1 lists the various types of litter disposed into the Mediterranean Sea according to Matthews; an annual input of garbage from ships is 660,000 mt. This is probably an underestimate because it is based on data from the early 1970s and does not include data concerning recreational activity. Furthermore, the data regarding fishing vessels came only from Greece and Italy.

Horsman (1982) computed the type and quantity of commodities used on board a merchant ship by comparing the store lists of

TABLE 1.1. Estimate of trash discarded from ships into the Meditterranean Sea.

Sources	Trash discarded (10^6 kg/yr)
Passenger lines	2.4
Merchant ships[a]	12.1
Merchant ships[b]	632.8
Fishing boats[c]	5.0
Military activity	10.0
Offshore oil production	0.3
Total	662.6

[a]Trash produced by crew.
[b]Cargo waste (pallets, wires, plastic covers, dunnage, etc.).
[c]Based on data from Italy and Greece only.
From Matthews (1975), with permission.

the ship on various dates. Assuming that all the trash generated from the used commodities was thrown into the sea, and using the data provided by Matthews (1975), Horsman estimated the annual input to the Mediterranean Sea from merchant ships to be $48.3–93.6 \times 10^6$ metal pieces, $3.0–4.5 \times 10^6$ glass items, and 4.5×10^6 plastic containers.

Bingel (1989) tried to estimate the quantity of fishing gear lost in the Mediterranean by the fishing fleet. He collected data in Turkey and related them to coastal length and to the continental shelf area. He then applied these statistics to the Mediterranean as a whole and obtained the following estimates (in mt/year):

According to number of vessels	3342
According to length of coastline	2803
According to shelf area	2637

As this review shows, information concerning the input of garbage in the Mediterranean is scanty and insufficient for estimating the amount of marine debris. The information is based only on ship-source debris, involving numerous assumptions, extrapolations, and data manipulations, and is outdated. It is doubtful that these approaches provide anything better than order-of-magnitude estimates, and even these are associated with great uncertainty.

Coastal Litter

Most of the quantitative data regarding litter in the Mediterranean Sea come from beaches.

It is the easiest environment to sample and is also the most important environment from both aesthetic and economic points of view. The earliest studies on litter on the Mediterranean coast are those of Shiber (1979, 1982, 1987) in which he reports on the presence of plastic pellets on the beaches of Lebanon and Spain. In a second debris survey conducted on the beaches of Lebanon in 1988, 10 years after his first survey, he provides semiquantitative data, using terms such as "rare," and "abundant" (Shiber and Barrales-Rienda 1991). During this 10-year period, the concentration of plastic pellets increased on three previously surveyed beaches and decreased on two others. Megalitter, however, increased on all five surveyed beaches during that period.

A coordinated effort to measure Mediterranean litter was initiated by the United Nations Environmental Program (UNEP) and was carried out in 1988 in Cyprus, Israel, Italy (Sicily), Spain, and Turkey (Gabrielides et al. 1991; UNEP/IOC/FAO 1991; Golik and Gertner 1992). Altogether, litter quantities were measured by transect along 61 beach profiles on 12 beaches at a frequency of once per month for about a year, for a total of 659 samples. Examination of these data shows a wide range of litter concentrations from a mean of 0.53–640 pieces/m in counts, or from 4.2 to 6628 g/m in weight (UNEP/IOC-/FAO 1991). Table 1.2 provides a comparison of the mean litter quantities in participating countries. However, this comparison must be made with care in view of large differences in the number of samples that were collected in each country and the high variability of the data. For example, the high values that were obtained in Sicily are undoubtedly biased because one beach near Palermo, Sicily, is probably used as a dumping ground for construction refuse.

Floating Litter

Very little information exists concerning the floating litter in the Mediterranean Sea. When Saydam et al. (1985) used a neuston net to collect floating litter in the northeastern Mediterranean, they found that litter quantities ranged from zero, in many of the tows, to a maximum of 7.2 kg/km^2. One must bear in mind that neuston nets are designed to collect plankton, and that this sampler may bias the results toward the smaller litter particles.

A similar approach was used by Marino et al. (1989) off the coast of Spain. Their net was larger, sampling a 16-m-wide band of surface water and covering an area of 0.4 km^2 at each sampling. The mean plastic concentration was found to be 2086 pieces/km^2 or 61.8 kg/km^2 in July 1988, while in March 1989 the values were only 352 pieces/km^2 and 14.7 kg/km^2, respectively. Marino et al. attributed the difference to a storm with onshore winds that occurred before the March 1989 sampling; a large portion of the floating litter was driven shoreward by these winds.

Morris (1980) and McCoy (1988) estimated the concentration of floating garbage pieces by visual observations from ships. Morris estimated a concentration of 2000 pieces/km^2 some 40 miles southwest of Malta; McCoy reported only 0.12 pieces/km^2 from the Ionian Sea and attributed the low concentration to the fact that the observations were carried out away from shipping lanes.

Seabed Litter

Information on litter found on the sea floor of the Mediterranean Sea is also scanty. Bingel et al. (1987) and Bingel (1989) reported on litter that was retained by trawl nets on the continental shelf in Mersin and Iskanderun

TABLE 1.2. Comparison of debris levels between various Mediterranean countries.

Study area	Spain[a]	Sicily, Italy[a]	Cyprus[a]	Israel[a]	Israel[b]
Number of samples	17	46	123	472	86
Mean counts (pieces/m)	33.2	102	10.4	7.3	8.7
Mean weights (g/m)	159.3	1595	87.1	–	–

[a]From UNEP/IOC/FAO (1991).
[b]From Bowman et al. (unpublished data).

bays off the coasts of Turkey and off Cyprus in 1983–1984 and in 1989. The mean litter concentration was found to be 49 kg/km^2 in Mersin Bay, 29 kg/km^2 in Iskanderun Bay, and 53 kg/km^2 off Cyprus.

In 1993, Galil et al. (1995) analyzed the litter content retained in a beam trawl net at 17 stations along a profile from Italy to the eastern shores of the Mediterranean Sea at depths ranging from 194 to 4614 m. The width of the beam trawl was 2 m, and as the trawling time and speed were recorded it was possible to estimate the litter concentration on the sea bottom. In 12 of the 17 stations (70%) litter was found; the concentration was 0.2–8.5 × 10^{-3} pieces/m^2.

The data concerning debris quantity in the Mediterranean Sea (on beaches, sea surface, and sea bottom) that are provided here only demonstrate the paucity of information presently available. Again, this information is insufficient to provide a debris baseline for the Mediterranean Sea as a whole for comparison with future studies. At best, only the data regarding coastal pollution for specific beaches could be used for this purpose. The data presented do provide, however, a preliminary estimate of the pollution level in terms of orders of magnitude, i.e., 10–100 pieces/m for the beach, 100–1000 pieces/km^2 for the sea surface, and 0–1000 pieces/km^2 for the sea floor. Both debris density and debris distribution data from the Mediterranean exhibit high variability.

Types of Litter

Figure 1.1 shows the composition of litter on teaches from five Mediterranean countries that participated in the UNEP project (UNEP/IOC/FAO 1991). The composition is given in terms of mean relative abundance of the various constituents of the litter for each country. The dominant constituent is plastic, which ranges from 60% to 80%. The only exception is Sicily, where one of the sampled beaches was probably used for dumping construction material; this reduced the relative abundance of plastics. Considering that the

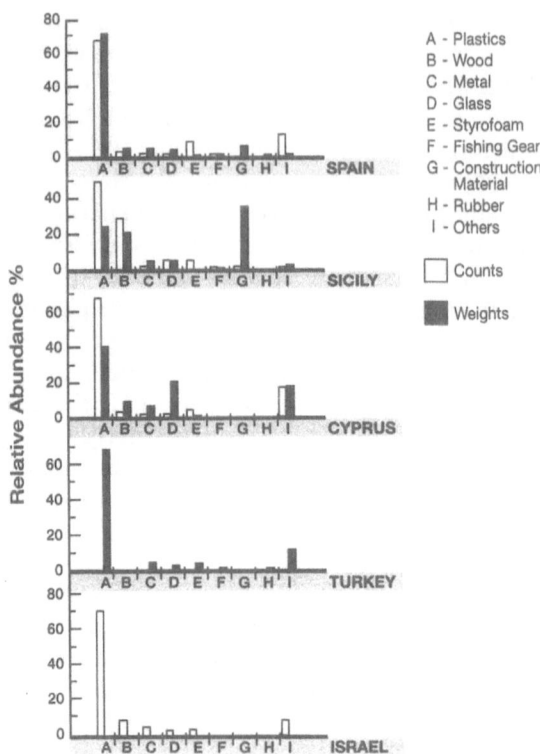

FIGURE 1.1. Relative abundance of various components of litter on the beaches of five Mediterranean countries (Gabrielides et al. 1991).

"others" category constitutes about 10%, its constituents (wood, glass, and metal) account for only a small overall percentage, although in some countries wood is the second most abundant item. Other constituents, such as Styrofoam and rubber, are also relatively rare. It is of interest to note that fishing gear is rather rare along the Mediterranean beaches (2.8% was the highest abundance recorded).

Within the plastics category, items consisted of fragments, bags, sheets, containers for beverages, food, cosmetics, engine oil, toys, and other products. The term "wood" includes driftwood, lumber, and sometimes remnants of crates and pallets. Most of the metal debris is tins used for beverages and aerosol cans. Likewise, most of the glass is represented by soft drink bottles, and, to a lesser extent, light bulbs. Other materials include pieces of garments, Styrofoam, foam rubber, paperboard and cardboard cartons, and paper.

The composition of the floating litter is also dominated by plastic. According to observations made by Morris (1980) near Malta, 60%–70% of floating marine debris were plastics, which consisted of bags, cups, sheets, and packaging material. The rest of the floating material consisted of timber, rubber, nylon, ropes, glass bottles, and paper. McCoy, who conducted similar observations in the Ionian Sea, did not provide quantitative data for the various components of the litter but reported that plastics (mostly containers) and wood were the most abundant materials in the litter. Marino et al. (1989) found that in the coastal waters off Barcelona, Spain, the floating litter consisted (in number of pieces) of 74.5% plastic, 15.2% Styrofoam, and 3.1% wood.

The data concerning marine debris found on the seabed are too scanty to characterize its nature. Loizides (Bingel 1989) reported that off Cyprus 80% (by weight) of the litter dredged from the bottom was metallic objects and 1.4% was plastics. However, in terms of number of pieces, 23% were metal and 45% were plastic. Bingel (1989) reported that off the coast of Turkey the most abundant material, in terms of weight, was wood at 43%, and plastics constituted 32%. Galil et al. (1995) found some 278 pieces of debris on the sea bottom of the eastern Mediterranean in 12 stations; these consisted of 43.8% paint chips, 35.9% plastic components, 9% metal, and 3% glass.

Sources of Debris

Identification of the sources of litter may prove to be the most important issue in searching for a solution to litter pollution, because measures that stop production of debris at their source can be very effective. In the case of marine debris, the immediate question is whether the major source for this litter is land based, i.e., derived from coastal garbage dumps, or from bathers or by drainage from land, or is marine based, from ships dumping litter into the sea.

In the case of the Mediterranean Sea, Golik and Gertner (1990, 1992) compared the composition of the litter found on the Israeli coast to that reported on the beaches of western Europe. Dixon and Cooke (1977), and Dixon and Dixon (1980, 1983) studied the composition of containers stranded as coastal litter in terms of their original use, i.e., before they were turned into debris and dumped. Their studies were carried out on beaches in Portugal, the English Channel, England, Western Cherbourg Peninsula, west Jutland, and the southern North Sea. They found that on all these beaches the most abundant containers were those that had contained cleansers and household detergents. The metal containers were mostly those of marine oil or grease, and the carton containers were for long-life milk. Elsewhere, on the beaches of Amchitka Island, Alaska, Merrell (1980, 1984) found that most of the litter was remnants of fishing gear. In contrast, on the Israeli beaches, Golik and Gertner (1990, 1992) were impressed by the high relative abundance (although they did not conduct actual counts) of containers used for beverages, food, and cosmetics (suntan lotion) and other objects, which are brought to the beach by bathers. They suggested, therefore, that the litter on the Israeli beaches results mostly from recreation activity and should be considered as land-based litter.

Shiber and Barrales-Rienda (1991) arrived at a similar conclusion for the beaches of Lebanon. They attributed the plastic pellets, which were found on the beaches, to local plastic factories as well as to the loss of cargo during unloading of raw material from ships. During the recent wars in Lebanon, massive population shifts have prevented an orderly collection of garbage and "even when street cleaners are able to sweep the streets and sidewalks, they usually wind up dumping the trash onto the beaches!" (Shiber and Barrales-Rienda 1991).

The impression that the Mediterranean shores are polluted mostly by land-based litter was shared by participants of the UNEP project and is supported by the low concentration of remnants of fishing gear, which has a maximum abundance of 2.8% in Turkey (UNEP/IOC/FAO 1991). The difference between the sources of debris on the Mediterranean coast and those on Europe's western shores is not surprising, in view of the fact that the bathing and recreation season on the Mediterranean shores is longer and more popular than on the western shores of Europe. In addition, at least in the English Channel, the maritime traffic must be heavier than in the Mediterranean, causing a higher proportion of marine-based litter in the English Channel.

While some land-based litter may be washed into coastal waters during storms, it seems unlikely that this litter will be carried great distances from shore. It is reasonable to speculate that most floating debris found in offshore waters, as well as most litter on the sea floor, in the Mediterranean comes from ships. Indeed the most abundant litter item found on the Mediterranean Sea bottom by Galil et al. (1995) was paint chips, which originate from ship chipping and is therefore definitely of marine origin.

Factors Controlling Litter Distribution in Space and Time

The data regarding litter distribution in the Mediterranean permit a preliminary investigation of the factors that control marine debris distribution. Gabrielides et al. (1991) and Golik and Gertner (1992) emphasized the effect of large population concentrations on the pollution levels of nearby beaches. In Sicily, the mean concentration of litter on the beach of Ficarazzi, which is near Palermo, the largest city in Sicily, was 231 pieces/m. On the beach of Eraclea, which is far away from any population concentration in Sicily, the mean pollution level was only 9 pieces/m. Similarly, in Israel significant statistical differences in debris levels were found between beaches close to Haifa, one of the largest cities in Israel, and remote beaches. Furthermore, Golik and Gertner (1992) reported that the most contaminated beach sampled in Israel (Akhziv) is located near the border be-

tween Israel and Lebanon. They attributed the high level of pollution on this beach to debris that was carried by the sea from coastal population concentrations in southern Lebanon, as evidenced by the inscriptions on the remnants of packaging material.

The type of debris found on a beach is also related to local human activities. For example, the litter on Ficarazzi Beach, near Palermo in Sicily, consisted of nearly 40% (by weight) of debris from constructive activities. In Israel, the litter found on the beach of Palmakhim, which is located a few kilometers north of Ashdod Harbour, consisted of 46% wood and lumber. This debris is mostly crates and pallets, drifting from the nearby harbor. Palmakhim is the only beach studied in Israel where plastic was not the most abundant debris type (Bowman et al. unpublished data).

Golik and Gertner (1992) demonstrated a seasonal effect on debris quantities. Figure 1.2 shows the monthly means of litter quantities found on six beaches in Israel. Two statistically significant minima were found during that period: one in July 1988 and the other in December 1988 through February 1989. The first one resulted from a beach-cleaning operation and is therefore artificial, but the other was a result of winter storms. During these storms, high waves transported litter from the upper shoreface to the back of the beach and even beyond, thus leaving the beach relatively clean. During the rest of the year, debris transported to the beach from the sea, or left by bathers, accumulates on the beach.

Litter Dynamics

Bowman et al. (unpublished data) conducted an experiment designed to investigate the dynamics of litter on the beach. They repeatedly measured litter on three beaches in Israel. On each visit to the beach, they recorded all the litter items present and spray-painted each item. On their next visit, they recorded all the remaining painted pieces of litter, as well as the new unpainted ones that

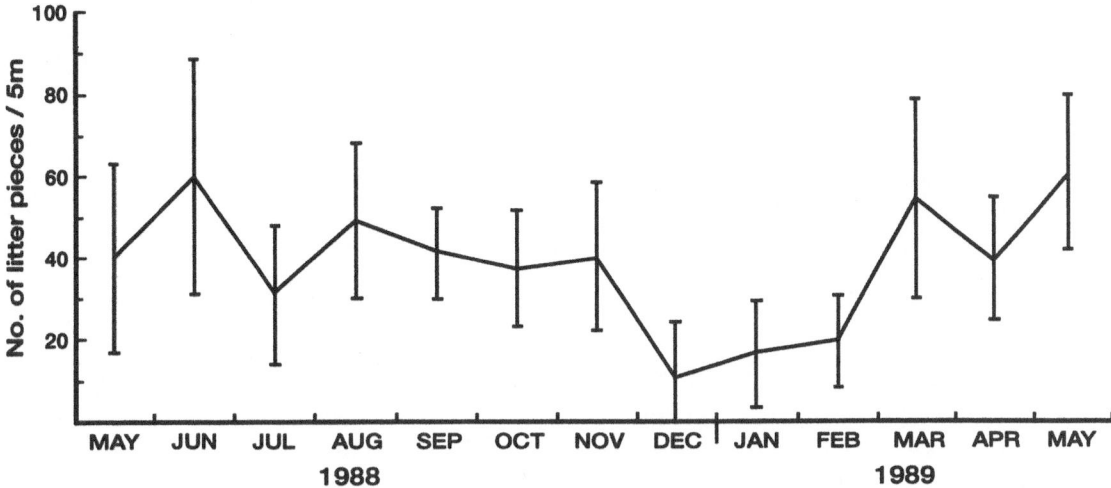

FIGURE 1.2. Variations of monthly means and standard deviations (*vertical bars*) of litter quantity (pieces/5 m) on six beaches in Israel (Golik and Gertner 1992).

were then painted with another color. This was repeated six times, and the results from one of the beaches are presented in Fig. 1.3. The quantity of beach litter decreased rapidly in the first week for each of the sampling sessions and more slowly thereafter. After the eleventh week, during September and October of 1991, another sharp decline in debris quantity occurred for all five sampling sessions. Bowman et al. proposed that a certain fraction of the litter is highly mobile and is transported away from the studied beach, probably by wind, only to be replaced by similar litter from the neighboring beaches. The less mobile fraction remained for a longer period of time until a strong storm removed it. Indeed, it was found that during the first week most of the items that disappeared from the beach were plastic containers and other plastic and Styrofoam debris; then, the lumber disappeared. The most stable items were glass bottles, metal cans, rope, and clothing.

Lines for Future Research

The information gathered so far on marine debris in the Mediterranean Sea may be helpful in defining future research objectives. The first objective should be to determine

whether there is a systematic change in litter quantity, either in a specific location or in the Mediterranean Sea as a whole. Do we experience an increase in marine debris because of continuous population and industrial growth as well as increased urbanization in the Mediterranean countries, or is there a decline in litter from preventive measures such as MARPOL Annex V and increasing public awareness? To answer this question, long-term programs that monitor litter on the beaches and in the sea must be adopted by the Mediterranean countries.

Another important question is the source of the litter, because identified sources will indicate where we have to expend our efforts and resources to control this problem. This is a difficult question to answer, because there are no clear characteristics that differentiate land-based from sea-based litter (with the exception of fishing gear or paint chips, which are definitely marine-based litter). The future study of empty containers may yield significant results leading in this direction. The inscription and imprinting on these containers provide information concerning their origin, and their original use may also be indicative as to their origin. Another approach to determine the characteristics of marine-based litter is to investigate litter on the beaches of desolate or sparsely inhabited

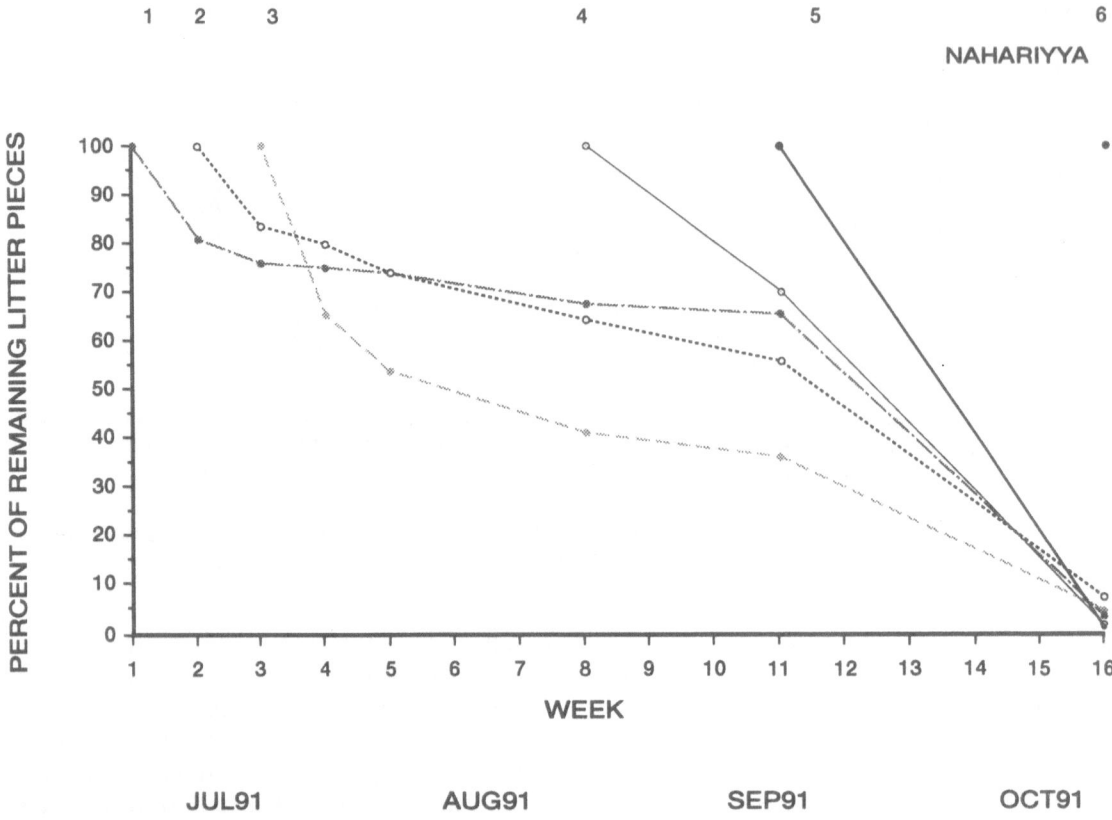

Figure 1.3. Reduction rate in litter quantity on the beach of Nahariya, Israel, is described in percentage of the original quantity found in each of the six sampling exercises graphed (Bowman et al. unpublished data).

islands in the Mediterranean Sea. If there is no garbage generated on the island, then whatever accumulates on the beach must be marine-based litter. A survey of marine debris on such islands may indicate particular characteristics. Surveys of this nature in the Southern Hemisphere, are described by Gregory and Ryan (Chapter 5, this volume).

Research on litter dynamics, once the litter is on the beach, is of great importance for designing a strategy to address the litter pollution problem. Answers to questions such as what is the lifetime of debris on a beach or in the sea, or what is the rate of input and outflow of litter into and from a certain area, are of extreme importance when plans are being drawn to manage the marine debris (solid waste) problem.

Because the major component of the debris is plastic, an effort should be directed to eliminate plastic items from the debris population. Efforts to produce plastic with a predetermined lifetime have already been started, with reported success (Scott 1990). If indeed the industry would produce plastic material for packaging that will disintegrate to benign elements after its useful life, marine debris and solid waste problems might be significantly reduced.

Acknowledgments. The author would like to thank D. Bowman and N. Manor-Samsonov for permission to use some of their unpublished data, and O. H. Osborne, who critically read the manuscript and made useful suggestions.

2.

Distribution of Floating Debris in the North Pacific Ocean: Sighting Surveys 1986–1991

Satsuki Matsumura and Keiichi Nasu

Introduction

The Intergovernmental Oceanographic Commission (IOC) and the Japan Meteorological Agency identified the necessity of monitoring marine debris in the North Pacific Ocean, and in its adjacent waters, in the early 1970s. In conjunction with the IOC's Global Investigation of the Pollution in the Marine Environment (GIPME), the Japan Meteorological Agency started a marine debris sighting survey in 1972.

As a result of the Japan Meteorological Agency's preliminary findings and decisions taken at the International North Pacific Fisheries Commission's (INPFC) 1986 annual meeting, the Fisheries Agency of Japan established a Pacific-wide sighting survey of floating marine debris. The areas of the survey included the North and South Pacific Oceans, the Sea of Japan, the Yellow Sea, the Bering Sea, the Sea of Okhotsk, the East China Sea, the Indian Ocean, the Tasman Sea, and the Antarctic Ocean. Mio et al. (1989) reported the results of the Fisheries Agency's 1986 sighting survey in regard to the North Pacific Ocean and its adjacent waters. Other reports, such as the Proceedings of the Workshop on the Fate and Impact of Marine Debris (Shomura and Yoshida 1985) and the Proceedings of the Second International Conference on Marine Debris (Shomura and Godfrey 1990), provided further detailed information on the sources and quantities of marine debris in the North Pacific Ocean.

This paper updates the Fisheries Agency's sighting survey results by presenting the 1987 through 1991 survey results relevant to the North Pacific Ocean and its adjacent waters. The survey's current methodology, the types of marine debris sighted, and estimates of densities of marine debris are described.

Method

The current methodology evolved from the 1986 preliminary survey work. Participants in that survey, which included 20 vessels and covered a total sighting distance of 80,546 nautical miles, counted 7458 pieces of floating marine debris. Using these results, a practical survey plan was formulated: an observation manual and data-recording formats were designed and distributed to observation vessels; videotaped training courses were prepared by the National Research Institute of Far Seas Fisheries, and navigation officers on ships belonging to the Fisheries Agency were nominated as sighting observers. On other ships, observers were appointed and then specially trained.

Observers used the distance between themselves and the sighted debris, and the angle of the debris from the bow, to calculate (and then record) the perpendicular distance of debris from the ship's trackline. Observers also recorded the types, numbers, and lengths (sizes) of marine debris, as well as

TABLE 2.1. Categories of marine debris sighted in the North Pacific Ocean, 1986–1991.

Type	Description
Manufactured objects	
Artificial	Net (e.g., trawl net and gill net), fishing gear other than nets (e.g., plastic and Styrofoam—including items such as floats and flag buoys)
Petrochemical related	Petrochemical products (e.g., plastic, vinyl, polyester), Styrofoam (excluding fishing gear), pieces of wood (e.g., boxes and boards), glass and metal products (e.g., bottles and cans)
Natural objects	Seaweed, logs
Unknown debris	All other items

other ancillary data. With the exception of active fishing gear, all types of marine debris (greater than 5 cm in size) were counted as part of the survey. The various types of marine debris were recorded within two main categories: manufactured objects and natural objects. A further categorization of marine debris, including objects identified by observers during the survey, is shown in Table 2.1.

When a number of objects of the same type were found at the same location and the number equaled 99 or less, the actual number was recorded. When the number of objects was estimated to be greater than 99, the number of objects was recorded as 99. A group of more than 99 occurred occasionally in coastal waters.

The maximum length of floating objects was estimated by eye. For example, when the sighted object was shaped like a box, and its size was approximately 20 × 30 × 40 cm, the object's length was assumed as 40 cm. The lengths of marine debris were recorded as S (small), M (medium), and L (large). These categories were defined by the following parameters: S, less than 50 cm; M, 50–200 cm; L, larger than 200 cm.

The angle from the course of the observation vessel, and the distance from the observation vessel, were estimated and recorded by the sighting observers. Other ancillary data that were also recorded every hour, on the hour, included date and time, location, number of researchers, visibility, wind force (by Beaufort scale), and surface temperature.

Sighting Effort

Table 2.2 shows the annual number of ships (voyages) that participated in the survey from 1986 through 1991, as well as the survey trackline lengths (nautical miles) for the same period. Excluding the 1986 preliminary survey, 64 different vessels conducted 204 survey voyages during the 1987–1991 period; observers surveyed approximately 926,000 nautical miles and counted 136,338 pieces of marine debris.

Figures 2.1A–C depicts the survey vessels routes (otherwise known as "tracks") during 1987 and are representative of all survey voyages during the 1987–1991 period. Not unlike the 1986 preliminary survey, survey vessels (during 1987–1991) included research ships, training ships, fisheries patrol boats, volunteer fishing boats, and cargo boats. The survey fleet was organized by the Fishing Ground Environment Conservation

TABLE 2.2. Effort of marine debris sighting survey, North Pacific Ocean, 1986–1991.

Year	Number of voyages	Sighted distance (nautical miles)
1986[a]	20	81,000
1987	34	139,000
1988	45	245,000
1989	47	206,000
1990	42	203,000
1991	36	133,000
Total	204 (64)[b]	926,000

[a]1986 was the preexperiment survey; only the data from 1987–1991 surveys were used for statistical analyses. [b]A total of 64 ships participated in the sighting surveys in 1987–1991.

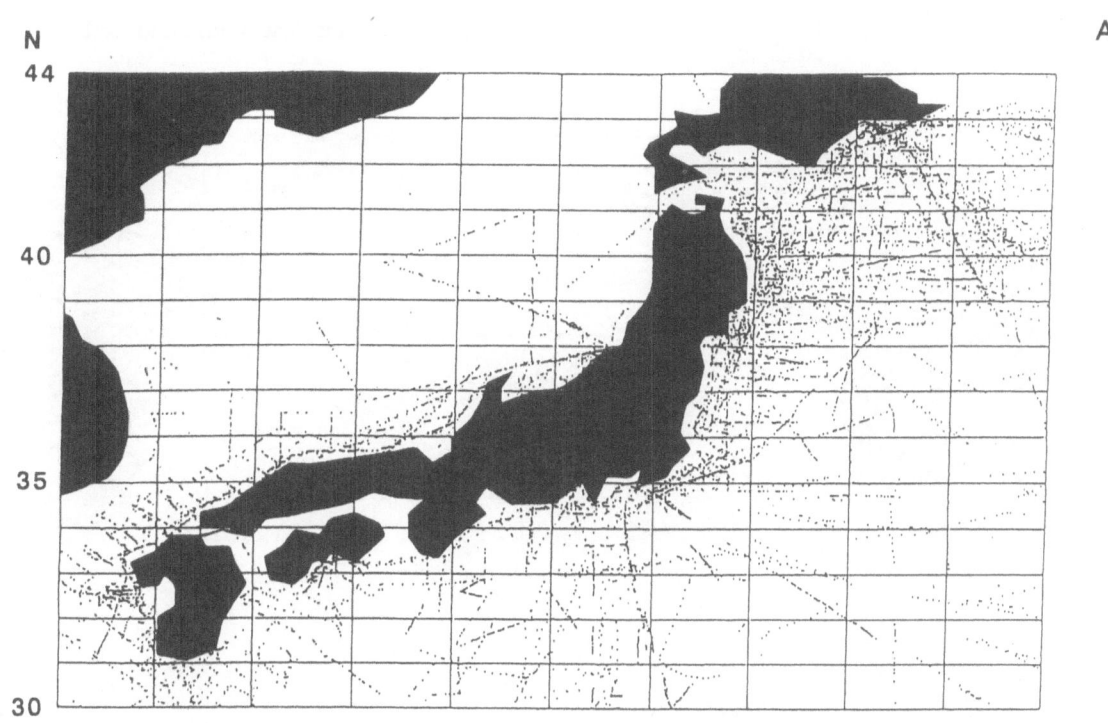

FIGURE 2.1A–C. Tracks of vessels engaged in marine debris sighting surveys. A An example during one year (1987) in coastal waters of Japan. B Tracks on northwest Pacific Ocean. C Tracks on Pacific Ocean.

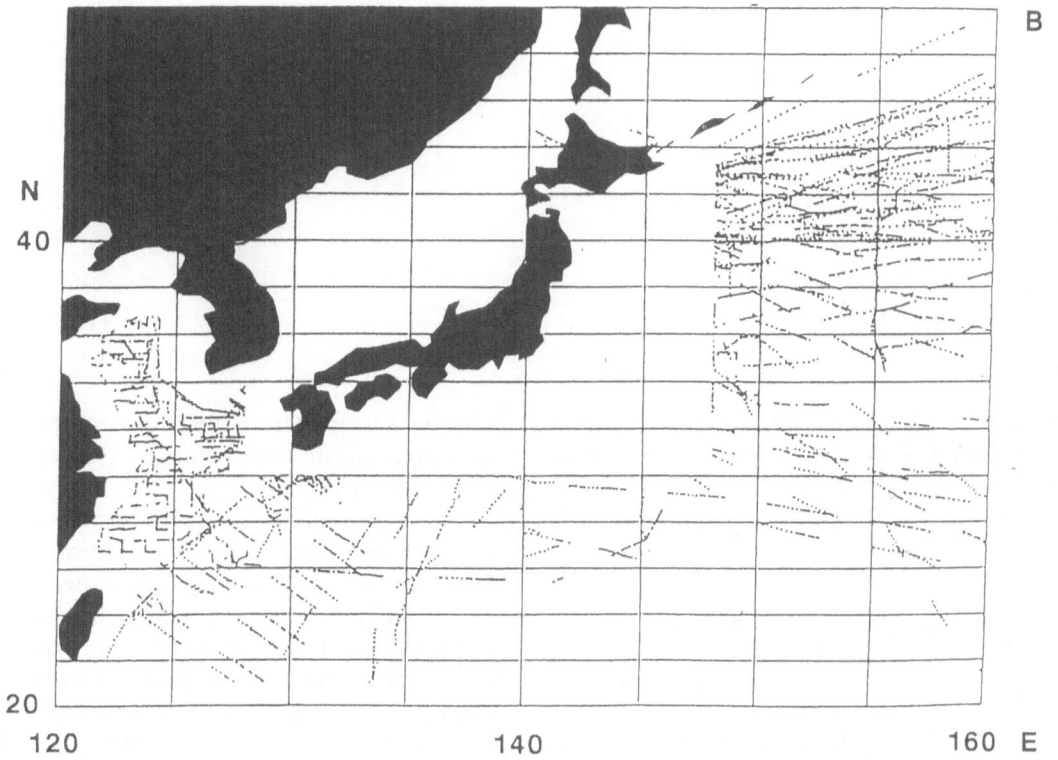

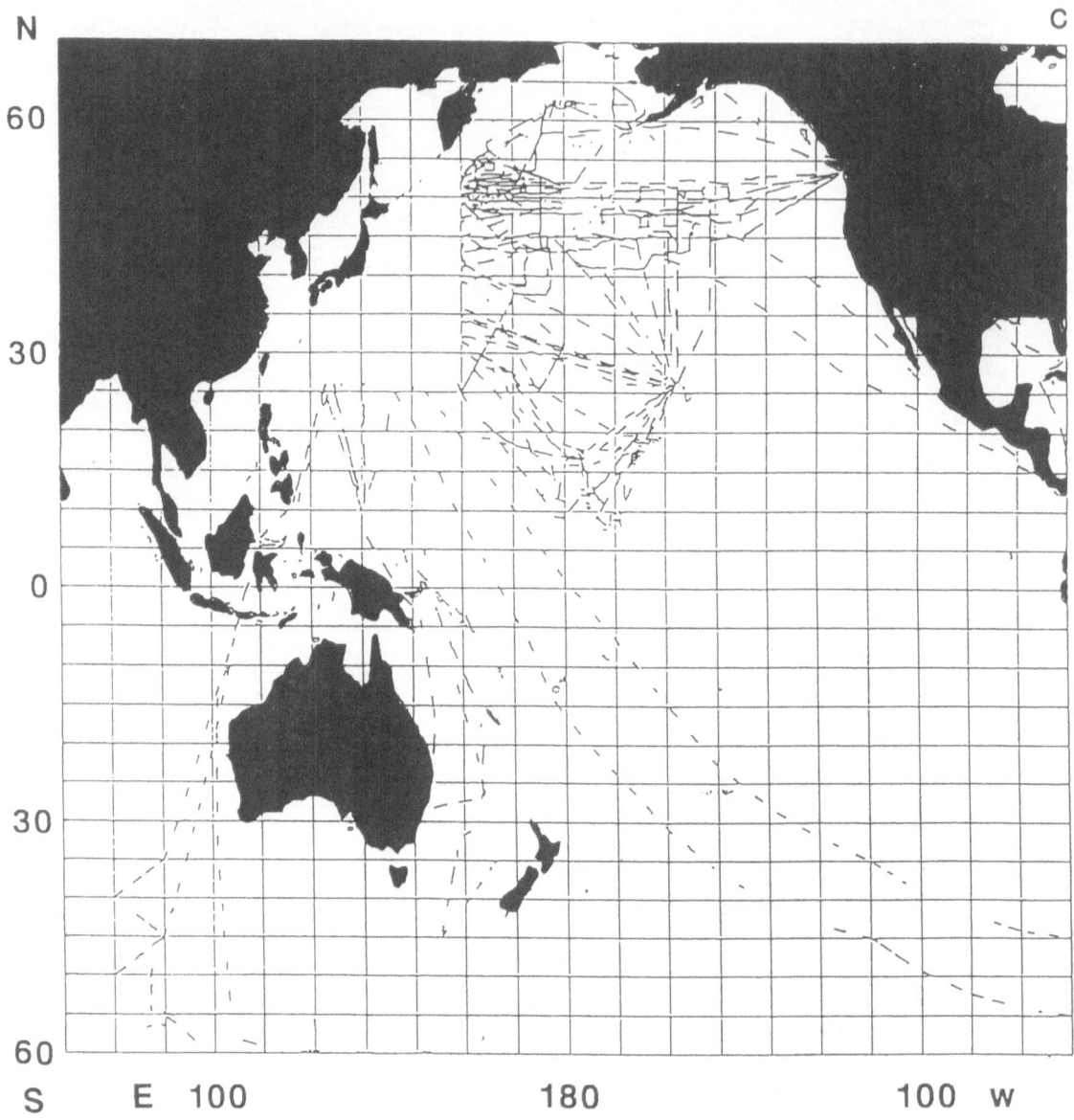

Division of the Fisheries Agency of Japan. The sighting effort, in terms of vessel track-miles, was greatest in Japanese waters and in the western Pacific Ocean.

The Composition of Marine Debris

Figure 2.2 shows the composition of sighted debris during each year, 1986 through 1991. With the exception of 1987, the composition ratio underwent little change. However, in the Bering Sea during 1987, the composition ratio was influenced by a high density area of seaweed that was more than twofold the seaweed density level during other years. Otherwise, the average marine debris composition, during 1986 through 1991, was 10% fishing gear, 60% total petrochemical (including fishing gear, Styrofoam, and other plastic debris), and 30% natural objects (e.g., logs and seaweeds).

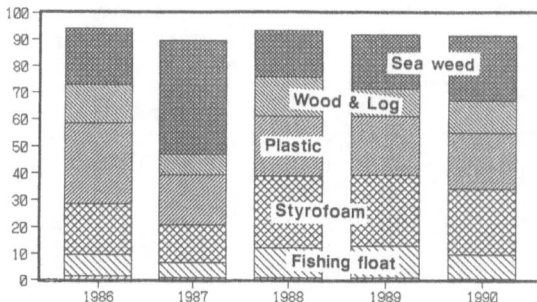

FIGURE 2.2. Composition of sighted debris, 1986–1990.

Amounts of Marine Debris

The highest proportion of debris sightings were between 10 to 20 m from the vessel's trackline (24%). Furthermore, the data indicate that 69% of debris was sighted at a distance within 40 m and 91% within 100 m. This relation between the amount of debris sighted and the perpendicular distance was similar to that found by Nasu and Shimadzu in their 1969 whale sighting survey.

Estimation of Density of Marine Debris

The line transect method was used for estimating the density distribution of marine debris based on the number of pieces of debris sighted, the perpendicular distance from the ship's track, and the trackline length. The number of pieces of debris per square nautical mile (density), D, is calculated by the following equation (Seber 1982):

$$D = \frac{n}{2(w/1852)L}$$

where n is the number of debris sighted, L is the surveyed trackline length (nautical miles), and w is the effective perpendicular distance (m).

The effective perpendicular distance is calculated by the following equation using the sighting probability by perpendicular distance, g(y), estimated from the data:

$$w = \int_0^c g(y)dy$$

where c is half the maximum track width of the observation (100 m was used for this paper) and y is the perpendicular distance of an object from the vessel's trackline.

For calculating a sighting probability of perpendicular distance, the hazard rate model was used. The hazard rate model is also used in whale sighting surveys (Kishino et al. 1989), and is expressed as the following equation:

$$g(y) = 1 - \exp[-(Y/A)^{(1-B)}]$$

where Y is the recorded perpendicular distance and A and B are parameters. The density was calculated for each 5° latitude by 10° longitude block within the sighting effort.

Density of Marine Debris by Type

Fishing Nets.

Figure 2.3A shows that the density of fishing nets was relatively higher in the midlatitudinal area of 20° to 30° N, 150° to 130° W of the eastern Pacific Ocean. This tendency was almost identical with the results obtained in previous years. In addition, the densities of fishing nets in waters on the Pacific Ocean side of Japan, between 30° to 40° N, 140° to 150° E, were also high. A significantly high density area formed northeast of Hawaii during previous years (Mio et al. 1989).

Fishing Gear Other Than Nets.

Figure 2.3B shows that the range of distribution of fishing gear was wider than that of fishing nets. The highest density area (larger than 120 pieces per 100 square nautical miles) was distributed between 25° to 35° N, 130° to 180° W. Densities of 50 pieces per 100 square nautical miles, or greater, were found in 17% of the total blocks surveyed.

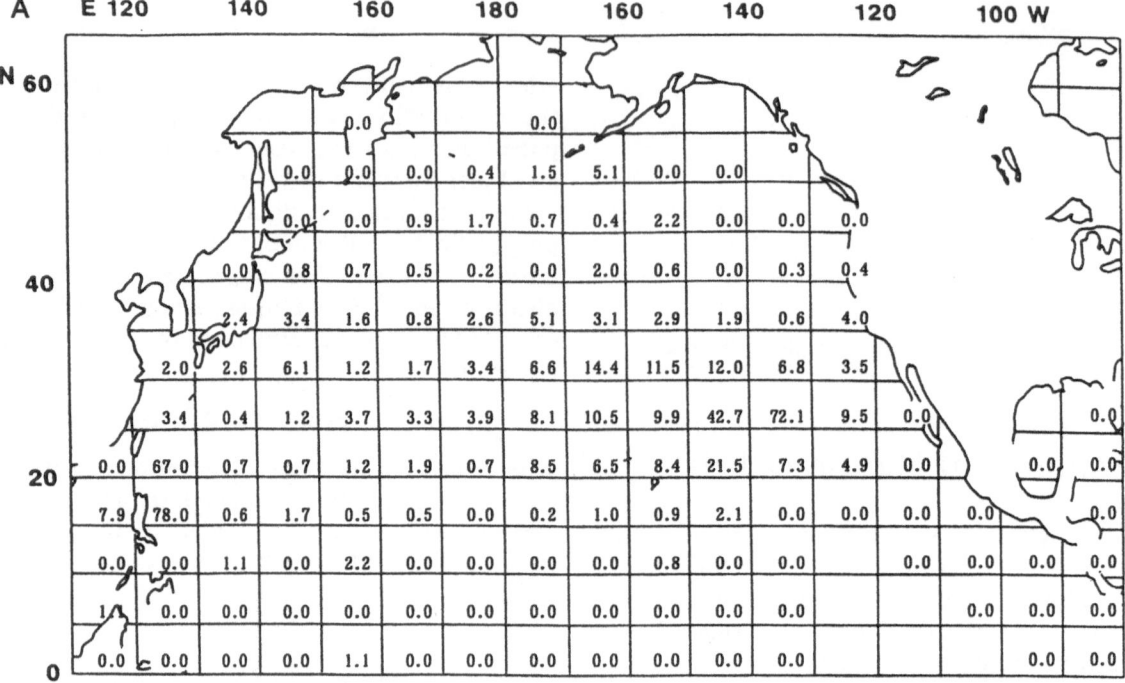

Figure 2.3. A Density of floating fishing nets/100 nm², 1987–1991. B Density of fishing floats and fishing gear other than nets/100 nm², 1987–1991. C Density of floating Styrofoam/100 nm²,1987–1991. D Density of floating plastic products/100 nm², 1987–1991. E Density of floating logs/100 nm², 1987–1991. F Density of floating seaweed/100 nm², 1987–1991. G Density of all floating objects/100 nm², 1987–1991.

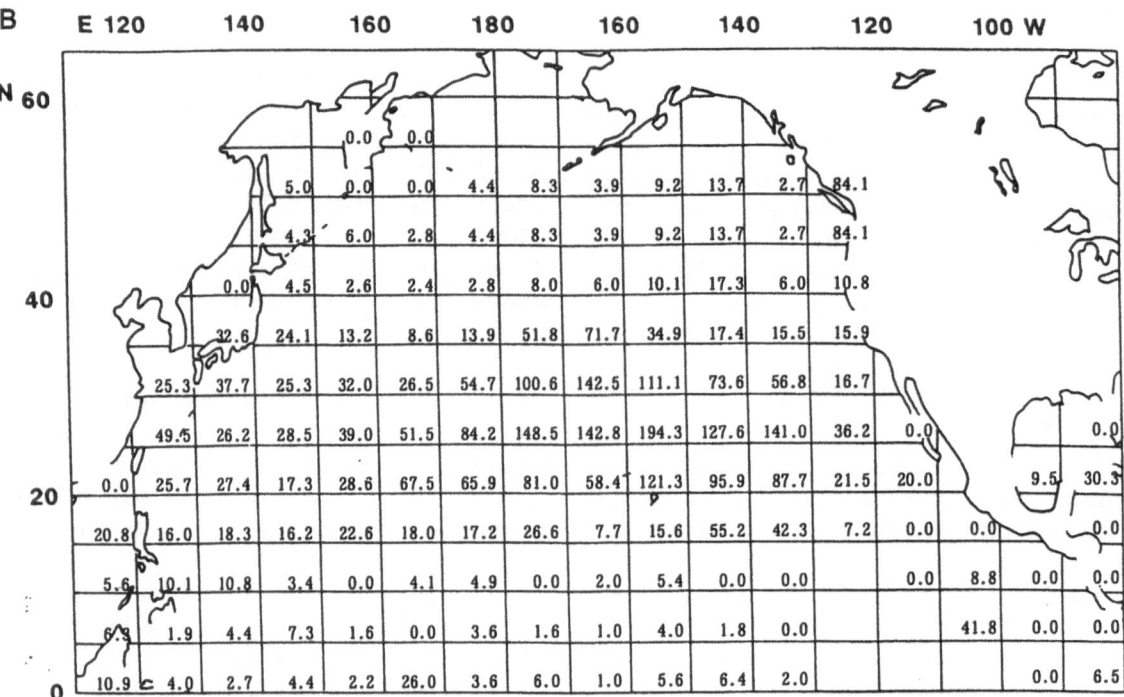

C

Longitude: E 120 — 140 — 160 — 180 — 160 — 140 — 120 — 100 W

	E 120	130	140	150	160	170	180	170	160	150	140	130	120	110	100	90
			0.0	0.0			0.0									
			0.0	0.0	0.0	4.5	2.2	10.1	4.8	0.0	14.4					
		2.3	4.4	1.0	3.0	63.3	10.4	3.9	13.8	6.0	0.0					
40		0.0	13.8	5.9	4.3	7.0	7.1	6.6	8.0	9.3	5.7	21.7				
		198.6	86.7	33.2	20.8	24.1	26.8	61.2	24.0	20.4	17.6	34.9				
	43.4	244.5	146.3	100.8	81.1	85.1	105.5	170.5	146.1	110.4	104.2	47.5				
	224.0	277.6	113.8	92.9	106.3	100.2	161.4	90.2	99.4	163.5	143.1	48.9	0.0			0.0
20	3178.5	225.2	73.1	58.2	48.3	73.0	57.6	30.7	29.3	34.1	36.9	33.4	52.5	55.0	563.0	539.4
	368.7	40.6	39.5	38.5	26.3	17.1	13.0	15.6	9.7	9.2	15.7	38.7	26.4	47.6	40.0	188.5
	18.3	33.2	17.2	8.6	2.3	2.7	5.3	0.0	4.9	3.4	0.0	0.0		8.4	19.2	256.5	151.2
	33.?	8.5	7.2	8.1	0.9	0.0	3.1	2.8	3.9	30.7	2.0	13.6			91.8	113.6	117.2
0	26.3	14.8	18.1	17.1	7.1	0.0	3.9	7.7	10.5	6.1	3.5	2.2			0.0	139.4

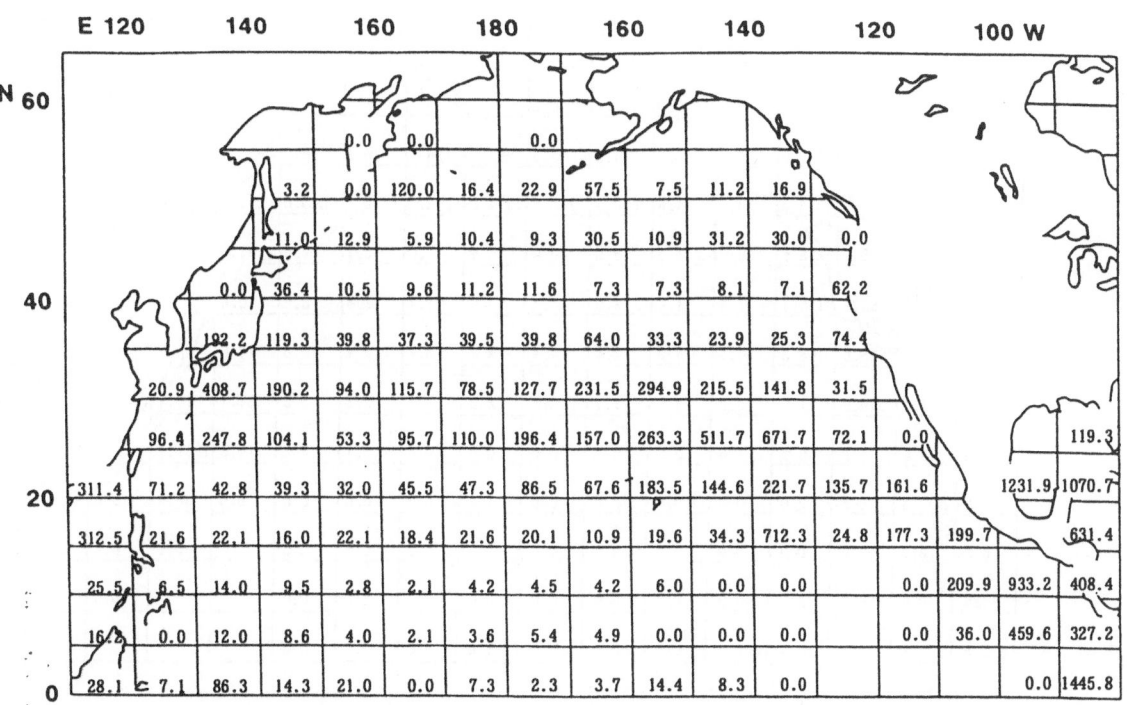

D

Longitude: E 120 — 140 — 160 — 180 — 160 — 140 — 120 — 100 W

	E 120	130	140	150	160	170	180	170	160	150	140	130	120	110	100	90
			0.0	0.0			0.0									
		3.2	0.0	120.0	16.4	22.9	57.5	7.5	11.2	16.9						
		11.0	12.9	5.9	10.4	9.3	30.5	10.9	31.2	30.0	0.0					
40		0.0	36.4	10.5	9.6	11.2	11.6	7.3	7.3	8.1	7.1	62.2				
		192.2	119.3	39.8	37.3	39.5	39.8	64.0	33.3	23.9	25.3	74.4				
	20.9	408.7	190.2	94.0	115.7	78.5	127.7	231.5	294.9	215.5	141.8	31.5				
	96.4	247.8	104.1	53.3	95.7	110.0	196.4	157.0	263.3	511.7	671.7	72.1	0.0			119.3
20	311.4	71.2	42.8	39.3	32.0	45.5	47.3	86.5	67.6	183.5	144.6	221.7	135.7	161.6	1231.9	1070.7
	312.5	21.6	22.1	16.0	22.1	18.4	21.6	20.1	10.9	19.6	34.3	712.3	24.8	177.3	199.7	631.4
	25.5	6.5	14.0	9.5	2.8	2.1	4.2	4.5	4.2	6.0	0.0	0.0		0.0	209.9	933.2	408.4
	16.?	0.0	12.0	8.6	4.0	2.1	3.6	5.4	4.9	0.0	0.0	0.0		0.0	36.0	459.6	327.2
0	28.1	7.1	86.3	14.3	21.0	0.0	7.3	2.3	3.7	14.4	8.3	0.0			0.0	1445.8

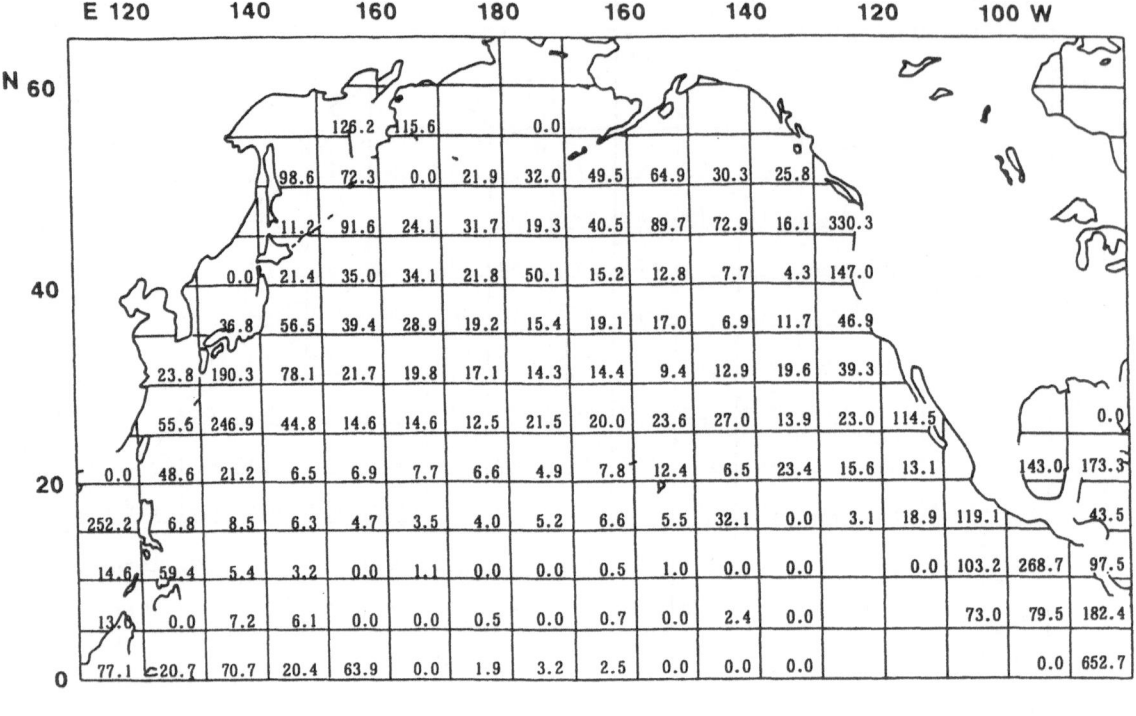

E

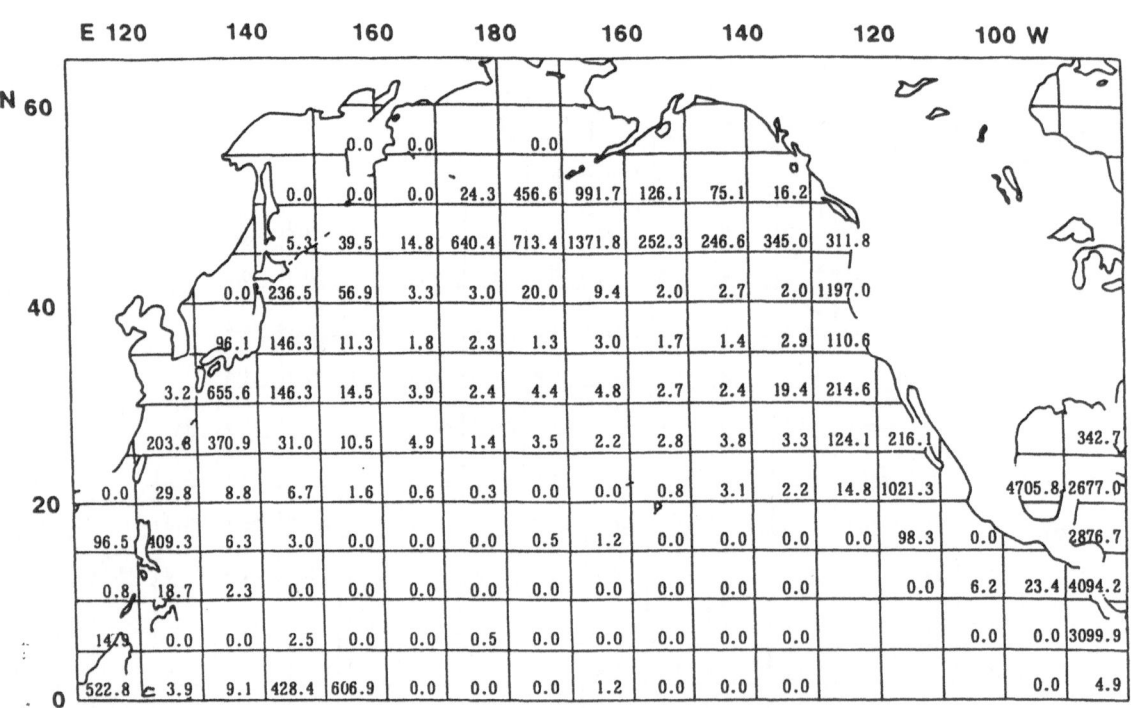

F

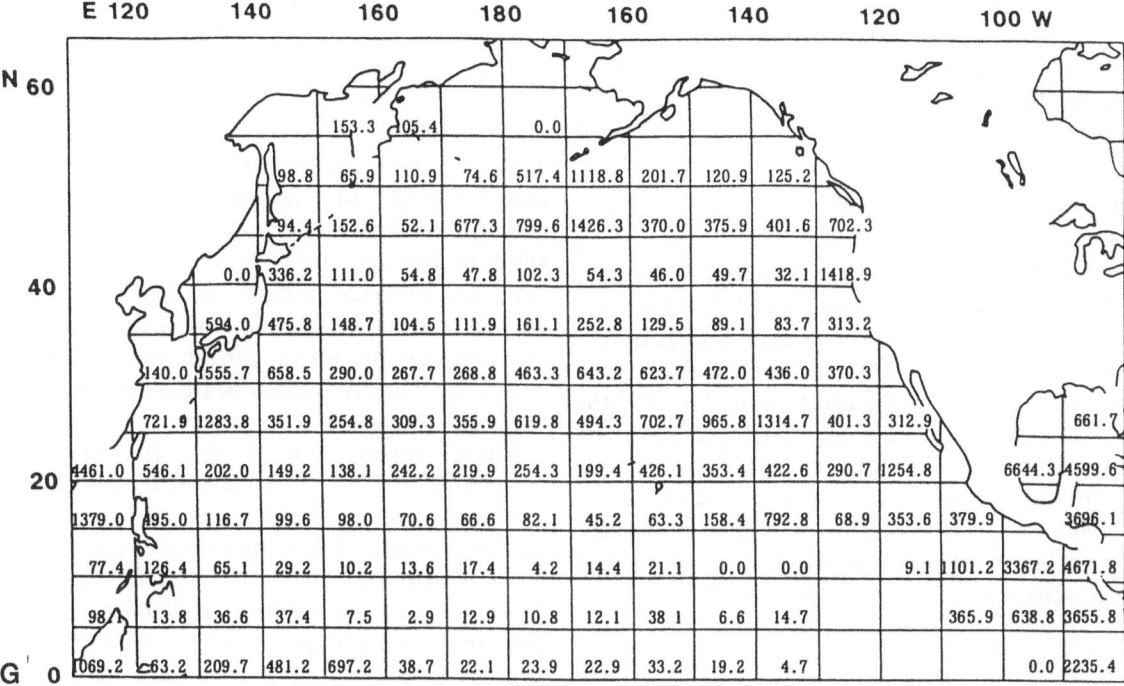

Styrofoam.

Figure 2.3C shows that high-density areas of Styrofoam (more than 200 pieces per 100 square nautical miles) were distributed from the East China Sea to the south of Japan. The highest density block was near the coast of China, in the East China Sea. The coasts of middle America and the Gulf of Mexico were also extremely high density areas. Elsewhere, a relatively high density of Styrofoam was found north of Hawaii (coincident with the high density of fishing nets). In general, high-density areas of Styrofoam are distributed along coastal regions, with the exception of waters north of Hawaii. This indicates that much of the Styrofoam observed was generated from land. Styrofoam was found in 77% of the total North Pacific Ocean blocks surveyed and is the most widely dispersed material.

Plastic Products.

Figure 2.3D shows that the highest density areas of plastic products were found near the middle America coast of the Pacific Ocean and in the Caribbean Sea. The next highest density areas were in the north of the South China Sea and along midlatitude 20° to 35° N. The waters northeast of Hawaii also included high-density areas of plastic debris. Generally, plastic debris was very widely distributed. Like Styrofoam, plastic products were observed in 77% of the total blocks surveyed in the North Pacific Ocean, and its high density in coastal waters also indicates significant land-based sources.

Floating Logs.

Figure 2.3E shows that the density of floating logs was relatively high in the area north of Hawaii, which was similar to that of other items. However, the density was generally low in the midlatitudinal area of the central Pacific Ocean. On the other hand, higher densities of floating logs were observed in areas from the Gulf of Alaska to northern California. Generally, the density of floating logs was higher in coastal waters throughout the Pacific Rim, from Central America to Indonesia. This phenomenon resulted from the sighting of floating logs that originated from the land.

Seaweed.

Figure 2.3F indicates that seaweed has a tendency to be more dense in areas near land. The highest densities were observed in areas of the middle America coasts, both in the Gulf of Mexico and in the Pacific Ocean. The Pacific Ocean coasts of Taiwan, Japan, the Aleutian Islands, and Alaska Bay also had high seaweed densities. Seaweed obviously comes from coastal waters, and most of it is so easily degradable that it will not persist long enough to be transported into midocean areas.

Conclusion

Figure 2.3G shows that the highest densities, of all combined floating objects, are in coastal waters. This results from the concentration in these waters of floating logs and seaweed, which accounted for almost half the debris sighted. Synthetic objects are distributed over a much wider range. Total debris densities in coastal waters were 20–40 objects per square nautical mile, while the density in the north equatorial current area (5° to 15° N, across the central Pacific) was about 0.2 objects per square nautical mile, and 1–3 objects per square nautical mile in the subarctic boundary area (35° to 45° N).

Based on the survey observations that the highest densities of marine debris form in coastal waters, we conclude that the majority of marine debris seems to originate from land discharge. Even after considering the contribution of seaweed density to this observation, and that Styrofoam originates from ships as well as land, and the effects of ocean currents, the need for mitigation of the land-based sources of floating marine debris is strongly indicated. Furthermore, because the types of marine debris observed in this survey float on the sea surface, at least for some period of time, it is obvious that the distribution, movement, and accumulation of marine debris are closely related to the general circulation of the sea surface layer.

These surface circulation patterns are caused by drift currents, which are produced by wind stress. Therefore, the distribution and movement of marine debris are assumed to depend on the dynamics of drift currents. As such, the accumulation and dispersion of marine debris should conform to the large time and spatial scale drift current factors such as the convergence and divergence zones of the oceans. This assumption is supported by the consistent high-density area north of Hawaii, where a large gyre current exists. At short time scales, distribution of floating debris, Styrofoam in particular, is assumed to depend on wind dynamics such as wind direction, force, and duration.

Acknowledgments. The survey fleet was organized by the Fishing Ground Environment Conservation Division, Fisheries Agency of Japan. We thank former Deputy Director Shuichi Takehama for planning and coordinating the first few years of this project. We also thank the captains and observers of the 64 vessels that voluntarily participated in the survey. And last, we thank Dr. K. Hiramatsu and Professor N. Kishino for helping us analyze our sighting data.

3∎
Marine Debris in the Caribbean Region

James M. Coe, Stefan Andersson, and Donald B. Rogers

Introduction

The Wider Caribbean Region is generally defined as the geographic area including the Gulf of Mexico, the Caribbean Sea, and the adjacent waters of the Atlantic Ocean from south of 30° N latitude and within 200 nautical miles of the Atlantic coast of the Caribbean States (Cartagena Convention 1983). The states of the Wider Caribbean Region include 12 continental states, 13 island states, the Commonwealth of Puerto Rico, 3 overseas departments of France, a territory (St. Maarten) shared by the Netherlands and France, and 11 dependent territories (WRI 1992) (Fig. 3.1)

The two major water masses in this region are the Gulf of Mexico and the Caribbean Sea, which are joined via the Yucatán Channel (located between the Yucatán Peninsula and Cuba). Water from the Caribbean Sea flows north into the Gulf and then exits to the east along the north coast of Cuba and through the Straits of Florida to the Gulf Stream. In the eastern Gulf, the Loop Current flows north from the Yucatán Channel then turns sharply to the southeast on its way to the Straits of Florida. In the western Gulf, currents are dominated by an anticyclonic (clockwise) gyre off the coasts of Louisiana, Texas, and northern Mexico. There is little exchange of water between the two areas of the Gulf with the exception of an occasional eddy that may pinch off from the Loop Current and enter the western Gulf (Heneman 1988).

The Caribbean Sea is influenced by two oceanic currents that drift parallel to the equator: the North Equatorial Current and the South Equatorial Current. Both these currents flow from east to west. Together these currents, known as the Caribbean Current, bend northwestward toward the Yucatán Channel. There are two smaller bodies of water with cyclonic (anticlockwise) flows in this region: the Bay of Campeche, and the southwest Caribbean gyre partially encircled by the coasts of Nicaragua, Costa Rica, and Panama (Heneman 1988).

It is estimated that by the year 2000 the population of the Wider Caribbean's coastal and island areas will reach 65 million people (WRI 1992). Within the last three decades, much of the Caribbean Region has undergone a dramatic sociocultural shift away from the more traditional rural and agricultural lifestyle toward urban-industrial and tourist-based employment. United Nations Environment Programme (UNEP) (1994) reported that the populations of coastal areas of the Wider Caribbean Region have been growing at a 3% annual rate since 1980. This shift not only brings more people and their wastes into the coastal zone, it greatly increases their dependence on processed, manufactured, and packaged goods that, in turn, results in increased persistent wastes. Not only are the coastal inhabitants and mariners contributors to the marine debris problem, the residents of the many watersheds draining into the Gulf of Mexico and Caribbean Sea as well as the tourist population also contribute.

The Caribbean Tourist Organization esti-

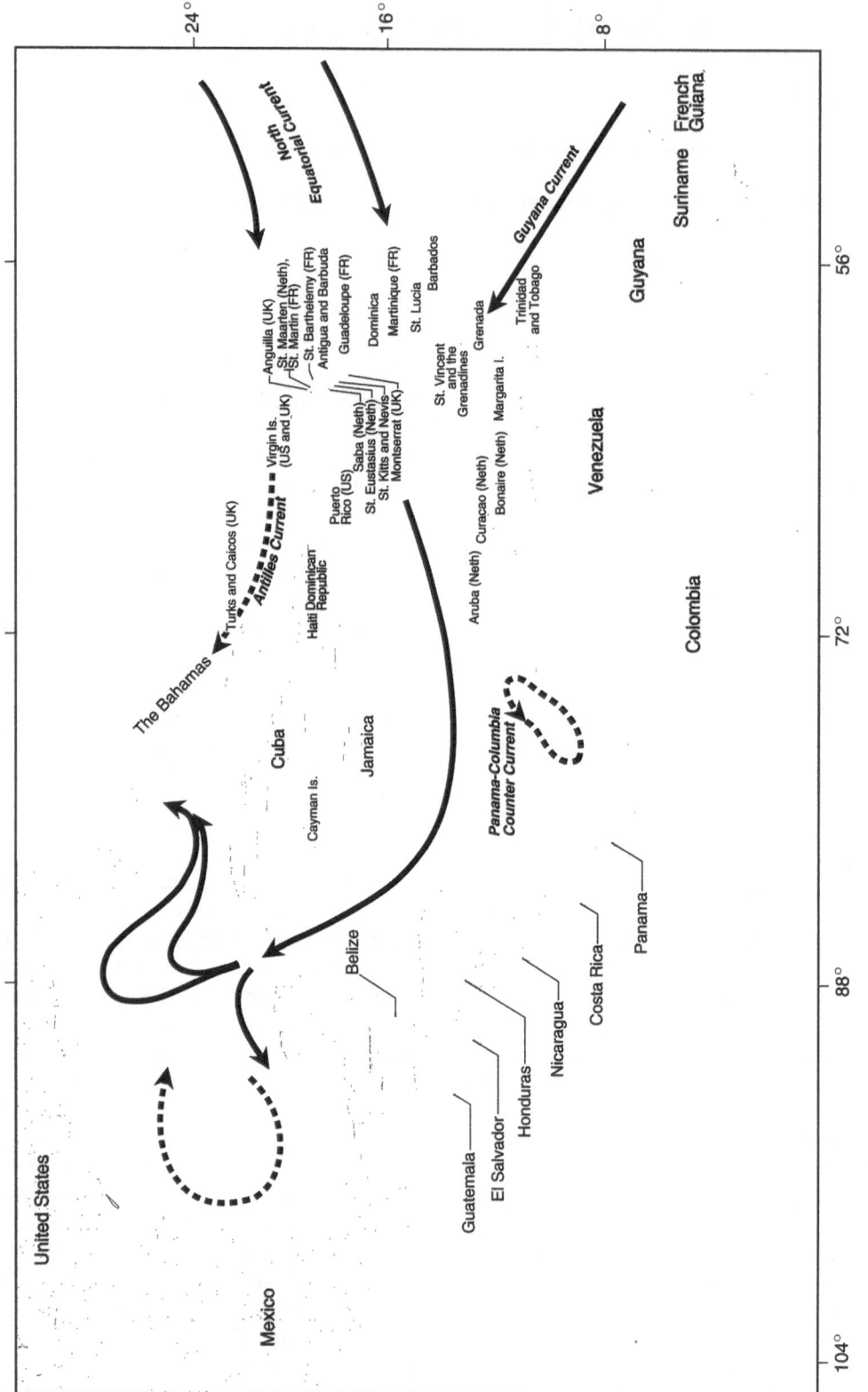

FIGURE 3.1. The Wider Caribbean Region and ocean currents.

mated that 12 million tourists visited the region in 1994 (UNEP 1994), with an annual growth rate of 9.0%. Meanwhile, cruise ships bring another 8 million tourists annually (UNEP 1992), with a growth rate of 7.5%. And although there is no watershed census, there are conservatively tens of millions living in the upland areas of the coastal watersheds of the Wider Caribbean Region whose persistent wastes may be carried into the marine environment. There are 20 major rivers (not including the Amazon) that contribute roughly one-sixth of the surface water that enters the Gulf of Mexico and the Caribbean Sea (UNEP 1992). The impact to the region by marine debris is further compounded when considering that the Guyana Current, which sweeps the northeast coast of South America and passes into the Caribbean Sea through the Lesser Antilles, distributes floatable debris from the Amazon and other southern rivers. Tropical storms and hurricanes passing through the region also affect the amounts, types, and distribution of marine debris. Torrential rainfall and storm surges along hurricane-lashed coasts generate large, periodic increases in the amounts of land-based marine debris. At sea, storm winds sweep the surface, driving floating debris ashore great distances from its origins.

This paper reviews results of two ongoing marine debris surveys in the Caribbean region [a review of the amounts, types, and distribution of marine debris in the Gulf of Mexico is provided in this section (Chapter 4) by Ribic et al.]. The principal data are provided by the IOC/UNEP Caribbean Marine Pollution Assessment and Control Programme (CEPPOL). In 1990 and 1991, the Intergovernmental Oceanographic Commission (IOC) and the United Nations Environment Programme (UNEP) jointly surveyed six localities within the region: Barbados, Colombia, Dominica, Medico, Puerto Rico, and St. Lucia. The Center for Marine Conservation (CMC)—a nonprofit, nongovernmental organization—provides additional data on these and other localities: CMC gathers data from, and provides guidelines on, international coastal cleanup events. When presented side-by-side, CEPPOL and CMC data provide a very prelim-

inary look at the amounts, types, and distribution of ocean- and land-based marine debris in the Caribbean region.

The CEPPOL Survey: 1990–1991

Method

The CEPPOL sampling design was derived from the IOC/FAO/UNEP (1989) pilot survey in the Mediterranean (MEDPOL). In an attempt to prove MEDPOL's recommended guidelines for beach surveys adequate, CEPPOL sampling stations were primarily homogeneous, undisturbed beaches of offshore islands. On each beach, a group of 12 randomly selected transects were sampled; each transect was 5 m wide. The data from the individual transects were pooled so as to have at least 30 different combinations of transects per group. Groups of 2–6 transects were plotted against the mean number of items per group. Individual data from each transect were also plotted, and the mean and SD of both group and individual data were calculated. From the plots, it was observed that a minimum of 4 transects were needed for the collection of data to produce a mean that was within ± 1 SD of the universal mean (which agrees with MEDPOL's recommended guidelines).

The transects were oriented perpendicular to the beach (from waterline to vegetation line). Beaches (stations) were selected according to the following criteria: (1) stations were not to be contaminated by beachgoers; (2) they were not to be subject to periodic cleaning; (3) they were not to exhibit a "steep" slope; (4) they should not be in the vicinity of rivers; (5) they should have homogeneous sedimentology; and (6) they should be located windward and be open to the sea. At each station, transects were sampled one time per month (for 1 year).

Samples collected within each transect were classified by material type, counted, and weighed. When samples were identified as being part of one fractured item (several parts of the same item scattered about), the pieces were counted as one item. The total weight per material type was recorded. Otherwise, data pertaining to the geographical location of the beach, as well as the beach's physical characteristics, were recorded. A list of items commonly associated with marine activities was included on the CEPPOL data sheet. The numbers of each of those items, as well as the total number of items collected at each transect, were also recorded.

Caribbean Coastal Cleanups

The Center for Marine Conservation (CMC) started coordinating Caribbean coastal cleanup events in 1989: Mexico was the first wider Caribbean country to participate. As of 1994, 25 Caribbean states participate in the annual event. The individual cleanups are conducted by volunteers who are organized by local coordinators cooperating with the CMC International Coastal Cleanup Program. CMC's role also includes data analysis and reporting data by country.

Method

Universal data cards are used by all coordinators participating in voluntary coastal cleanups. When an item is collected, it is tallied in one of eight categories on the data card: plastic, Styrofoam, glass, rubber, metal, paper, wood, and cloth. Weight of trash is determined locally, using methods that vary from site to site; therefore, CMC interprets and reports data based on item count rather than weight or volume (Sheavly 1995).

Generally, CMC provides data analysis for each individual cleanup in four ways: the country's 12 most abundant items collected (e.g., plastic pieces, plastic rope, plastic caps and lids, etc.), major characteristics of debris in each zone [a zone is normally one contiguous beach] (i.e., traceable debris reported, entangled wildlife report), the percent composition of each country's beach debris, and

the number of debris items associated with identifiable types and sources of debris. In the latter group, items associated with identifiable types and sources of debris, CMC further reflects item counts in seven categories: bottles, commercial fishing, recreational fishing, galley waste, (vessel) operational waste, sewage, and medical waste. Other than bottles, each source category has at least several indicator items. For example, indicator items for recreational fishing include plastic floats and plastic (monofilament) fishing line; indicator items for commercial fishing include synthetic rope, plastic lightsticks, and foamed plastic buoys. Operational waste includes plastic strapping bands, wooden pallets, and glass light bulbs; galley waste includes plastic trash bags and plastic milk jugs. And last, indicator items of sewage include plastic tampon applicators and rubber condoms, and medical wastes includes plastic syringes.

Amounts, Types, and Distribution

Puerto Rico.

CEPPOL surveyed two beaches in Puerto Rico: Cayo Turrumate and Bahia Sucia. Cayo Turrumate is a reef island 2 nautical miles southeast of the Puerto Rican mainland, and Bahia Sucia is located on the southwestern tip of Puerto Rico. Both beaches face the southeast, in the direction of the trade winds.

At Cayo Turrumate, the most abundant items collected were plastics, which accounted for 55.3% of the total items, 10.0% of which were from marine activities (Fig. 3.2A). The second most abundant items collected were processed wood and synthetic foam (each equaling 12.0% of the total items, representing 83.0% and 0.6% by weight, respectively). Likewise, at Bahia Sucia, plastics were the most abundant item (57.0%), while synthetic foam was second (33.0%) and processed wood was third (4.5%; however, 43.0% by weight) (Fig. 3.2A). The data from both Bahia Sucia and Cayo Turrumate

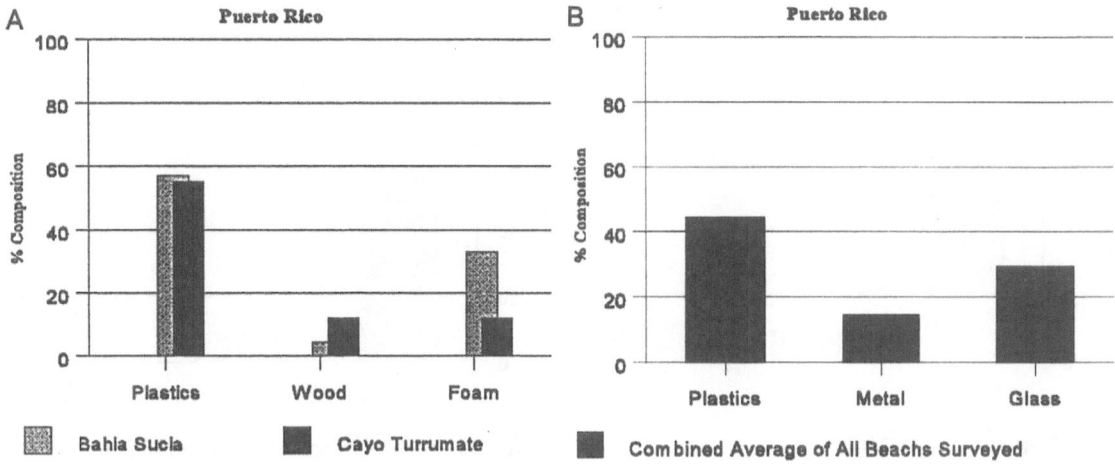

FIGURE 3.2. A CEPPOL (Caribbean Marine Pollution Assessment and Control Program) 1991–1992 survey results (Puerto Rico). B CMC (Center for Marine Conservation) 1994 coastal cleanup results (Puerto Rico).

identify fishing activities as a significant source of marine debris. The percentage of fishing-related debris in relation to the total amount of items collected ranged from 8.2% to 48.4%. The CEPPOL results identify many of the collected items with shore-based activities, as well as to inappropriate (or illegal) disposal practices.

In general, the voluntary beach cleanup data correspond well with the CEPPOL data. Of the 232,982 total items collected during Puerto Rico's 1994 coastal cleanup event, plastics were the most abundant item, accounting for 44.5% of the total (Fig. 3.2B). At Mayaguez, on the west coast of Puerto Rico, processed wood was the most prevalent single item reported by CMC, and plastics accounted for 48.2% of the total items collected. Otherwise, CMC's data, which encompass 24 beaches in Puerto Rico, show that glass beverage bottles were the most abundant item reported throughout Puerto Rico. Among the 78,515 debris items identifiable by type and source, there were 64,731 bottle items (glass and plastic), 5,561 galley items, and 3,421 commercial fishing items.

Colombia.

CEPPOL surveyed two isolated beaches in Colombia: Playa Blanca and Castilletes. Playa

Blanca is located south of Cartagena and Castilletes is located on the southeast side of Peninsula de Guajira. The composition of debris was similar at both beaches. The most abundant items were plastic (Fig. 3.3A), accounting for 59.0% of which 4.0% was related to marine activities. The second most abundant item was synthetic foam, averaging 12.0% of the items per transect. Processed wood accounted for 9.0% of total items collected, 13.0% by weight. The combined data from Playa Blanca and Castilletes show fishing activities as a significant source of marine debris. While the percentage of fishing-related debris in relation to the total amount of items collected ranges from 11.0% to 31.0%, the most abundant items collected were shampoo (and other personal care product) containers, plastic bottles, beverage cans, and plastic garbage bags. These items suggest significant debris input from shore-based activities.

Similar results are reported by CMC. In Colombia, CMC reported data from four beaches: Casimba Beach, Isla Rosario, La Boquilla Beach, and Punta Arena Beach. From these beaches, 8269 total items were collected during Colombia's 1994 coastal cleanup event. Plastics accounted for 57.2% of the total amount collected (Fig. 3.3B). The three most frequently identified types and sources were bottles (1062 items), galley

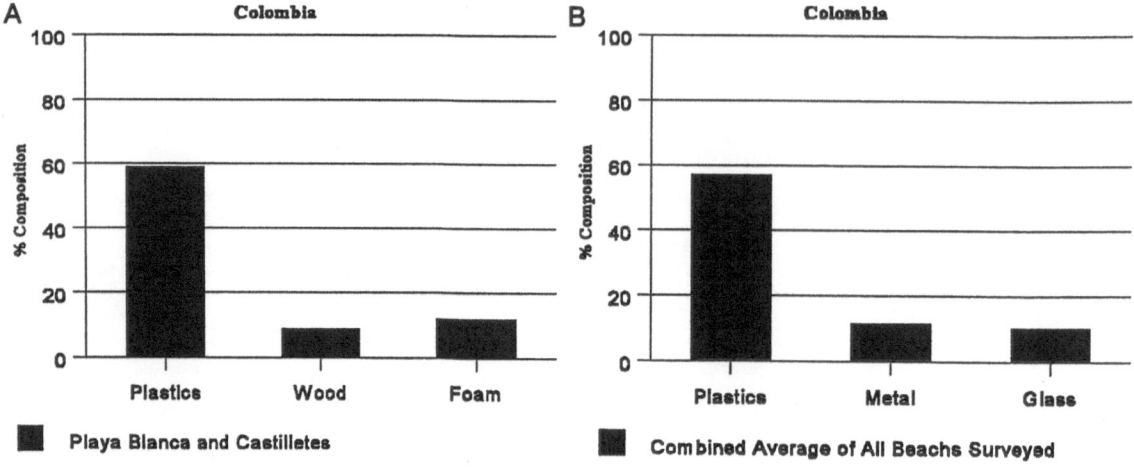

FIGURE 3.3. A CEPPOL 1991–1992 survey results (Colombia). B CMC 1994 coastal cleanup results (Colombia).

waste (790 items), and operational waste (235 items), suggesting a mix of land, ship, and recreational sources.

Mexico.

CEPPOL surveyed four beaches on the northwestern margin of Mexico's Yucatán Peninsula: Telchac, Chelem, Celestúm, and Chuburna. At Telchac, the average number of items per transect was 0.4, with an average weight of 51 g per transect; the most frequent

items were plastics and metals (each equaling 33.0% of the total items collected) (Fig. 3.4A). Also at Telchac, cloth accounted for 22.0% of the total items collected. The other surveyed beaches were more homogeneous in the amount of items collected, with a combined average of 4.0 items per transect. At Chuburna and Celestún, plastics were the most abundant items collected, accounting for 36.0% and 38.0% of the total, respectively; metals were the second most abundant items (32.0%) (Fig. 3.4A). However, at Chelem, metals were the most abundant items collected, accounting for 47% of the total,

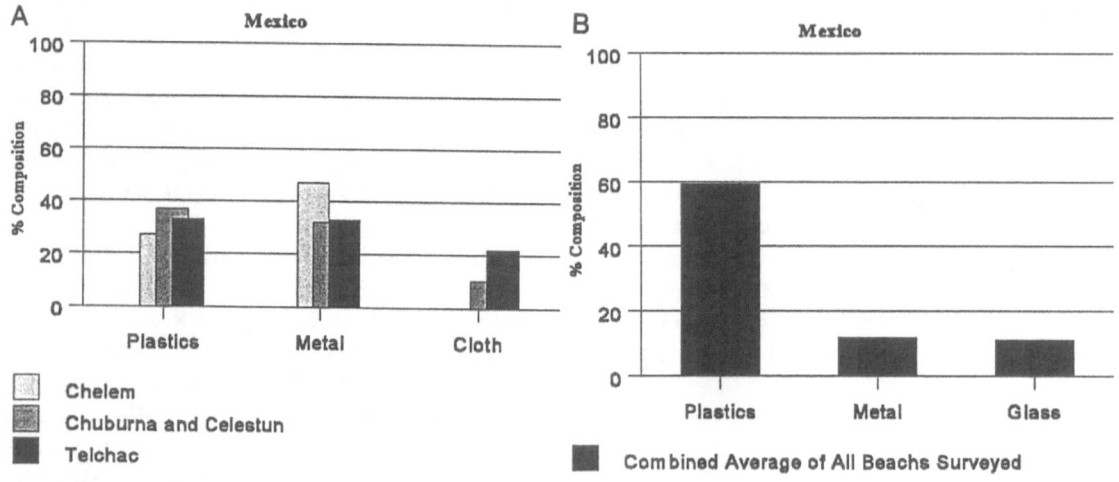

FIGURE 3.4. A CEPPOL 1991–1992 survey results (Mexico). B CMC 1994 coastal cleanup results (Mexico).

and plastics were the second most abundant items (27%) (Fig. 3.4A).

Among the data provided by CMC from Mexico's 1994 coastal cleanup event, there were four beaches that are on the Yucatán Peninsula or otherwise on the Gulf of Mexico: Campeche, Yucatán, Cozumel, and Tamaulipas. Campeche is located on the western margin of the Yucatán Peninsula, approximately 100 km south of Celestún; Yucatán is located on the northernmost margin of the Yucatán Peninsula, east of Telchac; Cozumel is located on the Isla de Cozumel, on the east side of the Yucatán Peninsula; and Tamaulipas is located on the Gulf of Mexico, near Tampico. On all the Mexican beaches surveyed, plastics were the most abundant items collected (59.5% of the total) (Fig. 3.4B). On Campeche, plastics accounted for 62.6% of the items collected, plastic trash bags, being the most prevalent items. On Yucatán, plastics accounted for 51.3%, and plastic straws were the most prevalent item. On Cozumel, plastics accounted for 68.4%, and plastic pieces were the most prevalent item. And on Tamaulipas, where plastic trash bags were the most prevalent item collected, plastics accounted for 49.3%. The three most frequently identifiable types and sources (of the 73,726 identifiable items collected) on Mexico's beaches (not including beaches on the Pacific Ocean) were bottles (11,767

items), galley (8,089 items), and commercial fishing (3,391 items).

Barbados.

CEPPOL surveyed two beaches in Barbados: Gays Cove and Long Beach. Gays Cove is located on the eastern margin and Long Beach is located on the southeastern margin. Plastics accounted for 70.0% of the combined totals of items collected at both beaches (Fig. 3.5A). Also at both beaches, fishing gear was the second most abundant item collected. In terms of weight, processed wood accounted for 80.0% (plastics, 9.0%) at Long Beach, and 32.0% (plastics, also 32.0%) at Gays Cove. There were also notable amounts of metal, cloth, and Styrofoam on both beaches.

Chancery Lane was the only beach involved in the Barbados 1994 coastal cleanup event. Chancery Lane is located on the western side of Barbados. Here, plastics accounted for 82.3% of the 2972 items collected (Fig. 3.5B). The three most common items collected were plastic pieces (20.2% of the total items collected), plastic rope (12.8%), and plastic fishing nets (8.7%). The three most frequently identifiable types and sources (of the 2972 identifiable items collected) were commercial fishing (675 items),

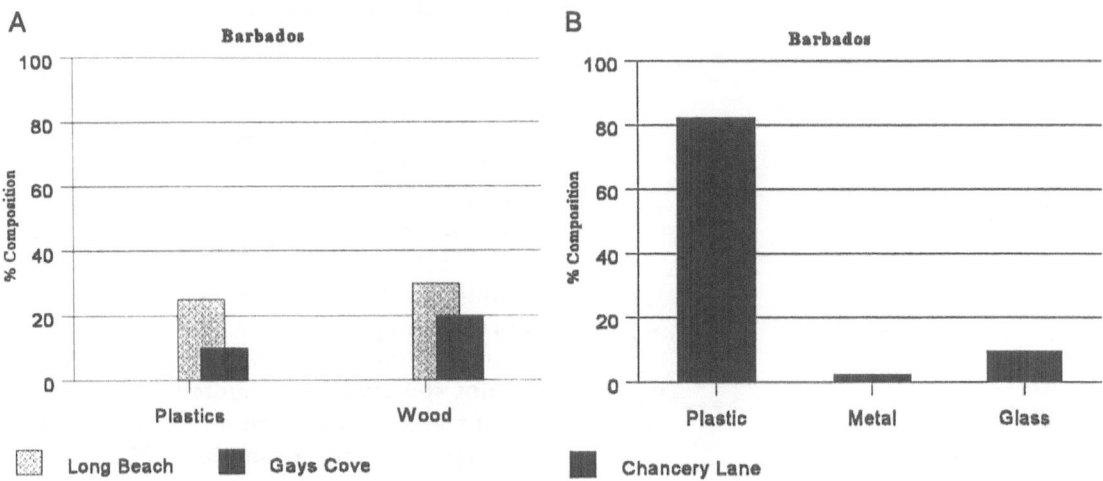

FIGURE 3.5. A CEPPOL 1991–1992 survey results (Barbados). B CMC 1994 coastal cleanup results (Barbados).

bottles (148 items), and operational (133 items).

St. Lucia and Dominica.

Coastal cleanup events were not conducted in St. Lucia and Dominica. In the CEPPOL survey, data on St. Lucia and Dominica are combined: a total of six beaches were surveyed, three on St. Lucia and three on Dominica. Although the names of these beaches are not provided, on Dominica all three beaches surveyed were west-coast recreational. Of the three beaches surveyed on St. Lucia, two were recreational beaches on the west coast and one was nonrecreational on the east coast. In St. Lucia, plastics accounted for 51.3% of the total items collected, and processed wood accounted for 1.2% (Fig. 3.6), while in Dominica, processed wood accounted for 36.0% of the total items collected and plastics accounted for 16.0%. It should be noted that the Dominica survey sites were close to a fishing village where wooden boats are often abandoned and sometimes drift to other shores.

It is also important to note that from the nonrecreational (east and windward-facing) beach on St. Lucia almost 100% more items were collected than from the other five beaches. On the nonrecreational beach, plastics accounted for 82.0% of the total items collected, and processed wood accounted for only 0.4%. Also, while the debris on the nonrecreational beach revealed diverse, off-island origins, the debris on the recreational beaches came primarily from local sources. This indicates the effects of winds and ocean currents on the amounts and distribution of marine debris and the importance of beach use and exposure on the composition of accumulated debris. Results similar to those from St. Lucia and Dominica have been obtained from Panama, the Cayman Islands, and Cuba. At these localities, plastics were the most abundant items collected, followed by Styrofoam, processed wood, or metal as the second most abundant items collected.

Discussion

Quantitative data on marine debris in the Wider Caribbean Region are limited. Scientists and other observers, such as Siung-Chang and Deane (1984), Heneman (1988), Heneman and Coe (1989), Corbin and Singh (1993), Sheavly (1995), plus papers in this volume by Wade (Chapter 15), Chaparro and Velez (Chapter 29), and Singh and Xavier (Chapter 30) have reported on the fate of persistent, land- and ocean-based marine debris in this region; however, various limitations—statistical and physical—often occur when gathering this type of information. For example, CMC's beach cleanups are conducted by volunteers from within each country, and the actual cleanup event is only conducted 1 day per year. Such cleanup results, although otherwise invaluable, lack the rigor and replication needed to fully characterize the amounts, types, and distribution of marine debris.

Additionally, the variabilities of point sources, non-point sources, and temporal and physical factors that influence distribution are not specifically accounted for in the design of any of the marine debris surveys in this region. For example, although the CEPPOL surveys use a transect method in their study (primarily on homogeneous, undisturbed beaches), their sample site selection criteria create some unknown bias against

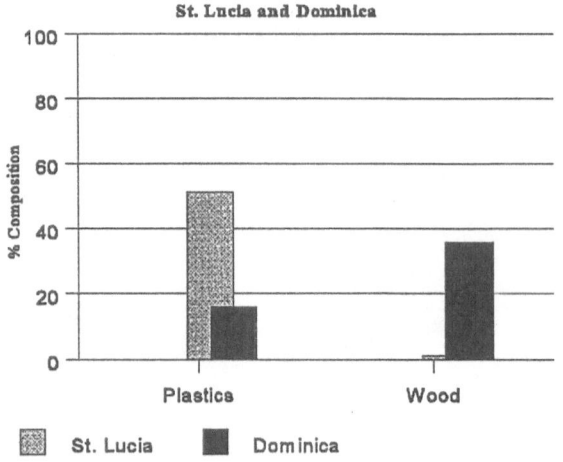

Figure 3.6. CEPPOL 1991–1992 survey results (St. Lucia and Dominica).

land-based sources of debris: selected stations are specifically not contaminated by beach-goers; they are not in the vicinity of rivers; and they are located on the windward sides of the surveyed islands. Comparisons based on the extrapolation of survey data on the amounts, types, and distribution of marine debris in the Wider Caribbean Region are speculative because of the general inability to assess, or account for, biases.

These limitations aside, the data from the CEPPOL marine debris surveys and from the CMC voluntary beach cleanups allow some preliminary observations. In each site where data were collected, plastic debris made up the majority or near-majority by count. Pro-cessed wood debris typically dominated when measuring by weight. Urban area and recreational beaches commonly showed a high level of local sources. Little-used beaches, open to the prevailing winds, com-monly had debris from diverse, far-flung sources. These remote, exposed beaches had typically higher proportions of fishing-related debris and plastics in general.

Although there are exceptions to each of these general observations, both off-shore and land-based sources of debris are obvious in the data. It appears that absent or inade-quate solid waste collection and disposal is a major contributor to the marine debris problem throughout the Wider Caribbean Re-gion (Siung-Chang and Deane, 1984; World Bank 1991). This effect is direct through the dumping of solid wastes on beaches and in estuaries and rivers, and indirect through the deterioration and escape of persistent wastes from inadequate disposal sites. Ships may deliver their garbage to ports only to see it return to the harbor via the local shoreside dump. Further, the lack of disposal capacity makes littering both necessary and socially acceptable, as observed in the recreational beach survey data.

Acknowledgments. The authors would like to thank Seba Sheavly, Kathy O'Hara, and Randy Burgess of the Center for Marine Con-servation for their support during the prepa-ration of this manuscript. Also, special thanks to Wendy Carlson with the Graphics Unit at the Alaska Fisheries Science Center for her preparation of Figure 3.1.

4∎
Distribution, Type, Accumulation, and Source of Marine Debris in the United States, 1989–1993

Christine A. Ribic, Scott W. Johnson, and C. Andrew Cole

Introduction

Marine debris as a "new" pollution problem was first identified in the United States in the 1980s during public education and awareness campaigns that addressed the condition of coastal beaches (Wallace 1985; Pruter 1987a; O'Hara 1990). Because coastal beaches collect debris from all sources, most studies evaluating marine debris have focused on beaches (O'Hara 1990; Ribic 1991; Cole et al. 1995).

This paper summarizes recent studies on the distribution, type, accumulation, and source of marine debris on coastal beaches and in select harbors of the United States, emphasizing plastic debris. We first present the problem by three coastal regions: Atlantic, Gulf of Mexico, and Pacific. We then provide a national overview and discuss trends in debris abundance from 1989 to 1993.

Methods

Marine debris data were primarily from national surveys coordinated by the Center for Marine Conservation (CMC) (Hodge et al. 1993) and localized surveys conducted by the National Park Service (NPS) (Cole et al. in press), the National Marine Fisheries Service (NMFS) (Johnson 1990a, 1990b, 1993), and the U.S. Environmental Protection Agency (EPA) (Battelle Memorial Institute 1990; Battelle Ocean Sciences 1992).

Surveys by CMC provided information on the composition and ubiquitous nature of marine debris on a national scale. In the CMC program, U.S. coastal beaches have been cleaned annually since 1986 and shorelines bordering the Great Lakes since 1991; in 1992, about 7500 km of beaches were cleaned in 29 states (Fig. 4.1) (Hodge et al. 1993). Data collected by CMC (on standardized data cards) were summarized into seven debris categories: plastic (including Styrofoam), glass, rubber, metal, paper, wood, and cloth. The CMC program is all volunteer, with a brief training session on identification of debris items. Volunteers walk the beach (including throughout and behind dunes), searching for, recording, collecting, and removing debris. Beaches are cleaned once a year, usually in September. We used data from the cleanups primarily to indicate presence or absence of debris items, not for abundance estimates. A complete description of the CMC program and annual state summaries are provided in O'Hara and Younger (1990), Debenham and Younger (1991), Younger and Hodge (1992), and Hodge et al. (1993).

In the NPS Marine Debris Monitoring Program, debris was sampled quarterly (December, March, June, and September) at eight national parks or seashores on the Atlantic, Gulf, and Pacific coasts from 1989 to 1992 (Fig. 4.1). Atlantic Coast locations were Cape Cod National Seashore (NS) in Massachusetts,

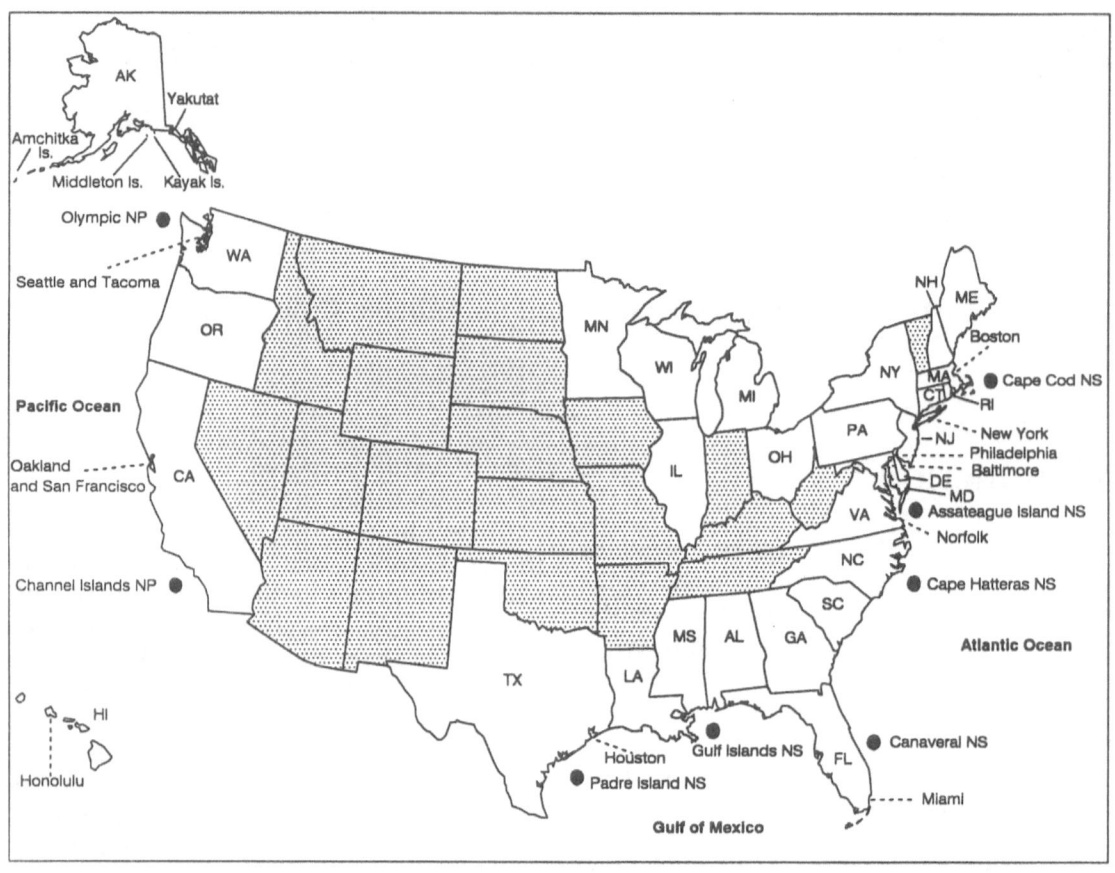

Figure 4.1. States participating in 1992 Center for Marine Conservation beach cleanups, eight national parks and seashores in the National Park Service Marine Debris Monitoring Program 1988–1992, select harbors (*dashed lines*) in the U.S. Environmental Protection Agency's Harbor Studies Program 1989–1991, and the National Marine Fisheries Service Alaska study sites 1989–1993. *NP*, National park; *NS*, national seashore.

Assateague Island NS in Maryland and Virginia, Cape Hatteras NS in North Carolina, and Canaveral NS in Florida. Gulf coast locations were Gulf Islands NS in Mississippi and Alabama and Padre Island NS in Texas. Pacific Coast locations were Channel Islands National Park (NP) in California and Olympic NP in Washington State.

At each NPS site, personnel were trained to identify and survey debris by a standardized procedure. Park personnel identified usually five 1-km sections of beach at each site where conditions were uniform and visitation light. All anthropogenic debris (except wood and stone masonry) visible from walking height was recorded on data sheets and collected. Debris was collected from water edge to the

edge of a dune line (if present) or vegetation line. Data were summarized into four primary categories: plastic (including Styrofoam), glass, metal, and other (e.g., paper, cloth, leather). Plastic items were further categorized by general source: fishing (e.g., net fragments, floats, fishing line), packaging (e.g., bottles, cups, bags), personal effects (e.g., hats, balloons, toys), and miscellaneous (e.g., fragments, sheeting, pellets).

Because debris was so abundant at Padre Island NS, six beach sections were randomly chosen and sampled with fixed 50-m-wide transects. Transect lengths varied between 35 and 100 m, depending on season. Because of the change in survey methods, only 2 years of data were available for Padre Island NS;

therefore, data from that location were not included in trend assessment. A complete description of NPS methods, including quality assurance procedures and annual summaries, are provided in Cole et al. (1990, 1992, 1995) and Manski et al. (1991).

Surveys by NMFS were conducted on approximately 32 km of outer coast beaches (mostly 1-km sections) at four locations in Alaska (see Fig. 4.1). Eight beach sections near Yakutat in the eastern Gulf of Alaska were surveyed twice a year, usually in March or April and again in September from 1989 through 1993; 7 beach sections on Kayak Island in the central Gulf of Alaska, once in August 1991; 6 beach sections on Middleton Island in the central Gulf of Alaska, once in June 1989 and July 1992; and 11 beach sections on Amchitka Island in the Aleutians, once in June 1993). Beach sections had been previously surveyed near Yakutat in September 1988, on Kayak Island in July 1988, on Middleton Island in July 1987, and on Amchitka Island in September 1987 (Johnson 1990a).

Survey methods in Alaska (Merrell 1985) were similar to those previously described for the NPS. Special emphasis, however, was placed on trawl web because major trawl fisheries for groundfish operate offshore (Low et al. 1985), and substantial amounts of trawl web are lost or discarded each year (Berger and Armistead 1987; Johnson 1989). Trawl web is also a predominant item that entangles northern fur seals (*Callorhinus ursinus*) on the Pribilof Islands (Fowler et al. 1985).

To determine deposition of trawl web and other entanglement debris (gill net, rope, and packing straps) on Alaska beaches, all items on select beach sections were usually counted and removed. At Yakutat, deposition rate (average number of pieces km^{-1} yr^{-1}) was determined by dividing the total number of "new" entanglement pieces counted each year (sum of March or April and September surveys) by eight (number of 1-km beach sections). To determine whether a declining trend existed, the average deposition rate of trawl web, gill net, rope, and strapping from 1989 to 1993 was examined by

curvilinear regression (Snedecor and Cochran 1967). At Kayak Island, all entanglement debris was counted and removed from six of seven beach sections in 1988 and 1991. At Middleton Island, trawl web was counted and removed from all beach sections in 1989 (Johnson 1990b) and in 1992. At Amchitka Island, trawl web and gill net were tagged or painted orange on five randomly selected 100-m sections of each 1-km study beach in 1987 (counts were extrapolated to pieces/km); the entire 1 km of each study beach was resurveyed in 1993. In addition to the removal of entanglement debris, on at least one beach section at each location all other plastics were recorded and plastic bottles were examined for country of manufacture to identify possible sources. A more detailed description, including quality assurance procedures, of NMFS methods is provided in Johnson (1990b, 1993, 1994).

Floating debris in several U.S. harbors was examined by the EPA. Sites included Boston, New York, Philadelphia, Baltimore, Norfolk, Miami, Houston, Seattle–Tacoma, San Francisco–Oakland, and Honolulu (see Fig. 4.1). Harbors were sampled at ebb tide with modified plankton nets towed from one of several vessels. Because of the nonrandom distribution of debris "slicks" in harbors, transects could not be used to estimate the amount of debris in each harbor. Debris items likely to come from urban sources, especially sewer overflows, were the focus of the study, and a modified CMC data form was used to record data. Items of concern to EPA included plastics that could pose a threat to human health (e.g., medical debris) or wildlife (e.g., rope) or could economically damage an area. A complete description of EPA items of concern and sampling methods, including quality assurance procedures, are provided in Battelle Memorial Institute (1990) and Battelle Ocean Sciences (1992).

Data from all studies were standardized to follow the categories of the NPS (Cole et al. 1992). Plastics, including Styrofoam and rubber, were classified as fishing gear, packaging, personal effects, miscellaneous, or medical. The medical category (e.g., syringes, cocaine "crack" vials) was added because those items are prevalent in harbors; in the NPS study, medical items were placed in the miscellaneous category because of their low incidence. Sewage items (e.g., condoms, tampon applicators) were put into the personal effects category. Select debris items were also divided into impact categories: wildlife (entanglement or ingestible plastics) and human health (medical and sewage). In all cases, cigarette butts were excluded from analyses. Fragments (small unidentified plastic pieces) were excluded from total debris because of inconsistent sampling but were included in discussion of ingestible plastics. Pellets, the raw material from which plastics are made, were included in all analyses. Types and composition of debris were summarized as percentages, and accumulation rates were standardized to number of pieces per kilometer.

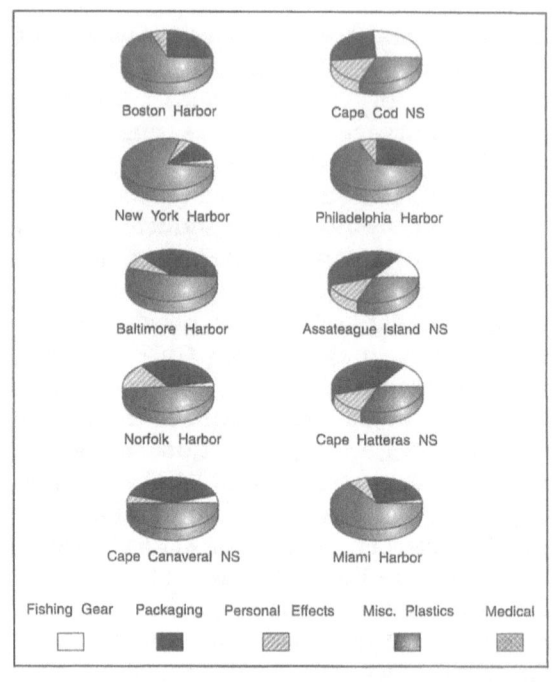

FIGURE 4.2. Percent composition of plastic debris (based on total number of plastics; excluding cigarette butts and fragments) by category for select harbors (1989–1991) and national seashores (1992) on the U.S. Atlantic Coast. Sewage is included in personal effects. NS, national seashore.

Atlantic Coast Results

Distribution and Type

Most man-made debris on beaches and in harbors of the Atlantic Coast was made of plastic. Surveys by CMC in 1992 documented that plastics as a percentage of the total debris varied by state from a low of 48% in North Carolina and Maine to a high of 89% in Virginia. On average, plastics constituted 90% of the total debris at Cape Cod NS, Assateague Island NS, Cape Hatteras NS, and Canaveral NS in 1992. Most debris in Atlantic Coast harbors (1989–1991) was also plastics (average, 72%; $n = 6$). Northern Atlantic harbors (Boston, New York, Philadelphia, and Baltimore) had the highest percentage (73%–92%) of plastics while less-urbanized sites (Norfolk) had the lowest (67%). Miami, heavily urbanized, had a high percentage (81%) of plastic, similar to northern Atlantic harbors.

Packaging and miscellaneous items comprised most of the plastics. At all NS sites in 1992, packaging made up 26%–40% and miscellaneous items 31%–50% of all plastics (Fig. 4.2). At the NS sites, the five most

abundant debris items usually included hard or foam plastic fragments, small plastic bags, and caps or lids (Table 4.1). The composition of plastics in harbors (1989–1991) was similar to that on beaches. In harbors, packaging ranged from 13% to 36% and miscellaneous items from 49% to 78% of the total plastics (Fig. 4.2). Plastic fragments, pellets, small pieces of plastic sheeting, and miscellaneous food wrappers were usually the most abundant items in harbors (Table 4.1). The greater abundance of miscellaneous items in harbors than on beaches was directly attributable to large numbers of pellets. Another difference was the presence in harbors of a small but consistent amount of medical debris (0.1%–2.0%). Medical items were present in only trace amounts at the NS sites.

Accumulation

The quarterly accumulation rate of plastics varied by location at the NS sites. In 1992, the

TABLE 4.1. The five most abundant plastic debris items (excluding cigarette butts) on national seashore (NS) beaches (1992) and in harbors (1988–1991) of the U.S. Atlantic Coast.

Item	Boston Harbor	Cape Cod NS	New York Harbor	Philadelphia Harbor	Assateague Island NS	Baltimore Harbor	Norfolk Harbor	Cape Hatteras NS	Canaveral NS	Miami Harbor
Pellets	2	—	1	1	—	1	—	—	—	4
Fragments[a]	1	2, 3	2	2	1, 5	2	1	1, 3	1, 2	1
Small sheeting	5	—	3	3	—	3	3	—	—	2
Miscellaneous food wrappings	4	—	4	—	—	4	3	—	—	5
Spheres	—	—	—	5	—	—	4	—	—	3
Cups, spoons, forks, straws[b]	—	—	5	—	4	—	—	—	3	—
Caps and lids	3	5	—	—	—	—	—	5	4	—
Packaging material	—	—	—	—	—	5	—	—	—	—
Cigarette wrappers	—	—	—	—	—	—	5	—	—	—
Beverage labels	—	—	—	4	—	—	—	—	—	—
Small bags	—	4	—	—	2	—	—	2	—	—
Bottles <3.8 l	—	—	—	—	—	—	—	—	5	—
Rope <1 m	—	1	—	—	—	—	—	—	—	—
Balloons	—	—	—	—	3	—	—	4	—	—

Items are ranked in descending order of abundance from 1 to 5. A dash indicates the item was not one of the five most abundant.
[a]Fragments separated into hard and foam fragments in the NS surveys.
[b]Straws only for the NS survey.

lowest average accumulation rate was at Assateague Island (317 pieces/km) and the highest at Cape Cod (1128 pieces/km). For all NS sites on the Atlantic Coast, the average quarterly accumulation rate of all categories of plastics was 716 pieces/km.

Source

The major source of identifiable plastics that washed ashore on the Atlantic Coast was the United States; Canadian items were second in frequency. Virginia had the most diverse foreign debris, with items from eight countries. Debris with cruise ship logos were particularly abundant on Florida beaches.

Entanglement Debris

At the NS sites, the average quarterly accumulation rate of all entanglement debris ranged from 8 to 29 pieces/km. In 1992, the accumulation rate was greatest for rope (3–14 pieces/km) but trawl web (0–5 pieces/km) and strapping (0–1 pieces/km) were consistently low; this pattern was evident in all survey years (1989–1992).

Ingestible Plastics

Plastics that can be ingested by wildlife were the most abundant debris on beaches and in harbors (see Table 4.1). At the NS sites in 1992, ingestible plastic made up 31%–56% of the total plastic debris with average quarterly accumulation rates of 142–352 pieces/km. In harbors, fragments and pellets made up 26%–71% of the total plastics. Northern Atlantic harbors had the highest percentage of pellets and fragments (average, 59.5%; n = 4) compared to other Atlantic Coast harbors (average, 29.1%; n = 2).

Medical and Sewage Debris

Medical and sewage debris were relatively uncommon on most Atlantic Coast beaches.

At the NS sites in 1992, average quarterly accumulation rate of medical debris ranged from 0 to 3 pieces/km while sewage debris ranged from 1 to 45 pieces/km. Most sewage debris was at Cape Cod NS and consisted of tampon applicators and cotton swabs. Medical and sewage debris were consistently found in Atlantic Coast harbors, averaging 1.6% of the plastics (excluding fragments).

Gulf of Mexico Results

Distribution and Type

Most debris on beaches and in harbors of the Gulf Coast was made of plastic. By state, plastics as a percentage of the total debris ranged from 56% in Florida to 68% in Louisiana. In 1992, plastics made up 82% and 94% of the total debris at Gulf Islands NS and Padre Island NS, respectively. Plastics were 99.9% of the total debris in the Houston Ship Channel.

Packaging and miscellaneous items dominated the plastics. At the NS sites in 1992, miscellaneous plastics made up 50%–70% of the total plastics and packaging 17%–32% (Fig. 4.3). Although the five most common

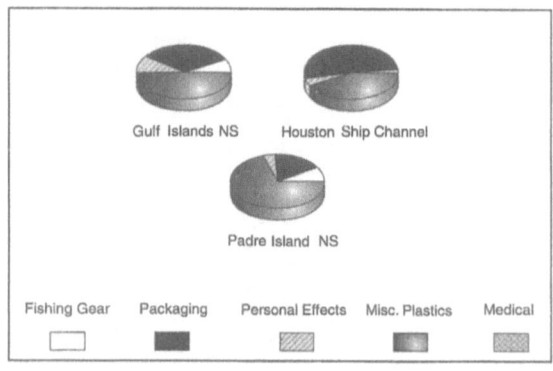

FIGURE 4.3. Percent composition of plastic debris (based on total number of plastics; excluding cigarette butts and fragments) by category for select harbors (1989–1991) and national seashores (1992) on the U.S. Gulf Coast. Sewage is included in personal effects. NS, National seashore.

plastic debris items at the NS sites varied, the two most abundant items were consistently hard and foam fragments (Table 4.2). Composition of plastics in the Houston Ship Channel was dominated by pellets (99.3%) (Table 4.2). Excluding the pellets, the remainder of the plastic debris was similar to that on the beaches: miscellaneous items made up 42% and packaging 51% of the debris (Fig. 4.3). Medical debris was uncommon in the channel (0.4%) and rare on most Gulf coast beaches.

Accumulation

In 1992, the average quarterly accumulation rate for all plastics was 544 pieces/km at Gulf Islands NS and 17,714 pieces/km at Padre Island NS. The high accumulation rate at Padre Island NS was directly attributable to numerous hard (4,883 pieces/km) and foam (3,972 pieces/km) fragments.

Source

Debris from more than 19 countries washed ashore on Gulf Coast beaches. On Texas beaches, Mexican debris was most prevalent. Daily surveys at Padre Island NS identified debris (salt bags, gloves, etc.) from commercial shrimping operations (Miller 1993).

Entanglement Debris

Although entanglement debris made up less than 5% of all plastics, a consistent amount washed ashore at both NS sites. In 1992, the average quarterly accumulation rate of all entanglement debris ranged from 21 to 260 pieces/km. Accumulation rate of trawl web (1–4 pieces/km) and strapping (1–13 pieces/km) was low compared to that of rope and gaskets, particularly at Padre Island NS. For example, at Padre Island NS, accumulation rate of rope and gaskets averaged 28 and 118 items/km, respectively. Entanglement debris made up only 2% of the plastics in the Houston Ship Channel.

Ingestible Plastics

Plastics that may be ingested by wildlife were the most abundant debris on beaches and in the Houston Ship Channel (Table 4.2). At the NS sites in 1992, ingestible plastics made up 34%–47% of the total plastic debris; average quarterly accumulation rates ranged from 186 to 8333 pieces/km. In the Houston Ship Channel, fragments and pellets made up more than 99% of the total plastics.

Medical and Sewage Debris

Medical and sewage debris were uncommon at Gulf Islands NS and in the Houston Ship

TABLE 4.2. The five most abundant plastic debris items (excluding cigarette butts) on national seashore (NS) beaches (1992) and in the Houston Ship Channel (1988–1991) on the U.S. Gulf Coast.

Item	Gulf Islands NS	Houston Ship Channel	Padre Island NS
Pellets	—	1	—
Fragments[a]	1, 2	2	1, 2
Filaments	—	3	—
Beverage labels	—	4	—
Spheres	—	5	—
Caps and lids	3	—	5
Small sheets	—	—	3
Bottles <3.8 l	4	—	—
Small bags	5	—	—
Rope <1 m	—	—	4

Items are ranked in descending order of abundance from 1 to 5. A dash indicates the item was not one of the five most abundant.
[a] Fragments separated into hard and foam fragments in the NS surveys.

Channel, but relatively common at Padre Island NS. In 1992, the average quarterly accumulation rate of medical debris was 1.0 piece/km or less at Gulf Islands NS and 26 pieces/km at Padre Island NS. Accumulation of sewage debris averaged 1 piece/km at Gulf Island NS and 11 pieces/km at Padre Island NS. Medical and sewage debris in the Houston Ship Channel made up only 1% of the total plastics (excluding pellets and fragments).

Pacific Coast Results (Including Alaska and Hawaii)

Distribution and Type

Plastics dominated the man-made debris on beaches and in harbors of the Pacific Coast, including Alaska and Hawaii. Based on CMC surveys in 1992, plastics as a percentage of the total debris varied by state from 48% in Washington to 70% in Oregon. In 1992, plastics made up 97% of the total debris at Olympic NP and 95% at Channel Islands NP. In Pacific Coast harbors ($n = 5$), plastics comprised 80%–96% of the total debris.

In most locations, plastic debris was dominated by packaging and miscellaneous items (Fig. 4.4). In Alaska, however, derelict fishing gear usually made up most of the plastic debris; fishing gear constituted 53% and packaging items 42% of the total number of plastic debris pieces at all locations sampled from 1989 to 1993 (Fig. 4.4). In Alaska, two of the five most abundant debris items were directly associated with commercial fishing: gill-net floats and rope (Table 4.3). The only location in Alaska where packaging debris dominated (56%) was at Yakutat, probably because residents occasionally recreate on these beaches. At the NP sites in 1992, the two most abundant debris items were consistently hard and foam fragments (Table 4.3). In Pacific Coast harbors, miscellaneous plastics dominated the debris (average, 69.8%; $n = 5$), followed by packaging (average, 22%) (Fig. 4.4). The most abundant debris in harbors was fragments, pellets, foam spheres, and small pieces of sheeting (Table 4.3). Most miscellaneous plastics in harbors were com-

TABLE 4.3. The five most abundant plastic debris items (excluding cigarette butts) on national park (NP) beaches (1992), Alaska beaches, and in harbors (1989) of the U.S. Pacific Coast.

Item	Alaska	Olympic NP	Seattle Harbor	Tacoma Harbor	San Francisco Harbor	Oakland Harbor	Channel Islands NP	Honolulu Harbor
Pellets	—	—	5	1	2	2	—	5
Fragments[a]	—	1, 2	1	2	1	1	1, 2	1
Small sheeting	—	—	3	4	4.5	4	—	2
Miscellaneous food	—	—	—	—	—	5	—	4
Spheres (1 cm)	—	—	2	5	3	—	—	—
Large spheres	—	—	—	3	—	3	—	—
Cups, spoons, forks, straws[b]	—	—	—	—	—	—	5	—
Caps and lids	3	3	—	—	—	—	4	—
Packaging material	—	—	4	—	4.5	—	—	—
Bottles <3.8 l	2	—	—	—	—	—	3	—
Cigarette wrappers	—	—	—	—	—	—	—	3
Small bags	5	—	—	—	—	—	—	—
Rope <1 m[c]	4	5	—	—	—	—	—	—
Gill-net floats	1	—	—	—	—	—	—	—
Foam packaging	—	4	—	—	—	—	—	—

Items are ranked in descending order of abundance from 1 to 5. Fragments not counted in Alaska surveys. A dash indicates the item was not one of the five most abundant.
[a]Fragments separated into hard and foam fragments in the NP surveys.
[b]Straws only for the NP surveys.
[c]Size not separated in Alaska.

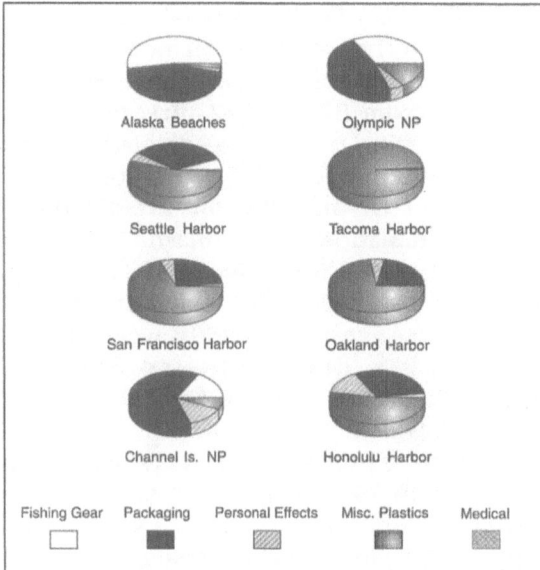

FIGURE 4.4. Percent composition of plastic debris (based on total number of plastics, excluding cigarette butts and fragments) by category for select harbors (1989), national park (NP) beaches (1992), and beaches in Alaska (1989–1993) on the U.S. Pacific Coast. Sewage is included in personal effects.

posed of pellets (average, 39%; $n = 5$), particularly in Tacoma, WA (88.9%).

Accumulation

Abundance of total plastics varied by location. In Alaska, plastics ranged from 203 pieces/km at Amchitka Island (Johnson 1993) to more than 1000 pieces/km at Middleton Island (Johnson 1989). In 1992, the average quarterly accumulation rate for plastics was 1729 pieces/km at Olympic NP and 813 pieces/km at Channel Islands NP.

Source

Major sources of plastics that washed ashore on the Pacific Coast were the United States and Asia (Japan, Taiwan, Korea). In Alaska, of bottles identifiable to country of manufacture, about 55% were from the United States,

37% from Asia, and 8% from other countries (e.g., Russia, Canada). Debris of foreign origin were also found on beaches in Oregon (Taiwan), Washington (Japan, Korea), California (Mexico), and Hawaii (Japan, China).

Entanglement Debris

Entanglement debris made up a greater percentage of the total plastics in Alaska than elsewhere on the Pacific Coast and Hawaii. For example, at Kayak and Amchitka islands, entanglement debris made up 11%–31% of the total plastic debris, compared to only 2% at Channel Islands NP and Olympic NP. In 1992, the average quarterly accumulation rate of trawl web (0–1 piece/km) and strapping (1–2 pieces/km) was consistently low at Olympic NP and Channel Islands NP; rope was the most abundant entanglement debris (5–13 pieces/km). In harbors, fishing debris was a small percentage of the total plastics (average, 2%; $n = 5$).

Ingestible Plastics

Plastics that may be ingested by wildlife were the most abundant debris on beaches and in harbors. At the NP sites in 1992, ingestible plastics made up 44%–74% of the total plastic debris; average quarterly accumulation rates ranged from 356 pieces/km at Channel Islands NP to 1276 pieces/km at Olympic NP. In Pacific Coast harbors, plastic pellets and fragments averaged 53% of the total plastics.

Medical and Sewage Debris

Medical and sewage debris was uncommon on beaches or in harbors on the Pacific Coast. In 1992 at each NP, the average quarterly accumulation rate of medical and sewage debris combined was 4.0 pieces/km or less. In Alaska, medical and sewage debris was rare. Medical and sewage debris combined averaged 0.5% of the total plastic (excluding pellets and fragments) in harbors. Washington state harbors were the only sites in the EPA program where no sewage debris was encountered.

National Overview

Several patterns emerged from the regional analysis of marine debris. Plastics were the most abundant man-made debris on beaches and in harbors of the United States. Packaging and miscellaneous categories were usually the dominant plastics. In all regions, common debris in harbors included pellets and fragments; on beaches, in contrast, common items included plastic bags, cups and lids, and fragments. Beaches in the Gulf of Mexico accumulated the most debris (average 9129 pieces km^{-1} quarter^{-1}), followed by beaches on the Pacific (1271 pieces km^{-1} quarter^{-1}) and Atlantic coasts (716 pieces/km).

Debris that could harm wildlife or affect human health was present in all regions. The most abundant entanglement item on all beaches was rope (also found in harbors); only beaches in Alaska and Hawaii had substantial amounts of trawl web or gill net (J. Henderson, NMFS, personal communication). Quarterly accumulation rate of entanglement debris was greatest on beaches in the Gulf of Mexico (average, 141 pieces/km), followed by the Pacific (20 pieces/km) and Atlantic (15 pieces/km) coasts. Ingestible debris had the highest accumulation rate on the Gulf (average, 4260 pieces/km) and Pacific (816 pieces/km) coasts, followed by the Atlantic (265 pieces/km) Coast. Typically, fragments made up a large percentage of the ingestible debris on beaches and pellets made up a large percentage of ingestible debris in harbors. Medical and sewage debris was most prevalent in Atlantic Coast harbors. Small amounts of medical and sewage debris were found on most beaches; exceptions were medical debris at Padre Island NS and sewage debris at Cape Cod NS. Pacific Coast beaches and harbors appeared to have the least medical and sewage debris.

Trends

Nationwide, no clear trend was evident in total plastic debris abundance from 1989 to

1993. The NPS study revealed that the average quarterly accumulation rate of debris (all parks combined except Padre Island NS) increased from 820 to 922 pieces/km from 1989 to 1992. Quarterly accumulation rates in individual parks varied substantially. For example, at Olympic NP the average quarterly accumulation rate of plastic debris increased from 533 pieces/km in 1989 to 1729 pieces/km in 1992. At Channel Islands NP, however, the average quarterly accumulation rate of plastic debris decreased from 961 pieces/km in 1989 to 813 pieces/km in 1992. On beaches near Yakutat, Alaska, total plastics decreased steadily from an average of 409 pieces/km in 1989 to 227 in 1991; in 1992, however, total plastics increased dramatically to 953 pieces/km. At Middleton Island, Alaska, total plastics decreased from about 1100 pieces/km in 1989 to 875 in 1992. Similar annual variability was observed in the quarterly accumulation rate of individual debris items at all NPS sites (Table 4.4).

Studies at Amchitka Island, Alaska, provided some long-term trends in plastic debris abundance (Table 4.5). The same beach sections on Amchitka Island have been surveyed periodically since 1972 (Merrell 1980, 1984). In 1993, an average of 203 plastic debris items/km were observed, the lowest amount since 1972 (Table 4.5). Fewer pieces of trawl web and strapping and fewer gill-net floats and bottles were found in 1993 than in 1982 (the last year all beaches were surveyed completely). The items that declined the most from 1982 to 1993 were gill-net floats (86%), strapping (72%), and trawl web (29%) (Table 4.5).

Entanglement debris decreased on some

TABLE 4.4. Average quarterly net accumulation rates (average number/km) for select debris items from seven National Park Service study sites, 1988–1992 (Padre Island NS not included).

Item	1988	1989	1990	1991	1992
Rope <1 m	46.2	47.0	53.0	92.6	55.7
Small bags	0.9	43.8	38.2	54.7	32.7
Tampon applicators	6.9	6.9	9.7	6.0	3.4
Caps and lids	45.5	51.8	68.4	91.3	55.8
Bottles <3.8 l	53.1	38.2	46.4	47.6	37.9

Data from Cole et al. (1990, 1992, 1995) and Manski et al. (1991).

TABLE 4.5. Trend in abundance (average number/km) of six of the most common plastic debris items on Amchitka Island, Alaska, 1972–1993.

Item	1972	1973	1974	1982	1987	1993
Trawl web	12	17	24	34	55	24
Rope	10	20	26	25	67	37
Gill-net floats	66	93	126	59	25	8
Bottles	13	23	45	38	39	33
Caps and lids	12	13	25	33	26	38
Strapping	30	32	71	58	34	16
Total[a]	160	219	361	284	309	203

The same ten 1–km beach sections were surveyed each year except 1987. In 1987, five randomly selected 100–m sections of each 1–km study were sampled; counts were extrapolated to number/km. Strapping includes open and closed bands.

[a]All plastic debris including items not in this table (excludes fragments).

Data from Johnson et al. (1990a) and Merrell (1984).

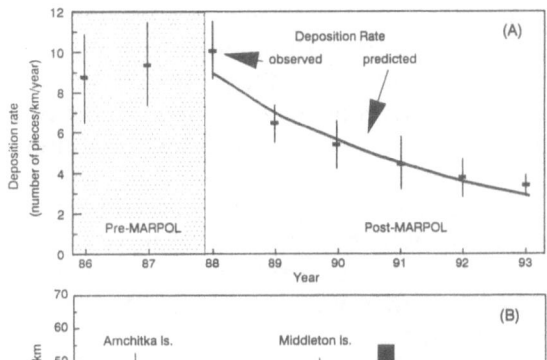

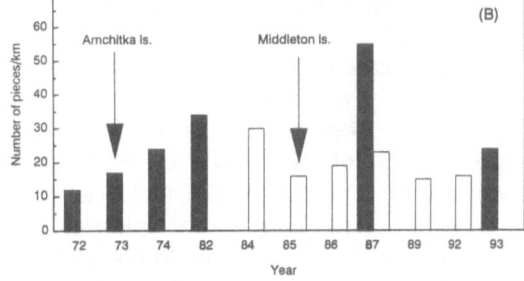

FIGURE 4.5. A Deposition rate of trawl web on eight 1–km beach sections year Yakutat, Alaska, before (*dotted area*) and after MARPOL Annex V legislation. Observed values, mean ± SE; logarithmic regression (*solid line*), $r^2 = .99$, $n = 5$, $P < .001$. Data from Johnson (1989, 1994). B Average deposition of trawl web on beach sections at Middleton ($n = 3$, 2.9 km total) and Amchitka ($n = 10$, 10 km total) Islands, Alaska. Data from Merrell (1984) and Johnson (1990a, 1993).

Alaska beaches, from 170 pieces in 1989 to 113 pieces in 1993 on beaches near Yakutat and from 205 pieces in 1989 to 175 pieces in 1992 on beaches at Middleton Island (Table 4.6). This decrease largely resulted from a decline in trawl web deposited ashore. Johnson (1994) reported that the deposition rate of trawl web at Yakutat declined significantly ($p < .01$) from 10.1 pieces km^{-1} yr^{-1} in 1988 to 3.8 pieces km^{-1} yr^{-1} in 1992. This decline continued in 1993: deposition rate was only 3.4 pieces km^{-1} yr^{-1} (Fig. 4.5). There was no significant ($p > .05$) trend in deposition rate of other entanglement debris on beaches near Yakutat from 1989 to 1993. Rope, gill net, and strapping averaged 10.0, 0.4, and 1.1 pieces km^{-1} yr^{-1}, respectively.

TABLE 4.6. Number of type of entanglement debris items deposited on eight 1–km beach sections near Yakutat and on six beach sections (5.5 km total) on Middleton Island, Alaska.

	Yakutat					Middleton Island	
Item	1989	1990	1991	1992	1993	1989	1992
Trawl web	52	43	36	30	27	48	55
Rope	106	84	82	61	80	139	104
Gill net	2	3	3	5	2	2	11
Strapping	10	16	7	7	4	16	5
Total	170	146	128	103	113	205	175

Beaches near Yakutat were cleared of entanglement debris twice each year. Beaches on Middleton Island were cleared of most entanglement debris (trawl web, gill net) in June 1989 and resurveyed in July 1992.

Trawl web also declined at other locations in Alaska. At Amchitka Island in 1993, trawl web averaged 24 pieces/km compared to 55 pieces/km in 1987 and 34 pieces/km in 1982 (Fig. 4.5). At Middleton Island, trawl web averaged 16 pieces/km in 1992, slightly more than in 1989 (15 pieces/km) but overall the second lowest quantity observed since 1985 (16 pieces/km) (Fig. 4.5).

Discussion

Laws have been enacted to reduce the amount of marine debris originating from both ocean- and land-based sources. In December 1988, Annex V of the International Convention for the Prevention of Pollution from Ships (MARPOL 73/78) entered into force. MARPOL was the first international agreement to prohibit dumping plastics and to regulate disposal of other garbage into the sea from ships. The Clean Water Act has jurisdiction over debris that enters U.S. water through storm drains. Regardless of source, the ubiquitous distribution and abundance of plastic debris of all types on coastal beaches and in urban harbors indicates that we cannot ascertain whether present laws, such as MARPOL Annex V and the Clean Water Act, are effective.

Plastics made up the majority (48%–95%) of the debris washed ashore in the 1990s. Similarly in 1988, national surveys by CMC reported that plastics made up 62% of the total debris (O'Hara 1990). On beaches in heavily populated states, most plastics were packaging and miscellaneous items, probably from land-based sources. In Alaska, however, because study beaches are distant from urban areas, most debris (including packaging) was from ocean-based (mostly fishing) and not land-based sources. Nationally, elimination of plastic bottles, caps and lids, and bags would reduce debris on beaches by at least 20%.

Although laws to reduce marine debris have been in effect for the last 5 years, no consistent decline in abundance of plastics was observed on beaches. Plastic debris was more abundant overall in 1992 than in 1989; however, abundance trends in some states can be obscured by annual variations in ocean currents, weather (hurricanes), or fishing effort and location. For example, the dramatic increase in total plastics at Yakutat, Alaska, in 1992 was largely the result of the 10-fold increase in gill-net floats. An El Niño in 1991–1992 may have influenced surface currents and winds (strong northward flow) in the North Pacific Ocean (Cannon et al. 1985; Reed and Schumacher 1985), resulting in greater deposition of debris on beaches near Yakutat than elsewhere in the Gulf of Alaska. At Middleton Island, for example, a nearly identical number of gill-net floats (~435/km) were found in 1989 (Johnson 1990a) and in 1992.

The exact source of most marine debris is difficult to determine. Few items can be accurately traced to a specific source except for commercial fishing gear (e.g,. trawl web,

lobster-claw bands, floats), offshore oil platforms (e.g., hard hats, write-protection rings), or the cruise ship industry (e.g., items with logos). Unless an item has a logo or other identifying mark, accurate source determination is impossible. The CMC regularly records items bearing a logo, but these are a minute fraction of the total collected. For example, in 1992, only 30 of approximately 4.5 million pieces collected had cruise ship logos.

Country of manufacture of plastics found on beaches or in harbors also does not necessarily identify the "true" source. Vessels that resupply in foreign ports undoubtedly purchase goods from those countries. Assuming, however, that most U.S.- and foreign-manufactured bottles are lost from like vessels, beach surveys indicate that the United States, Canada, Mexico, and Asian countries are probably major sources of debris on U.S. beaches. Most trawl web on Alaska beaches is probably from the United States because foreign trawl fisheries offshore of Alaska ended in 1987 (Kinoshita et al. 1993), while most monofilament gill nets and gill-net floats are from foreign sources because they are banned in Alaska (Uchida 1985; Johnson 1990a).

Categorizing debris may also make determination of source difficult. The CMC, NPS, and NMFS studies place debris in general source categories. The CMC categories (including all wastes, not only plastics) include recreational fishing and boating, commercial fishing, operational, galley, medical, and sewage wastes (Hodge et al. 1993). The NPS and NMFS categories are basically the same: fishing gear, packaging, personal effects, and miscellaneous (Johnson 1993; Cole et al. 1995). In all cases, however, debris classified into one category may originate from another. For example, lightsticks (chemically luminescent tubes) are used by the fishing industry in the United States and Caribbean countries (Morrell 1992), and also by divers off the coast of Florida. Thus, lightsticks could be placed in either the fishing or personal effects categories.

The high incidence of fragmented plastic in U.S. harbors indicates that fragmented plastics on beaches came not only from degradation on beaches but also from marine waters.

Fragmented plastics and pellets have also been observed in platform-of-opportunity studies on both the Atlantic and Pacific coasts (Colton et al. 1974; van Dolah et al. 1980; Day et al. 1990b).

Although plastics remained abundant on beaches and in some harbors, some debris showed a declining trend. The low incidence of medical and sewage debris on beaches and in West Coast harbors is probably the result of recent modernization of sewage treatment systems and reduced sewage overflow (Battelle Memorial Institute 1990). Some entanglement debris has also declined on beaches; specifically, trawl web declined in 1993 near Yakutat, Alaska, for the fifth consecutive year. The average deposition rate in 1993 (3.4 pieces km^{-1} yr^{-1}) was the lowest observed since studies began in 1985 (Johnson 1989). Declining trends in trawl web were also observed at Amchitka and Middleton Islands, Alaska. This decline in trawl web is consistent with the decline in entanglement of fur seals in trawl web (Fowler et al. 1993). Both studies indicate that the rate of loss and discard of net fragments in the North Pacific Ocean are decreasing.

The historical pattern of trawl web deposition is similar among Middleton Island, Amchitka Island, and Yakutat. Before approximately 1988 (pre-MARPOL), trawl web increased on beaches at all three locations. After 1988 (post-MARPOL), trawl web declined at all three locations, although fishing effort increased in Alaska. For example, in the Gulf of Alaska from 1988 to 1992, the number of trawl vessels that landed groundfish increased steadily from 122 to 234 (Fig. 4.6) and the domestic trawl catch increased from 109,000 to 202,000 metric tons (Kinoshita et al. 1993).

Declines in trawl web deposition and in the entanglement rate of fur seals in trawl web occurred after MARPOL was implemented. It should not be assumed, however, that MARPOL was the primary factor for the observed declines. In addition to MARPOL, increased public awareness through education, U.S. observers on many trawl vessels, the

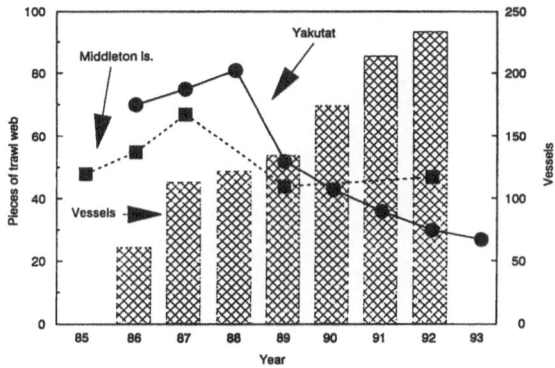

Figure 4.6. Total number of trawl web pieces (*circles*) deposited on select beach sections [2.9 km total at Middleton Island (*squares*) and 8 km total near Yakutat (*circles*)] and increase in number of trawl vessels in the Gulf of Alaska, 1985–1993. Data from Johnson (1989, 1990a, 1990b, 1993).

switch from foreign to domestic fisheries, and better port disposal facilities are probably responsible for the decline in trawl web discarded at sea (Johnson 1994).

Because many plastics (e.g., bottles, bags, caps and lids, pellets) are still abundant on beaches and in harbors, efforts must continue to reduce the disposal or loss of all plastics into the marine environment. Monitoring of beaches and harbors documents items attributable to various source categories (e.g., commercial fisheries, sewage) and identifies industries where further regulations or education efforts need to be concentrated.

Acknowledgments. We thank the NMFS Marine Entanglement Research Program, Alaska Fisheries Science Center, for partially funding the NPS program and the Alaska studies. We thank Wayne Trulli, Battelle Ocean Science, Duxbury, MA, and David Redford, U.S. EPA Office of Water for allowing use of unpublished data from Honolulu Harbor and for providing data summaries from other harbors in the EPA program. We thank CMC for providing information on the annual beach cleanups. We thank EPA staff and also thank J. Thedinga of the NMFS Auke Bay Laboratory for reviewing this manuscript.

5.

Pelagic Plastics and Other Seaborne Persistent Synthetic Debris: A Review of Southern Hemisphere Perspectives

M.R. Gregory and P.G. Ryan

Introduction

For centuries humans have indiscriminantly discarded their waste into, and on the margins of, oceans, lakes, and rivers. Seafarers traditionally disposed their garbage by simply heaving it overboard, and the practice continues to this day despite international agreements such as the London Dumping Convention (LDC) and the International Convention for the Prevention of Pollution from Ships (MARPOL). When quantities of mostly (bio) degradable waste were low, environmental and other consequences remained minimal. However, the advent of nondegradable synthetic materials has had profound biological and environmental effects (Laist 1987; Laist, Chapter 8, this volume) on shores and in oceanic and coastal surface waters (Pruter 1987a).

Awareness of the problems slowly developed through the 1950s. The magnitude of the marine debris problem was brought into the scientific and wider public arena by the *Ra* expeditions and Heyerdahl's (1971) observations of an equatorial Atlantic sea surface (p. 208) "... so filthy that we could not put our toothbrushes in it ... covered with clots of oil ..." (p. 298) "... scarcely a day passed without some form of a plastic container, beer-can, bottle ... and other rubbish drifting close by." Plastics and other synthetic materials are now recognized to be contaminants of marine waters of global significance, the extent of which can be gauged from the Proceedings of the first two International Conferences on Marine Debris (Shomura and Yoshida 1985; Shomura and Godfrey 1990).

The "health" of Northern Hemisphere seas, and particularly of inshore waters adjacent to heavily populated and industrialized regions, has been progressively deteriorating (GESAMP 1991). The impact is not so evident across the expanses of the Southern Ocean (Gregory 1990a). Oceans cover more than 80% of the Southern Hemisphere, and the pressures of population, industrialization, and shipping traffic intensity are far less than in the Northern Hemisphere where the cover is closer to 60% (Lutgens and Tarbuck 1992). The tyranny of distance has its advantages in dispersal and dilution of pollutants (Gregory 1990a). Furthermore, the strongly established ocean surface circulation patterns (Fig. 5.1) ensure that transequatorial mixing of pelagic contaminants between the hemispheres is minimized.

Amounts, Distribution, and Sources of Persistent Pelagic Debris

Shoreline surveys of pelagic plastic and other persistent litter have been carried out at many places around the Southern Hemisphere, in-

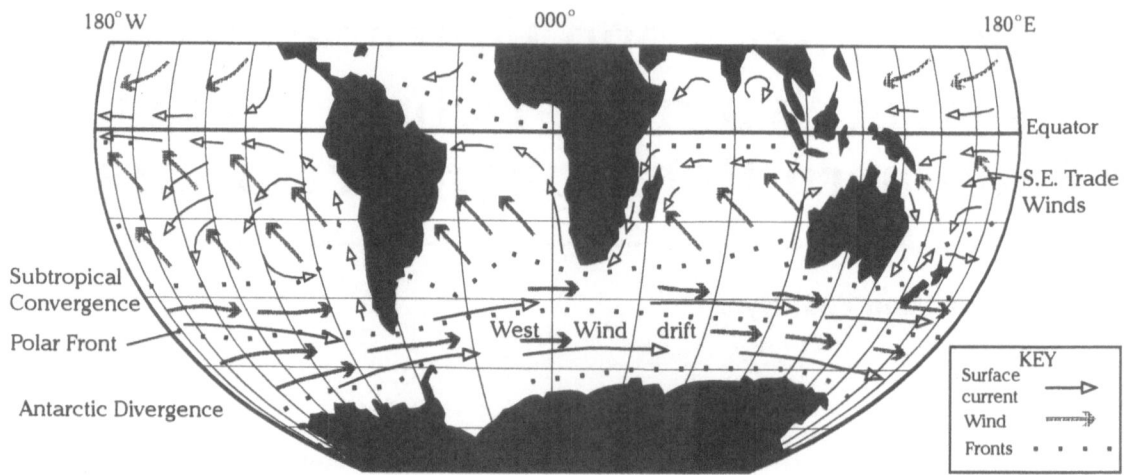

FIGURE 5.1. Oceans cover a little more than 80% of the Southern Hemisphere. Drift and dispersal of pelagic debris are controlled by the oceanic fronts, surface currents, and dominant winds.

cluding Argentina (Goodall 1990a, 1990b), South Africa (Ryan 1990a), Western Australia (Young 1993), Victoria, Australia (Heislers 1994), Tasmania, Australia (Slater 1991), New Zealand (Gregory 1978, 1991; Smith and Tooker 1990), and Antarctica (Gregory et al. 1984; Gregory 1990a; Goodall 1990c). It is also perhaps worthy of note that Harper and Fowler (1987) were recording plastic pellets in beached prions, *Pachyptila salvini* and *P. desolata,* from New Zealand as early as the late 1950s, long before recognition of the problem in neuston tows off the eastern seaboard of North America in the early 1970s (Carpenter and Smith 1972; Carpenter et al. 1972; Colton et al. 1974).

These surveys focused on metropolitan areas (O'Callaghan 1993), although some efforts have been made to visit more remote localities (Slater 1991; Wace 1990, 1991) including the island states of Oceania (Gregory 1990b), remote Subantarctic Islands (Gregory 1987; Ryan 1987a; Slip and Burton 1990, 1991), and Antarctica (Gregory et al. 1984; Torres and Gajardo 1985; Gregory 1990a; Greenpeace 1991). These regional studies include detailed commentaries on litter types, quantities, distribution patterns, and identification of possible sources. There are often accompanying remarks on environmental and ecological or biological impact that in general emphasize the hazards to life by en-

tanglement and ingestion and also on aesthetic factors. Many beach surveys are undertaken through government- or university-funded research agencies; other surveys, often of less scientific rigor, are associated with voluntary beach cleanups organized by public interest groups. Without standardized field techniques and data recording and processing (Ribic et al. 1992), comparisons and interregional evaluations are difficult. Both "standing crop" and "accumulation rate" approaches have been used with quantities expressed either by weight or by number per unit length (kilometer) of coastline. Count and weight measures do not necessarily correlate well when attempting to establish intensity of contamination.

Shipboard sighting surveys of drifting debris items in the North Pacific, such as those reported by Dahlberg and Day (1985), Mio and Takehama (1988), and Matsumura and Nasu (Chapter 2, this volume), have not been undertaken in Southern Hemisphere oceans. Baseline surveys of smaller debris items using surface tow (neuston) nets are limited to waters adjacent to New Zealand (Gregory 1990a, 1990b) and South Africa (Ryan 1988a) and are sporadic elsewhere around the region (Gregory et al. 1984; Grace 1994). The following sections provide a region-by-region review of published and unpublished survey data from the Southern Hemisphere.

Chile

We have little information on plastics and other anthropogenic debris on Chilean shores. Between latitudes 20° and 45° S, densities may reach 200 items/km but are generally less. Around Valparaiso and Valdivia to the south, macrolitter densities may be significantly greater (>300 items/km) while mesolitter (virgin plastic pellets and degradational chips) concentrations of more than 5000/m were noted in a brief 1985 survey. Bourne and Clark's (1984) observations of garbage quantities in the Humbolt Current off Valparaiso also suggest significant litter strandings along this coast.

Argentina and Uruguay

Figure 5.2 shows 11 beach sites in three widely separated regions from southern Uruguay to Tierra del Fuego (Goodall 1990a, 1990b), while Fig. 5.3 indicates the proportions and types of persistent man-made litter items larger than 10 mm around Uruguay and Northern Argentina. For the five northernmost localities (Fig. 5.2, inset A), quantities of man-made litter items ranged from 77 to 755 for each kilometer of beach (average, 403). Plastics were the most numerous of 12 identified litter categories, averaging 70%, with glass, cloth, and cardboard of lesser importance, except at Coronilla where glass reached 47%. The baseline survey sites are popular resort and recreational beaches, so it is not surprising that items commonly left by beachgoers (e.g., plastic soft drink bottles and cups) are conspicuous. Material directly attributable to fishing activities was not common (2% or less). The influence of the Rio de la Plata system, which receives domestic and household trash from the cities of Buenos Aires and Montevideo (Fig. 5.2, inset A), is acknowledged. Although the database is meager, densities of litter appear to decrease with distance from these two cities. Plastic debris composition at the two southern localities, Mar Chiquita (Fig. 5.2, inset A) and Puerto Valdés (Fig. 5.2, inset B), suggest the

primary sources were beachgoers and shore fishermen with lesser contributions from offshore.

Plastics also dominate the man-made beach litter at four of five survey sites around eastern Tierra del Fuego (Fig. 5.2, inset C). The exception was San Sebastian where leather, metal, glass, and cloth quantities all exceeded plastic. The San Sebastian survey site was influenced, however, by a "big" tide that effectively cleaned the beach a few days before the survey. The tide lifted plastics and other floatables from the shore but did not affect debris from offshore oil rig activity. Quantities varied from 71 to 535 items/km (average, 253) if Viamonte is ignored. Viamonte, with 5198 items/km, is excessively contaminated by a city dump some 35 km away.

Repeated surveys at sites other than Viamonte permit some preliminary estimation of accumulation rates (Goodall 1990a), which range from 12 to 140 items km^{-1} yr^{-1} (Table 5.1). Offshore sources, which include ship traffic, naval vessels, and local fishing activities around the Beagle Channel (e.g., Harberton and Moat) or oil rigs and tankers (San Sebastian), are important. Onshore sources are also recognized to be of significance (e.g., Viamonte). In addition to Argentina and Chile, countries of origin for individual items include Belgium, Greece, Holland, New Zealand, the United States, and Germany. In Goodall's surveys (1990a, 1990b), virgin plastic granules were not recorded from Argentinean shores.

South Africa

The plastic marine debris problem off southern Africa and the surrounding seas was reviewed by Ryan (1990a). Subsequently, considerable progress has been made toward organizing beach cleanup programs, largely through education programs (e.g., Dolphin Action and Protection Group's "Prevent Plastic Pollution" campaign). This activity recently has spread to Namibia, where there has been a survey (August 1992) and subsequent monitoring of macrolitter along the

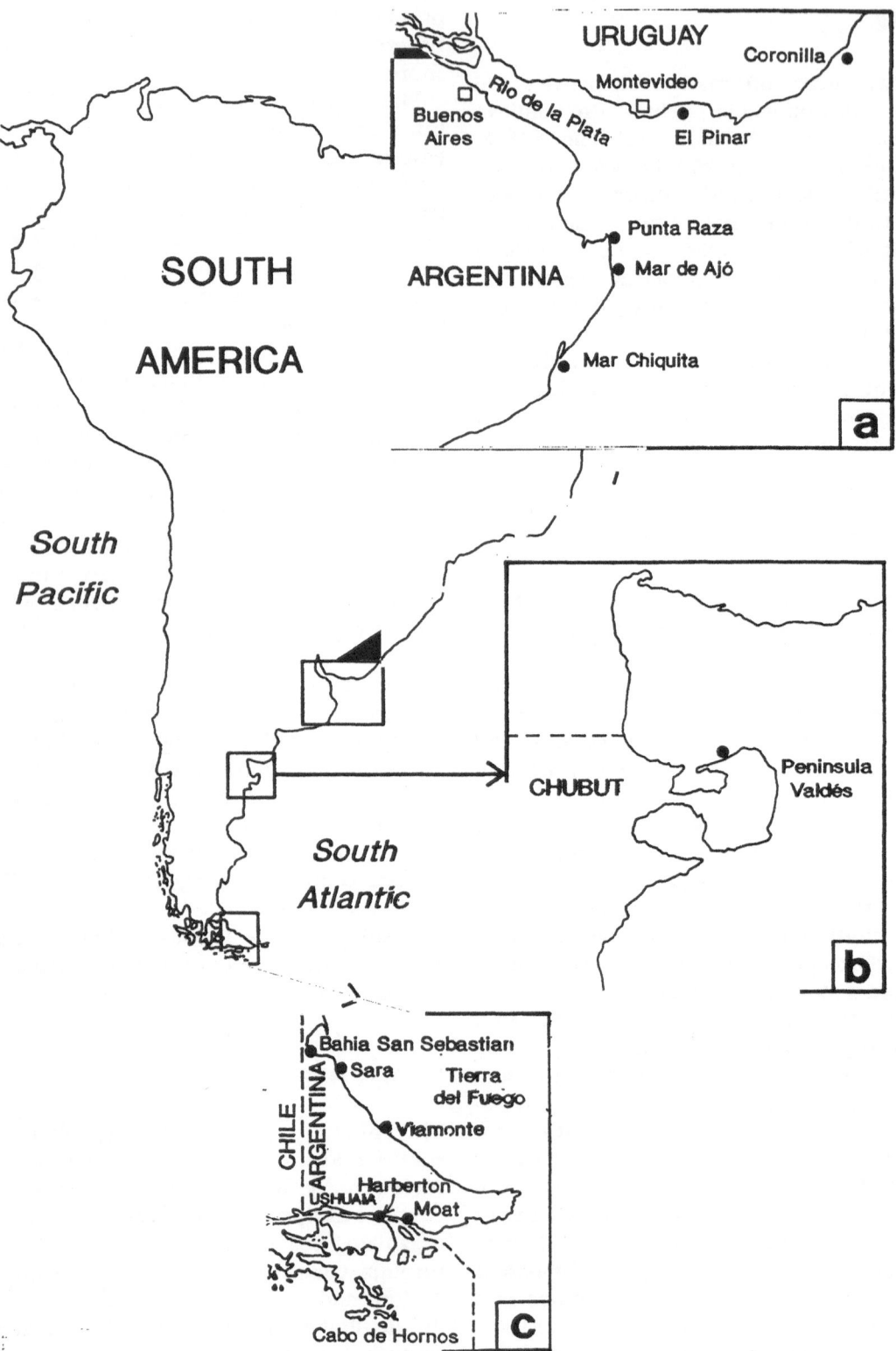

FIGURE 5.2. Sampling sites for surveys of beach debris on the shores of Uruguay and Argentina (Goodall 1990a, 1990b).

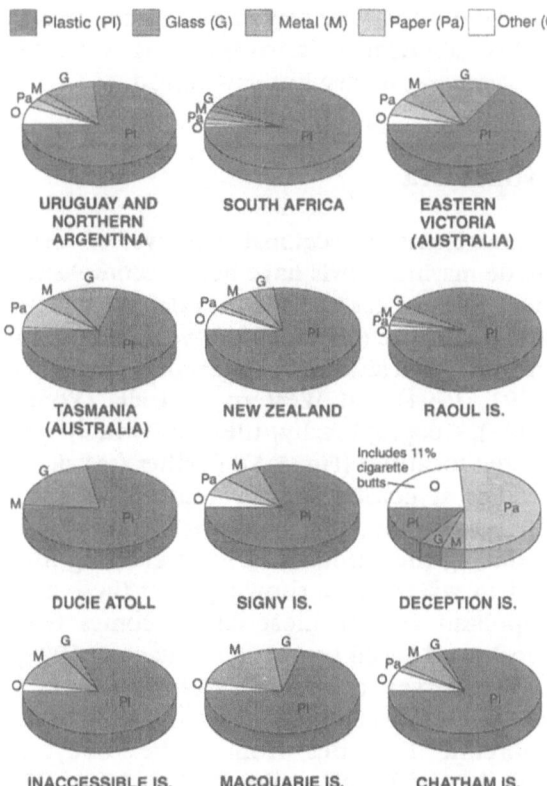

Plastic (Pl)　Glass (G)　Metal (M)　Paper (Pa)　Other (O)

URUGUAY AND NORTHERN ARGENTINA

SOUTH AFRICA

EASTERN VICTORIA (AUSTRALIA)

TASMANIA (AUSTRALIA)

NEW ZEALAND

RAOUL IS.

DUCIE ATOLL

SIGNY IS.

DECEPTION IS.
Includes 11% cigarette butts

INACCESSIBLE IS.

MACQUARIE IS.

CHATHAM IS.

Figure 5.3. Litter composition for representative localities around the Southern Hemisphere.

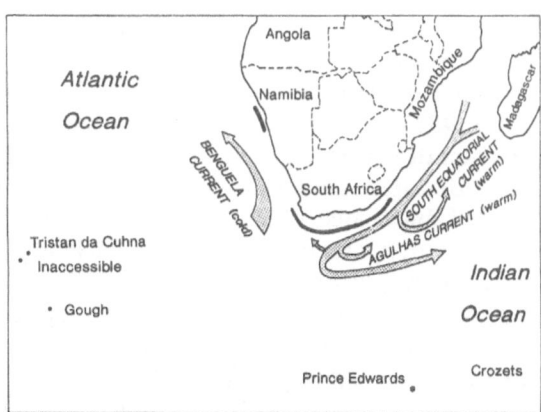

Figure 5.4. Principal marine debris sampling areas (*heavy line*) of South Africa and Namibia. Subantarctic Islands of the African sector and important oceanographic features are identified.

coast from Swakopmund to the Ugab River Mouth (Wildlife Society of Namibia, press release, 20 April 1993).

Quantitative estimates of both mesolitter and macrolitter stranded on 50 South African beaches between Saldanha Bay and the Kei River were made in 1984 and again in 1989 (Ryan 1990a; Ryan and Moloney 1990). Figure 5.4 indicates the general localities of the quantitative estimates, as well as influential regional currents. These counts showed

Table 5.1. Estimated annual accumulation rates of man-made debris items in eastern Tierra del Fuego.

Locality	Number of items $km^{-1} yr^{-1}$
Harberton	136
Moat	39
Sara	12
San Sebastian	140

that at matched sites, plastic litter increased significantly during the 5-year sampling interval. The average density of mesolitter increased from 491 to 678 items/m of beach (38% increase), while that of macrolitter increased from 1090 to 2990/km of beach (174% increase). The strong correlation between mesolitter and pumice (an indicator of oceanic dispersal rather than local sources) suggests that inshore currents are more important than local point sources in determining the abundance of this type of debris on beaches. There was a weaker correlation with macrolitter, indicating the combination of land- and sea-based sources of this material. This question is considered later in the section on debris sources.

Most of the beached mesolitter was composed of virgin plastic pellets, with the proportion decreasing from 1984 (80%) to 1989 (68%). Despite this proportional decrease, the absolute abundance of both virgin pellets and other mesolitter fragments increased from 1984 to 1989. Densities on beaches ranged from 4 to 7698 particles/m in 1984 and from 24 to 11472 particles/m in 1989. However, recent observations suggest that the abundance of virgin pellets has reduced substantially, coinciding with a decrease in the incidence of pellets in seabirds (Ryan 1995). Mean densities of virgin granules at 50 beaches in 1994 decreased 36% from 1989 levels and 24% from 1984 levels. Densities of other mesodebris were unchanged relative to the 1989 survey.

Plastics made up more than 90% of all artifacts (98% of mesolitter and 88% of macrolitter) on the 50 South African beaches. Other categories of macrodebris were, in decreasing order of abundance, objects made from wood, glass, metal, paper, cloth, and wax. Packaging made up the bulk of the plastic litter (65% and 69% of macrolitter in 1984 and 1989, respectively), with fishing materials constituting most of the remaining items.

In a study of macrodebris in Algoa Bay, a heart-shaped bay on the southeast coast of South Africa, Forsyth (1993) sampled seven 100-m strips of beach (spaced at 1-km intervals). An average density of 7420 articles/km of beach (including in the adjacent dune field), with an average mass of 319 kg/km, were recorded: plastic, glass, and metal items constituted 80%, 13%, and 2.3% by number, and 66%, 22%, and 3% by mass, respectively.

On the central Namibian coast in 1992, the overall density of macrolitter averaged 117 kg/km of beach (range, 22–308, based on 46 samples of 500 m each). The density of plastic ranged from 4 to 142 kg/km, glass (mainly disposable beer bottles) from 2 to 230 kg/km, paper and cardboard from 0 to 90 kg/km, and monofilament line from 0 to 41 kg/km. Of this litter, only 15% was deemed to have come from offshore sources; the majority came from shore-based recreational activities (primarily surf angling).

Little is known about the distribution and abundance of marine debris on the sea floor around South Africa. The only quantitative assessment is for False Bay, a large shallow bay abutting the metropolitan area of Cape Town (Rundgren 1992). Transect counts of all plastic debris (>20 mm diameter) were made by scuba divers at 18 sites around False Bay in spring (September) and in autumn (March). Packaging made up more than 80% of all plastic debris, with bags (65%), bottles (22%), and food wrapping (12%) constituting the bulk of items. Counts generally were higher in spring, apparently because offshore winds and stormwater runoff during the winter rainy season carry much of the debris into the sea. However, at a site sampled monthly there was a midsummer peak in debris abundance, corresponding with the December–January holiday period.

Australia

Comprehensive regional surveys of man-made marine debris have been accomplished for shores of eastern Victoria (Heislers 1994) and Tasmania (Slater 1991) with limited locally focused studies in South Australia (Wace 1991, 1994) and Western Australia (Young 1993). Geographically, these are widely separated localities (Fig. 5.5). Further, restricted studies at coastal sites around the cities of Melbourne, Sydney, and Brisbane aimed at reducing the amounts of litter entering Australian waters have shown that at these metropolitan centers most debris comes from local, land-based sources (O'Callaghan 1993).

The survey of eastern Victoria (Fig. 5.5, lower portion) included more than 700 km of shoreline. Densities, from count surveys of 179 km of beach, ranged from 11 to 273 items/km and weights, based on 69 km, from 2.6 to 16.4 kg/km (Table 2). Plastics accounted for more than 60% of the man-made litter items, with glass, metal, and timber of generally lesser significance (see Fig. 5.3). Litter density is greatest, and dominated by material from shipping and offshore fishing activities, for the coast west of Wilsons Promontory and facing into Bass Strait. East- and northward to the New South Wales (NSW) border, quantities decrease and are variably dominated by local onshore and recreational beach-user sources, with distinctive "plumes" around some access points. At more remote, less frequently visited places, offshore sources become significant.

Slater (1991), reporting on the results of 150 surveys between January 1990 and June 1991 at 88 separate sites and covering 177 km of the Tasmanian shoreline, noted that statewide there were approximately 300 debris items/km. This debris was 65% plastic (Fig. 5.3), 40% of which was from fishing activities. However, 30% of the debris could also be regarded as having its source in coastal recreation. In the southwest of the World

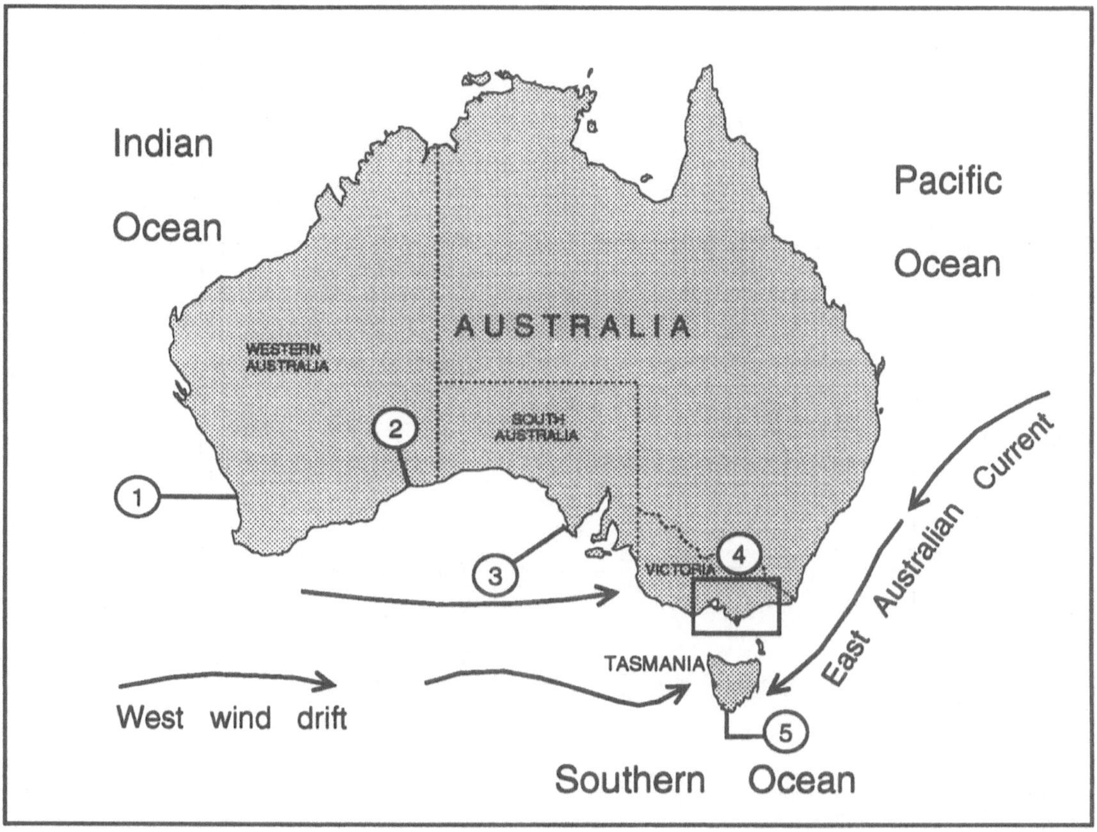

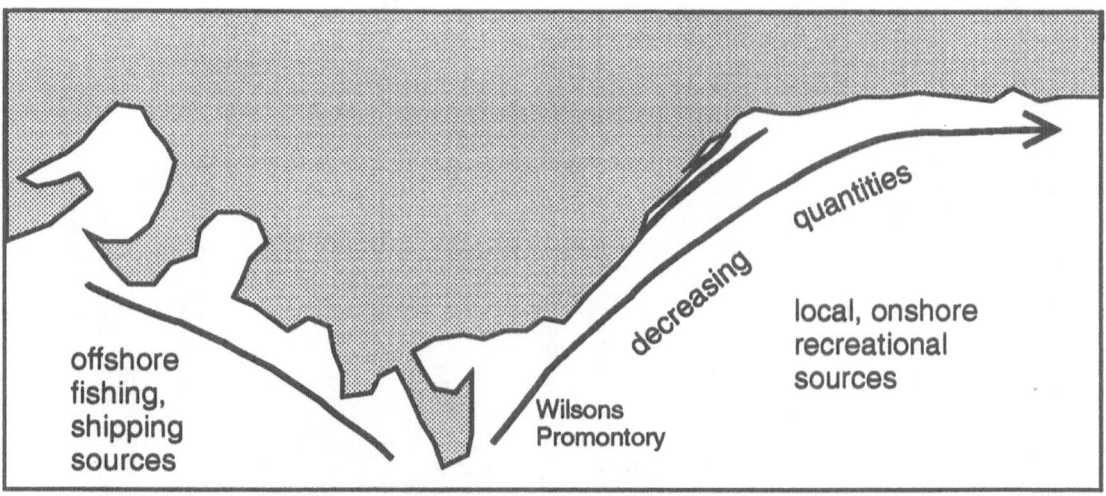

FIGURE 5.5. Australian regional survey localities (*circled numbers, top*): *1*, Marmion Marine Park; *2*, Eyre Bird Observatory; *3*, Anxious Bay; *4*, eastern Victoria (*enlarged, lower*); *5*, World Heritage Area, Tasmania.

TABLE 5.2. Summary of typical and representative macrolitter densities, composition, and variation around the Southern Hemisphere.

Country	Length surveyed (km)	Number of items/km		Proportion (%)[f]				Fisheries related (%)	Weight (kg/km)
		Range	Average	Plastic	Glass	Metal	Paper		
Chile[b]	0.3	(136–380)[c]	200	–	–	–	–	–	–
Argentina	9	71–755	336	37–72	<3–46	–	–	<2–>22	87–305
South Africa	27	480–30,000	4120	93	3	2	2	25–40	319–441
Namibia	23	–	–	–	–	–	–	15	22–308
Western Australia	0.8	(230–5120)	3600	–	>18	–	–	40	123
South Australia	25	–	–	62	31	5	–	(est)>80%	13–15
Victoria, Australia	179	11–273	76	62	14	9	10	40	2.6–16.4
Tasmania, Australia	177	–	300	71	12	6	3	40	–
New Zealand	467	<5–>60,000	997	75	4	6	–	23	–
Ducie Atoll, Oceania (25°S)	–	–	(240)	77	21	>1	–	>25	–
Raoul Island, Oceania (30°S)	<8	<200–>1500	450	>80	<10	<1	–	44	–
Subantarctic Tristan da Cunha (37°S) and Gough Island (40°S)	13.6	<10–2400	880	77	4	19	0	44	–
Subantarctic Prince Edward Island (46°30'S)	4.9	0–225	32	88	0	12	0	22	–
Subantarctic Heard (53°S)	24	–	13[d]	51	12	33	–	40	–
Subantarctic Macquarie Island (54°S)	~94	–	9[d]	71	5	20	–	29	–
Subantarctic Chatham Island (44°S)	<40	<40–>1000	450	70–>80	<2–<10	<1–>8	<2	>70	(39–91)
Subantarctic Auckland Island (50°30'S)	<3	p–<300[f]	100	78	8	7	2	50	–
Subantarctic Campbell Island (52°30'S)	<3	p–<100[e]	25	86	8	2	–	58	–
Subantarctic Bird Island (54°S)	<0.3	(<200–>1000)	–	100[?]	–	–	–	25–90	–
Antarctica Signey Island (60°30'S)	0.1	(>1000)	–	78	1	9	12	50[?]	–
Antarctica Livingston Island (62°30'S)	1	>150	–	80	2	11	2	>20[?]	–
Antarctica Deception Island (63°S)	1	175	–	14	5	5	51	–	4
Antarctica Scott Island (67°30'S)	0.1	Nil	–	–	–	–	–	–	–
Antarctica Ross Sea	~35	0–>30	–	Mostly lumber and tin cans				–	–
Antarctica Budd Coast	–	Nil	–	–	–	–	–	–	–

[a] Proportions generally based on item number.
[b] Plastics only recorded.
[c] Parenthetical values are extrapolated.
[d] Annual accumulation rate ignoring wood and timber.
[e] The annotation p represents a low or undetermined number.

Heritage Area, which occupies much of the remote western side of Tasmania, surveys at 10 localities showed 165 debris items/km, and although composition was similar (63% plastic), fishing-related debris was doubled to 80% (Slater 1992). The importance of local fisheries, as well as distant oceanic sources, was acknowledged: debris was transported from afar by the Southern Ocean's westwind drift and the East Australian Current.

From beach clearances in the austral springs of 1991, 1992, and 1993, Wace (1991, 1994) reported litter densities along the 26-km length of Anxious Bay on the west coast of Eyre Peninsula, South Australia (Fig. 5.5), reaching 13–15 kg/km. If wood is ignored, this debris is composed of plastic (62%), glass (31%), and metal (5%). Recreational and casual visitors are negligible on this remote beach, and the source of debris lies in local and distant offshore fishing activity. Comparable quantities are also figured

for a 1-km coastal stretch at the Eyre Bird Observatory, Western Australia (Wace 1991), which lies on the western side of the Great Australian Bight. Here a Southern Ocean source is claimed for most of the litter.

At Marmion Marine Park, north of Fremantle, Western Australia, litter levels in 1992 reached 3660 items/km and 123 kg/km (Edwards et al. 1992). These quantities were an increase of 133% (item count) and 228% (in terms of weight) from a similar study completed in 1985 (Cary et al. 1987). On this shore, 60% of the litter came from beach-users and the remainder had its source in the local rock lobster fishery (Young 1993).

Plastic pellets have been reported from the shores of eastern Australia with numbers exceeding 1000/m only around the ports of Melbourne and Sydney (Gregory 1990b, Table 1). Recent observations, partially shown in Table 5.3, indicate low densities along the Bass Strait shore of Tasmania (to a

TABLE 5.3. Summary of representative plastic mesolitter (virgin pellets and degradational flakes <5 mm across) densities on Southern Hemisphere shores.[a]

Country	Densities (m⁻¹)	Comments
Chile (30°S)	≈1000–4000	Virgin pellets predominate
Chile (40°S)	≈100	Virgin pellets predominate
South Africa (25°–34°S)	20–12,000	40%–90% pellets, proportional decrease from 1984 to 1989
Western Australia	p[b]	
South Australia	p->100	
Victoria, Australia	p->1000	High numbers only near Port of Melbourne
New South Wales, Australia	p->>2000	Numbers decrease abruptly with distance from Sydney
Queensland, Australia	p->5	Degradational flakes generally greater than pellets
Tasmania, Australia	p->10	Only present along shores facing Bass Strait
Northern North Island, New Zealand (35°30'S)	5->100	Quantities based on 1978 data; recent surveys indicate densities have significantly decreased. Limited evidence from remote,
Auckland, New Zealand (37°S)	>100->100,000	west coast North Island localities, suggest degradational flakes
Western North Island, New Zealand, (38°S)	5->100	greater than, or equal to, pellets. Near metropolitan areas, degradational flakes are not so conspicuous (many fewer than
Eastern South Island, New Zealand (43°30'S)	>10->>1000	pellets).
Southern South Island, New Zealand (46°30'S)	p->100	
Oceania	p->1000	Degradational flakes locally abundant, greatest concentrations
Subantarctic Tristan de Cunha and Gough Island	≈10	on shores facing southeast trade winds
Subantarctic Chatham Island	p->>1000	High numbers of pellets and degradational flakes restricted to northern coast [the Subtropical Convergence may be a factor in this]

[a]Based on surveys of suitable (sandy) beaches (Gregory 1978, 1990b, 1993; Gregory and Ryan, unpublished data). There are some suitable subantarctic beaches with no reports of mesolitter (e.g., Northeast Harbour and Campbell Island); this is surprising given the fact that pellets and other plastics are abundant in seabirds that seldom, if ever, feed north of the Subantarctic Convergence.
[b]The annotation p represents a low or undetermined number.

local maximum of >10/m at Burnie) and in northern Queensland between Townsville and Port Douglas (local maxima, <5/m).

New Zealand

Following earlier efforts by officers of the Department of Conservation (Macfie 1989; Matson 1989), during an 8-month period between October 1989 and May 1990 Greenpeace N.Z. conducted a countrywide "Adopt-a-Beach Campaign." During this effort, volunteers completed 1220 surveys at 388 beach localities from northernmost New Zealand to Stewart Island in the south. A total of 467 km of beach (less than 5% of the country's coastline) was surveyed. A distinction was made between items (= every single piece of debris collection) and articles (= identifiable objects of debris, excluding fragments) (Smith and Tooker 1990). It was shown that the average litter density on an item-count basis was almost 1000/km. Plastic articles were dominant at 75% (64% items); lesser litter categories included glass (5%), metal (12%), paper, cardboard, and timber (3%), and cloth and rubber (6%). On an item basis, the relevant values are glass (11%), metal (10%), paper, cardboard, and timber (12%), and cloth and rubber (3%). Smith and Tooker (1990) noted that packaging figured prominently at 40% of the total debris, and they also emphasized that some 40% of articles could be considered to have their source in the fishing industry. However, the latter value may be an overestimate (Gregory 1991).

Foreign sources of debris have been widely reported from New Zealand and its offshore islands, with plastics again dominant (Gregory 1978, 1990b, 1991, 1993; Smith and Tooker 1990). Most identifiable material (80%) is of Asian origin (Japan, Korea, Taiwan), but many other countries are minor contributors (e.g., Argentina, Australia, France, Germany, Mexico, Norway, Russia, and the United Kingdom). Where the country of origin can be determined, the ratio of domestic to foreign debris is greater than 10:1. Fishing-related debris is concentrated on shores lying down drift or adjacent, or both, from local and distant fishing grounds, i.e., the east and west coasts of northern North Island and the east (Canterbury) and west coasts of South Island (Fig. 5.6A).

The quantities of debris attributable to the fishing industry are not surprising given unpublished records and accounts kept by MAF (Ministry of Agriculture and Fisheries) observers on vessels of all nationalities working in the New Zealand Exclusive Economic Zone (EEZ). Domestic refuse, engine room waste, dirty rags, and plastic sheets, and packaging from processing floors were routinely thrown overboard together with discarded netting and broken trawl gear. It was noted that this took place on some vessels that had incineration and can-crushing facilities. There is a consensus that the worst offenders (to 1991) were New Zealand and Korean vessels. There are frequent accounts of trawl nets being fouled by wire rope and hawsers, abandoned cod-ends and other fishing gear, and metal objects (including a refrigerator). It is perhaps ironic that at the same time this debris is being swept from the seabed, the recovering vessel is dumping its own waste overboard.

Smith and Tooker's (1990) Greenpeace report overemphasized offshore sources at the expense of onshore sources. Coastal cities are a major contributor to marine debris. Quantities of this debris typically decrease rapidly with distance from its onshore source. At some localities, extrapolated item density may exceed 60,000/km (Gregory 1991). The importance of onshore sources has recently been demonstrated by Arnold et al. (1994) in a study in which stormwater and other drainage outfalls to Auckland Harbour have been covered by a mesh screen, trapping larger items. It is estimated that annually more than 10 million debris items are entering local waters by this route. Although industrial catchments were very important contributors, commercial and residential sources were also significant.

Longer term trends in changing beach litter composition have been recorded at two localities: Kawerau, a remote beach on the west coast of northern North Island, and Rangitoto and Motutapu Islands near the entrance of Auckland Harbour. At Kawerau, four surveys between 1974 and 1982 revealed an increase in the proportion of plastics (Hayward 1984), and this pattern continues. At Rangitoto and Motutapu Islands, annual beach cleanups by Island Care between 1989 and 1992 indicated increasing amounts of paper, cardboard, and confectionery and convenience food wrappings; a decrease in sheeting, fiber, metal, and foam plastic; and no change in quantities of hard plastic and glass. Overall, volume of debris increased at Rangitoto and Motutapu Islands during that time period (Lewis et al. 1994).

Paper, confectionery and convenience food wrappings, and plastic soft drink bottles are of negligible significance on remote shores of New Zealand. They are conspicuous on popular recreational beaches all around the country and particularly those near coastal cities (to more than 75%). A summer seasonal buildup in density of this material has been recognized at Tauranga (Smith and Tooker 1990) and at popular beaches near Auckland, which reflects the period of greatest recreational use.

Gregory recorded the accumulation and distribution of virgin plastic granules on New Zealand beaches in 1978. The greatest concentrations, which locally greatly exceeded 100,000/m, were found near the cities of Auckland, Wellington, and Christchurch. As Table 5.3 shows for Auckland and Christchurch, quantities generally decreased with distance from these important manufacturing centers. During the late 1980s, pellets were resampled at 262 of the original 326 sampling localities: of the 262 resampled localities, 12% increased, 45% decreased, and 44% showed no change. Additional evidence, partially shown in Fig. 5.6B, suggests a decrease in pellet numbers that may be attributable to improvements in the handling and transport of manufactured goods (Gregory 1993).

Oceania

Plastics and other man-made debris are common on both populated and unpopulated islands of Oceania (Gregory 1990b, 1991). Quantities are highly variable, and debris often originates from local communities. Litter composition is generally comparable to that reported from developed countries of the region (i.e., plastics >70%). However, significant amounts of litter have also been reported from several remote, uninhabited, and seldom-visited islands, e.g., Ducie Atoll and Henderson Island (Benton 1991, 1995) and Oeno Atoll (Preece, R.C., Department of Zoology, University of Cambridge, U.K. 1993, personal communication) in the Pitcairn Group. At Ducie, density reaches 397 items/km of which 23% can be related with certainty to fishing activity, based on the table in Benton (1991). Similarly on Raoul Island in the Kermadec Group some 1000 km northeast of New Zealand, which has a transient population of four manning a weather station, litter density varies from less than 200 items/km to pockets with more than 1500 items/km (Gregory 1991, Fig. 7).

The component of indisputable fishing-related debris is 20%–25%. However, much of the material on both Ducie Atoll and Raoul Island is household or domestic in character and could well have a source in fishing vessels or other shipping activity—it may have traveled from afar. J.R.E. Hager (UNESCO, Jakarta, Indonesia, 1991, personal communication) reported for Dravuni Island (Fiji) that 121 kg/km of beach debris had stranded on the windward side and only 1 kg/km on the leeward side. Gregory (1990b, 1991) noted a similar pattern on many other islands of Oceania and considers this dispersal pattern a reflection of the ever-present southeast trade-winds.

Accumulation rates on islands of Oceania are probably high. Repeated surveys of southwest-facing Denham Bay, Raoul Island, in 1990 and 1993 suggest an annual rate of at least 83 items/km may be appropriate. Virgin plastic granules are widely recorded from beaches of Oceania, and local concentrations may exceed 1000/m but in general are much less (Gregory 1990b, Table 2). On north-facing Bells Beach of Raoul Island, pellet numbers per meter have progressively decreased over surveys in 1988, 1990, and 1993

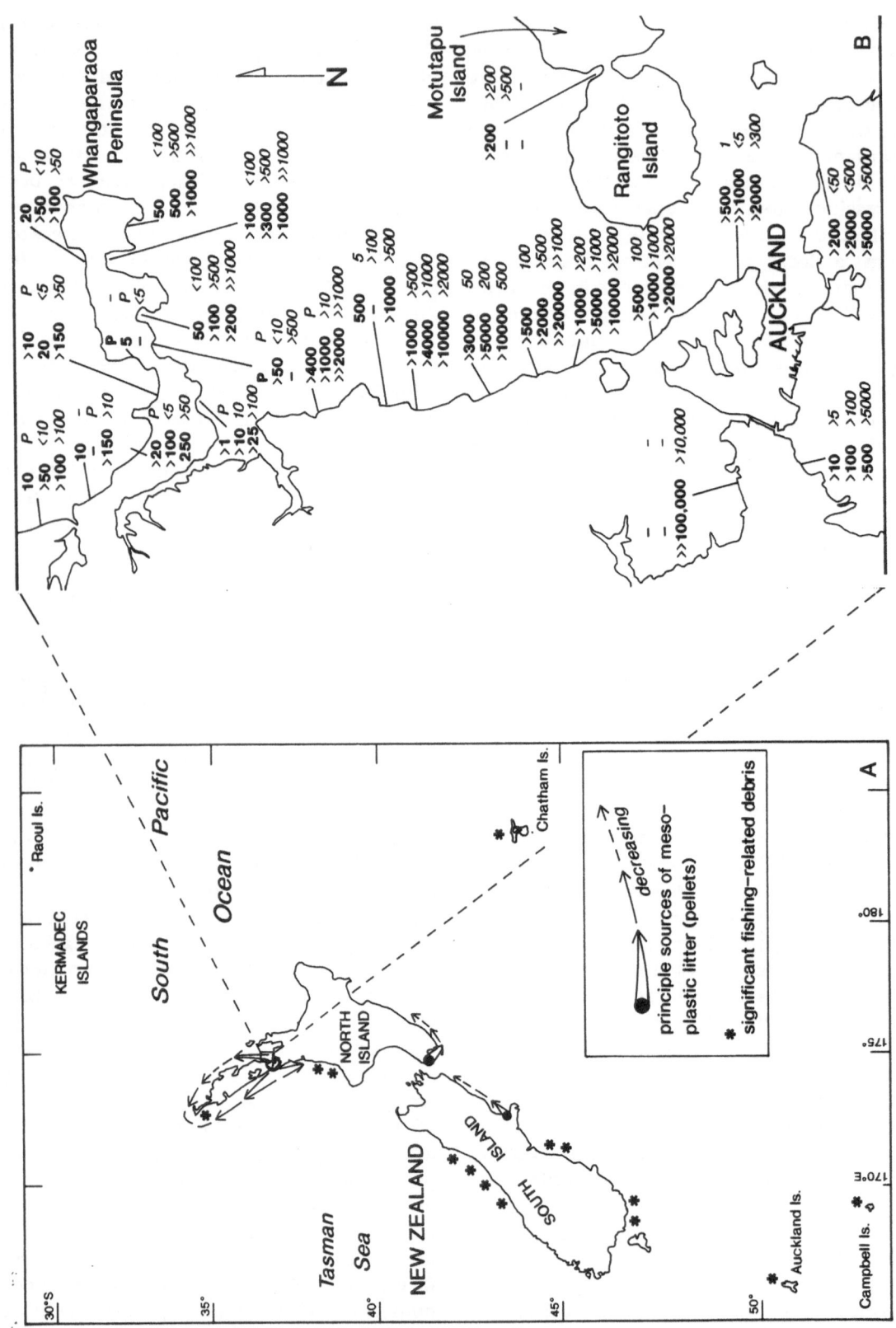

from more than 50 to less than 10 to less than 1. During this same period, quantities of megalitter and macrolitter have not decreased.

Subantarctic Islands

Plastics and synthetic man-made debris of all kinds, together with driftwood, pumice, and other natural flotsam, are conspicuous on remote and isolated Subantarctic shores, such as islands of the New Zealand (Gregory 1987, 1990b) and African (Ryan 1987a; Ryan and Watkins 1988; Ryan and Moloney 1993) sectors of the Southern Ocean as well as the Australia territories of Heard and Macquarie Islands (Smith et al. 1989; Slip and Burton 1990, 1991). These studies consistently reported plastic items at 70% or more of the anthropogenic litter and that much of it is fishing related (buoys, netting, rope hawsers, etc.). The mean density of litter on Heard and Macquarie Islands is 13 and 9 items/km, respectively, which is much less than that reported from Campbell Island and Auckland Island of the New Zealand sector, where it reaches 100 items/km (see Table 5.2). These values are comparable with those of Prince Edward Island in the African sector (Table 5.2), which lies at a similar latitude.

Chatham Island, approximately 1000 km east of the South Island, is generally included with the New Zealand Subantarctic islands. However, with a population of nearly 700, mostly fishermen and farmers, Chatham Island is atypical. Anthropogenic litter densities reach 1000 items/km near settlements on the west- and north-facing shores and only 37 items/km on isolated eastern shores (Fig. 5.7). Elsewhere in the Subantarctic Islands, levels of anthropogenic litter also increase near fishing communities and research stations. At

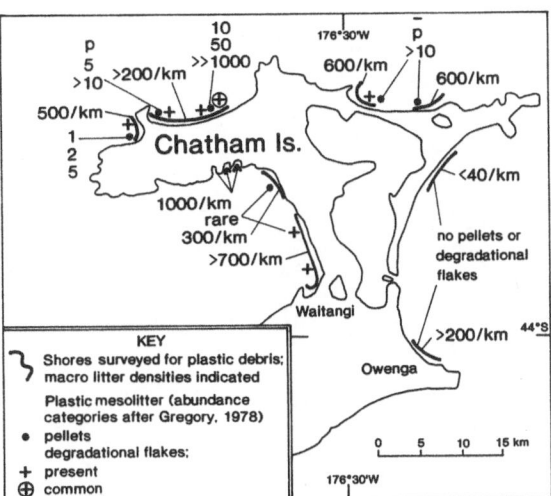

FIGURE 5.7. Debris quantities and distribution on Chatham Island.

Tristan da Cunha, litter density reaches 2400 items/km (70% is related to fishing activities), which is higher than regional subantarctic levels (Table 5.2). Note that both Tristan da Cunha Island and Chatham Island lie a considerable distance north of the Polar Front.

All these southern islands are influenced by the west wind drift and strong circumpolar circulation of the Southern Ocean (see Fig. 5.1). Quantities of debris are greatest on the windward shores (west-facing) and least on the leeward shores (east-facing). While some of the stranded debris is related to inshore fishing activity, some of it is truly "oceanic" and comes from distant sources (Fig. 5.8). Annual accumulation rates between 9 (Macquarie Island), 13 (Heard Island), and 800 (Inaccessible Island in the Tristan da Cunha Group) items/km have been determined. Six months of regular surveys on the west coast of Inaccessible Island, between October 1989 and March 1990, showed an annual accumulation rate of approximately 800 items/km.

←

FIGURE 5.6A,B. Maps showing New Zealand and the Auckland area. A Distribution patterns of mesoplastic litter (plastic pellets) and fishing-related debris around New Zealand and Chatham Island. B Decreases in plastic pellet populations (expressed as number/m) between surveys of the late 1970s (*bold numbers*) and early 1990s (*italic numbers*). Pellet numbers generally decrease with distance from metropolitcan Auckland.

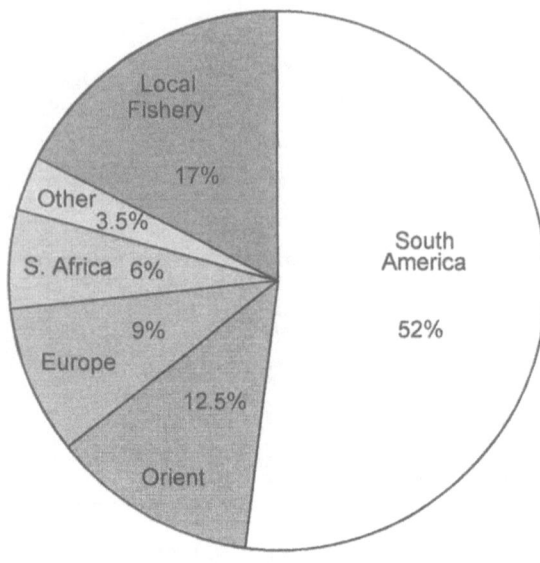

FIGURE 5.8. Origins of macrolitter items stranding on the western side of Inaccessible Island, 1989 and 1990.

Plastics made up 88.3%, glass 4%, metal 3.3%, wood 3.7%, and cork (float) and wax, 0.3% each. Packaging (51.7%) and fishery items (27.2%) made up the bulk of the plastic debris.

Beach debris surveys, following removal of all anthropogenic material, are in progress at Main Bay, on Bird Island, South Georgia (Rodwell 1990; Arnould 1993; Reid 1993), and Signy Island, in the South Orkney Islands (Chalmers et al. 1991; Shears et al. 1992). These and other surveys all emphasize the abundance of timber, much of which is bleached and abraded and has been beached for some time. Otherwise, the most common materials are fishing gear, such as floats and rope, packaging straps (most of which are cut), and the usual assortment of plastic bottles and sheeting.

Plastic pellets occur with some regularity on remote Chatham Island shores and reached more than 1000/m at one northern coast locality. This may be related to the Subtropical Convergence astride which the island lies (see Fig. 5.1). Pellets have not been recorded on other Subantarctic Islands of the New Zealand and Australian sectors (Gregory 1987, 1990a; Slip and Burton 1991); however, they have been noted on Tristan da

Cunha (Ryan 1987a), which, like the Chathams, lies in a more northerly latitude.

Antarctica

Remoteness and isolation have not left Antarctic shores immune from contamination by anthropogenic marine debris. There are several reports of material stranded on the shores of the Ross Sea (Gregory et al. 1984; Gregory 1990a), as well as around the Antarctic Peninsula and its adjacent islands (Torres and Gajardo 1985; Goodall 1990c; Greenpeace 1991). In some instances this material can be traced to nearby research stations (Arnould 1993), rubbish dumps of abandoned stations (Gregory et al. 1984), or other human activities (Torres and Aguayo 1993).

Similar materials are also reported from Whalers Bay, Deception Island (Goodall 1990c), Byers Peninsula, Livingston Island, and South Shetland Islands (Greenpeace 1991). Note that with increased tourist activity around the Antarctic Peninsula, confectionery items, film canisters, cigarette butts, and other evidence of the casual day visitor are appearing in the beach litter.

High Seas

Understanding of the dynamics of plastic pollution at seas remains limited (Gerrodette 1985). Of the Southern Hemisphere seas, the only region subject to reasonably intensive investigation is that around South Africa during the late 1970s (Ryan 1988a, 1990a). Limited surveys, using an otter-type neuston net and tows of generally 1-h duration, have been made around New Zealand and between New Zealand and the Ross Sea, Antarctica (60 stations were occupied over several surveys during the course of two austral summers) (Gregory et al. 1984; Gregory 1987, 1990a, 1990b). Some further data have come from ships of opportunity (e.g., Greenpeace's RV *Gondwana*) on passage across the Pacific sector of the Southern Ocean between New Zealand and Drake Passage and the Antarctic

Peninsula (10 stations in 1989 and a further 10 stations in 1991). The RV *Tangaroa,* on its delivery voyage, provided data from a crossing of the Indian Ocean and Tasman Sea from Cape Town to Tasmania (18 stations along the 38°S parallel). Grace (1994) also took neuston net samples during a crossing of the Indian Ocean, between Australia and the Gulf of Suez, including several in the area of the Whale Sanctuary, on the Greenpeace vessel RV *Rainbow Warrior* in 1993.

These neuston tows and associated sighting surveys have shown that, across the expanses of the eastern South Pacific and the Pacific Sector of the Southern Ocean, pelagic debris quantities are minimal. South of the Polar Front, the density is unlikely to exceed 1 or 2 items/km^2: between the Polar Front and the Subtropical Convergence, quantities may approach 20 items/km^2. North of the Subtropical Convergence amounts may reach 2000 items/km^2, but could be higher immediately down drift from urban centers (Gregory 1990b).

In the eastern Indian Ocean, north of Australia, up to 16 megalitter items were sighted per day, with fishing floats and plastic bottles conspicuous. Plastics and other man-made litter increased westward to more than 11 items/km in the eastern-central Indian Ocean then fell away dramatically to fewer than 2 items/km across most of the central and western parts. A region around the Seychelles was considered particularly clean and free of debris (0.3 items/km) (Grace 1994), although plastic and other man-made litter is not uncommon on the shores of Aldabra Atoll (Payet, R., Ministry of Foreign Affairs, Victoria, Seychelles, 1994, personal communication).

An extensive grid of neuston trawls made during the late 1970s provided detailed information on the seasonal abundance of plastic mesolitter around the southwestern Cape Province of South Africa. During that period, the mean density of plastic was 3640 particles/km^2 and 42.4 g/km^2 (Ryan 1988a). Since then there have been no assessments of the density of mesolitter at sea in coastal waters. A small number (39) of neuston trawls made in oceanic waters southwest of Africa in 1987 collected only two pieces of mesolitter, an average density of only 50 particles/km^2 (Ryan 1990a), which is considerably lower than the 1970s estimate of approximately 3000 particles/km^2 for oceanic water west of southern Africa (Morris 1980).

At sea, there is evidence that macrolitter densities (at least in the vicinity of large urban areas) decrease rapidly with distance offshore. Aerial transects counted 19.6 items/km^2 at 10 km offshore, compared with only 1.6 items/km^2 50 km offshore (Ryan 1988a, 1990a). Similar patterns were found in the density of mesolitter (Ryan 1988a).

Interregional Comparisons and Variations

Interregional comparisons are difficult to make because of differences and inconsistencies in sampling methodology and category identification. There is also the unresolved problem inherent in the mixing of materials from onshore and offshore sources. Domestic and household debris often cannot be separated into either category with any certainty. Nevertheless, the following general observations can be made:

The proportion of plastic is reasonably consistent, between 60% and 80%.

Density, in terms of item numbers, is highly variable. It is greatest near population centers where local land-based sources are often dominant and packaging materials are generally conspicuous.

Onshore sources, for local as well as remote recreational beaches, are dominated by paper, confectionery and convenience food wrapping, beverage containers, and straws.

The proportion and relative abundance of debris clearly attributable to the fishing industry and shipping activities increases with distance from population centers and proximity to important fishing grounds. It reaches a maximum at remote, unpopulated islands.

Foreign-source material becomes conspicuous and is relatively more abundant at remote localities where local sources are minimal.

Quantities floating on the high seas are higher around Southern Africa and perhaps the east-central Indian Ocean and in the waters of Oceania. Elsewhere, quantities are very low, reflecting dispersal and dilution with distance, and never approach those reported from the North Atlantic and North Pacific.

Evidence from remote oceanic islands (see Tables 5.2 and 5.3) suggests a southward-decreasing, strong latitudinal gradient in litter densities from subtropical and temperate waters through the Subtropical Convergence to the Polar Front and beyond. The major oceanic fronts are effective barriers to the passage of pelagic litter, which tends to concentrate along these fronts.

Sources and Sinks: Onshore or Offshore

Southern Hemisphere studies of marine debris, particularly those from remote localities and isolated islands, invariably stress offshore sources. Such sources are usually from the fishing industry or other shipping activities or both. It cannot be considered accidental that debris is conspicuous on shores adjacent to, or down drift from, many fishing grounds and shipping lanes. Indeed the significance of these sources may be underestimated because domestic or household litter, often attributed to onshore sources, may be of the same maritime origin.

It must also be recognized that some material is truly oceanic, coming from resident populations after lengthy periods afloat on the high seas. The evidence of distant sources and long-distance transport is incontestable; for example, the predominance of South American debris reaching the Tristan da Cunha Group (Ryan 1987a). Figure 5.8 reveals some of the sources, and source countries, that contributed to macrolitter on the westside of Inaccessible Island during 1989 and 1990. More than half the South American items came from Argentina; Uruguay and

Brazil were also key South American contributors. Furthermore, while most of the Asian debris reaching Inaccessible Island during that period belonged to Taiwan, other contributors included (in descending order) the USSR, UK, Spain, Scandinavia, Germany, France, Polland, Greece, Australia, New Zealand, United States, and West Indies. Elsewhere, Goodall (1990b) has documented New Zealand marine debris reaching Argentina.

There are several factors that greatly influence the wide dispersal of marine debris from all sources around the Southern Hemisphere. Around the Subantarctic Islands, numerous observers from the late nineteenth century (Russell 1896) to the present (Lutjeharms et al. 1988) have emphasized the west wind drift and strong circumpolar circulation of the Southern Ocean. Around the islands of Oceania, observations at Raoul Island (Gregory 1991) and Ducie Atoll (Benton 1991, 1995) indicate that marine debris is influenced by the southeast tradewinds.

There is little understanding or recognition of the "sinks" for marine plastic and other persistent synthetic debris. Williams et al. (1993) have drawn attention to the potentially damaging environmental impact of plastic and other litter accumulating on the sea floor. The extent of this problem for the Southern Hemisphere cannot be adequately assessed at present. For some time anecdotal evidence from scuba divers has suggested moderate amounts of plastic and other debris on the sea floor adjacent to popular recreational localities near Auckland and elsewhere around New Zealand. Cleanup programs by divers in South Africa harbors have collected vast amounts of debris, and these areas clearly are important point sources of nonfloating as well as floating debris. There is also the evidence, cited previously, of material recovered from the seabed of fishing grounds.

In regions of strong onshore winds, plastic and other debris once beached may be blown considerable distances inland. Observations at Raoul and Chatham Islands and around southern West Australia suggest the quantities hidden from view in vegetated dune fields may exceed those at the shoreline. Material disappearing and preserved through burial by drifting sand could be exhumed during later erosive episodes to rejoin the visible litter pollution (Gregory 1993). Forsyth (1993) has similarly noted significant quantities of blown debris in a dune field at Algoa Bay, South Africa. Coastal wetlands and mangrove swamps (Berry 1994) are also places where plastics and other litter can accumulate.

There has been some speculation over the degradation rates of pelagic plastics (Gregory 1978, 1990b). For beach-cast material a latitudinal gradient is evident. The proportion of chalky and embrittled materials is highest on low-latitude islands of Oceania (decreasing through middle latitudes of southeastern Australia and New Zealand), and on Subantarctic Islands it is low with physical or mechanical battering being more evident (Gregory 1987, 1990b). Degradation rates while afloat are likely to be much less than those for comparable items cast ashore and exposed to sunlight (Andrady 1990). Until the sinks for plastic and other persistent synthetic debris can be fully identified and quantified, the population dynamics of these materials, as envisaged by Gerrodette (1985), will elude us.

Discussion

The global oceanic flux of plastic and other persistent synthetic debris is difficult to measure and monitor from shipboard studies on the high seas because of logistics, time constraints, costs, and the individual commitment or effort demanded. Baseline debris surveys of beaches in the vicinity of urbanized and industrial regions, or popular recreational and tourist localities, are of questionable reliability because of the likelihood for contamination from local land-based sources.

Accessible but remote beach localities seldom frequented by casual visitors, like Anxious Bay in South Australia, are ideal places to conduct baseline surveys and hence monitor accumulation rates of pelagic plas-

tics (Wace 1991, 1994). West-facing Masons Bay, Stewart Island, New Zealand, from which almost 1 ton of debris/km was removed in 1988 (Macfie 1989), is a similarly located site and there are others in South America and southern Africa.

The Commission for the Conservation of Antarctic Marine Living Resources (CCAMLR 1993) has recognized that remote Subantarctic Islands, fully open to oceanic influences and free from seasonal ice and snow cover, provide appropriate monitoring sites from which the flux of marine debris released into, and circulating under, the west wind drift of the Southern Ocean can be evaluated. Long-term programs to monitor the incidence of marine debris and its impacts on marine living resources were initiated in the late 1980s at several island localities and on the Antarctic Peninsula. Because the surveys are following a standardized method, data collected will be statistically comparable and permit reliable assessment of the effects of MARPOL Annex V and other influences on the pelagic marine debris levels in the Southern Ocean.

We recognize the desirability of establishing a Southern Hemisphere "network" of key monitoring localities on comparable open, west-facing shores. From these, and given time, it may be possible to calibrate the present fluxes of anthropogenic pelagic litter around the region and determine variations in the longer term.

Marine debris in Southern Hemisphere surface waters, and on its shores, may have either land-based onshore or vessel-based offshore sources, but in either case these are local or regional in character. Artifacts of European (e.g., Norway, Poland, Russia) and Asian (e.g., Korea, Japan, Taiwan) manufacture at not uncommon on Subantarctic Islands: most artifacts can be attributed to their countries' fishing activities. These observations, together with the abundance of South American debris at Tristan da Cunha, emphasize the extent to which oceanic circulation patterns are a barrier to the transfer of pelagic litter and debris pollution between Southern and Northern Hemisphere surface waters.

With the exception of the Sao Paulo–Rio de

Janeiro axis, the Southern Ocean lies distant from major industrial and urban sources of pelagic litter pollution, as do contiguous surface waters for much of the Indian, South Pacific, and South Atlantic Oceans. The incidence of beached marine debris around the Southern Hemisphere, other than near population centers, is much less than that commonly reported from the Northern Hemisphere. Even at remote localities where much debris can be traced to the fishing industry or frequently traversed shipping lanes, or both, quantities are less. These observations should not lead to complacency, and there should be no lessening effort to reduce and control pelagic debris levels around the region.

Conclusions

The magnitude of contamination of surface waters and shorelines around the Southern Hemisphere by pelagic plastics, and other persistent synthetic and man-made materials, has not reached that of the Northern Hemisphere. For reasons of geography and demography it is unlikely ever to do so. Nevertheless, significant if minor quantities that have varying aesthetic and environmental impacts are widely reported. Recognizing the cumulative nature of the marine debris problem, even these minor levels must be considered unacceptable.

Geographically remote localities from around the region are ideal places for on-going monitoring surveys. Data from these isolated sites will allow determination of the variation in quantities and composition of marine debris and perhaps some measurement of the effectiveness of MARPOL V. On the other hand, full knowledge of the Southern Hemisphere flux and population dynamics of marine debris will have to await a level of understanding of the sinks of these materials that is at least as complete as our current understanding of their sources.

Acknowledgments. M. R. Gregory has had funding for research on pelagic debris for many years from the University of Auckland Research Committee. Recent support has come from the Research Agenda of the Ministry for the Environment. P. G. Ryan's synoptic debris surveys in South Africa are funded by the Plastic Federation of South Africa. Logistical support to visit islands in the African sector of the Southern Ocean was given by the South African Department of Environmental Affairs. The manuscript was typed by R. Bunker and the figures drafted by L. Cotterall. The assistance of K. Johnston is also acknowledged. In preparing this review we have benefited greatly from the generosity of, and freedom with which, colleagues and fellow workers have released unpublished results and observations to us. We thank the organizers for supporting our attendance at the Third International Conference on Marine Debris. M. R. Gregory also recognizes assistance from the U.S. Marine Mammal Commission.

SECTION II
Biological Impacts

SECTION II
Biological Impacts

Introduction

Of all the marine debris stories recorded over the years, none provides a more graphic demonstration of the kind of unexpected impact marine debris can have than the story of a British naval disaster in the 1982 Falklands War. It is reported that the Argentines were able to predict the location of the British frigate HMS *Sheffield* by correlating its trail of trash [dumped overboard at sea] with their knowledge of local ocean currents. The Argentines' prediction was correct: On May 4, 1982, their Exocet missile found its mark, virtually destroying the ship, causing more than 40 casualties, and forcing a general revision of naval warfare tactics.

Of course, while most of the impacts of marine debris are much less dramatic, they are nonetheless utterly logical even if unexpected. A predominant feature of the solid waste dilemma is that most people do not appreciate the ramifications of its omnipresence and are genuinely surprised when it impacts some facet of their lives: urban flooding from trash-clogged drainage systems, damaged boat motors, strangled animals on beaches. Yet, where can we look (except perhaps straight up) and not see some form of solid waste or potential solid waste? The point is, our wastes are increasing everywhere, potentially affecting every natural and man-made system. Rather than be surprised by this, we ought to expect every imaginable impact—even on warships—and use the simple logic of their occurrence to craft effective control and reduction measures. This, of course, requires information.

The need for impacts information, and its value in justifying work that addresses the marine debris problem, has powered a nonlinear, catch-as-catch-can approach to accumulating the impacts record. Much of the information on the impacts of marine debris has arisen from observations by biologists and oceanographers already in the field for other purposes. In fact, with few exceptions, the data on impacts of marine debris have been recorded by professionals whose principal work, or object of study, has been disrupted (usually unexpectedly) in some way by debris. Over time, this more-or-less haphazard collection of information has begun to provide the bases for answering (a) which types of impacts should be measured; (b) how much will it cost to measure the impacts of marine debris; (c) how are measurements of impacts best scaled against the measurements of amounts, types, and distribution of marine debris; and, perhaps most importantly, (d) what are the trends in biological and economic impacts caused by marine debris, and how should these trends influence societal response?

In the next two sections, these questions will be partially addressed. In Section Two, we look at the current factors contributing to the biological impacts of marine debris. Robards et al., Winston et al., Laist, and Carr and Harris all provide significant background for evaluating which impacts should be measured more systematically, as well as for designing appropriate investigations. We defer the discussion of social and economic considerations of the impacts and control of marine debris to Section Three.

6.

The Highest Global Concentrations and Increased Abundance of Oceanic Plastic Debris in the North Pacific: Evidence from Seabirds

Martin D. Robards, Patrick J. Gould, and John F. Piatt

Introduction

Plastic pollution has risen dramatically with an increase in production of plastic resin during the past few decades. Plastic production in the United States increased from 2.9 million tons in 1960 to 47.9 million tons in 1985 (Society of the Plastics Industry 1986). This has been paralleled by a significant increase in the concentration of plastic particles in oceanic surface waters of the North Pacific from the 1970s to the late 1980s (Day and Shaw 1987; Day et al. 1990a). Research during the past few decades has indicated two major interactions between marine life and oceanic plastic: entanglement and ingestion (Laist 1987). Studies in the last decade have documented the prevalence of plastic in the diets of many seabird species in the North Pacific and the need for further monitoring of those species and groups that ingest the most plastic (Day et al. 1985). Plastics may be consumed because particles resemble prey items (Day et al. 1985, 1990a), or by consuming prey with plastics in their gut (Kartar et al. 1976). In turn, adult seabirds may pass plastics on to chicks by regurgitation (Fry et al. 1987).

Two classes of plastic are commonly found in seabirds (Day et al. 1985; Ryan 1987b): pellets and fragments. Plastic "pellets" (also known as nibs, resins, or cylinders) are the raw product of the plastic industry. Pellets lost at the manufacturing plant or during transportation may enter the marine ecosystem directly or via drainage systems. Plastic "fragments" or "user" plastics are small, weathered pieces of larger manufactured items (e.g., fishing floats, buckets, and bottles) that are discarded or lost at sea, particularly from fishing boats and marine shipping (Scott 1975; Merrell 1980). Other forms of ingested debris include toys, Styrofoam, monofilament line, rubber, and plastic film (Baltz and Morejohn 1976; Day 1980; Robards et al. 1995).

Day (1980) completed the first comprehensive investigation of plastic ingestion by North Pacific seabirds. He analyzed 1968 stomach samples from 37 species of seabirds collected between 1969 and 1977 throughout a wide geographic area of subarctic coastal Alaska. Most specimens were collected at major seabird colonies in the Aleutian Islands (e.g., Buldir Island), and in the northern Gulf of Alaska (e.g., Shumagin, Semidi, and Kodiak Islands). He found plastic in 15 of the 37 seabird species collected. The frequency of plastic particle ingestion varied among species and was most prevalent in surface feeders such as fulmars, some shearwaters, petrels, and phalaropes (Day 1980), as well as some planktivorous diving species (e.g., auklets). Similar results have been observed in other studies (Ryan 1987b; Moser and Lee 1992).

Distribution and abundance of small plastic particles (< 25 mm) collected in neuston nets from the North Pacific and Bering Sea were reported by Wong et al. (1974, 1976), Shaw (1977), Shaw and Mapes (1979), Takatani et al. (1986), Day and Shaw (1987), and Shaw and Day (1994). These studies found that pelagic plastic is most abundant in the central subtropical and western North Pacific, suggesting an association with tanker and general ship traffic in the western Pacific and the "downstream effects" of pollutants entering the ocean near Japan and adjacent countries. Japan and southern California are the two major petrochemical and plastic manufacturing centers in the North Pacific (Guillet 1974; Wong et al. 1976). Plastic entering the ocean in southern California probably moves south away from the subarctic North Pacific in the California Current system. However, this current runs into the Equatorial Current, travels west across the Pacific, and enters the Kuroshio Current that moves north to Japan. Plastic entering the ocean in Japan (or reaching Japan in the Kuroshio Current) probably moves eastward in the North Pacific Drift Current (Day et al. 1985). Of 109 identifiable items ingested by laysan albatrosses on Hawaii, 108 originated in Japan (Pettit et al. 1981). The North Pacific Drift Current splits to form the California and Alaska currents. Of the plastic transported into the northern Gulf of Alaska by the Alaska Current, some apparently moves inshore and is consumed by seabirds. The highest incidence of ingested particles in the subarctic North Pacific (Day 1980) was in the Aleutian coastal waters. Densities of small plastic particles in the subarctic North Pacific and Bering Sea are 26 to 400 times lower, respectively, than in subtropical waters. Of small oceanic plastic particles found in the central North Pacific, 3.7% were pellets and 96.3% were user fragments (Day and Shaw 1987). In contrast, 19% of plastic particles found off southwestern Cape Province, South Atlantic in 1977–1978 were pellets and 39% were user fragments.

Available evidence suggests that plastics are damaging to seabirds when they are consumed in sufficient quantity to obstruct the passage of food or cause stomach ulcers (Fry et al. 1987; Ryan 1987b). Other effects may include bioaccumulation of polychlorinated biphenyls (PCBs) (Aldershoff 1982; Ryan et al. 1988), toxic effects of hydrocarbons (Carpenter et al. 1972), diminished feeding stimulus (Ryan 1988b), reduced fat deposition (Connors and Smith 1982), lowered steroid hormone levels, and delayed reproduction (Azzarello and Van Vleet 1987). However, at present acute effects of plastic ingestion are rarely observed, and chronic effects on body condition are generally equivocal (Day 1980; Ryan 1987c, 1990b; Moser and Lee 1992).

In this paper we present three major findings: (1) the frequency and number of plastic particles ingested by seabirds of the subarctic North Pacific have increased during the past two decades; (2) seabirds collected in the central North Pacific have a higher incidence and quantity of plastic particle ingestion than seabirds collected in the subarctic North Pacific or elsewhere in the world; and (3) experiments to assess the effects of ingested plastic on seabirds are probably underestimating what constitutes a large plastic load.

Methods

Seabirds for this study originated from two main sources. A collection was made in the subarctic coastal area of the North Pacific (1988–1990) in the same season and areas sampled by Day (1980) for temporal comparisons of plastic ingestion. A second collection was made in the central North Pacific (1990–1991) to assess geographic variation in incidence of plastic ingestion and types of plastic ingested.

1988–1990 Sample Collection

Stomach samples of 1799 seabirds comprising 24 species were collected during May 1988 through August 1990 for ongoing long-term studies of feeding ecology. Samples were collected at seven sites in Alaska ranging over a distance of about 2800 km: at Agattu and Buldir Islands in the western Aleutian

Islands, at Aiktak Island in the eastern Aleutian Islands, and Shumagin and Semidi Islands south of the Alaska Peninsula, in Kachemak Bay (lower Cook Inlet), and in Prince William Sound (Fig. 6.1).

To ensure that samples were large enough for statistical comparisons of key species, collections were more extensive for four species: parakeet auklets, tufted puffins, horned puffins, and black-legged kittiwakes (Table 6.1 provides Latin names of sampled species). These species were chosen for intensive study because of their large populations, widespread distribution, ease of collection, and because they had been the focus of the earlier study of plastic ingestion by Day (1980). Stomachs were dissected and contents were preserved in the field.

1990–1991 Sample Collection

Seabirds were collected as part of the National Marine Fisheries Service scientific observer program to assess catch and by-catch of driftnet fisheries. The use of seabirds collected from driftnets for research provides a nondestructive method of collecting large samples of many species.

Seabirds were salvaged from nets from May 1990 until November 1991 (one bird in the database was collected in 1987). The observers worked in the central North Pacific, ranging from 28°N to 47°N and from 145°W to 141°E. After collecting birds from nets, the observers identified, counted, and froze the birds for the duration of the cruise.

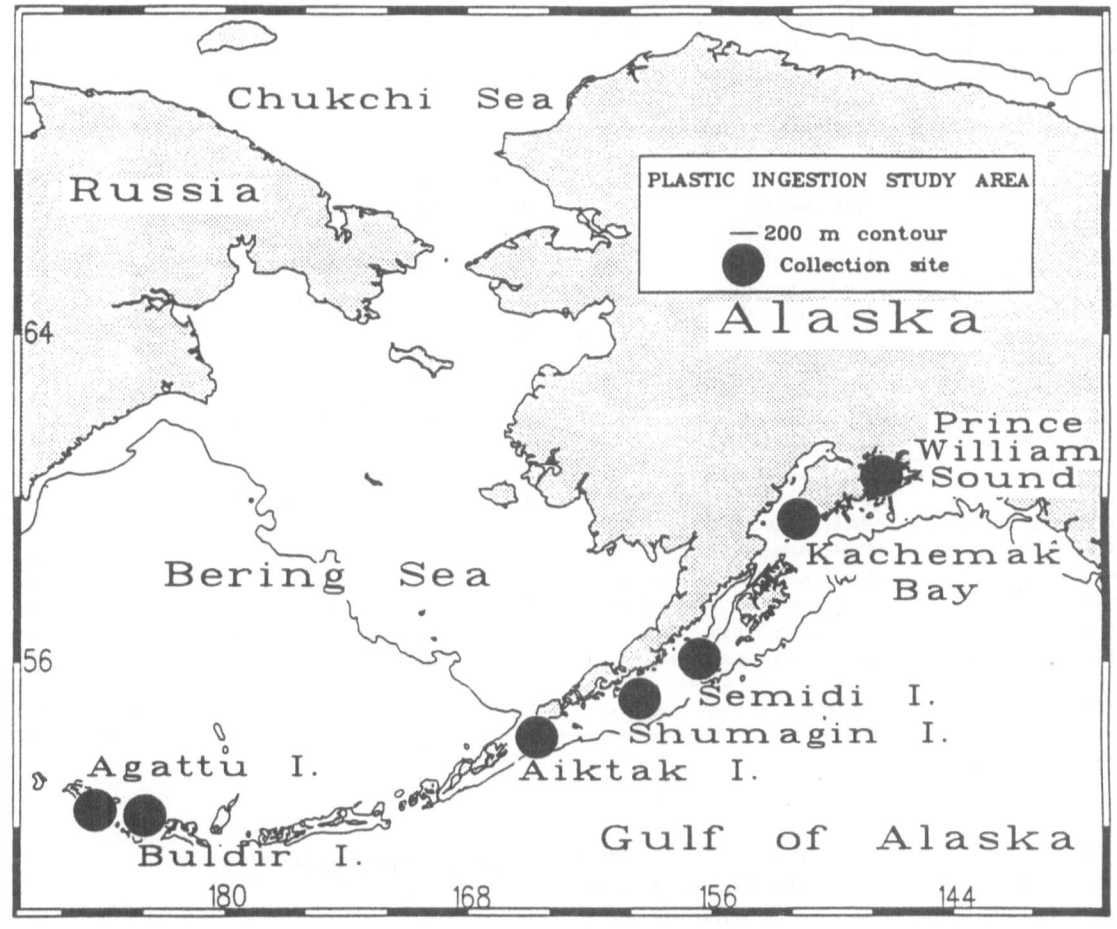

FIGURE 6.1. Location of Subarctic North Pacific sampling areas (*large circles*).

Martin D. Robards, Patrick J. Gould, and John F. Piatt

TABLE **6.1.** Percent incidence of plastic ingestion for species collected in the North Pacific.

Common name	Latin name	Subarctic North Pacific, 1969–1977[a]		Subarctic North Pacific, 1988–1990[b]		Central North Pacific, 1990–1991[d]	
		Percent incidence	n	Percent incidence	n	Percent incidence	n
Laysan albatross	Diomedea immutabilis	—	—	—	—	93	167
Black-footed albatross	Diomedea nigripes	—	—	—	—	45	110
Northern fulmar	Fulmarus glacialis	58	38	84	19	88	42
Sooty shearwater	Puffinus griseus	43	76	—	—	85	543
Short-tailed shearwater	Puffinus tenuirostris	84	200	80	5	88	200
Buller's shearwater	Puffinus bulleri	—	—	—	—	98	118
Pink-footed shearwater	Puffinus creatopus	—	—	—	—	100	1
Flesh-footed shearwater	Puffinus carneipes	—	—	—	—	95	83
Streaked shearwater	Calonectris leucomelas	—	—	—	—	0	2
Dark-rumped petrel	Pterodroma phaeopygia	—	—	—	—	50	2
Solander's petrel	Pterodroma solandri	—	—	—	—	50	2
Mottled petrel	Pterodroma inexpecta	—	—	—	—	60	5
Fork-tailed storm-petrel	Oceanodroma furcata	100	8	86	21	100	12
Leach's storm-petrel	Oceanodroma leucorhoa	25	4	48	64	67	3
Tristam's storm-petrel	Oceanodroma tristami	—	—	—	—	100	4
Red-tailed tropicbird	Phaethon rubricauda	—	—	—	—	0	1
Pelagic cormorant	Phalacrocorax pelagicus	0	3	20	10	—	—
Red-faced cormorant	Phalocrocorax urile	0	2	0	16	—	—
Bar-tailed godwit	Limosa lapponica	—	—	—	—	100	1
Red-necked phalarope	Pharlaropus lobatus	67	3	—	—	—	—
Red pharlarope	Phalaropus fulicaria	—	—	—	—	100	1
Pomarine jaeger	Stercorarius pomarinus	—	—	—	—	0	1
Long-tailed jaeger	Stercorarius longicaudus	—	—	—	—	0	2
South polar skua	Catharacta maccormicki	—	—	—	—	50	2
Mew gull	Larus canus	0	10	25	4	—	—
Glaucous gull	Larus hyperboreus	3	33	—	—	—	—
Glaucous-winged gull	Larus glaucescens	0	63	0	21	—	—
Slaty-backed gull	Larus argentatus	—	—	—	—	0	1
Black-legged kittiwake	Rissa tridactyla	5	188	8	256	0	5
Red-legged kittiwake	Rissa brevirostris	13	46	27	15	—	—
Common murre	Uria aalge	0	191	1	134	—	—
Thick-billed murre	Uria lomvia	1	138	0	92	100	1
Cassin's auklet	Ptychoramphus aleuticus	40	10	11	35	—	—
Parakeet auklet	Aethia psittacula	75	116	94	208	33	3
Crested auklet	Aethia cristatella	0	85	3	40	—	—
Least auklet	Aethia pusilla	1	89	0	13	—	—
Whiskered auklet	Aethia pygmaea	0	5	0	22	—	—
Ancient murrelet	Synthliboramphus antiquus	0	16	0	68	—	—
Marbled murrelet	Brachyramphus marmoratus	0	61	0	96	—	—
Kittlitz's murrelet	Brachyramphus brevirostris	0	5	0	17	—	—
Pigeon guillemot	Cepphus columba	0	18	3	43	—	—
Rhinoceros auklet	Cerorhinca monocerata	0	20	0	1	44	9
Horned puffin	Fratercula corniculata	37	148	37	120	57	28
Tufted puffin	Fratercula cirrhata	15	348	25	489	88	8

[a]Day (1980).
[b]Robards et al. (1995).
[c]Robards (1993).

After the cruise, the birds were sent to seabird biologists to be used for seabird ecology studies, including diet studies. A total of 1357 stomachs from 28 species were used in this study.

Plastic Analysis

The stomachs and gizzards of the collected birds were dissected and food items separated from nonfood items. Nonfood items included plastic objects and particles, rubber bands and gloves, cigarette filters, screws and nails, Styrofoam, fibrous wads, wire, stones, and pumice. Most plastic particles were found in the ventriculus (gizzard), and few were found in the proventriculus (stomach).

Nonfood items were analyzed in a manner consistent with previous studies (Day 1980) to allow for comparisons. The number of plastic particles in each stomach were recorded; particles were then weighed and classified according to their color, size (maximum dimension), shape, and type (pellets or user items). For this study, only the size, type, and number of particles were used. For comparison of incidence of ingestion and mean numbers of particles ingested between years and areas, we required a minimum sample size of 15 from each study to reduce potential bias from low sample sizes.

Results

Temporal Change in Subarctic North Pacific Plastic Ingestion

Of the 24 species of seabirds that were collected in both Day's (1969–1977) and our (1988–1990) study periods, the number of species that contained plastic particles increased from 12 in the 1970s to 15 in the 1980s.

Of the 17 species found to contain plastic in either study period, plastic ingestion increased in 12 species. Overall, there was a significant (χ^2 = 1100, 16 df, P < .001) increase in the frequency of plastic ingestion between the period of 1969–1977 and the late 1980s. Species with high frequencies of occurrence of plastics included all procellarids (fulmar, shearwaters, and storm-petrels), black-legged kittiwakes, parakeet auklets, and horned and tufted puffins. For species with adequate sample sizes, the frequency of plastic ingestion increased over time by 0.2%–26.3% (horned puffin and northern fulmar, respectively) (Fig. 6.2).

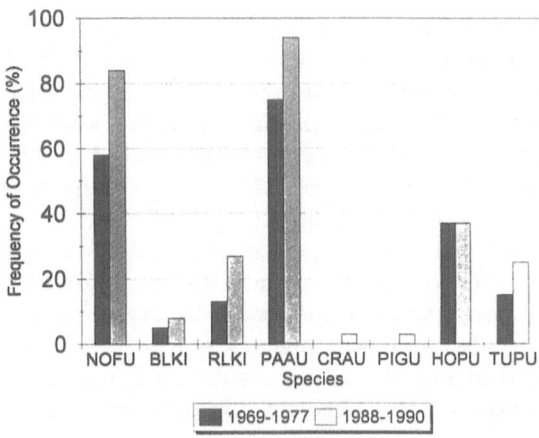

FIGURE 6.2. Temporal change in frequency of occurrence of plastic ingestion by subarctic North Pacific seabirds. *NOFU*, Northern fulmar (*Fulmarus glacialis*); *BLKI*, black-legged kittiwake (*Rissa tridactyla*); *RLKI*, red-legged kittiwake (*Rissa brevirostris*); *PAAU*, parakeet auklet (*Aethia psittacula*); *CRAU*, crested auklet (*Aethia cristatella*); *PIGU*, pigeon guillemot (*Cepphus columba*); *HOPU*, horned puffin (*Fratercula corniculata*); and *TOPU*, tufted puffin (*Fratercula cirrhata*). *Solid bars*, 1969–1977; *hatched bars*, 1988–1990.

Species that did not ingest plastic in any year of study included red-faced cormorant (n = 18), glaucous-winged gull (n = 84), whiskered auklet (n = 27), ancient murrelet (n = 84), marbled murrelet (n = 157), Kittlitz's murrelet (n = 22), and rhinoceros auklet (n = 21). Only 3 of 545 common and thick-billed murres that were examined over all years contained plastic. A low incidence of ingestion was also observed in glaucous gulls and least auklets.

The mean number of particles per bird increased significantly (Mann–Whitney U = 0.87, P < .05) between collections of the four key study species (parakeet auklets, tufted and horned puffins, and black-legged kittiwakes) made in 1969–1977 and those from 1988–1990. Parakeet auklets ingested, on average, the most particles of the 1988–1990 study with 17.1 particles per bird (versus 13.7 particles per bird in the 1969–1977 study). Similarly, the weight of plastic ingested by the four key study species increased significantly (Mann–Whitney U = 1.43, P < .05) between 1969–1977 and

1988–1990. Maximum numbers of particles varied widely among species (tufted puffin, 51; northern fulmar, 26; black-legged kittiwake, 15; horned puffin, 14; Leach's storm-petrel, 13; fork-tailed storm-petrel, 12). The largest number of particles per individual was 87 in a parakeet auklet.

Of 4417 plastic pellets examined from 15 species collected between 1988 and 1990, the majority were of the two main types: pellets (76.4%) and user plastics (21.5%). The remainder (2.1%) of items were unrecognizable plastic pieces. This was comparable with the 833 plastic particles analyzed by Day (1980), who found 70% pellets and 30% user plastic. Seven seabird species accounted for 98.9% of all particles recovered, and the composition of particles varied greatly among species. Pellets were ingested most by diving species such as tufted puffins and parakeet auklets., while user plastics were common in surface-feeding species such as storm-petrels and kittiwakes. Table 6.1 presents a summary of this study's results as well as the Latin names of each species sampled.

Geographic Variation in North Pacific Plastic Ingestion

Of the species (excluding species with only one sample) collected in the central North Pacific between 1990 and 1991, 5 species (Laysan albatross, Buller's shearwater, flesh-footed shearwater, fork-tailed storm-petrel, and Tristam's storm-petrel) had a plastic incidence of more than 90%, and 4 (northern fulmar, sooty shearwater, short-tailed shearwater, and tufted puffin) had an incidence greater than 80%. Of the 27 species collected in the subarctic North Pacific between 1988 and 1990, only 1 species had an incidence of plastic ingestion greater than 90% (parakeet auklet) and 2 an incidence more than 80% (northern fulmar and fork-tailed storm-petrel).

For the 11 plastic-ingesting species that were represented in both subarctic North Pacific and central North Pacific collections, 9 displayed a higher incidence of ingestion.

For the 4 species with adequately sized samples collected in both areas (Fig. 6.3), there was a significantly higher frequency of plastic ingestion in the central North Pacific ($\chi^2 = 9$, 3 df, $P < .05$).

The highest mean number of particles for a central North Pacific seabird species with adequate sample size was 18 for northern fulmars, followed by 16 for Buller's shearwaters and 14 for Laysan albatrosses. The only bird collected in the subarctic North Pacific containing similar mean numbers was the parakeet auklet, which contained 17 particles per individual. Three of the procellariiform species from the central North Pacific had individuals that ingested more than 100 particles. A short-tailed shearwater contained the highest number of particles found in any seabird (135 particles), followed by northern fulmars (114 particles) and sooty shearwaters (108 particles). The highest number of particles ingested by any seabird collected in the

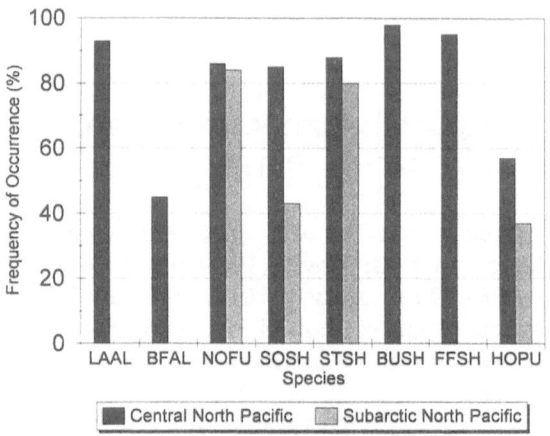

FIGURE 6.3. Frequency of occurrence of plastic ingestion by central North Pacific seabirds, with species collected in subarctic North Pacific added for comparison. *LAAL*, Laysan albatross (*Diomedea immutabilis*); *BFAL*, black-footed albatross (*Diomedea nigripes*); *NOFU*, northern fulmar (*Fulmarus glacialis*); *SOSH*, sooty shearwater (*Puffinus griseus*); *STSH*, short-tailed shearwater (*Puffinus tenuirostris*); *BUSH*, Buller's shearwater (*Puffinus bulleri*); *FFSH*, flesh-footed shearwater (*Puffinus carneipes*); and *HOPU*, horned puffin (*Fratercula corniculata*). *Solid bars*, central North Pacific; *hatched bars*, subarctic North Pacific.

subarctic North Pacific was by a parakeet auklet that contained 87 particles.

As in other studies (Moser and Lee 1992; Ryan 1987b), we found that the weight of particles ingested by central Pacific seabirds to be related to body size. Laysan albatrosses ingested the highest mean loads (1.94 g per bird with a maximum of 39.2 g), and horned puffins the least (0.18 g per bird with a maximum of 1.83 g). However, black-footed albatrosses were an exception to this. They ingested the second highest mass of plastic (11.26 g) but ingested only on average 0.32 g of plastic.

User plastic was the most abundant type of plastic ingested by seabirds in the central North Pacific (24% pellets, 76% user plastic). In contrast, seabirds in the subarctic North Pacific ingested mostly pellets (76% pellets, 22% user plastic, 2% unrecognizable plastic particles). This is consistent with the availability of different types of plastic, most (95%) of which is user type (Day and Shaw 1987). Some of the recognizable plastic objects are consistent with debris originating from dumping as opposed to fishing activities. These objects included plastic cars, game dice, a bubble gum toy case (still with enclosed toy doll), and refrigerator magnets. Debris items of unknown origin included cigarette lighters, plastic wrapping, gloves, cigarette filters, and plastic fittings. Larger plastic particles and objects were found in the larger seabirds such as fulmars and albatrosses, which was also the case in the North Atlantic (Moser and Lee 1992).

Discussion

Temporal Variation

Our study corroborates Day's (1980) finding of widespread ingestion of plastic particles by seabirds in the subarctic North Pacific. If we assume that differences in sampling intensity among species and geographic areas between Day's (1980) and our study do not bias our analyses, then it appears that although the type of ingested plastic has changed little over a 10- to 15-year period, the frequency and quantity of plastic ingestion has increased significantly.

This increase in plastic particle ingestion parallels an observed increase in plastic pollution in the North Pacific during the same time period (Day and Shaw 1987). In subarctic waters and the Gulf of Alaska, levels of small plastic pollution rose from 0–132 particles/km^2 in the mid-1970s to 12,800 particles/km^2 in the mid-1980s (Day et al. 1990b). Similar increases over these time periods were observed in the Bering Sea (68 and 600 particles/km^2, respectively) and in the subtropical waters (100 and 726 g/km^2, respectively).

Similar temporal observations of plastic pollution and ingestion of plastics by seabirds in the North Atlantic have recently been reported by Moser and Lee (1992). They analyzed the stomach contents of 1033 seabirds collected off the coast of North Carolina from 1975 to 1989. Procellariiform birds contained the most plastic, and the frequency of occurrence of plastic increased in 7 of 8 procellarid species during the study period. This increase was attributed to increasing levels of plastic particle pollution in the North Atlantic (Moser and Lee 1992).

Geographic Variation

Incidence and quantity of plastic ingested by different species are influenced by feeding method (Day 1980; Azzarello and van Vleet 1987). Numerous species throughout the world feed to a varying degree by scavenging at the sea surface (Ashmole 1971). This method of feeding is difficult to quantify and may lead to a bias in comparing the quantity of plastic ingested between different species assemblages, depending on the availability of scavengable plastic. However, a higher incidence of plastic ingestion in similar species in specific areas indicates a greater availability of plastic and thus more highly polluted oceanic waters.

Several species that we collected in the central North Pacific for this study were also collected in the same general area by Ogi

(1990). He found an 89% incidence of plastic ingestion in sooty shearwaters ($n = 193$), and an 82% incidence in short-tailed shearwaters ($n = 265$) collected in the central North Pacific between 1979 and 1987. This compares closely with our results of an 85% incidence of plastic ingestion in sooty shearwaters ($n = 543$) and 88% incidence in short-tailed shearwaters ($n = 200$).

If collection of seabirds from drift nets does not significantly bias the sample toward seabirds with higher plastic loads, then it appears that seabirds collected in the central North Pacific ingest plastic at a higher frequency than seabirds collected in the subarctic coastal area of Alaska. The composition of plastic in central North Pacific seabirds indicates a strong correlation between plastic ingested and availability of plastic in the oceanic surface waters, fragments of user plastic being the most common form of ingested plastic as well as the most common form of plastic in neuston trawls of central North Pacific surface waters (Day and Shaw 1987).

Apart from plastic ingestion studies involving only a few species, several recent studies have investigated large species assemblages around the world. The most comprehensive studies examined 36 species collected in the tropical Pacific (Ainley et al. 1990b; Spear et al. 1995); 38 species collected off North Carolina in the North Atlantic between 1975 and 1989 (Moser and Lee 1992); 15 species collected from Gough Island in the central South Atlantic Ocean (Furness 1985a); 60 species collected in the

southern hemisphere, mostly off South Africa (Ryan 1987b); and 23 species collected in the Antarctic (Ainley et al. 1990a).

Four species collected in the North Pacific were also collected as part of these other studies, allowing for a direct comparison of plastic ingestion between areas (Table 6.2). These results indicate that seabirds in the central North Pacific have the highest frequency of plastic ingestion. Northern fulmars have also been collected on the European coast (Van Franeker 1985), with an incidence of plastic ingestion of 92% ($n = 96$), the highest known incidence of plastic ingestion for fulmars. However, these birds were collected dead off beaches, which probably biased results when comparing to collections of live fulmars. These fulmars contained a mean number of 11.9 plastic particles per bird (maximum, 96), which was less than the mean number of 18 particles (maximum, 114) found in birds collected in the central North Pacific.

Taxa at the family level were used for comparisons between areas with different species assemblages. Highest known incidences of plastic ingestion from around the world are compared with incidences in seabirds from the North Pacific in Table 6.3. For five species of South Atlantic albatross collected by Ryan (1987b), the highest incidence was only 11%, by a sample of 18 black-browed albatrosses (*Diomedea melanophris*). The species with highest incidence of ingested plastic in the seven recent studies of large species assemblages are presented in Table 6.4 The subarctic North Pacific study

TABLE 6.2. Percent incidence of plastic ingestion for species found in a variety of global areas.

Species	Central North Pacific (n)[a]	Subarctic North Pacific (n)[b]	Tropical Pacific (n)[c]	North Atlantic (n)[d]	South Atlantic (n)[e]
Northern fulmar	88 (42)	84 (19)	—	86 (44)	—
Sooty shearwater	85 (543)	43 (76)[f]	75 (36)	40 (5)	51 (63)
Leach's storm-petrel	67 (3)	48 (64)	20 (354)	38 (8)	—
Black-legged kittiwake	0 (5)	8 (256)	—	10 (41)	—

[a]Robards (1993).
[b]Robards et al. (1995).
[c]Spear et al. (1995).
[d]Moser and Lee (1992).
[e]Ryan (1987b).
[f]Day (1980) (no sooty shearwaters were collected between 1988 and 1990).

TABLE 6.3. Highest percent incidence of plastic ingestion and mean number of particles ingested for different seabird families collected in the North Pacific and other areas of the world.

	North Pacific[a]			Outside the North Pacific			
Family	Species (n)	Percent incidence	Mean particle number	Species (n)	Area	Percent incidence	Mean particle number
Diomedeidae	Laysan albatross (167)	93	14	Black-browed albatross (50)[b], *Diomedea melanophris*	South Africa	11	0.2
Procellariidae	Buller's shearwater (118)	98	16	Blue petrel (74)[b], *Halobaena caerulea*	South Africa	92	9.7
Phalacrocoracidae	Red-faced cormorant (16)	0	0	Bank cormorant (167)[b], *Phalacrocorax neglectus*	South Africa	1	<0.1
Laridae	Red-legged kittiwake (15)	27	1	Bonaparte's gull (32)[c], *Larus philadelphia*	North Atlantic	19	3.0
Alcidae	Parakeet auklet (208)	94	17	"Auks" (37)[d]	Scotland	5	–

[a]Robards (1993) and Robards et al. (1995).
[b]Ryan (1987b).
[c]Moser and Lee (1992).
[d]Bourne (1976).

contained five procellariiform species, but is unique in being the only study in which a nonprocellarid (parakeet auklet) is the most common consumer of oceanic plastic.

The family Laridae contains many surface-scavenging species. However, the maximum incidence of plastic ingestion was only 27% by red-legged kittiwakes. Glaucous-winged gulls, with a sample size of 84, displayed a zero incidence of plastic ingestion. Elsewhere in the world, the highest incidence of plastic ingestion was 19% by Bonaparte's gulls (*Larus philadelphia*). These results suggest a behavioral difference for this taxa in their selectivity at identifying food items, or a short residency time for plastic particles in the gut.

Seabirds collected in the tropical Pacific by Spear et al. (1995) indicated seabirds moving south from the North Pacific generally contained more user plastic. In contrast, seabirds migrating north from the South Pacific generally contained more pellets. This corresponds with the high volume of user plastic found in North Pacific surface waters (Day

TABLE 6.4. Highest incidence of plastic ingestion by seabirds in seven recent studies of large species assemblages.

Location	Species displaying greatest percent incidence of plastic ingestion (n)	Percent Incidence
Central North Pacific[a]	Buler's shearwater, *Puffinus bulleri* (118)	98
Subarctic North Pacific[b]	Parakeet auklet, *Aethia psittacula* (208)	94
South Africa[c]	Blue petrel, *Halobaena caerulea* (74)	92
North Atlantic[d]	Northern fulmar, *Fulmarus glacialis* (44)	86
Central South Atlantic[e]	Great shearwater, *Puffinus gravis* (13)	85
Tropical Pacific[f]	Sooty shearwater, *Puffinus griseus* (36)	75
Antarctica[g]	Blue petrel, *Halobaena caerulea* (62)	56

[a]Robards (1993).
[b]Robards et al. (1995).
[c]Ryan (1987b).
[d]Moser and Lee (1992).
[e]Furness (1985a).
[f]Spear et al. (1994).
[g]Ainley et al. (1990a).

and Shaw 1987) and in seabirds of the central North Pacific.

In conclusion, some of the highest global incidence of plastic ingestion occurs in central North Pacific seabirds, and this corresponds to high plastic concentrations in oceanic surface waters. Input of plastic to the North Pacific Ocean from mainland and shipping sources, coupled with the currents and convergencies of the region, concentrate marine debris on the ocean surface at levels that appear higher than for any other oceanic region of the world.

The only solution to remedy this problem is to reduce the density of plastic particles at sea. Annex V of the International Convention for the Prevention of Pollution from Ships (MARPOL), which came into force in 1988 and prohibits dumping of all plastic at sea, may have already reduced the input of plastic to the North Pacific. Preliminary data from work completed in the early 1990s by Spear et al. (1994) indicate that increased frequency of ingestion may be dropping slightly. Further study of procellarid species will help track continued changes in levels of oceanic plastic debris.

Effects of Plastic Ingestion

There have been several studies using birds artificially loaded with plastic particles to assess the effects of plastic ingestion (Ryan 1990b, Sileo, L., U.S. Fish and Wildlife Service, unpublished data). These investigations showed some evidence of detrimental health effects from ingested plastic particles, particularly in relationship to growth and feeding desire. However, smaller plastic loads were used than for equivalent species or for similarly sized birds found wild in the central North Pacific. To produce more conclusive evidence of chronic health effects, these experiments need to be repeated using plastic loads closer to the maxima found in wild populations of similar species. Results can then be applied to wild populations of seabirds that have known levels of plastic ingestion.

A search for correlations between plastic load and health indices for wild populations of seabirds has been generally unsuccessful in producing any more than indirect evidence of chronic health effects (Day 1980; Connors and Smith 1982; Ryan 1990b). These investigations used collections of wild birds, with maximum numbers of ingested plastic particles less than that observed for similar species in the central North Pacific. Spear et al. (1994) is the only investigation at present to show a statistically significant negative correlation between plastic loads and body weight. However, it may be impossible to demonstrate direct cause-and-effect relationships between plastic ingestion and body condition in wild seabirds because of natural variability in the environment and the fact that affected birds may quickly disappear (die) from sampled populations. Statistical confirmation of impacts using carefully controlled experiments is of primary importance for future study.

Acknowledgments. We thank Jay Pitocchelli, and the crew of the M/V *Tiglax* (Alaska Maritime National Wildlife Refuge), for logistic support and help in collection and processing of the subarctic birds [all specimens were collected under state and federal permits issued to J. Piatt]. Assistance in the field was also provided by John Wells, Andrea MacCharles, Alan Springer, Gus van Vliet, Bay Roberts, Leslie Pulcher, Mark Holmgren, John Lang, and Vernon Byrd. We thank Chris Babcock, Karen Laing, Bay Roberts, and Gerry Sanger for help during project planning and data analysis. The Marine Mammal Laboratory of the National Marine Fisheries Service and Region 7 of the U.S. Fish and Wildlife Service provided financial support for this project. We especially thank Linda Jones and Kenton Wohl of those organizations. We thank all the people who cooperated in the 1989–91 High Seas Driftnet Observer Programs for their help in obtaining the data used in this study, and especially Shannon Fitzgerald, Greg Morgan, and Dan Waldek. Larry Spear of Point Reyes Bird Observatory sorted the stomachs. Finally, thanks go to Peter Ryan, Gerry Sanger, Carolyn Gove, Mary Cody, and Cott Mignery for help with analysis of plastic particles and the preparation of this manuscript.

7∎
Encrusters, Epibionts, and Other Biota Associated with Pelagic Plastics: A Review of Biogeographical, Environmental, and Conservation Issues

Judith E. Winston, Murray R. Gregory, and Leigh M. Stevens

Introduction

Entanglement, ingestion, and ghost-fishing are well-documented biologically damaging effects of marine debris. Debris may also smother benthic communities on soft and hard bottoms (Parker 1990). For a number of organisms, however, plastic debris provides a positive opportunity, creating new habitats in the form of numerous, semipermanent floating islands, which are driven by winds and currents around the world's oceans. Although these epibiotic assemblages seem to be most common in warm-water regions, biologically encrusted plastic items have already been found at sites ranging from the Subantarctic to the Equator (Gregory et al. 1984; Gregory 1990a, 1990b). This paper focuses on studies by the three authors at sites in the Western Atlantic and the Southern Pacific, with findings of worldwide relevance.

Western Atlantic Drift Plastic Studies: Islands in the Stream

Three studies of drift plastic epibionts were carried out at North Beach in Fort Pierce, Florida (approximately 27° 29′ N, 80° 25′ W). These Florida studies have been supple-mented by observations and collections made on the shores of Bermuda in 1978 (Gregory 1983) and in October 1990 (Gregory, unpublished data). Bermuda beaches are also strongly influenced by the Gulf Stream and gyral circulation of the Sargasso Sea.

The Florida study site (Fig. 7.1) is a gently sloping, east-facing barrier island beach, stretching north from the Fort Pierce Inlet along the Atlantic coast of North Hutchinson Island. Inshore currents generally flow southward. The north-flowing Gulf Stream current (as indicated by water color, temperature, and presence of *Sargassum*) is usually encountered between 15 and 20 miles offshore (Worth and Hollinger 1977; Hugh Reichardt, Smithsonian Marine Station, personal communication). Prevailing winds are from the southeast, or, in the late fall and winter, from the northwest (Worth and Hollinger 1977). Material washed up on the beach may include sea grasses, mangrove leaves, and mangrove propagules carried out of the adjacent Indian River Lagoon through Fort Pierce Inlet as well as locally derived recreational debris (e.g., fast-food containers and drink bottles). However, at any time of year (but most frequently in fall and winter) passing storm systems may cause several days of onshore easterly winds, which increase wave action and cast *Sargassum, Physalia,* and other members of the Gulf Stream *Sargassum* community onto the beach, along with large amounts of mostly plastic marine debris.

could be encountered year round and that several calcified taxa besides bryozoans were consistently found attached to plastic items. Examination of freshly beached material also showed that noncalcified organisms like algae and hydroids were a part of this encrusting biota. It also seemed, at least on casual observation, that pieces beaching in summer were greener than those arriving in winter. Thus, when opportunities arose, two additional collections were made to address the following questions: (1) What were the dominant organisms on these plastic islands? (2) Were there consistent patterns of encrustation? (3) Were noncalcified organisms a

FIGURE 7.1. Drift line at North Beach, Fort Pierce, Florida, shows pelagic *Sargassum,* seagrass, and plastic trash.

Florida Study One

The initial Florida study (December 1980) was undertaken as part of a survey of the marine bryozoans of the Indian River area (Winston 1982a). In this first study, plastic items encrusted by bryozoans (Fig. 7.2) were collected along 0.8 km of beach from the Inlet breakwater northward. It was concluded that the bryozoan *Electra tenella,* previously known from natural flotsam such as floating wood and sea beans, was the dominant organism on the trash, and further speculation is that this species might be increasing its abundance and distribution via the increasing amount of plastic entrained in Caribbean and Gulf Stream currents (Winston 1982b).

· Repeated visits to this beach over the next few years showed that stranded plastic trash

FIGURE 7.2. Gallon plastic container with large colonies of *Electra tenella,* foraminiferans, and crustose algae. North Beach, Fort Pierce, FL, study 1. *Scale bar* = 5 cm.

significant component of the community? (4) Were there any seasonal differences in the biota found on the trash? and (5) Could the flora and fauna reveal anything about the origin of the trash itself?

Florida Studies Two and Three

A summer sample was collected in July 1988, when several stormy days (E and E-NE winds up to 20 knots and 5–8 ft seas) prevented work on another project offshore. The collection was made along a 0.4-km stretch of the beach, following the high-ride drift line between the seaward ends of two public access paths. To document the occurrence of algae and other noncalcified encrusters, the collection was made during a receding tide while the freshly washed-up trash and associated flotsam was still wet from the waves. Natural flotsam cast up with the trash included *Sargassum* bunches and *Physalia* jellyfish, indicating an offshore Gulf Stream source.

The aim was to collect all fresh material brought in by the storm. Probably most of the sample was cast up by the high tide, but the

transect might have also included some refloated material from previous tides. We collected 335 pieces (all those more than 2 cm in size) in plastic garbage bags to preserve moisture. On return to the laboratory, 250 pieces were measured, noted as to type and origin (if ascertainable), and examined under a dissecting microscope for attached organisms. Specimens were preserved dry, or, when necessary for later identification, saved in the appropriate preservative (formalin or alcohol). Sampling was repeated in February 1994, after 3 days of strong east winds. In this third study 126 items were collected and processed as were the summer samples.

The Western Atlantic Drift Biota

The groups of organisms found in the second and third Florida studies as well as their frequency of occurrence are summarized in Table 7.1 At least 64 taxa, representing 9 phyla, were found on the 376 items sampled. Many of these taxa were also present on collections made on Bermudian shores. Five

TABLE 7.1. Organisms encrusting Florida drift plastic (two studies).

Major group	Estimated number of taxa	Rank (summer)	Rank (winter)
Foraminiferans	4+	1	1
Coelenterates			
Hydroids	2+	4	2
Colonial Scyphozoa	1		
Millepora	1		
Cup corals	1		
Bryozoans	26+	2	3
Algae			
Soft (red, brown, green)	8+	3	4
Calcareous	2+	–	–
Polychaetes			
Serpulids	2+	5	5
Spirorbids	1	–	–
Other tubes	3	–	–
Sponges	2+	–	–
Colonial ascidians	1	–	–
Mollusks	6+	–	–
Crustaceans			
Acorn barnacles	2+	7	7
Gooseneck barnacles	2+	6	6
Total	64+		

groups of encrusting organisms dominated in both localities.

Foraminiferans

The most abundant encrusters on North Beach plastics were benthic foraminiferans, amoeboid protists, whose delicate protoplasm is protected by a calcareous shell or one constructed of cemented sediment. On North Beach plastics at least four taxa occurred, but most individuals belonged to one species of the genus *Acervulina* (Fig. 7.3). They may well have been the first organisms to recruit onto the plastic, as they were frequently the only organisms present on an item. Their brown-and-white or bleached white specks (about 0.5–1 mm in diameter) were numerous, occurring on 59% of encrusted items in summer and 89% in winter. They were not so conspicuous on Bermuda plastics, where occasional *Rosalina* and a *Cibicides* together with rare miliolids were identified. The presence of planktonic foraminiferans on several plastic items from Bermuda was surprising. Little is known about the ecology of encrusting foraminiferans, even though they may be important framebinders and sediment producers in reef settings. Foraminiferans have been reported from fouling studies (Woods Hole Oceanographic Institution 1952), but as microfauna, they have seldom been considered in ecological or distributional studies of marine macrofauna from either fouling or benthic habitats.

Bryozoans

Bryozoans are suspension-feeding marine invertebrates (about 5000 species worldwide) whose calcified or soft-bodied colonies can be found in almost all marine habitats, although they most commonly encrust hard substrates in shallow to shelf depths. In terms of number of items encrusted, bryozoans were the second most abundant group on North Beach plastics in summer and the third most abundant group in winter. Bryozoans were also the most diverse group in terms of numbers of species represented (28). In all

three Florida studies, *Electra tenella* (see Fig. 7.2) was the most abundant bryozoan on plastic trash. The development of *E. tenella* has never been described; however, like other membraniporines, the species is presumed to be nonbrooding, with eggs spawned into seawater and developing into a triangular-shelled feeding larva (cyphonautes) that can remain in the plankton for as long as 2 months (Hyman 1959). Next in abundance on plastic trash were colonies of *Membranipora tuberculata*. *Membranipora tuberculata* has been reported from warmwater habitats around the world and is a dominant organism in the pelagic *Sargassum* community (Hentschel 1922; Friedrich 1969; Pestana 1985). It also occurs on other species of brown algae and occasionally on other substrates (Winston 1982a).

Three other membraniporine species, *Membranipora savartii*, *Membranipora arborescens*, and *Membranipora* sp., were also common (Winston 1982a), which was not surprising as their cyphonautes larvae could be expected to be carried long distances by ocean currents. It was more surprising that 23 species of bryozoans with nonfeeding or brooded larvae or both were also found on plastic, some of them more than once (Figs. 7.3 and 7.4). Larvae of these species, at least in laboratory studies, settle within a few days.

Bryozoans were similarly common in collections made from around Bermuda. *Electra tenella* and *M. tuberculata* were also frequent epibionts on pelagic plastics from Bermuda (Gregory 1983) together with at least another dozen unidentified species.

Hydroids

Hydroids, sessile colonial coelenterates (about 2000 species worldwide) whose colonies are also commonly found in fouling and hard substratum communities and whose feeding polyps are carnivorous upon zooplankton, were the second most abundant group on North Beach plastics in winter and the fourth most abundant group in summer. Hydroids were one of the first groups of organism ever reported from floating plastic

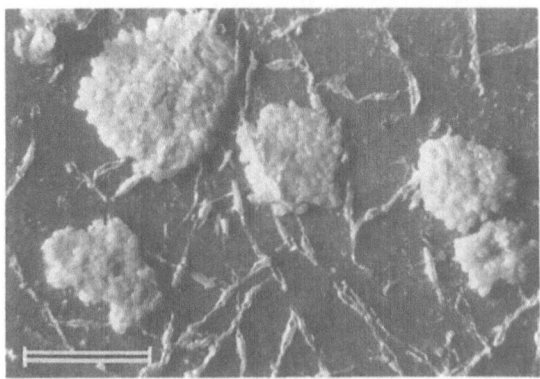

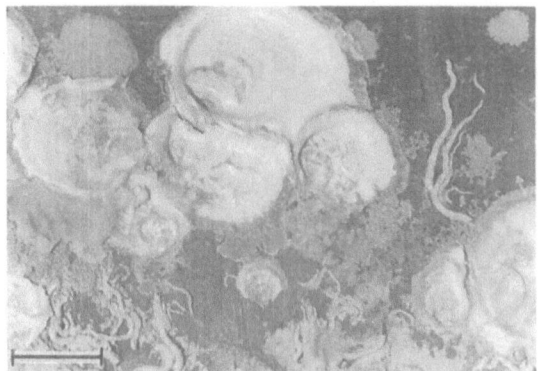

FIGURE 7.3. *Acervulina* sp. (foraminiferan) and bases of *Aetea* (bryozoan) colonies. North Beach, Fort Pierce, FL, study 2, item No. 184. *Scale bar* = 1 mm.

FIGURE 7.4. Portion of plastic container encrusted by mollusks, serpulids, bryozoans, and crustose algae. North Beach, Fort Pierce, FL, study 1. *Scale bar* = 1 mm.

debris (Carpenter and Smith 1972). However, because hydranths and hydroid colonies themselves are short lived and noncalcified, only the more heavily chitinized stolon regions usually remain to be found on beached debris. On dry or wet plastic their delicate traceries were barely discernible, but examination under the microscope showed that there were few pieces of plastic debris without them. It was clear that the stolons found represented more than one species; however, because hydranths and gonophores (reproductive chambers) are necessary to identify hydroids, no species identifications were possible. The few intact colonies seemed to belong to *Clytia* or *Halecium* species known as minor members of the *Sargassum* community. Larvae of these species are nonfeeding ciliated planulae that would normally attach in a few hours or days (Calder 1986).

Hydroid colonies, including species of *Clytia, Halecium,* and probably *Obelia* and *Sertularia,* were also commonly encountered on Bermuda plastics but few specimens were intact, and complete colonies were generally restricted to freshly beached material.

Algae

Algae, plants from a number of different major groups lacking certain structures (true roots, stems, leaves, and flowers) found in other plants, were the next most common encrusters on Florida plastics (Figs. 7.2 and 7.4). At least two crustose (calcareous) species occurred. Noncalcareous forms displayed morphology ranging from films to filaments and even included a young *Sargassum* recruit. Algae are common fouling organisms (Woods Hole Oceanographic Institute 1952), and are found in the *Sargassum* community. For example, calcerous *Fosliella* algae covered 1%–2% of the surface of *Sargassum* washed up on a Bermuda beach (Pestana 1985). They are also important in shallow benthic habitats and as framework builders and binders on reefs.

Algae were similarly common on the pelagic plastic stranding on Bermuda beaches. Of coralline taxa, *Fosliella* was most frequent (Gregory 1983), although *Jania, Mesophyllum, Lithophyllum,* and *Amphiroa* were tentatively identified also.

Tube-Building Polychaetes

Serpulids and spirorbids were also common encrusters of North Beach (Fig. 7.4) and Bermuda drift plastics. Sessile calcareous tube-building, filter-feeding polychaetes of the families Serpulidae and Spirorbidae are common and widely distributed in fouling and shallow water communities. Most ser-

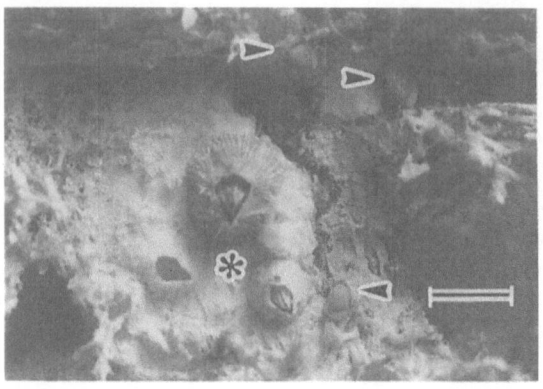

Figure 7.5. Plastic debris encrusted by acorn (*asterisk*) and goose (*arrows*) barnacles. North Beach, Fort Pierce, FL, study 3, item no. 61. *Scale bar* = 1 cm.

Figure 7.6. Plastic debris encrusted by hidden-cup corals (*Phyllangia americana*) and sponge (*arrow;* only dried fibrous skeleton remaining). North Beach, Fort Pierce, FL, study 3, item no. 61. *Scale bar* = 1 cm.

pulid tubes from the North Beach locality probably belonged to a species of *Hydroides,* but species identification was not possible from tubes alone. Tube forms of Bermuda material suggested that the most common taxon there was *Spirorbis corrugatus,* with *Spirorbis spirillum, Spirorbis borealis,* and *Hydroides dianthus* of lesser abundance. These species are also a conspicuous, if minor, element of the *Sargassum* community (Butler et al. 1983).

Barnacles

Next most abundant were cirripede crustaceans or barnacles (Fig. 7.5). Barnacles are a diverse and varied marine group with more than 1400 species, many cosmopolitan. Adults are shelled, sessile, and attached to hard substrata or other organisms. They feed on plankton or other particles captured or filtered by cirri. Generally, egg masses are brooded to the first nauplius stage larva, then shed into seawater where later larval stages are planktonic before settling and metamorphosing into sessile adults (Southward 1986).

Stalked lepadomorph or goose barnacles were the barnacles most frequently encountered on drift plastics from both Florida and Bermuda. This is not surprising because as a group they prefer floating substrates. The

most frequently identified species was *Lepas anatifera.* The taxon considered typical of drifting *Sargassum, Lepas pectinata,* was not identified at North Beach and was quite rare on Bermudian artifacts. Stalkless acorn barnacles attach to both fixed and floating objects. At least two species of acorn barnacle, *Balanus eburneus* and *Balanus amphitrite amphitrite,* have also been recorded.

In addition to these dominant organisms, members of other taxa more rarely encountered on North Beach and Bermuda drift plastics included sponges, corals (Fig. 7.6), fire coral (*Millepora* sp.), octocorals, gorgonians, ascidians, colonial scyphozoans, and bivalve mollusks (including *Chama, Pteria, Anomia, Pinctada, Isognomon,* and *Crassostrea*). Membranous sabellid polychaete tubes and mud tubes that could have belonged to either polychaetes or amphipods were also noted. Occasional crabs (*Planes minutus*) were associated with larger plastic artifacts (e.g., crates) in the Bermudian wrack.

Community Recruitment

The epibiota encrusting North Beach plastics is recruited from several sources. Obviously, the most significant source of recruits should be organisms from the closest natural habitat whose larvae are drifting in the same currents. The four or five dominant species, including *Electra tenella* and *Lepas* spp., are all species common to natural flotsam, a habitat which, although long known, seldom receives ecological attention (Friedrich 1969). The *Sargassum* community is probably another major source of recruits. The second most common bryozoan, *Membranipora tuberculata,* is a dominant organism in this community (Hentschel 1922; Friedrich 1969; Pestana 1985). Several organisms frequently encountered on plastic (e.g., hydroids and serpulids) may also be recruited from the *Sargassum* community. The remaining epibionts may be recruited from any and all environments that voyaging plastic "islands" pass over, including shallow coastal and fouling communities (e.g., *Crassostrea* or *Balanus amphitrite amphitrite*), and coral reefs (e.g., corals, sponges, *Millepora*). Drift substrata are most likely to be settled by species with long-lived larvae; however, the data show that species with short-lived larvae may also recruit, perhaps when their larvae are carried to the surface by storms or during periods of upwelling such as on the Florida shelf during summer (N. Smith 1981; Worth and Hollinger 1977). While most of the normally benthic species appear to have originated in Florida or the Caribbean, at least one exotic bryozoan species was found. A *Thalamoporella* species never before reported from Florida or the Caribbean was collected in studies one and three; it most closely resembles *Thalamoporella evelinae,* a species described from Brazil (Marcus 1941).

Seasonal Differences

The same five groups of organisms dominated the Florida community in summer and winter, although in a slightly different order. One seasonal difference was in the relative abundance of the two most common bryozoans. In summer, *Electra tenella* encrusted 14% of the collected items and *Membranipora tuberculata* encrusted only 4%, while in the winter sample, *M. tuberculata* was slightly more abundant (10%) than *E. tenella* (8%). It had been thought that the greener appearance of summer drift plastic might result from the presence of more locally derived trash (perhaps from outflow from the Indian River coastal lagoon). This hypothesis was not upheld; however, the samples did show clear seasonal differences both in levels of encrustation by organisms and in the amount of tar present. In the summer sample, 60% of plastic items were encrusted by at least one kind of organism, versus 37% in the winter. The pattern for tar was reversed with only 16% of plastic items having noticeable amounts of tar on them in July versus 52% in February.

Reproduction

It is clear that at least some of the encrusting organisms found were able to reproduce on their plastic islands. Although reproduction of the membraniporine bryozoans, most common on the trash, was not demonstrable because they spawn directly into the sea, the large colony sizes found indicated that they had attained sexual maturity. Ovicelled colonies of two brooding bryozoans were found in the Florida samples; one was the previously unrecorded *Thalamoporella evelinae* and the other was *Bugula minima,* a Florida–Caribbean species (Winston 1982a). Crustose algae with reproductive structures (conceptacles) were observed on both Florida and Bermuda material, and a large serpulid tube, with several similar small tubes aggregated around it, was noted in one instance. Gastropod egg cases were also found on some plastic items, although adult gastropods were never present.

Biotic Interactions

Presumably, motile fauna (other than rare crabs) abandon their host item before or when it reaches the surf zone. However, there was ample evidence of grazing in the form of numerous teeth marks, scrapes, and gouges left on the plastic. The teeth of sharks make distinctive serrated impressions (Fig. 7.7). Older pieces, with crazed or partially degraded surfaces or both, often had large numbers of tooth marks. Whether fish were attracted to the plastic because of its encrusting biota is not known, but barnacles, one of the abundant groups, are certainly eaten by many groups of fish, including wrasses, parrotfish, and wreckfish (Southward 1986). The predation intensity suggested by tooth marks is supported by the numerous observations of increased fish numbers associated with floating objects of all kinds (Hunter and Mitchell 1966; Hunter 1968; Fedoryako 1988).

The smooth hard surfaces of the plastic may make it less hospitable to smaller invertebrate grazers and predators than a layered multibranched clump of *Sargassum* or a spongy, easily penetrated piece of drifting wood, but there is some indirect evidence that invertebrate predators may occasionally be present. Membraniporine bryozoans produce spined zooids in response to nudibranch predation (Harvell 1984). In one instance, three large circular *Electra tenella* colonies were found with an inner (older) zone of unspined zooids and an outer (younger) zone of spined zooids. This change suggests the arrival of a predator to their habitat island. However, it must be acknowledged that a sudden change to production of spined

zooids has also been linked with a sharp increase in salinity (Hutchins 1941), possibly occurring when plastics passed from a river delta to the sea.

There was no indication of any biological succession on the plastic, nor was there any evidence of increased species richness with increased size of the host artifact. It has been shown for *Sargassum* clumps that age is as important as size (Fine 1970). This may also have been true for plastic islands; unfortunately, there is no good way to age [time in water] plastic. Based on observations such as brittleness and surface crazing, it appeared that large to medium-sized, middle-aged (somewhat faded, scraped, and bitten versus extremely cracked or brittle) pieces had the richest assemblages.

Biogeography and Oceanography

Plastic artifacts from up on the North Beach site have three potential sources: (1) local Florida debris (including merchant shipping and recreational cruises, commercial and recreational fishing, and inappropriate or illegal household trash disposal), in which case objects may have been in the water a few days to a few weeks; (2) sources from other countries (or states), associated with medium-distance transport (Caribbean, Brazil, or the Gulf of Mexico), in which case objects traveled several weeks or months in ocean currents; and (3) sources from other countries, associated with transoceanic movement, in which case items must have survived months or years at sea in the North Atlantic. Analysis and reflotation of encrusted plastic items showed that many (e.g., bottles, containers, washtubs, plates, cups, bowls, sandals) could be categorized as consumer or household trash from coastal or inland domestic sources and that most had floated at the surface.

Many items probably had a local source; that is, they were discarded within 500 miles of where they washed ashore. A few items seemed to have a cruise- or fishing vessel origin—Japanese containers of various types,

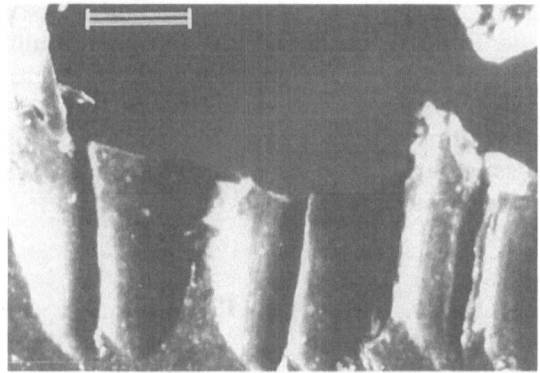

Figure 7.7. Shark bite in drift plastic container. North Beach, Fort Pierce, FL, study 2, item no. 132. *Scale bar* = 1 mm.

a cruise line bag, etc. The country of manufacture of 339 of the 452 items examined in the three studies however was not identifiable because items were fragmentary or lacked labeling or imprints. Of the remaining 113 items examined, 75% had a U.S. or probable U.S. origin, and an additional 7% had labeling in English, but no further identification.

The second most important source appeared to be the Caribbean, especially the Eastern Caribbean–Lesser Antillean region, as 8% of the examined items were manufactured in Venezuela. Another 5% had labeling in Spanish; however, the exact country of origin could not be established. Items manufactured in Columbia, Mexico, Honduras, Puerto Rico, and Brazil were also represented in the three collections. Other collections made at various times along the same stretch of North Beach yielded items manufactured in Barbados and Jamaica, confirming earlier impressions (Winston 1982b) that a substantial amount of the plastic flotsam reaching this beach originated in the eastern Caribbean. Regional trading is extensive, however, and the country of manufacture was not necessarily the country of origin.

The occurrence of material from eastern Caribbean sources agrees with what is known about prevailing current patterns. Old drift studies made by the U.S. Coast and Geodetic Survey, cited in Armstrong (1994), showed that a bottle put in the sea near Caracas could reach the Florida Keys in 4 months. More recent estimates indicate a minimum time for travel from the eastern Caribbean to the North Beach latitude of the Florida coast to be about 2 months (Winston 1982b). The three largest rivers affecting the Caribbean Sea are the Orinoco, the Magdalena, and the Amazon; together they supply about 20% of the total freshwater discharge into the world's oceans. Discharge from these rivers is greatest during the rainy season (from August to October), and work by biological oceanographers using remote sensing techniques has shown that the plume of nutrients produced by their maximum outflow reaches Puerto Rico by September or October (Müller-Karger et al. 1989). It is likely that such discharge pulses add large new cohorts of plastic islands to the population already afloat, as well as affecting the amounts, types and distribution of other marine debris in the Caribbean region (Coe et al.; see Chapter 3, this volume).

Although not impossible, sources associated with transoceanic movement are least likely in the North Beach area. Two items from Portugal were collected in North Beach studies (although they could, of course, have been discarded in Brazil and reached Florida via the Guyana Current), as well as a number of items from Greece and Scandinavia. As *Sargassum* drifts into cool water, it dies and sinks to the bottom (Schoener and Rowe 1970); plastic trash does not. Sea beans, the drift seeds of tropical plants, as well as other members of the Gulf Stream community (e.g., *Velella* jellyfish and *Janthina* snails), are occasionally cast up on western Atlantic beaches; sea beans also voyage from West Africa to the Caribbean via the Equatorial Current (Turk 1982; Armstrong 1994). Given a potential life span of at least several years, there is no reason why plastic items could not make similar voyages.

As in Florida, some of the plastic debris on Bermuda's shores has a local and often recreational origin. However, it is evident that many of the larger items come from offshore if not distant sources. These sources could lie in both the heavy trans-Atlantic shipping traffic and regional cruise vessel activities. Gregory (1983) has previously emphasized the importance of the gyral circulation of the Sargasso Sea in trapping man-made litter, some of which ultimately strands on Bermuda. The Gulf Stream is also an important factor, delivering plastics from Brazil, Venezuela, and several islands of the Caribbean (e.g., Trinidad, Jamaica, Puerto Rico) to Bermudian shores.

Pacific Drift Studies: Islands Near and Far

During the same time period that North Beach trash was being surveyed, studies of drift plastic and its biota were also being carried out in the Southern Ocean and the Pacific. The first of these (Gregory et al. 1984; Gregory 1987) showed that plastic debris could be found on even remote subantarctic islands, and beyond the Polar Front into the Southern Ocean itself. Although Southern Hemisphere debris inputs are minor compared to those of the much more heavily populated Northern Hemisphere, they are still significant (Gregory and Ryan, Chapter 5, this volume). Debris adrift in these waters can be transported around the globe in as little as 3–4 years via the West Wind Drift and Circum-Antarctic Current.

Further work by Gregory (1990a, 1990b) at numerous mainland and island sites ranging from tropical to subantarctic latitudes related patterns of debris abundance and distribution to surface currents and prevailing wind regimes. Surveys carried out on isolated and unpopulated islands of the southwest Pacific showed that much of the identifiable marine debris came from oceanic sources, often distant-water fishing or shipping activities. On populated remote islands, local sources were always important, but there was also an oceanic source consisting of objects that, by their brittle character and level of encrustation, were considered to have been adrift for some time. In contrast, on urban New Zealand shorelines and closely adjacent islands much of the debris came from local sources.

Pacific Encrusting Biota

There were striking similarities between the community found encrusting drift plastic in warm waters of the Western Atlantic (Carpenter and Smith 1972; Winston 1982b; Gregory 1983) and that in the Pacific. In warm and temperate water areas of the Pacific, plastic debris is also encrusted by Bryozoans, especially *Membranipora tuberculata* (which has been found on beached debris at sites from Northern New Zealand, Australia, Norfolk Island, Raoul Island, Fiji, Rarotonga, and Tongatapu), crustose algae, serpulids,

and acorn and goose barnacles (Fig. 7.8). A lone hermatypic coral of the kind often found attached to floating wood has been recorded from a plastic debris item on Raoul Island (29° S). The common reddish-pink encrusting reef-dwelling formaminiferan, *Homotrema rubra,* has also been noted on drift plastics from Raoul Island and Tuvalu (Gregory 1990b, Fig. 12). The extent of encrustation was less on subantarctic items, with only lepadomorph barnacles and spirorbid polychaetes recorded (Gregory 1990a). Fresh specimens of the well-known tropical Indo-Pacific oyster *Lopha cristagalli,* which had a solitary previous record from the northernmost New Zealand (Parengarenga, 34° 30' S) is a recent exotic discovery from a remote southern South Island beach (Fig. 7.9); in both instances it was attached to a tangled rope mass.

In a detailed study Stevens (1992) examined more than 3000 items from 37 km of coastline in northern New Zealand. On the 228 items of drift plastic that were encrusted by one or more organisms, a biota dominated by sessile calcified organisms was found which included bryozoans, barnacles, tubeworms, calcareous algae, foraminiferans, and hydroids. The community was similar to, but richer in species, than its Western Atlantic counterpart. The species richness, frequency of occurrence, and diversity of epibionts on these 228 items is summarized in Tables 7.2 and 7.3 and further illustrated in Fig. 7.8 (for New Zealand in general). As in the Western Atlantic, bryozoans dominated in terms of number of species (58). However, *Membranipora tuberculata* was the most abundant and *Electra angulata* (the Pacific counterpart of *Electra tenella*) was much less common. There were 26 bryozoan species new to the region, including 3 undescribed species, 2 of them only known from drift plastic. Many colonies were sexually mature.

Numerous epibionts overgrew others. In general, bryozoans were able to cover most other taxa, including other bryozoans. The most successful overgrowth was by *Celleporina bemiperistomata,* with *Watersipora subtorquata, Electra angulata,* coralline al-

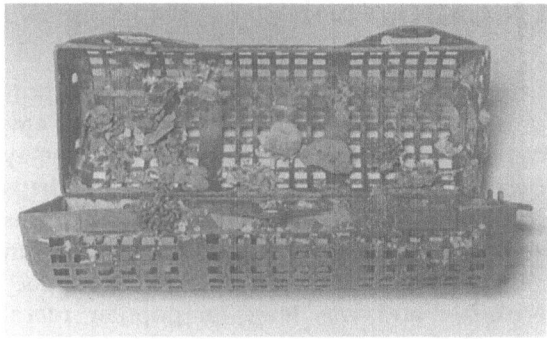

FIGURE 7.8. Heavily encrusted bait box (length, 33 cm) with bryozoans, mollusks, worm tubes, foraminiferans, and crustose algae. North Beach, Chatham Island, New Zealand.

FIGURE 7.9. *Lopha cristagalli*, a common tropical Indo-Pacific mollusk species; one of many attached to a tangled mass of plastic rope. Coal Island, Preservation Inlet, New Zealand. (Collected by J. Lindquist.)

gae, and the tubeworm *Hydroides elegans* of progressively decreasing significance. Two species, the barnacle *Lepas anatifera* and the bryozoan *Cryptosula pallasiana*, were present exclusively on plastic items with positive buoyancy, while four bryozoans (*Eurystomella forminigera, Foveolaria cyclops, Galeopsis porcellanicus,* and *Smittoidea maunganuiensis*) were restricted to negatively buoyant material. More than 40 species from all taxonomic groups were present on both buoyant and nonbuoyant debris.

Community Development and Biotic Interactions

Beach-cast material preserves only sessile organisms with hard skeletons (with perhaps a few desiccated remnants of soft-bodied forms). It gives little indication whether soft-bodied or motile animals were originally present in the community. To address that problem, Stevens (1992) moored 87 plastic bottles—polyethylene terephthalate (PET) and high density polyethylene (HDPE)—in seawater at depths of 0 and 10 m near the Leigh Marine Laboratory, on the east coast of New Zealand's North Island, 50 km north of Auckland. The development of communities fouling the plastic was followed over a period of 10 months. Sixty-seven sessile taxa were found on moored plastic items. Substrata were dominated by algae and hydroids, with algae cover increasing and hydroid cover decreasing during the 10-month experiment. Serpulid tubeworms and barnacles were also common. In addition, 32 motile taxa, mostly

TABLE 7.2. Species richness and frequency of occurrence of epibiontic taxa recognized on drift plastic items along northern New Zealand coastline (Stevens 1992).

Type of organism	Number of species (+ unidentified)	Single occurance of species	Rare (<5)	Few (5–10)	Common (11–30)	Abundant (31–50)
Bryozoans	58 (4)	23	20	10	4	—
Mollusks	8 (1)	6	2	—	—	—
Tubeworms	5 (3)	3	—	—	2	1
Barnacles	5 (0)	1	—	2	1	1
Foraminiferans	4 (0)	3	—	—	1	—
Ascidians	1 (0)	1	—	—	—	—
Hydroids	0 (1)	—	—	—	1	—
Algae	1 (1)	—	—	—	—	1
Total	82 (10)	37	22	12	9	3

TABLE 7.3. Frequency of species (bryozoans and total encrusting epibionts) found on drift plastic items along the northern New Zealand coastline (Stevens 1992).

Number of species per plastic item	Number of occurrences (bryozoans only)	Number of occurrences (all species)
1	74	121
2	1	39
3	2	25
4	1	10
5	1	8
6	0	4
7	0	6
8	0	4
9	0	3
10	0	3
>10	0	5
Total	79	228

crustaceans (crabs and amphipods) and polychaetes, were recorded. No bryozoans occurred on surface-moored bottles, and *Membranipora tuberculata* was not among those bryozoans occurring on the bottles at the 10-m depth.

This study of moored items demonstrated that the presence of larvae in the water column was the most influential factor controlling community development. Successional stages (which have sometimes been described from fouling communities) did not occur on the moored plastic and hence could not be used to age an item. However, growth rates were determined for organisms common on moored plastic over several months, and it was clear that similar studies could serve to establish a minimum time in the sea for a drift item. Stevens suggested that barnacles and bryozoans might be best for this purpose.

Stevens observed juvenile fish living inside some of the bottles and other fish grazing on bottle surfaces. Invertebrate grazers (amphipods, gastropods, chitons) were also common. During the 10-month experimental period the organisms encrusting the plastic actually changed the level at which bottles floated. Algae-covered bottles floated higher in the water than clean bottles, and encrustations of barnacles, mollusks, ascidians, and sponges forced bottles down.

Oceanography

In Pacific beach-cast plastic items, the portion of material with calcified and other epibionts varied geographically (locally it may exceed 20%; see Table 7.4), and perhaps seasonally, although this latter claim needs elaboration. Frequency of encrustation, in general, was highest at rarely visited remote localities and was least on popular recreational shores near urban centers and other upland sources.

Discussion

Encrusting Epibiota

In tropical, subtropical, and warm-temperate waters a distinct community of organisms encrusts drift plastic. The same groups of organisms, and apparently even some of the same species, have been found in widely scattered sites in the Caribbean, the Western

TABLE 7.4. Percent frequency of occurrence of encrusted, beach-cast plastic items found in selected North Atlantic and southwest Pacific Ocean localities.

Locality	Proportion of encrusted items (%)
Bermuda	
Grenadier Bay	5
Nonsuch Island	16
West Whale Bay	15
United States	
Fort Pierce (Florida)	37–60
Raoul Island	
North Coast	7–30
Denham Bay	13
Boat Cove	10
New Zealand ([+], remote localities–*, recreational and upland localities)	
Northeast (North Island)	7*–14[+]
West Coast (North Island)	5–10
Southland (South Island)	<1–5
Subantarctic Islands	0–<1
Australia	
Queensland (north of Townsville)	<5–>10
Tasmania	<3

Atlantic, and the Pacific. The community is dominated by sessile animals and plants known to occur on natural floating substrata such as tropical seeds, logs, and pumice, which are also known to drift many thousands of miles (Bryan 1971; Jokiel 1989; JMB Smith 1990, 1992), as well as on pelagic sea turtles and sea snakes. The epibionts of plastic also include species known from the pelagic *Sargassum* community. Probably next in importance on drift plastic are cosmopolitan warm-water fouling species, normally found in harbor and coastal environments, whose larvae tend to settle at or near the water surface and that have already been spread around the world by shipping vessels. Limited information on the fouling of pelagic plastic litter and trash in colder waters indicates fewer taxa may be involved.

Potential for Dispersonal of Marine Organisms

Individual items also have the potential to pick up segments of whatever benthic community they pass over on their voyages. In this way, drift plastics increase the potential for dispersal for a number of groups of marine organisms including those with short-lived larvae, as is illustrated by the finding that species of several groups in which brooding takes place reach reproductive condition on floating plastic (e.g., bryozoans, gastropods, and crustose algae). As a transoceanic dispersal mechanism, plastic substrata are more likely significant for colonial groups (in which one colony can become a founder population) than they are for solitary organisms. In bryozoans, cross-fertilization between colonies is normal in sexual reproduction, but when grown in isolation, a high proportion of species have the ability to self-fertilize and produce daughter colonies (Maturo 1991).

Drift plastic substrata can and do serve to bring in new species. This is shown by two examples from the limited work done so far with *Lopha cristagalli* in the Pacific (see Fig.

7.9) and *Thalamoporella evelinae* in the Atlantic. Compared to the numbers of larvae dumped into a harbor by discharge of ballast water (Carlton 1987; Carlton and Geller 1993), the contribution of floating plastic to environmental problems associated with the introduction of aggressive alien taxa is probably low. However, the following points should be kept in mind.

Dispersal by plastic debris is most likely to affect closely adjacent coastal regions (e.g., spread of an exotic species from a site of introduction like a populated harbor to nearby islands). Trash can be rapidly dispersed along a shoreline by currents. For example, a mooring lost at one New Zealand locality was beached 3 weeks later 30 km away (Stevens 1992).

The total habitat is still increasing and is semipermanent. Individual items appear to have the potential to make one or more circuits of the world oceans (e.g., the Gulf Stream or the Southern Ocean) before fragmenting, sinking, or being driven ashore.

Even when the number of larvae of a particular species colonizing plastic was low, a favorable set of circumstances could change that picture. For example, in regions like the Caribbean where seasonal outflow from rivers has a strong influence on hydrography, a chance combination of the appearance of a large cohort of plastic islands (carried from coastal and upland sources by the rivers) with an especially large cohort of larvae in any of the coastal benthic habitats (e.g., reefs and shelfs) that plastics pass over could result in a transport event large enough to affect the biota of distant shores.

More than 200 plants and animals are known from fouling communities (Woods Hole Oceanographic Institution 1952), and many of them have already effected a cosmopolitan distribution by traveling on ships. Travel by vessel fouling moves organisms rapidly through environments that are novel and often hostile in terms of temperature and salinity, yet some may survive. For example, a collection made of the fauna from the sides of an old ship that had come from Bermuda and was being broken up in Copenhagen harbor (where the water was colder and much less saline) yielded gravid jewel box clams (*Chama* sp.) as well as live crabs and blennoid fish (Bertelsen and Ussing 1936). A slow voyage on plastic (especially when conditions are favorable) would give encrusting biota much better chance of survival.

Potential for Dispersal: Terrestrial Organisms

As unlikely as it may seem, transport of terrestrial organisms (plants, invertebrates, and vertebrates) possibly has a greater probability of significant biological impact (Fig. 7.10). A number of studies have shown that insects, snails, isopods, millipedes, and plants can survive transport, for example, on rafts of vegetation or logs or both (Wheeler 1916; Heatwole and Levins 1972). Because of its smooth, nonporous surface, plastic is not as well suited to their survival as natural vege-

tation, but the report of viable seeds being transported in a plastic toy boat (West 1981) strongly suggests the possibility that insects or other hard-shelled invertebrates could be passively carried in the same manner.

Rafting long distances may be extremely unlikely for vertebrates (e.g., lizards, rats) because the very large masses of trash noted in some urban harbors are unlikely to stay together in stormy seas. The greatest potential for impact is local, i.e., the introduction of pests like rats from populated coasts to nearby islands where attempts are being made to preserve an endangered native biota. For example, beach cleanups around New Zealand show plastic regularly constitutes 70% or more of beach litter (Smith and Tooker 1990), and this suggested to Gregory (1991) that metropolitan and land-source aggregations of marine debris could end up on nearby islands. It is conceivable that islands of high conservation value and meriting heritage or scientific reserve status, with biotas supposedly protected from disturbance, could unknowingly be placed at risk. Events of this kind could lead to the introduction of exotic and aggressive alien taxa, thus eroding

Figure 7.10. Artist's conception of rafts of plastic debris as a dispersal agent for marine and terrestrial organisms.

or rapidly destroying years of environmental effort and progress. It is a danger of which those who manage or have stewardship over protected coastal commons should remain cognizant (Gregory 1991).

Oceanography

There is demonstrable need for collaborative research efforts among marine biologists, geologists, physical oceanographers, and students of man-made persistent pelagic debris (including plastic). The interactions could be scientifically rewarding and also have practical relevance to marine environmental management.

Lack of biological succession on plastic islands shows that their time afloat cannot be determined by community structure as Gregory once envisaged (1990b). However, colony size and growth rates of key taxa may permit minimum age estimates. Through timing of stranding events, marine debris research may assist oceanographers in relating peak river outflow in source regions to pollution or contamination events (or both) some distance away. A comparable example is the study of commercial fishery point sources and litter reaching the Texas coast (Miller et al. 1995). On the other hand, studies of epibionts may identify discrete and different water masses through which plastic debris has passed and hence assist in hindcasting currents and circulation patterns. An example is the forensic study of *Lepas* on an overturned yacht hull found to the east of New Zealand in 1989, which confirmed drift from north to south—a direction contrary to accepted ideas of the prevailing current sets (Foster, n.d.).

Final Fate

The community that develops on a piece of debris can make it float higher in the water or sink below the water surface, potentially affecting its trajectory with respect to ocean currents (whether primarily wind-driven or current-driven), and its total survival time, as well as whether it ends up on the bottom in shallow, continental shelf, or abyssal depths. Encrusting organisms may aid in breakdown of plastic or conversely may protect it from sun damage and desiccation effects. These aspects of the problem still need to be addressed, for very few studies have done so (Sibley and Strickland 1989). Plastic debris settling to the sea floor may also provide substrata suitable for colonization by a varied biota. Harms (1990) found 20 species inhabiting plastic debris dredged from the Elbe estuary. Rundgren (1992) studied plastic debris on the bottom at 18 sites around False Bay, near Cape Town, South Africa. Colonization of plastic shopping bags attached to the bottom at 5 m depths by benthic organisms (barnacles and mussels) took place within 2 weeks. Other observations included entanglements of sea urchins and sea fans with plastic, encrustation of plastic by an anemone and calcareous algae, and the presence of anoxic black sediment under a partly buried bag. Benthic ecologists have been putting plastic into the ocean for the last 30 years also, either as fouling substrata (Walters 1992) or as frames and supports for benthic panels, collectors, or other experiments. These submerged plastics have become experiments themselves, if unintentional ones, and input from ecologists should be sought.

In some areas plastic may be detrimental to benthic communities and demersal fisheries (Nash 1992). It was first noticed to impact fisheries almost 20 years ago (Hollström 1975), and the problem has not gone away. Nor is its effect restricted to urbanized coasts, for debris is moved and sorted by winds and currents (Bingel et al. 1987; Williams et al. 1993; Ribic et al., Chapter 4, this volume; Coe et al., Chapter 3, this volume; Matsumura and Nasu, Chapter 2, this volume; Gregory and Ryan, Chapter 5, this volume). In other areas, like the deep sea, sunken plastic might be a boon, providing food and substrate. It has been suggested that "whale falls," dead bodies of whales whose remains persist for several years, may provide a food resource and serve as a dispersal pathway for deep-sea vent organisms (CR Smith 1992; Ward 1994). We have begun to study the potential impact of benthic organisms whose larvae hitch a ride at the ocean surface, but it may be that our sunken trash is leaving a Hansel and Gretel-like trail of plastic crumbs through the ocean basins to guide benthic organisms to new settlement areas.

The epibionts of pelagic plastics may also be contributing increasingly to carbonate sediment budgets. Epibionts of sea grasses (Land 1979) and *Sargassum* (Pestana 1985) are known to be important contributors to the carbonate budget of tropical and subtropical, shallow marine sediments. Also, significant calcification rates, for bryozoans rapidly colonizing plastic substrates fixed close to the bottom in shallow temperate water near the site of Stevens' (1992) observations, have recently been recognized (Smith and Nelson 1994).

Conclusions and Recommendations

From this review a number of general conclusions and recommendations can be made. The scope of biological interactions with marine plastic debris is broad (see Fig. 7.10). The environmental problems receiving the most publicity so far are ingestion and entanglement, together with those of aesthetic values. The aggregation of fishes around floating objects, and the curiosity-driven attraction of some smaller marine mammals to flotsam are well known. However, it is only recently that the attention of the scientific community has been drawn to the significance of an encrusting (and pseudo-planktonic) biota and to the potential that pelagic plastics offer as vectors in both short- and long-distance dispersal of marine and possibly terrestrial organisms. Four primary aspects warrant further research.

1. Fully establish all facets of the biological significance of plastic both afloat and once it has settled to the sea floor. The latter is likely to be linked to fishery resources and their management.
2. Networking with beach cleanup groups to accumulate data on distributions of en-

crusting organisms. Study and determination of growth rates of indicator taxa may permit aging of pelagic plastic and compilation of guidebooks or brochures (or both) of common epibionts on drift plastic.
3. There is urgent need for study of biofilms on pelagic plastics and the role of microbial processes in the degradation of pelagic plastics.
4. Studies of existing pest exotics should be used to pinpoint characteristics of other species that might become problems, assess the likelihood of their doing so, and predict when and where their impact would be. For example, the zebra mussel (*Dreissena polymorpha*), a native of the Black and Caspain Seas, was accidentally introduced into the Great Lakes in 1986, probably via ballast water, and has become a major problem in North American freshwater habitats. Its damaging effects result from its life history characteristics: rapid growth, high fecundity, ability to attach securely to aquatic structures, and efficiency in filter feeding that can deplete the food resource for native organisms (Marsden 1992). An exotic serpulid introduced into New Zealand waters caused similar problems in a more limited area because of its ability to form massive growths on boats and submerged structures (Read and Gordon 1991). Such characteristics could be assessed for other potential invaders.

Acknowledgments. M. R. Gregory has had funding support for his ongoing studies of marine debris over many years from Auckland University Research Committee. Recent support has come through the Research Agenda of the Ministry for the Environment. Gratitude is also expressed to the Bermuda Biological Station for Research Publication number 1425 and the Starr Fellowship, which facilitated his field work on the island in late 1990. Acknowledgment is also made to the technical assistance of R. Bunker, L. Cotteral, S. Courtney, and K. Johnston. The contribution of L. M. Stevens to this paper comes from his M.Sc. research undertaken as part of the Environmental Science Programme at the University of Auckland. The enthusiastic support of Associate Professor J. Hay and the Leigh Marine Laboratory is acknowledged. His graduate research studies were financially supported by a Kippenberger Memorial Fellowship and a grant from the Department of Conservation. M. R. Gregory and L. M. Stevens also wish to express their appreciation to the late Brian Foster, for his friendship, advice, and guidance in the development of their ideas on the epibionts of drift plastics. J. E. Winston wishes to thank the Smithsonian Marine Station (contribution number 394), the American Museum of Natural History, and the Virginia Museum of Natural History for support of field work for North Beach studies, and to express her appreciation to Sherry Reed, Mary Rice, and Eliza Winston for help with collections.

8■

Impacts of Marine Debris: Entanglement of Marine Life in Marine Debris Including a Comprehensive List of Species with Entanglement and Ingestion Records

David W. Laist

Introduction

Lost and discarded marine debris, particularly items made of persistent synthetic materials, is now recognized as a major form of marine pollution. This perception was a seminal finding of the 1984 International Workshop on the Fate and Impact of Marine Debris (Shomura and Yoshida 1985). A major factor leading to this conclusion was information on the nature and extent of interactions between marine debris and marine life gathered by researchers working independently in different ocean areas during the 1970s and early 1980s. Compiled for the first time at the 1984 workshop, the information highlighted two fundamental types of biological interactions: (1) entanglement, whereby the loops and openings of various types of debris entangle animal appendages or entrap animals; and (2) ingestion, whereby debris items are intentionally or accidentally eaten and enter the digestive tract. Collectively, the review demonstrated that such interactions were far more widespread and common than previously thought, and that both types of interactions can cause the injury and death of individual animals of many different species.

Since 1984, much additional information has been collected on the nature and extent of interactions between marine debris and marine life. This paper attempts to review all information now available on marine debris entanglement interactions. Entanglement of marine life incidental to active fishing gear while affecting many of the same species, is considered a separate problem and issue and is not considered here. The principal objectives are to (1) compile a list of all species known to become entangled in marine debris and compare information on the occurrence of entanglement and ingestion of debris, (2) provide a guide to available information on entanglement of marine life in marine debris, (3) review factors affecting the magnitude of entanglement-related interactions with marine debris, (4) describe the effect of entanglement on individual marine animals, (5) review information on population-level effects of entanglement, and (6) identify research and mitigation measures that should be taken to address entanglement.

This analysis is based on a review of articles in peer-reviewed scientific journals and reports and publications prepared by or for government agencies. Where appropriate and possible, unpublished observations and data by individuals working directly with species or populations affected by marine debris have been considered. To help gather material, letters requesting relevant papers, reports, publications, or data were sent to more than 225 individuals in 21 countries actively engaged in studying marine debris effects or working to control sources of marine debris in the marine environment.

99

Problems Associated with Collecting and Analyzing Entanglement Data

For many reasons, efforts to characterize the occurrence and effect of entanglement with marine debris are difficult. As an awareness of these difficulties is important in interpreting available data, it is appropriate to briefly review some of the inherent problems before examining the reported data. These difficulties may be grouped into three categories: detection of entangled animals; biases in sampling and reporting; and determination as to whether entangled animals reflect interactions with marine debris (Table 8.1).

The most basic limitation in assessing marine debris entanglement is the difficulty in detecting entangled animals. Most animals vulnerable to entanglement are highly migratory (e.g., seabirds, sea turtles, and marine mammals) and tend to be scattered across wide ocean areas. Entangling debris also is scattered over broad areas, making interactions possible almost anywhere in a species range. When dispersed throughout their ocean ranges, animals are visible only for brief instances at or above the sea surface. The fleeting glimpses of wildlife afforded from the decks of ships or plane windows

does not provide a reasonable opportunity to detect entangled animals at sea either living or dead. Animals are usually visible only at great distances and often are only partially visible. Moreover, animals that become entangled and die may quickly sink or be consumed by predators at sea, thereby eliminating them from potential detection and sampling during limited survey periods. Also, those that die and float in entangling debris are often concealed within a mass of debris that itself is hard to see. Together, these factors frustrate systematic attempts to detect entangled animals at sea. As a result, most data on entangled animals at sea are opportunistic anecdotal records. When systematic sampling efforts have been attempted, small sample sizes have precluded statistically meaningful analyses.

Most entanglement records have, therefore, been gathered by land-based observers examining animals that strand on beaches or congregate seasonally on shorelines to nest, breed, molt, etc. Reliance on such land-based sampling, however, introduces a number of common sampling biases that force investigators to contend with fundamental assumptions and uncertainties that have not been tested or resolved. Most important, live entangled animals returning to shore include only those survivors entangled in debris light enough or close enough to shore to allow

TABLE 8.1. Factors complicating the analysis of marine entangelement trends.

Detection	Sampling and reporting biases
Entanglements occur as isolated events scattered over wide areas.	Virtually no direct, systematic at-sea sampling has been done and there are few long-term surveys.
Entangling debris is not easily seen on live animals at sea because animals may only be partially visible at great distances.	Sampling methodologies are inconsistent.
Dead animals are difficult to see because they float just beneath the surface and may be concealed within debris masses.	Strandings represent an unknown portion of total entanglements.
Dead entangled animals may disappear quickly because of sinking or predation.	Shore counts of live entangled animals are biased toward entanglement of survivors carrying small debris.
	Entangled animals spend less time ashore and more time foraging at sea.
	Some entanglements reflect interactions with active rather than derelict fishing gear.
	Many entanglement records may remain unpublished or are anecdotal and cannot be compared geographically or temporally.
	Few data are available for periods before 1980.

them to swim or fly to land. Invariably remaining unknown and unsampled are the animals that die at sea from entanglement in large debris items far from shore.

Moreover, animals that are entangled in small debris and get to shore do so at an increased metabolic cost. This imposes added food requirements, resulting in more time spent at sea feeding. It also increases the risk of predation because of decreased mobility. Thus, by some uncertain factor, shore-based counts of entangled animals underrepresent the number of animals caught in small debris. More importantly, land-based surveys offer no measure of the number of animals that die in large debris at sea. By the same token, dead entangled animals that strand on shore represent an unknown proportion of entangled animals that die from entanglement at sea.

The incidence of entanglement in debris also may be confused with the incidental "entanglement" or "by catch" of marine mammals, seabirds, turtles, and fish caught in active fishing gear. While active fishing gear clearly is not marine debris, lost and discarded fishing gear is. Accordingly, data on animals found stranded, free swimming, or otherwise caught in derelict fishing gear or fishing gear debris are considered victims of marine debris. In this regard, it is recognized that some animals found with net fragments attached may have been cut out of active fishing gear as unwanted bycatch when gear was retrieved. Also, some animals may tear free from active gear by themselves and carry small gear fragments with them. Thus, some unknown proportion of entanglement records almost certainly is caused by active gear rather than marine debris.

Unfortunately, it is rarely possible to determine after the fact if animals found carrying fishing debris became entangled in active fishing gear or in lost gear or gear fragments. Because lost and discarded fishing gear is a significant component of marine debris and poses obvious entanglement hazards, it clearly is inappropriate to assume that all animals found entangled in fishing-related materials became entangled in actively fished gear. For the present, therefore, investigators

and managers must work within the constraints of this conundrum.

Overview of Entanglement Records

Notwithstanding difficulties in obtaining entanglement data and their limitations, one way to assess the scope of entanglement interactions is by compiling a synoptic list of marine species worldwide for which individuals have been documented as being entangled in marine debris. To date, such lists have been prepared only for sea turtles (Balazs 1985) and pinnipeds (Fowler 1988). To assess the full range of biological effects caused by marine debris, it also is helpful to include species known to ingest marine debris. With regard to ingestion, Day et al. (1985) listed seabirds known to ingest plastics, Balazs (1985) reviewed ingestion by sea turtles, Walker and Coe (1990) compiled ingestion records for toothed whales, Hoss and Settle (1990) reviewed ingestion by bony fishes, Ryan (1987b) listed seabirds found to ingest plastics in the Southern Ocean; Ainley et al. (1990b) listed seabirds in the eastern equatorial Pacific found with plastics in the digestive tract, and Moser and Lee (1992) listed seabirds in the western North Atlantic with ingested plastics.

To combine and build on these efforts, Appendices 1 through 3 present synoptic lists of sea turtles (Appendix 1), seabirds (Appendix 2), and marine mammals (Appendix 3) reported entangled or containing ingested marine debris or both. The appendices update and expand earlier compilations by adding new or overlooked records, combining regional findings, and including reports of both entanglement and ingestion. Also, a list of fish and crustaceans known to be entangled or to ingest marine debris has been compiled (Appendix 4). In addition to providing a comprehensive listing of species known to be affected by marine debris, the appendices indicate the types of debris involved, the ocean basins in which the inter-

David W. Laist

actions have been reported, and, where possible, a subjective evaluation of whether the reviewed literature suggests that interactions are infrequent or more than infrequent for at least some locations.

Overall, the lists of affected species indicate that marine debris is a broad-scale pollutant affecting individuals of a significant percentage of the world's marine species (Table 8.2). Considering both entanglement and ingestion-related records, marine debris is known to affect individuals of at least 267 species worldwide (Table 8.2), including 86% of all sea turtle species, 44% of all seabird species, 43% of all marine mammal species, and numerous fish and crustacean species. For most of these species groups, significant numbers of species are subject to entanglement or ingestion of marine debris, although not necessarily both for individual species. For all species groups combined, entanglement records only were found for 90 species, ingestion records only were found

for 132 species, and both entanglement and ingestion records were found for 45 species. This suggests that a relatively small number of species are vulnerable to the combined effects of both entanglement and ingestion. For crustaceans, feeding mechanisms limit interactions exclusively to entanglement.

Entanglement appears to be a far more likely cause of mortality than ingestion-related interactions. For virtually all species with documented entanglements, at least some deaths are attributed to entangling debris. For at least some species (see following), there is evidence that significant levels of entanglement death occur. In contrast, for many species known to ingest marine debris there is equivocal or no evidence of ingestion-related mortality. In these cases, ingested debris is documented from living and dead animals but there is either no evidence of mortality or it is not clear that the dead animals died as the result of the presence of debris in their digestive tracts. Even in cases

TABLE 8.2. Number and percentage of marine species worldwide with documented marine debris entanglement and ingestion records by species group.

Species group	Total number of species worldwide	Number and percentage of species with entanglement records	Number and percentage of species with ingestion records	Number and percentage of species with entanglement or ingestion records or both
Sea turtles (see Appendix 1)	7	6 (86%)	6 (86%)	6 (86%)
Seabirds (see Appendix 2)	312	51 (16%)	111 (36%)	138 (44%)
Sphenisciformses (penguins)	16	6 (38%)	1 (6%)	6 (38%)
Podicipediformes (grebes)	19	2 (10%)	0 (0%)	2 (10%)
Procellariiformes (albatrosses, petrels, and shearwaters)	99	10 (10%)	62 (63%)	63 (64%)
Pelicaniformes (pelicans, boobies, gannets, cormorants, frigatebirds, and tropicbirds)	51	11 (22%)	8 (16%)	17 (33%)
Charadriiformes (shorebirds, skuas, gulls, terns, auks)	122	22 (18%0)	40 (33%)	50 (41%)
Other birds (see Appendix 2)	—	5	0	5
Marine mammals (see Appendix 3)	115	32 (28%)	26 (23%)	49 (43%)
Mysticeti (baleen whales)	10	6 (60%)	2 (20%)	6 (60%)
Odontoceti (toothed whales)	65	5 (8%)	21 (32%)	22 (34%)
Otariidae (fur seals and sea lions)	14	11 (79%)	1 (7%)	11 (79%)
Phocidae (true seals)	19	8 (42%)	1 (5%)	8 (42%)
Sirenia (manatees and dugongs)	4	1 (25%)	1 (25%)	1 (25%)
Mustellidae (sea otter)	1	1 (100%)	0 (0%)	1 (100%)
Fish (see Appendix 4)	—	34	33	60
Crustaceans (see Appendix 4)	—	8	0	8
Squid (see Appendix 4)	—	0	1	1
Species total		136	177	267

where there is evidence of frequent ingestion of debris (particularly certain species of seabirds, fish, and marine mammals), ingestion-related mortality has often not been confirmed or is reported rarely. In all cases in which entanglement is reported frequently or commonly, evidence of some mortality is apparent.

Sea Turtles

The six species of sea turtles for which entanglement records exist (Appendix 1) represents 86% of the world's sea turtle species (6 of 7 species). The records suggest a regular pattern of occurrence by all six affected species with entanglements documented in most ocean areas. Sea turtles also actively approach and ingest all types of floating debris, including entangling line, as if it were food. This behavior may precipitate many if not most entanglements. The flatback sea turtle (*Natator depressus*), whose distribution is limited almost exclusively to the Arafura Sea, the Gulf of Carpentaria, and the Coral Sea north and northeast of Australia, is the only sea turtle for which neither entanglement nor ingestion records were found.

The most thorough review of sea turtle entanglement records is by Balazs (1985), who listed some 60 cases involving five species from the mid-1970s to mid-1980s. Many if not most sea turtle entanglement cases involve animals that either died because of entanglement, would have died without human intervention, or had gangrenous flippers caused by tightly wrapped lines. The vast majority of sea turtle entanglement records involve monofilament line (likely from commercial or perhaps recreational fishing), rope, and commercial trawl and gillnet webbing. For at least some records, it is suspected that entangling material was encountered as active fishing gear. Other items known to entangle sea turtles include anchor lines, kite string, cloth strips, burlap, and plastic bags.

Coastal and Marine Birds

Reports of entangling debris were found for 56 species of marine and coastal birds (see Table 8.2 and Appendix 2), including 16% (51 of 312) of the world's seabird species listed by Harrison (1983) and 5 other coastal species. When considered from a taxonomic perspective, seabird entanglements appear to be most common in pelecaniformes (e.g., pelicans and gannets) and a few charadriiformes (e.g., coastal gulls), and less common in procellariiformes (e.g., albatrosses, petrels, and shearwaters), and least common in sphenisciformes (penguins) and podicipediformes (grebes). Although entanglements have been reported for 38% of penguin species (6 of 19 species), the records for each species involve only one or two birds. No entanglement records were found among gaviiformes (loons).

Overall, entanglements have been reported for far fewer species than ingestion (Table 8.3), and the pattern of occurrence among species tends to be opposite that for ingestion records. That is, many investigators note that ingestion of plastics is most common among procellariiformes, less frequent among charadriiformes, and least common among pelecaniformes (Day et al. 1985; Furness 1985a; Ryan 1987b; Ainley et al. 1990a; Sileo et al. 1990b; Moser and Lee 1992). For 57% of the seabird and coastal bird species with entanglement records (32 of 56 species), no reports of marine debris ingestion were found. Moreover, both entanglement and ingestion records were found in only 17% of the seabird species found to interact with marine debris (24 of 138 species). The records suggest that seabirds become entangled accidentally when seeking natural prey items associated with entangling debris, such as pelicans plunging for small fish that may be near floating line. By far, the debris items most frequently reported in seabird entanglement records are monofilament line and fishing net. Other entangling items reported commonly, particularly for some species, are fishing hooks, six-pack yokes, wire, and string.

Marine Mammals

Entanglement records were found for 28% (32 of 115 species) of the world's marine

Table 8.3. Number and percentage of seabird species with entanglement records only, ingestion records only, and both entanglement and ingestion records.

Species group	Number and percent of species with entanglement records only	Number and percent of species with ingestion records only	Number and percent of species with entanglement and ingestion records
Sphenisciformes (penguins)	5 (83%)	0 (0%)	1 (17%)
Podicipediformes (grebes)	2 (100%)	0 (0%)	0 (0%)
Procellariiformes (albatrosses, petrels, and shearwaters)	1 (2%)	53 (84%)	9 (14%)
Pelicaniformes (pelican, cormorants, boobies, gannets, frigatebirds, and tropicbirds)	9 (53%)	6 (35%)	2 (12%)
Charadriiformes (shorebirds, skuas, gulls, terns, and auks)	10 (20%)	28 (56%)	12 (24%)
All seabird species	27 (20%)	87 (63%)	24 (17%)
Other coastal birds	5 (100%)	0 (0%)	0 (0%)

mammal species (Table 8.2 and Appendix 3). For marine mammals, entanglement in marine debris appears to be most common among seals and sea lions (pinnipeds), particularly the eared seals (otarids), less common in baleen whales (mysticetes) and manatees (sirenians), and rare among toothed whales (odontocetes) and otters (mustellids). Using a comprehensive list of the world's marine mammal species compiled by Rice (1977), the entanglement records include 60% of baleen whale species (6 of 10 species), 58% of seal and sea lion species (19 of 33 species), 25% of sirenians (1 of 4 species), 8% of toothed whales and dolphins (5 of 65 species), and the only exclusively marine otter, the sea otter *Enhydra lutrus*. Entanglement records among toothed whales that are clearly not related to bycatch in active fisheries are almost absent. Also, as discussed next, entanglement data for baleen whales (primarily scaring from ropes) may reflect a high interaction rate with active fishing gear rather than marine debris. Among pinnipeds, entanglement reports are more prevalent both in number of species and frequency of occurrence for fur seals and sea lions (Otariidae) than true seals (Phocidae) (Laist 1987; Fowler 1988).

As in seabirds, there is relatively little overlap between marine mammal species known to become entangled and those that ingest marine debris. Of the 49 marine mammal species documented to interact with

marine debris, interactions for 47% (23 species) were limited to entanglement, 35% (17 species) were limited to ingestion, and only 18% (9 species) included reports of both entanglement and ingestion (Table 8.4). Also like seabirds, the taxonomic pattern of marine mammal entanglement records tends to be opposite that for ingestion. That is, pinnipeds, which have the highest incidence of entanglement records, have the fewest ingestion records. Conversely, the species group with the lowest percentage of species with entanglement records (toothed whales) is the group with the highest percentage of species ingestion records. The West Indian manatee is the only species of marine mammal for which there is evidence that both entanglement and ingestion occur regularly (Beck and Barros 1991).

Available data indicate that entanglement is a far greater threat to marine mammals than ingestion in the number of species affected, the frequency of occurrence, and the effect on individuals. While there are a few records of mortality from ingestion, almost all species for which entanglement records exist include reports of dead or seriously injured animals or both. The debris items most often identified in marine mammal entanglement records are trawl net and gillnet fragments and monofilament line. Other entangling debris commonly identified includes rope and line of unspecified origin and strapping or packing bands.

TABLE 8.4. Number and percentage of marine mammal species with entanglement records only, ingestion records only, and both entanglement and ingestion records.

Species group	Number and percent of species with entanglement records only	Number and percent of species with ingestion records only	Number and percent of species with entanglement and ingestion records
Mysticete (baleen whales)	4 (66%)	0 (0%)	2 (33%)
Odontocete (toothed whales)	1 (5%)	17 (77%)	4 (18%)
Phocidae (earless or true seals)	7 (88%)	0 (0%)	1 (12%)
Otariidae (sea lions and fur seals)	10 (91%)	0 (0%)	1 (9%)
Sirenia (manatees and dugongs)	0 (0%)	0 (0%)	1 (100%)
Mustellidae (sea otters)	1 (100%)	0 (0%)	0 (0%)
All marine mammal species	23 (47%)	17 (35%)	9 (18%)

Fish, Crabs, and Squid

Entanglement records also were found for 34 species of fish and 8 species of crabs, but no species of squid (Appendix 4). Virtually all entanglement records involve dead animals, and most of these were caught in derelict fishing gear, principally set- and drift-gillnets and crab traps. This list seems particularly incomplete because entanglement in lost or discarded fishing gear seem possible for virtually all species caught in active commercial fishing gear.

Factors Influencing Entanglement Rates

As indicated, evidence of entanglement in marine debris has been reported for a significant number of species. The magnitude of impact for each species, however, is different and depends on the frequency with which individuals interact with and become entangled in marine debris. This, in turn, depends primarily on factors affecting the amount and density of entangling debris likely to be encountered and on biological factors that predispose some species to entanglement.

Amount and Density

The amount and density of entangling debris is a function of disposal or loss patterns and

the physical processes by which it is moved, reconfigured, and eventually deposited. Many investigators have collected information on the types, amounts, and distribution of entangling marine debris; see, for example, Shomura and Yoshida (1985), Alverson and June (1988), Shomura and Godfrey (1990), and Chapter 1, this volume. In general, the amount and distribution of debris have been related to probable sources (e.g., urban centers, commercial fishing areas, and shipping corridors) and to surface currents and wind patterns.

Hazardous debris often enters the marine environment and is most concentrated in fishing grounds, coastal waters, and beach areas that are particularly important habitats for various marine species (Laist 1987). For example, Brothers (1989) reported high concentrations of lost gillnets on fishing grounds off Newfoundland; Henderson (1984, 1988, 1990) reported accumulations of entangling debris on remote atolls in the northwest Hawaiian Islands used by Hawaiian monk seals; A. Carr (1987) noted massive concentrations of hazardous floating debris along drift lines and current margins used by sea turtles and seabirds for feeding; and the National Marine Fisheries Service (1993) suggested that the subarctic boundary in the North Pacific is an area where both northern fur seals and oceanic debris tend to concentrate.

Animal Behavior

While patterns of debris disposal and loss and physical forces concentrate marine debris in certain important marine habitats, an array of animal behaviors, such as those related to feeding, play, and nest building, bring individuals of certain species into direct physical contact with entangling debris. In effect, these behavior patterns predispose certain species to entanglement.

Sea turtles, for example, often forage along drift lines where prey as well as floating entangling debris accumulates (A. Carr 1987). Data on ingestion of plastic by sea turtles also indicate that sea turtles are unable to distinguish synthetic materials from natural prey items (Fritts 1982; A. Carr 1987). While plastic bags, sheeting, and fragments are the predominant debris items found in turtle stomachs, derelict rope and monofilament line also are ingested (Balazs 1985). This suggests that turtles approach virtually any floating debris, including entangling items, as potential prey. Studies of ingestion by captive sea turtles also illustrate this behavioral tendency (Lutz 1990). Once attracted to entangling material, it is not difficult to imagine webbing, rope, or line enwrapping flippers and snagging shells.

Foraging strategies and feeding behavior also may be related to entanglement rates among seabirds. Several gulls, such as the herring gull, the black-backed gull, and the black-headed gull, often feed in garbage dumps (Ryan 1990a; Onions and Rees 1992), near fishing vessels (Ryan 1991), or in the wake of ships where concentrations of entangling debris are likely to be greatest. Like sea turtles, some seabirds known to become entangled, such as gannets, also feed along current margins (Ashmole 1971) where debris concentrates. In general, seabirds that feed by scavenging (e.g., herring and black-headed gulls) and plunging (e.g., pelicans and gannets) are among the species with the highest entanglement rates. This might be expected given that scavenging birds often pick through garbage to obtain food and that prey of plunging birds may associate with floating debris for cover. Entanglement records are usually least common among species that feed by pursuit diving, surface seizing, and dipping.

The collection of debris items for nest building is another behavior that increases entanglement risks for adults and chicks of certain birds. Two species for which this behavior is common are northern gannets and double-crested cormorants. Bourne (1976) first reported the use of plastic debris in nest construction by gannets, and more recently Montevecchi (1991) reported that virtually all gannet nests sampled at colonies in eastern Newfoundland (97%, 722 of 741 nests) have plastics incorporated into them. Gannets collect nesting material almost exclusively from offshore areas, and the most common debris reported from their nests were scraps of fishing net, rope, and line. For double-crested cormorants, Podolsky and Kress (1989) reported 37% of examined nests (188 of 497) in the Gulf of Maine contained plastic materials, principally plastic bags, lobster pot lines, and fishing net fragments.

Among marine mammals, the filter-feeding strategies of baleen whales may cause some entanglements. Right whales (Kraus 1990), humpback whales (National Marine Fisheries Service 1991; Wiley et al. 1995), bowhead whales (Philo et al. 1992), and gray whales (Heyning and Lewis 1990) have been observed with rope and line trailing from their mouths or entangled around flippers and flukes. Numerous reports of baleen whales swimming off with float lines to crab pots and gillnets (Heyning and Lewis 1990) suggest that many, if not most, baleen whales that become entangled or scarred by fishing gear encounter this material as active gear. At least some animals, however, probably become entangled in derelict rope and netting. Occasional interactions would seem likely during skim feeding, a strategy sometimes used by humpback whales, right whales, and bowhead whales to collect prey from the uppermost layer of water where floating rope or line also could occur.

For seals and sea lions, curiosity and play appear to be important factors causing animals to seek out and interact with entangling debris (Laist 1987). Most interactions involve pups and juveniles. Studies of young and adult captive northern fur seals in pools containing net fragments and strapping bands demonstrate that young animals are more likely to become entangled than adults (Yoshida and Baba 1985; Yoshida et al. 1985) and that young seals repeatedly approach, interact with, and become entangled in such material (Bengtson et al. 1988; Feldcamp et al. 1988). Some pups become entangled in beach-cast debris before ever entering the water. These findings fit well with field observations of entangled fur seals on the Pribilof Islands that report the highest rates of entanglement in juvenile age classes (Scordino et al. 1988). Higher entanglement rates among younger age classes also are reported for Hawaiian monk seals (Henderson 1984), Antarctic fur seals (Croxall et al. 1990), Australian fur seals (Pemberton et al. 1992), and California sea lions (Stewart and Yochem 1987, 1990). Declining rates of entanglement in older age classes may reflect decreased interest and curiosity in floating debris based on accumulated experience with such items over time (Feldcamp et al. 1988). It also is possible that older animals become too large to be caught in small mesh sizes, thereby decreasing the proportion of debris hazardous to them.

Most fish and crustacean entanglements occur in lost or discarded fishing gear specifically designed to exploit the normal behavior patterns of such species. For example, gillnets are designed to exploit fish swimming patterns and traps are designed to exploit fish and crab food preferences and feeding behavior. While the catch efficiency of lost gear declines as nets collapse, and traps may incorporate corrodible time-release panels to minimize ghost-fishing, it seems likely that virtually all target and nontarget species taken in commercial fisheries are also killed in lost or discarded gear. As animals become trapped in lost gear, they can lure other animals that in turn become trapped in a self-perpetuating cycle (Kruse and Kimker 1993).

In considering factors that influence the magnitude of the effect of marine debris on marine life, it is interesting to compare marine debris with other pollutants. For example, although the effects of marine debris and contaminants such as pesticides, heavy metals, and hydrocarbons are influenced by their concentration in the marine environment, chemical pollutants, even those magnified through the food chain, may be diluted to a point at which they pose no threat to individuals. Marine debris items, however, retain a potential to injure or kill individual animals independent of their concentration in the environment. That is, the ability of a strapping band to injure or kill a seal is not diminished by decreasing numbers of strapping bands in the ocean. Considering the behavioral attraction of some endangered species to entangling debris, this is an important distinction from other forms of marine pollution. It also is important to recognize that behavioral patterns that draw individual animals to entangling debris amplify biological effects. This phenomenon is functionally analogous to the biomagnification of other contaminants through the food chain, in that both mechanisms effectively amplify the impact of dispersed contaminants on marine life.

The Effect of Entanglement on Affected Individuals

While many different types of marine life are subject to entanglement, the effect of entangling debris is essentially the same for all species and is primarily mechanical (Laist 1987). Animals that become entangled may exhaust themselves and drown, have their mobility impaired to a point where they can no longer catch food or avoid predators, become hung up on rocks or other fixed objects by trailing rope or line, or incur wounds and infections from the abrasion or constriction of attached debris.

In many cases, animals that interact with entangling debris do not become entangled or become entangled only briefly with little or no apparent harm. During studies of captive northern fur seals, animals frequently freed themselves after only brief periods of entanglement in netting and strapping bands introduced into their pools (Yoshida et al. 1985). Also, entanglement data on northern right whales (Kraus 1990) suggest that baleen whales are able to free themselves of entangling debris with some degree of success. There are only a dozen records of right whales seen entangled in fishing gear between 1976 and 1989, but 57% of appropriately photographed northern right whales in the right whale photoidentification catalog (68 of 118 animals) bear scars on their peduncles and 17% (28 of 168 animals) have scars along their mouths indicative of chaffing by ropes and lines.

For animals unable to free themselves quickly, survival prospects are poor. Increased drag from attached debris imposes high energy requirements and restricts movements. Studies on a captive California sea lion (Feldcamp 1985) showed that entangling debris can cause a severalfold increase in both drag and power required for swimming. Feldcamp concluded that small sea lions experience relatively higher drag and greater power requirements for a given size of debris than larger adults and that entanglement of younger animals could cause proportionally higher mortality. Similar findings were reported by Feldcamp et al. (1988) from energetic studies on captive juvenile northern fur seals.

Attached debris also causes changes in behavior. Balazs (1985) suggested that entangled sea turtles are unable to function normally in feeding, diving, surfacing to breathe, or other essential behaviors. Quantitative studies of behavioral changes induced by entanglement have been done on free-ranging and captive northern fur seals. Bengtson et al. (1989) compared at-sea foraging behavior of free-ranging entangled and unentangled juvenile fur seals on the Pribilof Islands using radio tags equipped with depth-of-dive recorders.They showed that animals entangled in small debris items spend twice as much time foraging at sea as do unentangled animals and that their average diving depth is much shallower. Other studies have shown that the pups of entangled female fur seals have higher mortality rates and lower body weight than pups of unentangled females (DeLong et al. 1988) and that, after a year, entangled fur seals are resighted, at half the rate of unentangled fur seals (Stewart et al. 1989). Feldcamp et al. (1988) reported a marked de-

crease in the time captive fur seals spend swimming when the weight of entangling debris increased from 200 g to 250 g. Together, the studies suggest a vicious cycle in which entangling debris imposes added energy requirements while at the same time impairing foraging efficiency, leading to eventual starvation and death.

While gradual starvation may be the fate of animals entangled in relatively small pieces of debris, those that become entangled in larger items probably die quickly. Fowler (1987) listed northern fur seal entanglement records for areas other than rookeries and haul-outs from 1967 to 1986. The list includes 13 animals found dead; of those, 9 were entangled in debris weighing more than 500 g, 1 was entangled in debris weighing less than 500 g, and for 4 there was no weight estimate for the debris. In contrast, nearly 75% of the debris removed from entangled fur seals found on the Pribilof Islands between 1983 and 1992 was less than 150 g and more than 90% was less than 500 g (Fowler et al. 1993). The data suggest that northern fur seals entangled in debris weighing more than 500 g are not likely to survive long enough to make it back to rookeries or haul-outs unless they become entangled close to shore or on shore.

Death of entangled animals also may occur quickly from injury or predation. Feldcamp (1985) reported that when a net fragment was placed in a pool with a captive California sea lion, the animal inserted his head into a mesh opening and reacted immediately with a violent twisting action that further entangled the animal and tightened the netting around the neck. He concluded that if such a reaction is common among seals, drowning or ensuing wounds may be more immediate causes of death than starvation. Potentially lethal injuries such as necrosis and loss of appendages caused by the constriction of entangling rope and line, have been reported for West Indian manatees (Beck and Barros 1991), sea turtles (Balazs 1985), baleen whales (Heyning and Lewis 1990), and seabirds (J. Miller, National Park Service, Corpus Christi, Texas, personal communication).

Death also may come quickly for entangled birds. Trailing line or other debris on entangled birds returning to roosts can snag on branches or other fixed objects. In a matter of minutes, thrashing pelicans struggling to take flight after becoming hung up in trees can sustain massive injuries and die (R. Heath, personal communication). Fish and crustaceans caught in lost traps or immobilized in lost gillnets may die from cannibalism, predation, starvation, and suffocation as gear is buried by sand (Muir et al. 1984; Kruse and Kimker 1993; Paul et al. 1993).

Each of these observations has important implications for interpreting entanglement records. They suggest that animals entangled in debris above a certain size threshold may die at sea quickly and, unless entangled close to shore, rarely wash up on beaches. Also, while animals entangled in relatively small debris may survive somewhat longer and make it back to shore, the amount of time spent ashore tends to be less than for unentangled animals because of their greater food requirements and shorter life spans. In both cases, entangled animals are less likely to be seen by researchers on land. Also, entangled animals that die for reasons other than predation may be detectable for only short periods of time because of scavenging and decomposition or be concealed in ways that frustrate detection (e.g., resting on the sea floor or floating at sea just below the sea surface).

Impacts on Species and Populations

At least four approaches have been used to assess the rates and effects of entanglement in marine debris for particular populations of marine life: (1) seasonal field surveys to count and compare entangled and unentangled animals in a given population; (2) long-term necropsy and photoidentification studies of entanglements and entanglement wounds and scars; (3) population modeling; and (4) compilations of independent anecdotal reports. The first three approaches offer quantitative analyses and have been done in only a few cases, almost all of which involve marine mammals (Table 8.5). In most all cases, these

quantitative estimates of entanglement rates are put forward as minimum estimates or indexes. In general, definitive proof of population-level effects is lacking even for those species thought to be most affected by entanglement. However, because of the aforementioned sampling constraints, conclusions that suggest population impacts are low or insignificant should be treated with caution. Indeed, indirect analyses for some species offer convincing evidence that effects of entanglement are great enough to limit population growth or accelerate population declines.

Northern Fur Seals

The first systematic analyses of marine debris entanglement were done on the population of northern fur seals (*Callorhinus ursinus*) that breed and haul out during summer on the Pribilof Islands in the southeast Bering Sea. These entanglement studies, which are also the most extensive and in-depth work done on the subject to date, have played a major role in bringing the problem of marine debris to light. In many respects, this entanglement work has been made possible by the extensive amount of research done on the species. Indeed, because of a cooperative international management regime established early in the 1900s, this population is one of the most thoroughly studied marine mammal populations in the world. The purpose of the cooperative management program was to oversee a commercial harvest of juvenile male fur seals for their pelts. Under the program, harvesting was limited to a portion of the population's 2- to 3-year-old juvenile males that haul out in segregated groups on the breeding islands each summer.

Early in the 1950s, the fur seal population on the Pribilof Islands numbered an estimated 2.1 million animals. Its size at that time is believed to have been roughly equal to the preexploitation population size before episodes of intense harvesting in the late 1700s and 1800s (National Marine Fisheries Service 1993). From the 1960s to about 1970, the population size declined to about 1.2 million for reasons apparently related to a change in

TABLE 8.5. Observed entanglement rates for selected species.

Species	Observed percentage entangled	Highest percentage observed per colony	Location(s)	Study period	Source
Northern fur seal	0.15–0.71[a]	—	Pribilof Islands	1967–1992	Fowler et al. 1993
Cape fur seal	0.11–0.12[a]	0.66	South Africa and Namibia	1977–1979	Shaughnessy 1980
Antarctic fur seal	0.11	0.39	Bird Island, South Georgia	1988–1989	Croxall et al. 1990
Australian fur seal	1.9 ± 0.7	—	Tasmania	1989–1990	Pemberton et al. 1992
Hawaiian monk seal	0.18–0.85	7.5	Northwest Hawaiian Islands	1985–1988	Henderson 1990; T. Gerrodette, U.S. National Marine Fisheries Service, 1994, personal communication
California sea lion	0.08–0.16	—	Southern California Bight	1983–1988	Stewart and Yochem 1990
Northern elephant seal	0.10–0.16	—	Southern California Bight	1983–1988	Stewart and Yochem 1990
West Indian manatee	1.7 and 3.6[b]	—	Florida	1974–1985	Beck and Barros 1991
Northern right whale	12 and 57[b]	—	Western North Atlantic	1976–1989	Kraus 1990
Northern gannet	2.6	—	Helgoland, German Bight	1984–1985	Schrey and Vauk 1987
Loggerhead sea turtle	6	—	Western North Atlantic	1990–1992	Bjorndal et al. 1994

[a]Percentage is for juvenile male segment of the population only.
[b]The first estimate is based on carcass salvage data and the second is based on scars and entanglements from photoidentification catalog data.

harvesting practices between 1956 and 1968 when female, as well as juvenile male, seals were taken. After the end of the female harvest, the population began increasing in the early 1970s. However, the period of the increase was brief, and in 1974 fur seal numbers again began to decline at an average rate of 4%–8% per year. By 1983, the population was estimated to number 877,000 animals. Since then, the population size has remained relatively stable at about 900,000–1,000,000 (National Marine Fisheries Service 1993).

The effect of the female harvest does not appear adequate to explain either the magnitude or the persistence of the observed population decline after 1974. And while the cause of the post-1974 decline is still debated, there is good reason to believe that entanglement was and remains a significant factor in shaping the population trend. Even for this extensively studied population, however, the evidence remains largely circumstantial, illustrating the difficulty in documenting and measuring the effects of marine debris entanglement.

As with other pinnipeds, virtually all records of entangled fur seals are from periods when they haul out on land. On the Pribilof Islands, sightings of entangled fur seals date back to at least the 1930s (Fowler 1987). Because of increasing observations of entanglement, counts were begun in 1967 to record the number of entangled animals found in the annual harvests of juvenile males returning to the islands for the first time after spending their first 2–3 years at sea. When commercial harvests ended in 1984, comparable counts were continued by the National Marine Fisheries Service as part of its northern fur seal research program. The percentage of entangled juvenile males seen in the commercial harvests and research drives from 1967 through 1992 has ranged from 0.15% to 0.71% (Fig. 8.1). As noted, some 90% of the animals were entangled in debris weighing less than 500 g.

While observed entanglement rates seen on land are far too low to account for the unexplained portion of the fur seal population decline, the unrecorded number of animals

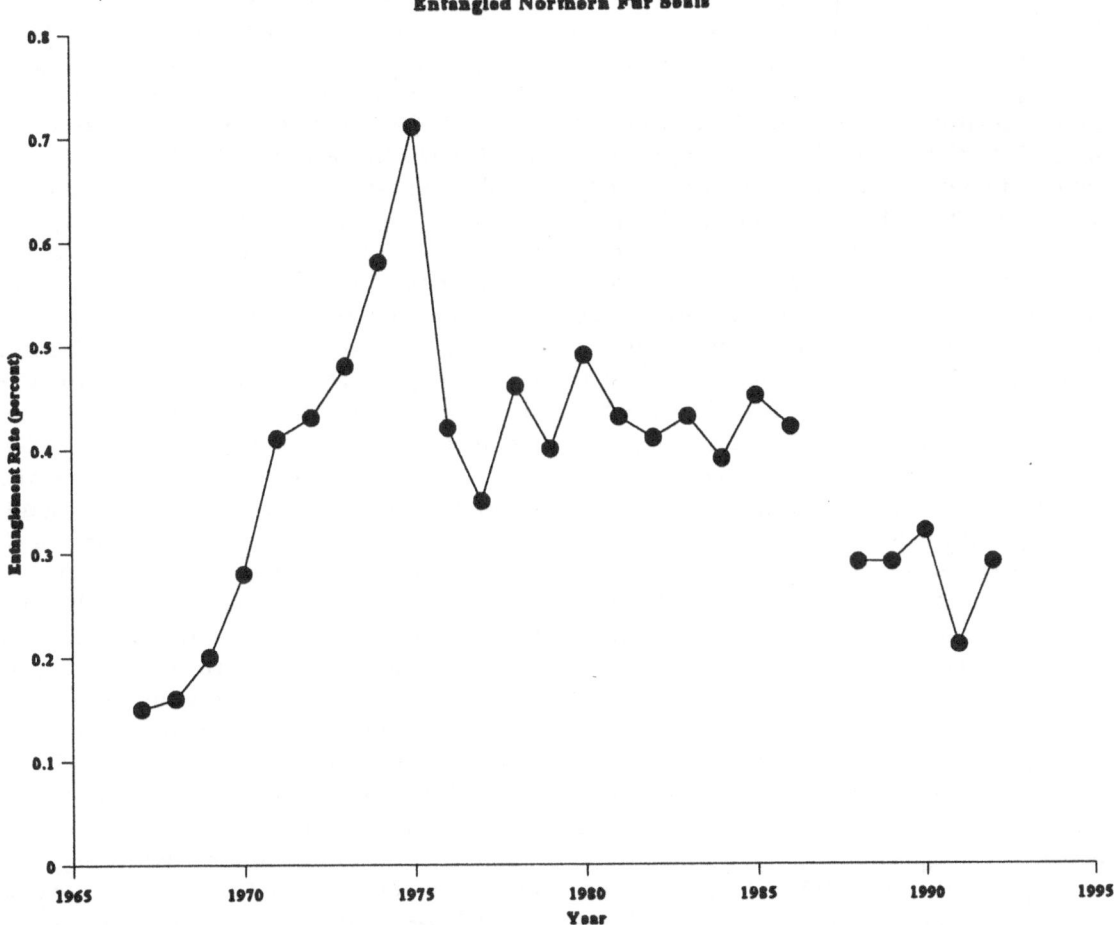

FIGURE 8.1. The percentage of juvenile male northern fur seals found entangled in the commercial harvest from 1967 to 1984 and in research roundups from 1985 to 1992 on St. Paul Island, Alaska (Fowler et al. 1993).

entangled and killed at sea in large debris was recognized as a potentially significant unknown factor. Therefore, to assess whether undetected entanglement at sea might be a cause of the decline, investigators turned to population modeling. The results show a statistically significant relationship between fluctuations in entanglement rates on land and changes in the numbers of pups born and an unexplained high mortality among juvenile animals before their initial return to the Pribilof Islands (Fowler 1982, 1985, 1987, 1988; Fowler et al. 1990; French and Reed 1990; Swartzman et al. 1990).

When combined with information on the amount of large net debris found on Alaska beaches (more than 500 g) that could entangle and prevent fur seals from returning to

land, the propensity of juveniles to become entangled, and the effect of debris on entangled individuals, the correlative analyses offer compelling indirect evidence that entanglement was a major factor in the population decline during the 1970s. Based on the size of the Pribilof Islands fur seal herd in the mid- to late-1970s, it was suggested that 50,000 Pribilof Island fur seals were killed annually at that time by entanglement in debris (Fowler 1982).

When indirect evidence offers a compelling explanation for the decline, direct supporting evidence has proven elusive despite the large number of entanglement-related deaths implied. In part, this is because it is still unclear where weaned pups go during their first years at sea. Moreover, detecting entangled animals at sea is exceedingly difficult. Also unclear is the extent to which entanglement is a function of inquisitiveness in young versus older animals and the size of animals relative to the size of openings in entangling debris. If the former factor is dominant, entanglement might be limited largely to pups and juveniles; however, if attraction to debris does not differ with age, a larger proportion of female fur seals may become entangled because their growth rate is less and their adult size is smaller than males.

Direct evidence of entanglement at sea is limited to a small number of records. Fowler's (1987) compilation of northern fur seal entanglement records from locations other than rookeries and haul-outs includes only 28 sightings between 1967 and 1987. These sightings confirm that both male and female animals become entangled and die at sea in relatively large pieces of debris (more than 500 g), but the sample size is far too small to verify or refute asserted mortality levels suggested by indirect analyses.

As shown in Fig. 8.1, counts since the late 1980s show a decline in juvenile seal entanglement rates observed on the Pribilof Islands through most of the 1970s and 1980s. In the late-1980s and early-1990s, observed entanglement on land declined to between 0.2% and 0.3%. This decline has resulted largely from a decrease in the number of seals entangled in trawl webbing, which may reflect a decrease in the discard of trawl net fragments by fishermen; there is no apparent decline in entanglement in strapping bands or other debris items (Fowler et al. 1993). Another possibility is that the decline reflects a change in fishing patterns and associated debris dispersal relative to the distribution of juvenile fur seals at sea. Given the modest signs of recovery of the fur seal population in recent years, it is possible that continuing entanglement is among the density-dependent factors impeding population recovery.

Cape Fur Seals

Like northern fur seals, young Cape fur seals along the Atlantic coast of South Africa and Namibia have been harvested for their pelts. Also like northern fur seals, efforts have been made to record the incidence of entangled animals seen in harvests at various rookeries. These studies, however, were limited to the 1977–1979 harvest seasons. The overall entanglement rates reported for surveyed Cape fur seal colonies were 0.12%, 0.11%, and 0.12% in 1977, 1978, and 1979, respectively (Shaughnessy 1980). For each year, about two-thirds of the observed entanglements were at the Cape Cross colony in Namibia even though that colony accounted for less than one-fifth of total annual harvests. Entanglement rates at Cape Cross ranged from 0.56% in 1977 to 0.66% in 1979. Cape Cross is close to a major purse seine fishing ground, and monofilament line was the most common entangling item at that location.

Shaughnessy (1980) cautioned against direct comparisons of entanglement rates for Cape fur seals and northern fur seal, in part because most harvested Cape fur seals were less than 1 year old while most harvested northern fur seals are 2–3 years old. However, given reported entanglement rates and the shorter amount of time Cape fur seals spent at sea exposed to debris, he suggests entanglement rates for Cape fur seals may be as high or higher than northern fur seals. Given more recent analyses of the effect of entanglement on northern fur seals, debris may be a significant source of mortality for at least the Cape Cross colony of Cape fur seals. Unfortunately, more recent entanglement surveys of Cape fur seal colonies have not been done.

Australian Fur Seals

Entanglement surveys of Australian fur seals were undertaken recently in Tasmania shortly after regional trawl fisheries began expanding. During observations of haul-out beaches in the 1989–1990 and 1990–1991 breeding seasons, 75 entangled animals were counted, 96% of which showed obvious signs of physical injury from attached debris (Pemberton et al. 1992). Juveniles and subadults comprised 66% of the entangled animals, and entangling material included trawl net (40%), plastic straps (30%), monofilament nets (15%), and nylon rope (15%).

Using counts of entangled and unentangled seals at a subset of survey sites in southern Tasmania, Pemberton et al. (1992) calculated a mean entanglement rate of 1.9% ± 0.7%. For reasons similar to those of Croxall et al. (1990), they put forward their calculated entanglement rate for Australian fur seals as a minimum estimate and added that conclusions regarding any influence entanglement might have on population trends at Tasmania's fur seal colonies will not be apparent for several years.

Antarctic Fur Seals

Studies to quantify Antarctic fur seal entanglement were done on Bird Island, South Georgia, during the 1988–1989 pup-rearing season by Croxall et al. (1990). Entangling debris was seen on 208 animals, representing at least 0.11% of the island's total estimated fur seal population. In only 19% of the entangled seals was debris considered loose enough to have any chance of coming off naturally. With regard to age and sex, 71% of the entangled animals were males, 88% of which were 4 years of age or less. Most entangled females (64%) were more than 4 years old. Unlike other seal entanglement studies in which fishing gear is the item most commonly seen, 75% of the Antarctic fur seal entanglements at Bird Island involve plastic straps and string and only 13% involve trawl web. Researchers were able to remove debris from 170 of the 208 entangled animals.

Croxall et al. (1990) consider the 0.11% entanglement rate a minimum rate for several reasons. They noted that a portion of the population (i.e., many juveniles) does not return to shore annually and thus are not available for detection of debris. They also noted that the number of animals entangled at sea and unable to return to Bird Island is unknown. In this regard, the high incidence of strapping bands and string seem noteworthy. If these items are representative of entangling debris in the species oceanic range

and heavy trawl net webbing is rare, more entangled Antarctic fur seals may be able to return to land than other species, such as northern fur seals, because entanglement rates in heavy items is comparatively low.

The calculated entanglement rate also is assumed to be low because some beaches were surveyed only occasionally, increasing the possibility that entangled animals may have been present but not counted. At beaches where there was confidence all entangled animals were recorded (i.e., those visited daily), entanglement rates were significantly higher, just under 0.40%. Considering such factors, Croxall et al. (1990) suggested a likely maximum entanglement rate might approach 1%. Extrapolated to the entire South Georgia fur seal population of 1.2 million animals, they estimated 5,000–10,000 animals may have been entangled in 1988–1989. Noting that Antarctic fur seals have been increasing at about 4%–15% annually, an order of magnitude greater than the highest observed entanglement rate, they concluded that entanglement is not likely to be significant for South Georgia's Antarctic fur seal population at present. They caution, however, that persistent high entanglement levels could cause a decline in the future as the population reaches equilibrium in response to density-dependent factors (Croxall et al. 1990).

Hawaiian Monk Seals

Entanglement records for endangered Hawaiian monk seals have been compiled by Henderson (1984, 1985, 1988, 1990). Between 1974 and 1984 he lists 27 entanglements and 8 additional reports of seals with scars indicating prior entanglement (Henderson 1985, 1990). From 1985 to 1988, Henderson lists 34 entanglements, including 3, 4, 12, and 15 in 1985, 1986, 1987, and 1988, respectively (Henderson 1990). The 1985–1988 entanglements include 4 animals found dead. Debris on most other entangled seals was described as being loose and was removed whenever possible. Most accounts involve pups. Fishing net and monofilament line constitute the most commonly reported debris.

Researchers visit the five major monk seal breeding islands for only a few days to a few months each year. Therefore, to examine trends between years and locations with different amounts of sighting effort, Henderson calculated observed entanglement incidents in units of 100 field camp-days per 100 seals. Between 1985 and 1988, entanglements increased eightfold from 0.06 to 0.48 incidents per 100 camp-days per 100 seals. Concurrent with this increase was an increase in the amounts of debris found and destroyed on haul-out beaches (Henderson 1990). Considering pups alone, variation in annual entanglement rates among the major breeding islands between 1982 and 1988 ranged from 0.37 to 4.4 incidents per 100 camp-days per 100 pups at French Frigate Shoals and Lisianski Island, respectively (Henderson 1990).

Based on the total number of Hawaiian monk seal entanglements from 1985 through 1988 (Henderson 1990) and estimates of total population size (including pups) for those years (T. Gerrodette, U.S. National Marine Fisheries Service, Honolulu, HI, 1994, personal communication), observed entanglement rates are 0.18%, 0.20%, 0.62%, and 0.85%, respectively. For individual colonies, entanglement rates in some years are much higher. Based on observed entanglement numbers and the total estimated population sizes for colonies at Kure Atoll and Laysan Island in 1988, observed entanglement rates are 7.5% (5 of 67 animals) and 2.1% (7 of 331 animals), respectively. These rates do not include correction factors to account for limited survey time or seals entangled and killed at sea and thus are undoubtedly low. Potential effects are mitigated by efforts to remove entangling debris not only from seals but also from haul-out beaches.

Pinnipeds in California

Stewart and Yochem (1987, 1990) reported on entanglement surveys of California sea

lions, northern elephant seals, harbor seals, and northern fur seals at haul-out sites on San Nicolas Island and San Miguel Island off the coast of California. Based on observations from 1983 through 1988, they concluded that entanglement contributes to the mortality of some animals, but that it is not a significant factor in population trends for any of the pinniped populations in the Southern California Bight (Stewart and Yochem 1990).

For California sea lions, reported entanglement rates combined for both islands range from 0.08% to 0.16% between 1983 and 1988 (Stewart and Yochem 1990). Monofilament gillnet was the most common entangling material. For northern elephant seals, reported entanglement rates over that period range from 0.10% to 0.15% with strapping bands the most commonly seen entangling debris. Entangling debris on harbor seals and northern fur seals was seen on only a very few occasions during the study.

West Indian Manatees

Beck and Barros (1991) reviewed entanglement records for endangered West Indian manatees in Florida. Their analysis suggests entanglements occur regularly and involve a small but increasing percentage of the population. From 1974 to 1985, 1.7% of the manatees salvaged in the Southern United States (16 of 940 carcasses) had one or both flippers scarred, missing, or entangled in monofilament line, rope, or crab trap lines. In 11 cases (1.2%), entanglement in line or netting was the identified cause of death. Also, 3.6% of free-ranging manatees in the manatee photoidentification catalog (26 of 729 individuals) bear scars or missing flippers from previous entanglements (Beck and Barros 1991). In recent years, sightings and rescues of entangled manatees have become more frequent. Because an unknown portion of these entanglement records result from interactions in active commercial and recreational fishing gear, mortality and injury rates caused by debris may be lower than those reported here.

Right Whales

Scars indicative of chafing and abrasion by rope and line have been identified on a much larger percentage of northern right whales in the western North Atlantic Ocean. Analyses of photographs in the right whale photoidentification catalog indicate 57% of appropriately photographed animals (67 of 118 whales) have rope scars on their peduncles and 17% (28 of 161 whales) have such scars around the mouth (Kraus 1990). Kraus (1990) attributed these scars to entanglement in fishing gear. Given a population estimate of 350 animals for the North Atlantic right whale population (Kraus 1990), at least 20% of the population (67 of 350) appear to have entanglement-related scars on peduncles alone. Also, 12% (3 of 25) of all documented right whale deaths recorded from 1970 to 1989 were attributed to entanglement in fishing gear. Given overlap between fishing grounds and the distribution of right whales, many if not most scars and deaths may be caused by interactions with active rather than derelict gear.

Sea Turtles

Until recently, information on sea turtle entanglement was limited almost exclusively to compilations of anecdotal records. The most thorough evaluation of such records is that by Balazs (1985), who compiled 60 entanglement records worldwide, 38% of which involve dead animals. The records cover five species and almost all are of single animals. Balazs concluded that sea turtles are affected to an unknown but potentially significant degree by entanglement in debris. As noted, sea turtles approach and attempt to eat all types of debris, including entangling material. Given information since Balazs' review on the use of drift lines where debris tends to accumulate (A. Carr 1986, 1987), it seems reasonable to assume that significant numbers of turtles may become entangled.

More recent quantitative assessments of sea turtle entanglement also suggest entanglement rates may be high. Based on sea turtle stranding records collected from 1980 to 1992 along the U.S. Atlantic and Gulf of Mexico coasts, entangling debris was found on 0.8% (142 of 16,327) loggerhead turtles, 6.6% (123 of 1,874) green turtles, 6.8% (66 of 970) leatherback turtles, 14% (36 of 258) hawksbill turtles, and 0.8% (18 of 2,140) Kemp's ridley turtles (W. Teas, personal communication). Data on entanglement rates among free-swimming sea turtles seen at sea also have been collected. Bjorndal and Bolton (1994) reported entangling debris was found on 6% of more than 800 loggerhead sea turtles caught in waters around the Azores from 1990 to 1993 and 5% of more than 1500 sea turtles of various species observed at sea worldwide. Assuming a high mortality rate among entangled sea turtles and the potential for a high percentage of female turtles to become entangled before reaching sexual maturity, the data suggest that entanglement mortality and reduced recruitment potential for at least some sea turtle species could cause population declines.

Northern Gannets

The only study found in this review presenting quantitative data on entanglement among seabirds is a report by Schrey and Vauk (1987) on northern gannets along part of their migratory route in the German Bight. They reported that 2.6% (8 of 313 sightings) of the gannets sighted at Helgoland in 1984 and 1985 were entangled in derelict fishing gear. Assuming that some of the 313 sightings were repeat sightings of individuals, they concluded entanglement rates in the area were definitely higher. They also reported finding 3 dead entangled birds (13% of all dead gannets found on Helgoland) and freeing 5 other entangled birds between 1976 and 1985.

Because most gannet colonies in the northeast Atlantic are in the British Isles, western France, Iceland, and Norway, where no entanglement surveys have been done, the significance of entanglements reported from the German Bight is unclear but may indicate problems elsewhere. As noted, almost all nests in some gannet rookeries contain plastic debris, indicating that gannets routinely collect such material for nest building. In 8 days of work at one rookery, Montevecchi (1991) reported finding 3 adults and 15 chicks entangled and concluded that a small percentage of adult and nesting gannets die at rookeries from entanglement in debris woven into nests. Despite finding a high percentage of double-crested cormorant nests with plastic debris in the Gulf of Maine, Podolsky and Kress (1989) reported finding no entangled cormorants at the rookeries they studied.

Pelicans

Although no statistical data on entanglement of pelicans are available, Ralph Heath (personal communication), founder of a seabird rescue and rehabilitation facility in southwest Florida, has found hundreds of pelicans that died in the past decade in south Florida from entanglement in line, string, and similar material. In addition, staff at the facility have removed entangling line from thousands of other pelicans. Because line removed from live birds is often unsnarled, interaction with active recreational fishing gear is suspected as a major source of the entangling material. In at least some cases, however, entanglement in lost or discarded lines seems probable. The tendency of these species to occur around docks and other areas where debris is likely to enter coastal waters and its plunging feeding behavior would suggest a high potential for becoming entangled in lost or discarded line.

Fish and Crabs

The death of fish and crabs in lost fishing gear (ie., ghost-fishing) is recognized as a serious concern for several commercial fisheries (Breen 1990; Kruse and Kimker 1993). Using interviews with fishermen to derive an 11% annual trap loss rate for the Dungeness crab fishery in the Fraser River Estuary of British Columbia, Breen (1987) estimated ghost fishing losses for that fishery in 1984 at 7% of the landed catch or about 21,100 kg. Based on other reported trap loss rates, Breen (1990) estimated an annual loss of about 10%–20% of the traps used in the coastal Dungeness crab and American lobster fisheries and concluded that associated ghost-fishing is a cause for concern.

Other estimates cited by Breen (1990) for annual ghost-fishing losses in lost pots and traps are 670 metric tons of American lobster in 1978 off the Northeast United States (Smolowitz 1978) and 300 metric tons of sablefish (7.5%–30% of landings) in the sablefish trap fishery off British Columbia from 1977 to 1983 (Scarsbrooke et al. 1988). Kruse and Kimker (1993) estimated 31,600 pots were lost in the Bristol Bay king crab fishery in 1990 and 1991. If each pot killed just one legal-sized crab per year, annual losses would be 205,400 pounds. Regulations for many crab and pot fisheries now include provisions for marking traps and using time-release devices on panels to minimize ghost-fishing (Breen 1990).

Lost coastal gillnets also may catch significant quantities of commercially important finfish and shellfish. Brothers (1989, 1992) reviewed a series of lost gillnet retrieval projects off Newfoundland, Canada, and reports recoveries of 148 gillnets in 20 days in 1975, 176 gillnets in 24 days in 1976, 16.5 gillnets in 15 days in February 1984, and no gillnets in 5 days in November 1984. Catch in the 148 lost nets retrieved in 1975 included 3,047 kg of groundfish and 1460 kg of crab. Total catch in the 176 nets recovered in 1976 included 4,813 kg of groundfish and 2,593 kg of crab. No fish were found in the 16.5 nets retrieved in 1984. Off Massachusetts, H.A. Carr and Cooper (1987) reported finding 9 lost nets in 15 submersible dives searching 40.5 ha of bottom. The findings suggest high densities of lost nets occur in at least some demersal gillnet fishing grounds off the northeast United States. Some of the nets were revisited over a 3-year period during which they continued to catch fish and shellfish (H. A. Carr et al. 1985; H. A. Carr 1986; H. A. Carr and Cooper 1987). Given the ability of lost nets to continue catching fish and crabs over a period of years, lost gillnets could be responsible for significant losses of some commercially valuable species.

Conclusions and Recommendations

Available information indicates that individuals of at least 135 marine species, including a significant percentage of the world's sea turtle, marine mammal, and seabird species, become entangled in marine debris. At least some entanglement-related deaths have been reported for most of these species. In almost all cases, a direct, absolute measure of the extent to which entanglement occurs or affects species at the population level does not exist. There are two principal reasons for this. First, most data have been gathered on beaches where animals haul out, roost, or strand. As a result, records are limited to animals that survive long enough to swim ashore or that become entangled close to shore. Because of logistical and practical constraints, at-sea surveys for entangled animals are rare and the number of animals entangled and killed offshore is unknown. Second, many entanglements involve fishing nets and line, and it is rarely possible to determine after the fact if entangled animals encountered their burden of gear when nets or line were active or after the gear was lost. There are no immediate prospects for resolving these uncertainties, and thus definitive information on the magnitude of entanglement mortality for affected species or populations seems unlikely in the near future.

Notwithstanding these limitations, some conclusions regarding levels of impact seem reasonable. For most seabirds (particularly procellariiform seabirds, penguins, grebes, and loons), toothed whales, and fish, evidence of entanglement is lacking or is based on isolated or infrequent reports. For these species, entanglement appears unlikely to cause effects at a population level. For certain other species, entanglement appears to be a chronic low-level source of mortality. Some of these species, such as gray whales, California sea lions, northern elephant seals, northern gannets, herring gulls, and shags, consist of large or increasing populations that appear to be unaffected by entanglement. However, other species subject to low levels of entanglement (e.g., West Indian manatees, Steller sea lions, and some sea turtles) are endangered or threatened. Although entanglement-related mortality for these species may be low compared to other sources of mortality, it constitutes an added obstacle to recovery and therefore merits attention.

For a few species, such as Hawaiian monk seals, northern fur seals, some sea turtles, certain fish and crabs caught commercially, and perhaps northern right whales, entanglement in marine debris appears to occur often enough to affect population abundance directly. In these cases, marine debris is an important research and management concern.

The types of marine debris most commonly associated with entanglement are fishing nets, monofilament line, lost crab traps and fish pots, rope, and strapping bands. The greatest source of this material is commercial fishing operations, although cargo vessels, recreational fishing, and land-based sources also may be significant contributors.

Given these points, resolution of marine debris entanglement problems would be best addressed through a combination of species management and source-reduction programs. For endangered or threatened species affected at low levels and for species for which marine debris may be a substantial source of mortality, the following actions are recommended:

Research should be undertaken to better document and monitor entanglement rates. Particularly promising in this regard are prospects for monitoring entanglement rates for sea turtles at sea. Periodic on-land surveys for entangled seals and seabirds that come to shore to nest, breed, and molt should be continued in cases in which such information provides the only source of knowledge about the extent of such interactions.

In conjunction with such studies and other field work, researchers should take steps to free entangled animals and to remove entangling debris from key habitats, such as rookeries.

Because of the predominance of fishing-related debris in entanglement incidents, source-reduction efforts should focus on incorporating new management measures into fishery management programs to avoid losses and to increase recovery of such items. In this regard, the following actions are recommended:

Require fishermen to report when, where, and the circumstances under which nets or traps are lost. Analyses of such information should help in devising ways to minimize losses and perhaps mounting clean-up efforts.

Investigate the feasibility of implementing pilot programs to clean up lost fishing gear from the sea floor in areas where such debris is likely to be concentrated.

Undertake studies to better document the magnitude of ghost fishing by derelict fishing gear through surveys to locate and record catch in lost gear and to record the catch in lost gear recovered incidentally during commercial fishing operations.

Require fishermen to retain all plastic or entangling gear caught incidentally during fishing operations for on-land disposal.

Institute incentives for retaining and returning used fishing gear for on-land disposal (e.g., bonuses for returning used gear or gear deposits).

Ensure that convenient reception facilities are available in fishing ports to receive material returned to shore.

Evaluate new technologies to decrease the potential for gear loss (e.g., automatic float releases that would allow floats to stay submerged where they would not be cut or towed by passing ships), and enhance the likelihood that lost gear could be recovered (e.g., attachment of sonic devices to help relocate lost gear).

Encourage or require alternative technologies and products that reduce the likelihood of loosing hazardous entangling items (e.g., using nonplastic straps or bait containers).

With regard to recreational fishing, steps should be taken to investigate the feasibility of degradable monofilament fishing line (e.g., line with a 1-year service life). As a general matter, targeted education and awareness programs should be developed and maintained for operators of commercial and recreational fishing vessels and cargo vessels. Such programs should describe the problems caused by lost items commonly used by each user group, relevant legal requirements, and provisions for properly disposing of plastics.

Acknowledgments. A review paper of this sort would not be possible without the generous assistance of many colleagues. At the risk of omitting many who provided help in compiling information used in this paper, I express special thanks to David G. Ainley, Sarah Allen, Raymond V. Arnaudo, Jason D. Baker, George Balazs, Cathy A. Beck, John L. Bengtson, Karen A. Bjorndal, Alan B. Bolton, W. Nigel Bonner, William R. P. Bourne, Gerald Brothers, Jeff Brown, Alan R. Bunn, Ruperto Chaparro, Susan J. Chivers, John Clary, James M. Coe, Anne Collet, John P. Croxall, Trevor R. Dixon, David E. Gaskin, William P. Gregg, Murray R. Gregory, Ralph T. Health, Jr., John R. Henderson, Robert J. Hofman, Karl W. Kenyon, Gordon H. Kruse, David M. Lavigne, Burney J. Le Boeuf, Lloyd F. Lowry, Peter L. Lutz, Theodore R. Merrell, Jr., Kenneth D. McDermond, Jeffrey Miller, John E. Miller, William A. Montevecchi, Daniel K. Odell, Rodney J. Paterson, Villere C. Reggio, Christine A. Ribic, Martin D. Robards, Betsy Schrader, David Sergeant, Peter D. Shaughnessy, Louis Sileo, Janet Slater, Wendy G. Teas, Fritz Trillmich, Jan Andries van

Franeker, Dean Wilkinson, Charles D. Wood-house, and Steven T. Zimmerman. For their great care in reviewing and commenting on early drafts and assistance in gathering information, I am particularly grateful to Charles W. Fowler, Burr Heneman, and most of all Peter G. Ryan. Finally, for the opportunity and support to work on this paper, I thank John R. Twiss, Jr. and the Marine Mammal Commission.

APPENDIX 1. Sea turtle entanglement (E) and ingestion (I) records.

Species	I/E	Material	Location	Source
Green sea turtle, *Chelonia mydas*	I+/E+	I: Plastic bags, plastic sheeting, line, synthetic cloth; plastic fragments; E: Monofilament line, trawl net, gillnet, anchor line, cloth strip, kite string	N. Pac.: Japan and Hawaii	Uchida 1990; Balazs 1980, 1985
			N. Pac.: Japan, Hawaii, and California	Henderson 1984; Mooney and Naughton, 1981; Balazs 1985
	I+/E−	I: Plastic banana bags, plastic fragments and cloth, unidentified plastic, rubber, aluminium foil, monofilament line; E: Fishing line	N. Atl.: Caribbean, Gulf of Mexico, Florida	Meylan 1978; Balazs 1985; Plotkin and Amos 1990; Bjorndal et al. 1994
			N. Atl.: Texas, Florida	Hildebrand 1980; Balazs 1985; Teas and Witzell 1994
	I−/E+	I: Plastic bags, vinyl film; E: Rope tied to anchor	S. Pac.: Peru, Australia	Hirth 1971; Brown and Brown 1982; Balazs 1983, 1985
			S. Pac.: Australia	
Loggerhead turtle, *Caretta caretta*	I+/E−	I: Styrofoam, plastic, plastic bags, monofilament line, plastic bottle, balloons, iron bolt, champagne cork, nylon thread; E: Monofilament line, gillnet, trawl netting, plastic woven sack	N. Atl.: Texas, Florida, Georgia, Virginia, Azores, and Mediterranean	Brongersma 1968; Salvador 1978; Van Nierop and den Hartog 1984; Balazs 1985; Gramentez 1988; Shoop and Ruckdeschel 1989
			N. Atl.: Caribbean, Texas, Florida, and Azores	Rabalais and Rabalais 1980; Balazs 1985; Bolten and Bjorndal 1991; Bjorndal and Bolten 1994; Plotkin and Amos 1990
	I−/E−	I: Plastic pellets in hatchings, plastic sheeting and bags, glass; E: Polypropylene rope	S. Atl.: South Africa	Hughes 1970, 1974b
	I−	Plastic plastic sheeting, line, and other debris	S. Atl.: South Africa	P.G. Ryan, personal communication
			N. Pac.: Japan	Balazs 1985; Uchida 1990
Hawksbill turtle, *Eretmochelys imbricata*	I+/E+	I: Plastic sheeting, plastic bags, plastic fragments, pellets, styrofoam, paper; E: Monofilament line, plastic onion bag	N. Atl.: Costa Rica, Florida, and Texas	Carr and Stancyk 1975; Hildebrand 1980; Hartog 1980; Meylan 1984; Balazs 1985
			N. Atl.: Florida and Texas	Broadrick 1982; Fletcher 1982; Wolf 1982; Balazs 1985; Plotkin and Amos 1990; Teas and Witzell 1994
	I+/E−	I: Numerous plastic particles; E: Monofilament gillnet	N. Pac.: Japan, Hawaii	Balazs 1985; Uchida 1990
			N. Pac.: Hawaii	Balazs 1978
Olive ridley turtle, *Lepidochelys olivacea*	E+	Trawl net, monofilament and nylon webbing, synthetic line	N. Pac.: Hawaii and Costa Rica	Afelin and Pulelos 1982; Balazs 1985

APPENDIX 1. *Continued.*

Species	I/E	Material	Location	Source
Kemp's ridley turtle, *Lepidochelys kempi*	I+/E+	I: Milk carton, plastics, burlap;	N. Atl.: Texas and Georgia	Balazs 1985; Plotkin and Amos 1990; Shoop and Ruckdeschel 1989
		E: Shrimp trawl, fishing line, rope	N. Atl.: Texas	Plotkin and Amos 1990; Balazs 1985
Leatherback turtle, *Dermochelys coriacea*	I+/E+	I: Plastic bags, plastic sheeting, plastic-coated prescription label, plastic fragments, nylon line;	N. Atl.: Netherlands, England, and eastern U.S.	Brongersma 1972; Schoelkopf 1981; Duguy et al. 1980; Fritts 1982; Balazs 1985, Sadove and Morreale 1990
		E: Lobster pot line, rope, nylon line	N. Atl.: England, Bermuda, France, and Texas	Duron and Duron 1980; Lee and Palmer, 1981; Duguy and Duron 1981, 1982; Duguy 1983; Hartog and Van Nierop 1984; Plotkin an Amos 1990
	I−/E−	I: Plastic sheeting;	S. Atl.: South Africa	Hughes 1974a
		E: Nylon rope	S. Atl.: South Africa	Balazs 1985
	I+/E−	I: Twisted vinyl fiber, large plastic sheets;	N. Pac.: Japan	Balazs 1985; Uchida 1990
		E: Monofilament gillnet, parachute anchor, rope	N. Pac.: Hawaii	Balazs 1985
	I−	Plastic bags, plastic sheeting	S. Pac.: Peru and New Zealand	Fritts 1982; Cawthorn 1985

Atl., Atlantic; Pac., Pacific; U.S., United States.

Minus (−), evidence of interactions based on rare or infrequent reports of isolated individuals.

Plus (+), some reports of interactions more than infrequent and/or evidence of a regular pattern of occurrence at low to high levels.

APPENDIX 2. Seabird entanglement (E) and ingestion (I) records.

Species	I/E	Material	Location	Source
Sphenisciformes (penguins):				
Gentoo penguin, *Pygoscelis papua*	E –	Plastic line	Ind. O.: Marion Is.	Ryan 1987a
Adeline penguin, *Pygoscelis adeliae*	E –	Fishing net, wire	S. Ocean	Slip 1990
Chinstrap penguin, *Pygoscelis antarctica*	E –	Trawl netting	S. Ocean: Seal Is.	United States of America 1991
Rockhopper penguin, *Eudyptes crestatus*	E –	Plastic	S. Atl.: Gough Is.	Ryan 1986
Jackass penguin, *Spheniscus demersus*	E –	Not identified	S. Atl.: South Africa	Ryan 1990a
Fairy penguin, *Euclyptula minor*	E – /I –	E: Fishing line; I: Bottle cap	S. Pac.: Tasmania	Slater 1992; Slater 1994
Podicipediformes (grebes):				
Great crested grebe, *Podiceps cristatus*	E –	Fishing line	N. Atl.: North Sea	Federal Republic of Germany 1985
Western grebe, *Aechmorphorus occidentalis*	E –	Monofilament gillnet fragment	N. Pac.: California	Jameson 1986
Procellariiformes (albatrosses, petrels, and shearwaters):				
Wandering albatross, *Diomedea exulana*	I	Plastic fragments	S. Atl.: S. Pac.	B.L. Furness 1983; Day et al. 1985
Royal albatross, *Diomedea epomophora*	I	Manufactured pieces of plastic	S. Pac.: Chatham Is.	M. Imber (cited et al. in Day et al. 1985)
Black-footed albatross, *Diomedea nigripes*	I+/E	I: Plastic fragments and bags; E: Not identified	N. Pac.: Hawaiian Is.	Conant 1984; Sileo et al. 1990a, 1990b; S. Fefer (cited in Henderson 1988)
Laysan albatross, *Diomedea immutabilis*	I+/E –	I: plastic fragments, light sticks, cigarette lighter, cellophane, paint chips; E: Monofilament gillnet	N. Pac.: Hawaiian Is. N. Pac.: Hawaiian Is.	Sileo et al. 1990a, 1990b; Fry et al. 1987; Robards et al., Chapter 6, this volume
Grey-headed albatross, *Diomedea chrysostoma*	I	Pieces of plastic	N. Pac.: Hawaiian Is. Ind. O.: Marion Is.	DeGange and Newby 1980; B.L. Furness 1983
Northern fulmar, *Flumarus glacialis*	I+/E –	I: Pellets, plastic pieces and sheeting, elastic thread, cigarette filters, tea bag, balloon, syringe tip, foil; E: Monofilament net	N. Pac.: Alaska N. Atl. N. Pac.	Day 1980; Robards et al., Chapter 6, this volume; Bourne 1976; Van Franeker 1985; Moser and Lee 1992; DeGange and Newby 1980

APPENDIX 2. *Continued.*

Species	I/E	Material	Location	Source
Southern fulmar, *Flumarus glacialoides*	I	Plastics	S. Atl.: S. Africa	Ryan 1987b
			S. Ocean: Antarctica	Van Franeker and Bell 1988; Ainley et al. 1990a
Giant fulmar, *Macronectes giganteus*	E –	Longline hooks, nylon line, string	S. Ocean: Antarctica	Anonymous, 1992
			Ind. O.: Marion Is.	Ryan 1987a
Dark-rumped petrel, *Pterodroma phaeopygia*	I –	Plastics	N. Pac.	Robards et al., Chapter 6, this volume
Great-winged petrel, *Pterodroma macroptera*	I –	Pellets	S. Pac.: New Zealand	M. Imber (cited in Day et al. 1985)
Solander's petrel, *Pterodroma solandri*	I –	Plastics	N. Pac.	Robards et al., Chapter 6, this volume
Mottled petrel, *Pterodroma inexpecta*	I	Plastics	N. Pac.	Robards et al., Chapter 6, this volume
White-winged petrel, *Pterodroma leucoptera*	I +	Plastic	Equatorial Pacific	Ainley et al. 1990b
Stejneger's petrel, *Pterodroma l. longirostris*	I +	Plastic	Equatorial Pacific	Ainley et al. 1990b
Pycroft's petrel, *Pterodroma l. pycrofti*	I	Plastic	Equatorial Pacific	Ainley et al. 1990b
Black-winged petrel, *Pterodroma nigripennis*	I –	Plastic	Equatorial Pacific	Ainley et al. 1990b
Thaiti petrel, *Pterodroma rostrata*	I –	Plastic	Equatorial Pacific	Ainley et al. 1990b
Collared petrel, *Pterodroma brevicpes*	I	Pellets	Equatorial Pacific	Spear et al., 1995
Murphy's petrel, *Pterodroma ultima*	I +	Pellets, nylon line	Equatorial Pacific	Spear et al., 1995
Kerguelen petrel, *Pterodroma brevirostris*	I +	Pellets	S. Pac.: New Zealand	S. Reed 1981
			S. Ocean	Ainley et al. 1990a
Cook's petrel, *Pterodroma cookii*	I +	Pellets	S. Pac.: New Zealand	M. Imber (cited in Day et al. 1985)
			Equatorial Pacific	Ainley et al. 1990b
Bonin's petrel, *Pterodroma hypoleuca*	I +	Plastic	N. Pac.: Hawaiian Is.	Sileo et al. 1990b
Atlantic petrel, *Pterodroma incerta*	I –	Pellets	S. Atl.: Gough Is.	R.W. Furness 1985a

APPENDIX 2. *Continued.*

Species	I/E	Material	Location	Source
Soft-plumaged petrel, *Pterodroma mollis*	I –	Pellets	S. Atl.: Gough Is.	R.W. Furness 1985a
Black-capped petrel, *Pterodroma hasitata*	I –	Pellets	N. Atl.: N. Carolina	Moser and Lee 1992
Juan Fernandez petrel, *Pterodroma e. externa*	I –	Plastic	Equatorial Pacific	Ainley et al. 1990b
White-necked petrel, *Pterodroma e. cervicallis*	I –	Plastic	Equatorial Pacific	Ainley et al. 1990b
Snow petrel, *Pagodroma nivea*	I –	Pellets	S. Ocean: Antarctica	Van Franeker and Bell 1988; Ainley et al. 1990a
Antarctic petrel, *Thalassoica antarctica*	I –	Plastics	S. Ocean: Antarctica	Ainley et al. 1990a
Blue petrel, *Halobaena caerules*	I +	Pellets	S. Atl.: Gough Is. S. Pac.: New Zealand S. Ocean	Ryan 1985, 1987c S. Reed 1981 Ainley et al. 1990a
Broad-billed prion, *Pachyptila vittata*	I +	Pellets	S. Pac. S. Atl.: Gough Is. S. Ocean	Bourne and Imber 1982 Ryan and Fraser 1988 Ainley et al. 1990a
Narrow-billed prion, *Pachyptila belcheri*	I +	Plastic	Equatorial Pacific S. Pac.: New Zealand	Ainley et al. 1990b Bourne and Imber 1982
Salvin's prion, *Pachyptila salvini*	I	Pellets	S. Pac.: New Zealand	Harper and Fowler 1987
Antarctic prion, *Pachyptila desolata*	I +	Pellets	S. Pac.: New Zealand	Harper and Fowler 1987; Ryan 1990a
Fairy prion, *Pachyptila turtur*	I	Pellets	S. Pac.: New Zealand	M. Imber (cited in Day et al. 1985)
Bulwer's petrel, *Bulweria bulwerii*	I	Plastic	N. Pac.: Hawaiian Is.	Harrison et al. 1983
White-chinned petrel, *Procellaria aequinoctialis*	I +/E –	I: Plastic fragments E: String Pellets	S. Atl. S. Ocean S. Pac.: New Zealand	B.L. Furness 1983 Ryan 1987a M.Imber (cited in Day et al. 1985)
Parkinson's petrel, *Procellaria parkinsoni*	I			

APPENDIX 2. *Continued.*

Species	I/E	Material	Location	Source
Pintado or Cape petrel, *Daption capense*	I+	Plastic particles, paint flakes	Equatorial Pacific S. Ocean: Antarctica S. Atl.: South Africa	Ainley et al. 1990b Ainley et al. 1990a Van Franeker and Bell 1988; Ryan 1987b, 1990a
Common diving petrel, *Pelecanoides urinatrix*	I	Plastic	S. Atl.: Gough Is.	Ryan 1986
Pink-footed shearwater, *Puffinis creatopus*	I	Pellets and plastic fragments	N. Pac.: California	Baltz and Morejohn 1976; Robards et al., Chapter 6, this volume
Flesh-footed shearwater, *Puffinis carneipes*	I+	Pellets and plastic fragments	N. Pac.	Robards et al., Chapter 6, this volume
Greater shearwater, *Puffinis gravis*	I+/E–	I: Pellets, polystyrene, thread E: Packing strip Pellets, styrofoam	N. Atl. S. Atl. S. Atl.	Bourne 1976; B.L. Furness 1983 Ryan 1987 Ryan 1991
Audubon's shearwater, *Puffinis lherminieri*	I	Pellets, styrofoam	N. Atl.: N. Carolina	Moser and Lee 1992
Buller's shearwater, *Puffinis bulleri*	I+	Plastic fragments	Equatorial Pacific N. Pac.	Ainley et al. 1990b Robards et al., Chapter 6, this volume
Wedge-tailed shearwater, *Puffinis pacificus*	I+	Pellets and plastic fragments	N. Pac.: Hawaiian Is.	Fry et al. 1987; Sileo et al. 1990b
Sooty shearwater, *Puffinis griseus*	I+/E–	I: Pellets and plastic fragments E: Trawl net	N. Pac. S. Pac. N. Atl. N. Pac.: Aleutian Is.	Ogi 1990; DeGange and Newby 1980 Ainley et al. 1990b Bourne 1976 Manville 1990
Short-tailed shearwater, *Puffinis tenuirostris*	I+/E–	I: Pellets and plastic fragments	N. Pac. S. Pac. N. Pac.	Day 1980 Ogi 1990; Ainley et al. 1990b; DeGange and Newby 1980
Newell's shearwater, *Puffinis auricularis*	I+	E: Monofilament gillnet Plastic	N. Pac.: Hawaiian Is.	Sileo et al. 1990b
Christmas shearwater, *Puffinis nativitatis*	I+	Plastic	N. Pac.: Hawaiian Is.	Sileo et al. 1990b
Little shearwater, *Puffinis assimilis*	I–	Pellets	S. Atl.: Gough Is.	R.W. Furness 1985a

APPENDIX 2. *Continued.*

Species	I/E	Material	Location	Source
Manx shearwater, *Puffinus puffinis*	1+	Pellets	N. Atl.: N. Carolina	Moser and Lee 1992; R.W. Furness 1985b
Cory's shearwater, *Calonectris diomedea*	I	Pellets	N. Atl.: N. Carolina	Moser and Lee 1992
Wilson's storm-petrel, *Oceanites oceanicus*	I+	Plastic fragments, styrofoam Plastic particles	N. Atl.: N. Carolina S. Ocean: Antarctica	Moser and Lee 1992 Van Franeker and Bell 1988; Ainley et al. 1990a
White-faced storm-petrel, *Pelagodroma marina*	I+/E	I: Pellets E: String	S. Pac.: Chatham Is. Equatorial Pacific Ind. O.: Marion Is.	Bourne and Imber 1982 Ainley et al. 1990b Ryan 1987a
British storm-petrel, *Hydrobates pelagicus*	I	Plastic	N. Atl.: Scotland	Van Franeker 1983
Leach's storm-petrel, *Oceanodroma leucorhoa*	I+/E−	I: Plastic fragments E: Monofilament line	N. Pac.: Alaska Equatorial Pacific N. Atl.: Newfoundland	Day 1980 Ainley et al. 1990b W.A. Montevecchi, personal communications; R.W. Furness 1985b
Markham's storm-petrel, *Oceanodroma markhami*	I	Plastic	N. Pac. Equatorial Pacific	G.V. Byrd (cited in Manville 1990) Ainley et al. 1990b
Wedge-rumped storm-petrel, *Oceanodroma tethys*	I −	Plastic	Equatorial Pacific	Ainley et al. 1990b
Sooty storm-petrel, *Oceanodroma tristrami*	I+	Plastic fragments	N. Pac.: Hawaiian Is.	Harrison et al. 1983; Sileo et al. 1990b; Robards et al., Chapter 6, this volume
Fork-tailed storm-petrel, *Oceanodroma furcata*	I	Plastic fragments	N. Pac.: Alaska	Day 1980; Robards et al., Chapter 6, this volume
White-belly storm-petrel, *Fregetta grallaria*	I	Plastic particles	S. Atl.: Gough Is.	R.W. Furness 1985a
Grey-backed storm-petrel, *Garrodia nereis*	I	Plastic particles	Gough Is., S. Atl.	R.W. Furness 1985a
Pelecaniformes (pelicans, boobies, gannets, cormorants, frigatebirds, tropicbirds):				
Red-tailed tropicbird, *Phaethon rubricauda*	I+	Plastic	N. Pac.: Hawaiian Is.	Sileo et al. 1990b
Brown pelican, *Pelecanus occidentalis*	E+	Monofilament line, string, fishhooks	N. Atl.: Florida N. Pac.: California	R. Heath, personal communication, Centaur Assoc. 1986 U.S. Fish and Wildlife Service 1980

APPENDIX 2. *Continued.*

Species	I/E	Material	Location	Source
White pelican, *Pelecanus erythrorynchos*	E	Monofilament line, fishhooks	N. Atl.: Florida	R. Heath, personal communication
Great frigatebird, *Fregata minor*	I+	Plastic	N. Pac.: Hawaiian Is.	Sileo et al. 1990b
Pelagic cormorant, *Phalacrocorax pelagicus*	I	Plastic	N. Pac.	Robards et al., Chapter 6, this volume
Shag, *Phalacrocorax aristotelis*	E+	Fishing net, fishing line, wire, six-pack yokes	N. Atl.: Brit. Isl.	Onions and Rees 1992
Little pied cormorant, *Phalacrocorax melanoleucos*	E–	Gillnet	S. Pac.: Tasmania	Slater 1990
Cape cormorant, *Phalacrocorax capensis*	E–	Fishing line	S. Atl.: South Africa	P.G. Ryan, personal communication
Great cormorant, *Phalacrocorax carbo*	E–	Fishing line	S. Atl.: South Africa	P.G. Ryan, personal communication
Australasian gannet, *Sula serrator*	E	Plastic strapping band	S. Pac.: Tasmania	Slater 1992
Northern gannet, *Sula bassana*	I–/E+	I: Metal spike E̲: Nets, rope, line, strapping bands	N. Atl.: Orkney Is. N. Atl.: Newfoundland German Bight	Bourne 1976 Montevecchi 1991 Schrey and Vauk 1987
Cape gannet, *Morus capensis*	E–	Orange bag	S. Atl.: South Africa	P.G. Ryan, personal communication
Red-footed booby, *Sula sula*	I–	Plastic	N. Pac.: Hawaiian Is.	Sileo et al. 1990b
Blue-footed booby, *Sula nebouxii*	I	Pellets	N. Pac.: Hawaiian Is.	Anonymous 1981
Masked booby, *Sula dactylatra*	I/E–	I: Plastic E̲: Trawl net scrap M̲onofilament line	N. Pac.: Hawaiian Is. N. Pac.: Hawaiian Is. N. Atl.: Texas	Sileo et al. 1990b Conant 1984
Ruddy turnstone, *Arenaria interpres*	E–			J.E. Miller, personal communication
Red-necked phalarope, *Phalaropus lobatus*	I	Plastic pieces	N. Atl.: N. Carolina N. Pac.: Alaska	Moser and Lee 1992 Day 1980; Robards et al., Chapter 6, this volume
Charadriiformes (shorebirds, skuas, gulls, terns, and auks):				
Bar-tailed godwit, *Limosa lapponica*	I	Plastic pieces	N. Pac.	Robards et al., Chapter 6, this volume

APPENDIX 2. *Continued.*

Species	I/E	Material	Location	Source
Red phalarope, *Phalaropus fulicaria*	I+	Plastic fragments, polystyrene, plastic particles	N. Atl.: N. Carolina N. Pac.: California	Moser and Lee 1992 Conners and Smith 1982; Robards et al., Chapter 6, this volume
Antarctic skua, *Catharacta maccormicki*	I−/E?	Fishing line in mouth and around neck	S. Ocean S. Pac.	Slip 1990 Robards et al., Chapter 6, this volume
Tristan skua, *Catharacta skua hamiltoni*	I−	Plastic particles	S. Atl.: Gough Is.	R.W. Furness 1985a
Great skua, *Stercorarius skua*	I+	Plastic particles	S. Atl.	Ryan and Fraser 1988
Long-tailed jaeger, *Stercorarius longicaudus*	I−	Plastics	N. Atl.: N. Carolina	Moser and Lee 1992
Parasitic jaeger, *Stercorarius parasiticus*	I−	Plastics	N. Atl.: N. Carolina	Moser and Lee 1992
Pomarine jaeger, *Stercorarius pomarinus*	I	Plastics	N. Atl.: N. Carolina	Moser and lee 1992
Laughing gull, *Larus atricilla*	I/E	I: Plastics E̲. Monofilament line, six-pack yokes	N. Atl.: Florida N. Atl.	Below 1979
Black-headed gull, *Larus ridibundus*	E+	Fishing line, fishhooks, string, wire, fishing net	N. Atl.: Brit. Isl.	Onions and Rees 1992
Bonaparte's gull, *Larus philadelphia*	I+	Plastic fragments Plastic particles	N. Atl.: N. Carolina Eastern Canada	Moser and Lee 1992 Braune and Gaskin 1982
Heermann's gull, *Larus heermanii*	I	Pellets and plastic fragments	N. Pac.: California	Baltz and Morejohn 1976
Mew gull, *Larus canus*	I	Plastic fragments	N. Pac.: California	Baltz and Morejohn 1976; Robards et al., Chapter 6, this volume; Vauk-Hentzelt 1982
Herring gull, *Larus argentatus*	I/E+	I: Plastic bags, stryofoam, cellophane E̲: Fishing line, string, fish hooks, wire, six-pack yokes	N. Atl.: North Sea N. Atl.: Brit. Isl.	Day et al. 1985; Vauk-Hentzelt 1982 Onions and Rees 1992
Western gull, *Larus occidentalis*	I−/E	I: Plastic E̲: Fishing line	Not identified N. Pac.: California	H. Ogi (cited in Day et al. 1985) Jameson 1986
Glaucous-winged gull, *Larus glaucescens*	I−	Plastic fragments	N. Pac.	Baltz and Morejohn 1976

APPENDIX 2. *Continued.*

Species	I/E	Material	Location	Source
Glaucous gull, *Larus hyperboreus*	I –	Plastic	N. Pac.	Day 1980
Great black-backed gull, *Larus marinus*	I/E	I: Plastic bags, styrofoam, etc. / E̲.: Monofilament line, six-pack yoke	N. Atl.: Maine / N. Atl.: Maine, New Jersey	Day et al. 1985 / Day et al. 1985; Montevecchi, personal communication
Lesser black-backed gull, *Larus fuscus*	E	Fishing line, string, six-pack yoke	N. Atl.: Brit. Isl.	Onions and Rees 1992
Silver gull, *Larus novaehollandiae*	E –	Six-pack ring	S. Pac.: Tasmania	Slater 1990
Kelp gull, *Larus dominicanus*	I +	Plastics	S. Atl.: South Africa	Ryan 1990c
Hartlaub's gull, *Larus hartlaubii*	E –	Fishing line	S. Atl.: South Africa	P.G. Ryan, personal communication
Black-legged kittiwake, *Rissa tridactyla*	I/E	I̲: Pellets, plastic fragments, nylon thread / E̲: Fishing line	N. Pac.: California / N. Atl.	Baltz and Morejohn 1976; Robards et al., Chapter 6, this volume / Bourne 1976; Moser and Lee 1992; Onions and Rees 1992
Red-legged kittiwake, *Rissa brevirostris*	I +	Pellets	N. Pac.: Alaska	Day 1980
Sabine's gull, *Zema sabini*	I	Plastic	N. Atl.: N. Carolina	Moser and Lee 1992
Common tern, *Sterna hirundo*	I/E	I: Plastic particles / E̲: Fishing line	N. Atl.: Canada, Brit. Isl.	Braune and Gaskin 1982 / Onions and Rees 1992
Sooty tern, *Sterna fuscata*	I+/E	I̲: Plastic / E̲: Not identified	N. Pac.: Hawaiian Is. / N. Pac.: Hawaiian Is.	Sileo et al. 1990b / Fefer (cited in Henderson 1988)
Bridled tern, *Sterna anaethetus*	I –	Plastic	N. Atl.: N. Carolina	Moser and Lee 1992
Crested tern, *Sterna bergii*	E –	Clear plastic bag	S. Atl.: S. Africa	Ryan 1990a
Sandwich tern, *Thalasseus sandvicensis*	E	Fishing line, string	N. Atl.: Brit. Isl.	Onions and Rees 1992
Black tern, *Chlidonias niger*	I –	Plastic	N. Atl.: N. Carolina	Moser and lee 1992
White tern, *Gygis alba*	I +	User plastics	Equatorial Pacific	Spear et al., 1995
Brown noddy, *Anous stolidus*	I +	Plastic	N. Pac.: Hawaiian Is.	Sileo et al. 1990b
Black noddy, *Anous minutus*	I	Plastic	N. Pac.: Hawaiian Is.	Sileo et al. 1990b

APPENDIX 2. *Continued.*

Species	I/E	Material	Location	Source
Black skimmer, *Rynchops nigra*	E+	Monofilament line	N. Atl.: New York	Gochfeld 1973
Razorbill, *Alca torda*	E	Fishing line, fishing net	N. Atl.: Brit. Isl.	Onions and Rees 1992
Dovekie, *Alle alle*	I	Plastic	Canadian Arctic	Bradstreet (cited in Day et al. 1985) Van Franeker 1983
Pigeon guillemot, *Cepphus columbia*	I	Plastic	N. Pac.	Robards et al., Chapter 6, this volume
Black guillemot, *Cepphus grylle*	E	Fishing net, fishing line, string	N. Atl.: Brit. Isl.	Onions and Rees 1992
Guillemot/common murre, *Uria aalge*	I – /E	I: Plastic	N. Pac.	Robards et al., Chapter 6, this volume
		E: Nylon fishing net, some kind of ring	N. Atl.: Norway, North Sea	Bourne 1977; Federal Republic of Germany 1985
Thick-billed murre, *Uria lomvia*	I –	Pellets	N. Pac.: Alaska	Day 1980
Murres (unidentified)	E	Net, line, & bottle fragments	N. Atl.: Newfoundland	R. Elliott (cited in Heneman 1988)
Cassin's auklet, *Ptychoraphus aleuticus*	I	Pellets, plastic fragments	Western N. Pac.	Jones and Ferrero 1985
			N. Pac.: Alaska	Day 1980; Robards et al., Chapter 6, this volume
Parakeet auklet, *Cyclorrhynchus psittacula*	I+	Pellets, plastic particles	N. Pac.	Day 1980; Pettit et al. 1981
Least auklet, *Aethia pusilla*	I	Plastics	N. Pac.	Day 1980; Robards et al., Chapter 6, this volume
Crested auklet, *Aethia cristatella*	I	Pellets, plastic particles	N. Pac.	Robards et al., Chapter 6, this volume
Rhinoceros auklet, *Cerorhinca monocerata*	I	Plastic fragments	N. Pac.: California	Baltz and Morejohn 1976; Robards et al., Chapter 6, this volume
Auklets (several unidentified species)	E+	Lost gillnet	Western N. Pac.	Jones and Ferrero 1985
Common/Atlantic puffin, *Fratercula arctica*	I+/E	I: Elastic thread E: Fishing net, fishing line	N. Atl.: Brit. Isl. N. Atl.: Brit. Isl.	Parslow and Jefferies 1972 Onions and Rees 1992
Tufted puffin, *Fratercula cirrhata*	I+/E –	I: Pellets E: Lost gillnet	N. Pac.: Alaska Western N. Pac.	Day 1980 DeGange and Newby 1980; Jones and Ferrero 1985; Robards et al., Chapter 6, this volume

APPENDIX 2. *Continued.*

Species	I/E	Material	Location	Source
Horned puffin, *Fratercula corniculata*	I+/E	I: Pellets — E: Lost gillnet —	N. Pac.: Alaska Western N. Pac.	Day 1980; Robards et al., Chapter 6, this volume; Jones and Ferrero 1985
Coastal birds other than seabirds:				
Osprey, *Pandion haliaetis*	E –	Monofilament line, six-pack yokes, fishing nets	N. Atl.: Maryland	Centaur Assoc. 1986
Canada goose, *Branta canadensis*	E –	Monofilament line, six-pack yokes	N. Atl.: Maryland	Centaur Assoc. 1986
Snow goose, *Chen hyperborea*	E –	Monofilament line, six-pack yokes	N. Atl.: Maryland	Centaur Assoc. 1986
Laysan finch, *Telespyza cantans*	E –	Drowned in plastic cooler	N. Pac.: Hawaiian Is.	Morin 1987
Common eider, *Somateria mollissima*	E –	Six-pack yokes, wire	N. Atl.: North Sea	Federal Republic of Germany 1985

Ind., Indian; O, ocean; Atl., Atlantic; Pac., Pacific.; Brit. Isl, British Islands.

Minus (−), evidence of interactions based on rare or infrequent reports of isolated individuals.

Plus (+), some reports of interactions more than infrequent and/or evidence of a regular pattern of occurrence at low to higher levels.

APPENDIX 3. Marine mammal entanglement (E) and ingestion (I) records.

Species	I/E	Material	Location	Source
Mysticete whales (baleen whales):				
Bowhead whale, *Balaena mysticetus*	I –/E –	I: Plastic sheeting E: Rope	Alaska, Beaufort Sea Alaska, Beaufort Sea	Lowrey 1993 Philo et al. 1992
Northern right whale, *Eubalaena glacialis*	E +	Gillnet, trawl net, rope, loster pot lines	N. Atl.	Kraus 1990 Marine Mammal Commission 1993
Southern right whale, *Eubalaena australis*	E –	Rope and floats	N. Pac. S. Pac.: New Zealand	Cawthorn 1985
Humpback whale, *Megaptera novaeangliae*	E +	Gillnets, seines, rope and line	N. Atl. N. Pac.	Humpback Whale Recovery Team 1991 Humpback Whale Recovery Team 1991
Gray whale, *Eschrichtius robustus*	E +	Gillnet, crabpot float lines, rope	N. Pac.	Mate 1985; Hare and Mead 1987; Heyning and Lewis 1990
Minke whale, *Balaenoptera acutorostrata*	I –/E –	I: Polythene bag, plastic sheeting E: Rope	N. Atl.: Texas, Gulf of Mexico S. Pac.: New Zealand	Anonymous 1988; Hare and Mead 1987 Cawthorn 1985
Odontocete whales (toothed whales):				
Sperm whale, *Physeter macrocephalus*	I +/E –	I: Trawl net, rope, mylar balloon, cups, newspaper E: Bouy line for longline	N. Atl.: Newfoundland, Denmark Strait N. Atl.: Bahamas	Mate 1985; Martin and Clarke 1986; U.S. Natl. Mus. (cited in Walker and Coe 1990) J. Brown, personal communication
Dwarf sperm whale, *Kogia breviceps*	I –	Bread wrapper, rubber hospital gloves	N. Atl.: Florida, N. Carolina	Barros et al. 1990; U.S. National Museum (cited in Walker and Coe 1990)
Pigmy sperm whale, *Kogia simus*	I –	Plastic bags, plastics	N. Atl.: S. Carolina, Florida, Texas	Charleston Museum and U.S. Natl. Mus. (cited in Walker and Coe 1990); Tarpley 1990
Baird's beaked whale, *Berardius bairdii*	I +	Longline, plastic sheeting, cellophane, glass, cigarette filters, fishhooks, roofing tile, cigarette lighter top	N. Pac.: Japan	Walker and Coe 1990
Cuvier's beaked whale, *Ziphius cavirostris*	I –	Plastic bags, plastic straw, asphalt	N. Atl.: Virginia N. Pac.: California	U.S. Natl. Mus. (cited in Walker and Coe 1990)
Gerval's beaked whale, *Mesoplodon europaeus*	I –	Plastic bags	N. Atl.: New Jersey, N. Carolina	U.S. Natl. Mus. (cited in Walker and Coe 1990)
Blainville's beaked whale, *Mesoplodon densirostris*	I –	Plastic bottle cap	N. Atl.: New York	U.S. Natl. Mus. (cited in Walker and Coe 1990)
		Plastic thread	S. Atl. Brazil	E. Secchi, Personal communication

APPENDIX 3. *Continued.*

Species	I/E	Material	Location	Source
Short-finned pilot whale, *Globicephala macrorhynchus*	I –	Plastic container	N. Atl.: N. Carolina	U.S. Natl. Mus. (cited in Walker and Coe 1990)
Long-finned pilot whale, *Globicephala melaena*	I –	Plastic	N. Atl.: France	A. Collet, personal communication
Rough-toothed dolphin, *Steno bredanensis*	I –	Plastic bag, heavy black plastic, fishhook	N. Pac.: Hawaii N. Atl.: Virginia	U.S. Natl. Mus. (cited in Walker and Coe 1990)
False killer whale, *Pseudorca crassidens*	I –	Remains of a jug	N. Atl.: Southeast U.S.	Barros et al. 1990
Killer whale, *Orcinus orca*	E –	Rope and floats	S. Pac.: New Zealand	Cawthorn 1985
Pacific white-sided dolphin, *Lagenorhynchus obliquidens*	I –	Plastic bags, plastic bottle caps, plastic drinking straw, waxed paper, fishhook	N. Pac.: California	Caldwell et al. 1965; Cowen et al. 1986; Walker and Coe 1990
Common dolphin, *Delphinus delphis*	I –	Plastic bag, cellophane, black plastic sheeting, balloon, fishhook	N. Pac.: California	Walker and Coe 1990; Los Angeles Cty. Museum (cited in Walker and Coe 1990)
Bottlenose dolphin, *Tursiops truncatus*	I + /E –	I: Plastic sheeting, plastic bag, bungie cord, shoelace, bottle cap, spring, cloth, fishhook E: Monofilament line	N. Pac.: California N. Atl.: Florida Ind. O.: Western Australia	Walker and Coe 1990; Schwartz et al. 1992 Barros et al. 1990 Mann et al. 1995
Risso's dolphin, *Grampus griseus*	I –	Plastic bag, balloon	N. Atl.: Massachusetts, France N. Pac.: Hawaii, Clifornia	U.S. Natl. Mus. (cited in Walker and Coe 1990); A. Collet, personal communication Los Angeles Cty. Mus. (cited in Walker and Coe 1990); E. Nitta (cited in Henderson 1988)
Striped dolphin, *Stenella coeruleoalba*	I –	Plastic bag	N. Atl.: N. Carolina	U.S. Natl. Mus. (cited in Walker and Coe 1990)
Northern right whale dolphin, *Lissodelphis borealis*	I –	Plastic bag, vinyl plastic, metal bottle cap	N. Pac.: California	Walker and Coe 1990
Harbor porpoise, *Phocoena phocoena*	I – /E	I: Cloth, platic bag, plastic E: Fishing line	N. Atl.: N. Carolina	U.S. Natl. Mus. (cited in Walker and Coe (1990); Hare and Mead 1987 Kastelein and Lavaleije 1991
Dall's porpoise, *Phocoenoides dalli*	I – /E –	I: Plastic sheeting, plastic gags, soda straw, bottle cap, cardbboard E: Lost gillnet	N. Atl.: North Sea N. Pac.: California Western N. Pac.	Los Angeles Cty. Mus. and Santa Barbara Mus. (cited in Walker and Coe 1990) Jones and Ferrero 1985; DeGange and Newby 1980

APPENDIX 3. Continued.

Species	I/E	Material	Location	Source
Franciscan dolphin, Pontoporia blainvillei	I	Unidentified debris	S. Atl.: Brazil	E. Secchi, personal communication
Tucuxi dolphin, Sotalia fluviatilis	I –	Clear plastic sheeting	S. Atl.: Brazil	E. Secchi, personal communication
Phocidae (earless or true seals):				
Hawaiian monk seal, Monachus schauinslandi	E +	Trawl net, monofilament line, plastic ring, plastic cup, plastic band, wire, fishhook	N. Pac.: Hawaiian Islands	Henderson 1984, 1990
Mediterranean monk seal, Monachus monachus	E –	Fishing gear, rubber hoop	Mediterranean Sea	Gots et al. 1992
Gray seal, Halichoerus grypus	E	Driftnet and trawl net fragments	N. Atl.: Nova Scotia	W.D. Bowen and S.S. Anderson (cited in Fowler 1988)
Harp seal, Phoca groenlandica	E	Monofilament gillnet and trawl net and strapping bands	N. Atl.	W.D. Bowen (cited in Fowler 1988)
Harbor seal, Phoca vitulina	E	Strapping bands and other debris	N. Pac.: California; N. Atl.: Nova Scotia, Brit. Isl.	Stewart and Yochem 1987; W.D. Bowen and S.S. Anderson (cited in Fowler 1988)
Northern elephant seal, Mirounga angustirostris	I –/E	I: Styrofoam cup; E̱: Monofilament line, trawl net & gillnet, packing bands	N. Pac.: Oregon; N. Pac.: California	Mate 1985; Stewart and Yochem 1987
Southern elephant seal, Mirounga leonina	E –	Rope, wire, plastic packing hoops (strapping bands)	S. Atl.: Argentina	Ramirez 1986
Leopard seal, Hydrurga leptonyx	E –	Monofilament fishing line, fishhooks	S. Pac.: Tasmania	Jenkin 1990; J. Slater, personal communication
Otariidae (sea lions and fur seals):				
Northern fur seal, Calorhinus ursinus	E +	Trawl net, strapping bands, rope, monofilament line, six-pack holder, metal headlight ring, lawn chair material	N. Pac.: Bering Sea	Scordino 1985; Laist 1987; Bengtson et al. 1988; Baba et al. 1990; Fowler et al. 1990, 1992
Guadalupe fur seal, Arctocephalus townsendi	E –	Unidentified line	N. Pac.: Mexico	B.J. Le Boeuf, personal communication
Cape/Australian fur seal, Arctocehpalus pusillus	E +	Trawl net, monofilament net fragment; monofilament line, plastic straps, nylon rope, rubber 'O'-rings, wire	S. Atl.: South Africa; S. Pac.: Tasmania	Shaughnessy 1980; Pemberton et al. 1992

APPENDIX 3. *Continued.*

Species	I/E	Material	Location	Source
Juan Fernandez fur seal, *Arctocephalus philippii*	E	Packing bands	N. Pac.: Chile	Cardenas and Cattan (cited in Wallace 1985)
New Zealand fur seal, *Arctocephalus fosteri*	E+	Strapping bands, fishing net, rope	S. Pac.: New Zealand	Cawthorn 1985; Fowler 1988
Antarctic fur seal, *Arctocephalus gazella*	E+	Nylon string, trawl net, plastic bags, rubber 'O'-ring, plastic packaging tape, twine, cloth strip, hard hat liner	S. Atl.: South Georgia	Bonner and McCann 1982; Croxall et al. 1990
South American fur seal, *Arctocephalus australis*	E	Fishing net fragments, plastic hoops plastic or rubber ring	S. Atl.: Argentina; S. Pac.: Peru	Ramirez 1986; Fowler 1988
California and Galapagos sea lions, *Zalophus californianus*	E+	Monofilament fishing line, gillnet, trawl net, packing bands, rope, rubber bands, other debris	N. Pac.: California; S. Pac.: Galapagos Is.	Stewart and Yochem 1987; F. Trillmich, personal communication
Steller sea lion, *Eumatopias jubatus*	I–/E	I: Styrofoam cup Ē: Trawl net, rope, strapping bands N̄et	N. Pac.: Oregon; N. Pac.: Bering Sea; S. Pac.: Aukland	Mate 1985; Calkins 1985; Loughlin et al. 1986; Cawthorn 1985
New Zealand sea lion, *Phocarctos hookeri*	E			
South American sea lion, *Otaria flavescens*	E	Rope, wire, plastic packing bands	S. Pac.: Argentina	Ramirez 1984, 1986
Sirenia (manatees and dugongs): West Indian manatee, *Trichechus manatus*	I+/E+	E: Rope, monofilament line Ī: Monofilament line, plastic bags, string twine, rope, fishhooks, wire, paper, cellophane, rubber bands	N. Atl.: Florida, Antilles	Beck and Barros 1991
Carnivora (sea otters and polar bears): Sea otter, *Enhydra lutris*	E –	Monofilament gillnet	N. Pac.: Aleutian Islands	DeGange and Newby 1980

Minus (−), evidence of interactions based on rare or infrequent reports of isolated individuals.

Plus (+), some reports of interactions more than infrequent and/or evidence of a regular pattern of occurrence low to higher levels.

Appendix 4. Fish, crab, and squid entanglement (E) and ingestion (I) records.

Species	I/E	Material	Location	Source
Fish:				
Parrotfish, *Scarus* sp.	E	Derelict barrier net	N. Pac.: Hawaii	Henderson 1988
Grubby, *Myoxocephalus aenus*	I	Polystyrene spherules	Eastern N. Atl.	Carpenter et al. 1972; Day 1988
Striped seasnail, *Liparis liparis*	I	Plastics	Western N. Atl.	Kartar et al. 1976
Flounder, *Platichthyes flesus*	I	Polystyrene spherules	Western N. Atl.	Katar et al. 1976; Day 1988
Winter flounder, *Pleuronectes americanus*	I/E	I: Polystyrene spherules; E: Lost gillnets	Eastern N. Atl.	Carpenter et al. 1972; Day 1980;
American plaice, *Hippoglossoides platessoides*	E+	Lost gillnets	N. Atl.: Cape Cod Bay; N. Atl.: Newfoundland	H.A. Carr et al. 1985; Brothers 1989
Witch flounder/greysole, *Glyptocephalus cynoglossus*	E	Lost gillnets	N. Atl.: Newfoundland	Brothers 1989
Pacific halibut, *Hippoglossus stenolepis*	E+	Lost crab pots	N. Pac.: Alaska	High and Worlund 1979
White perch, *Roccus americamus*	I	Plastic pellets, polystyrene spherules	Eastern N. Atl.	Carpenter et al. 1972; Day 1988
Striped bass, *Morone saxtatilis*	I	Plastic cigar holder	Western N. Atl.	Manooch 1973
Silversides, *Menidia manidia*	I	Polystyrene spherules	Eastern N. Atl.	Carpenter et al. 1972; Day 1988
Sand goby, *Gobius minutus*	I	Polystyrene spherules	Western N. Atl.	Kartar et al. 1976; Day 1988
Cunner, Tautog, *Tautogolabrus adspersus*	I	Polystyrene spherules	Eastern N. Atl.	Day 1988
	I	Plastic particles	Western N. Atl.	Kartar et al. 1976
Atlantic croaker, *Micropogonias undulatus*	I	Plastics	Western N. Atl.	Govoni (cited in Hoss and Settle 1990)
Tautog, *Tautoga onitus*	E+	Lost gillnets	N. Atl.: Cape Cod Bay	H.A. Carr et al. 1985
Herring, *Clupea harengus*	I/E	I: Polystyrene spherules; E: Lost gillnet	Eastern N. Atl.; Eastern N. Pac.	Day 1988; Bren 1990
Gulf menhaden, *Brevoortia patronus*	I	Plastics	N. Atl.: Gulf of Mex.	Govoni (cited in Hoss and Settle 1990)
Five-bearded rockling, *Ciliata mustela*	I	Plastics	Eastern N. Atl.	Kartar et al. 1976
Atlantic pollock, *Pollachius virens*	I	Polystyrene spherules, plastic cups	Eastern N. Atl.	Anonymous 1975; Day 1988

APPENDIX 4. *Continued.*

Species	I/E	Material	Location	Source
Atlantic cod, *Gadus morhua*	I/E+	I: Plastic cups; Ē: Lost gillnets	Eastern N. Atl.; Western N. Atl.	Anonymous 1975; Brothers 1989; Carr et al. 1985
Pacific cod, *Gadus macrocephalus*	E+	Lost gillnets	N. Pac.	High 1985
Blue whiting or paur, *Micromesistius pautassou*	I	Plastic cups	Eastern N. Atl.	Anonymous 1975
Turbot (unidentified)	E+	Lost gillnets	N. Atl.: Newfoundland	Brothers 1989
Wolffish (unidentified)	E	Lost gillnets	N. Atl.: Newfoundland	Brothers 1989
Lumpfish (unidentified)	E	Lost gillnets	N. Atl.: Newfoundland	Brothers 1989
Redfish (unidentified)	E	Lost gillnets	N. Atl.: Newfoundland	Brothers 1989
Sculpin (unidentified)	E	Lost gillnets	N. Atl.: Newfoundland	Brothers 1989
Chum salmon, *Oncorhynchus keta*	E+	Lost gillnet	Western N. Pacific	DeGange and Newby 1980
Coho salmon, *Oncorhynchus kisutch*	E+	Lost gillnet	Western N. Pacific	DeGange and Newby 1980
Pink salmon, *Oncorynchus gorbuscha*	I	Blue plastic fragment	N. Pac.	Day 1988
Salmon (unidentified)	E	Gill-net fragments	N. Pac.	Jones and Ferrero 1985
Pacific pomfret, *Brama japonica*	I	Plastic fragments	N. Pac.	Day 1988
Longnose lancetfish, *Alepisaurus ferox*	I+	Vinyl and polyethylene fragments, rubber, plastic soft drink bottles	Western N. Pac.	Kubota 1990
Unidentified billfish (Istiophoridae)	E	Mass of trawl and gillnet	N. Pac.: Hawaii	Henderson 1988
Dolphin, *Coryphaena hippurus*	I	Nylon rope, bottle, packaging, plastic fragment	Western N. Atl.	Manooch et al. 1984
Unidentified tuna (Scombridae)	E	Mass of trawl net and gillnet	N. Pac.: Hawaii	Henderson 1988
Little tunny, *Euthynnus alleteratus*	I	Plastic packaging	Western N. Atl.	Manooch et al. 1985
Blackfin tuna, *Thunnus atlanticus*	I	Plastic bag, plastic fragments	Western N. Atl.	Manooch and Mason 1983
Yellowfin tuna, *Thunnus albacares*	I	Plastic fragments	Western N. Atl.	Manooch and Mason 1983
Wahoo, *Acanthocybium solanderi*	I	Plastic sheeting	Western N. Atl.	Manooch and Hogarth 1983
Unidentified barracuda (Sphryaenidae)	E	Mass of trawl net and gillnet	N. Pac.: Hawaii	Henderson 1984

APPENDIX 4. *Continued.*

Species	I/E	Material	Location	Source
Ulua, *Caranx* sp.	E	Mass of net	N. Pac.: Hawaii	Henderson 1988
Sea raven, *Hemitripterus americanus*	E	Lost gillnet	N. Atl.: Cape Cod Bay	H.A. Carr 1986
Sea robin, *Prionotus evolans*	I	Plastic pellets, polystyrene spherules	Eastern N. Atl.	Carpenter et al. 1972; Day 1988
Ragfish, *Icosteus aenigmaticus*	E	Lost gillnet	Western N. Pac.	DeGange and Newby 1980
Catfish (unidentified)	E	Lost gillnets	N. Atl.: Newfoundland	Brothers 1989
Dusky shark, *Carcharhinus obscurus*	I -- /E --	I: Packaging Ē: Packing straps	S. Atl.: South Africa	P.G. Ryan, personal communiation
Blacktip shark, *Carcharhinus limbatus*	I -- /E --	Ī: Packaging Ē: Packing straps	S. Atl.: South Africa	P.G. Ryan, personal communication
Great white shark, *Carcharhinus carcharias*	I -- /E --	Ī: Gumboot Ē: Plastic straps	S. Atl.: South Africa	P.G. Ryan, personal communication
Copper shark, *Carcharhinus brachyurus*	E --	Plastic straps	S. Atl.: South Afria	P.G. Ryan, personal communicaton
Spinner shark, *Carcharhinus brevipinna*	E --	Plastic straps	S. Atl.: South Africa	P.G. Ryan, personal communication
Sandbar shark, *Carcharhinus plumbeus*	E --	Plastic straps	S. Atl.: South Africa	P.G. Ryan, personal communication
Tiger shark, *Galeocerdo cuvier*	I+/E --	I: Packaging, fishing net, plastic sheeting Ē: Plastic straps	S. Atl.: South Africa	P.G. Ryan, personal communication
Shortfin mako shark, *Isurus oxyrinchus*	I --	Packaging	S. Atl.: South Africa	P.G. Ryan, personal communication
Smooth hammerhead shark, *Sphyrna zygaena*	I --	Packaging	S. Atl.: South Africa	P.G. Ryan, personal communication
Scalloped hammerhead shark, *Sphyrna Lewini*	I --	Packaging	S. Atl.: South Africa	P.G. Ryan, personal communication
Spotted raggedtooth shark, *Carcharias taurus*	I -- /E --	I: Packaging Ē: Twine	S. Atl.: South Africa	P.G. Ryan, personal communication
Salmon shark, *Lamna ditropis*	E	Gillnet debris	N. Pac.	Jones and Ferrero 1985
Manta ray (Mobulidae)	E	Monofilament line	Not identified	Waterman (cited in Wallace 1985)

APPENDIX 4. *Continued.*

Species	I/E	Material	Location	Source
Skate (unidentified)	E+	Lost gillnet	N. Atl.: Cape Cod Bay, Newfoundland	H.A. Carr et al. 1985; Brothers 1989
Dogfish, *Squalis acanthias*	E+	Lost gillnet	N. Atl.: Cape Cod Bay	H.A. Carr et al. 1985; H.A. Carr 1986
Crabs (crustaceans)				
American lobster, *Homarus americanus*	E+	Lost gillnets and lobster traps	N. Atl.: Cape Cod Bay	H.A. Carr et al. 1985; H.A. Carr 1986; Sheldon and Dow 1975; Breen 1990
Northern cancer crab, *Cancer borealis*	E+	Lost gillnets	N. Atl.: Cape Cod Bay	H.A. Carr et al. 1985; H.A. Carr 1986
Dungeness crab, *Cancer magister*	E+	Lost crab pots	N. Pac.	High 1976, 1985; Breen 1987
Red king crab, *Paralithodes camtschatica*	E+	Lost crab pots	N. Pac.	High and Worlund 1979; High 1985
Snow crab (unidentified)	E+	Lost crab pots	N. Pac.	High 1985
Tanner crab, *Chionoecetes bairdi*	E+	Lost crab pots	N. Pac.	Kimker 1992
Red rock crab, *cancer productus*	E	Lost gillnet	N. Pac.	High 1985
Kelp crab, *Pugettia producta*	E	Lost gillnet	N. Pac.	High 1985
Crab (unidentified)	E+	Lost gillnets	N. Atl.: Newfoundland	Brothers 1989
Squids (mollusks)				
Flying squid, *Ommastrephes bartrami*	I	Plastic line	N. Pac.	Day 1988

Minus (−), evidence of interactions based on rare or infrequent reports of isolated individuals.
Plus (+), reports of interactions more than infrequent and/or a regular pattern of occurrence. Entanglement reports usually reflect due to ghost fishing by derelict fishing gear.

9.

Ghost-Fishing Gear: Have Fishing Practices During the Past Few Years Reduced the Impact?

H. Arnold Carr and Jessica Harris

Introduction

Fish stocks are depressed worldwide. The resultant economic hardships on many fishermen have forced changes in fishing operations and effort. In general, gear use and losses in trap or pot fisheries appear to be increasing concurrent with a shift to more durable gear and designs. Modifications in trawl construction and operations have resulted in increased net damage and loss. Evidence from a number of locations around the world indicates that recent demersal gillnetting practices are leaving more gear per fishing unit in the oceans; in some cases this lost gear is heavily concentrated on productive fishing grounds. These changes in traditional gear types and fishing methods are certainly increasing the potential for loss of commercial and noncommercial species because of ghost-fishing.

With emphasis on New England area fisheries, this paper reviews fishing operations and gear construction for trap (or pot), mobile trawl, demersal gillnet, longline, jigging, weir, and pound net fisheries with regard to the ghost-fishing impact on marine fish, the benthos, and other maritime activities.

Crustacean Traps

Two trap fisheries are considered here. First is the lobster trap fishery in the Northwest

Atlantic from Atlantic Canada to the mid-Atlantic region of the United States, and second is the Dungeness crab fishery off the U.S. Pacific Coast, including Alaska. Several other crab fisheries that have gear loss and ghost-fishing problems but which are not addressed in this paper are the crab fishery in the Gulf of St. Lawrence, Canada, and tanner and king crab fishery in the Bering Sea, and the new king crab fishery in the Bering Sea area off Russia. Gear loss and ghost-fishing by fish traps are also not considered here. Along the U.S. Atlantic Coast, few fish traps are used in open water; however, a recent resurgence of interest in fish trapping may change that. Managers are concerned with this because it may lead to increased interfishery gear conflicts and resultant gear loss.

Lobster Traps

The Northwest Atlantic lobster fishery occurs in both state and federal waters along the northeast coast of the United States and Atlantic Canada. In the United States, the fishery is managed by both state and federal management plans. Traditionally, lobster traps were made of wooden laths secured to a semicircular or trapezoidal frame with several compartments. Funnels constructed of webbing lead the lobsters to a baited center chamber, or parlor, from which escape is difficult. Traps are set individually or in "trawls" [trawl is a colloquialism used in the

New England lobster fishery to refer to a string of traps]. A trawl, in terms related to stationary fishing gear, is the number of traps connected in series with a marking buoy at each end. Depth of set, duration of a set ("soak time"), and number of traps fished are factors that vary with season, location of the fishery, and size of the vessel.

In New England, the lobster fishery has undergone several dramatic changes. In the mid-1900s the fishing industry mainly fished during the period from late April to November. The fishing season now extends throughout the year. Larger and more seaworthy vessels, as well as improved weather forecasting and navigational equipment, are the most significant factors enabling year-round fishing. Although inclement weather precludes lobstermen from fishing for extended periods, when they are able to fish, high seasonal prices can make even a small catch profitable (e.g., ex-vessel price of $4.50/lb).

Synthetic materials have replaced wood and other natural materials in the lobster trap system. Lobster traps are now constructed from plastic-coated wire and plastic skids, replacing the traditional wooden skids on the bottom of the traps. Buoys are plastic instead of wood and even the stick that passes through the buoy is available in plastic. The cost of the plastic sticks is about three times that of the wooden sticks, and even though this may appear to be a deterrent for the use of the plastic stick, many lobstermen prefer it because it resists marine borers and is much less likely to unexpectedly break where the stick enters the buoy.

Many lobstermen have switched from wooden traps to coated-wire traps because they are lighter and last longer. In 1992, 77% of all lobster traps fished in Massachusetts waters were the coated-wire type (McCarron and Hoops 1991, 1992). An increasing number of lobstermen who use coated-wire traps and are not actively fishing during the winter are "wet storing" their traps at sea during the winter months. This situation sharply increases gear loss because it exposes the gear to winter storms and conflict with mobile gear. Mobile trawlers have tradition-ally expected certain fishing areas to be more accessible to them in winter months; however, wet storage is encroaching on these areas.

Federal law mandates escape vents, a device designed to release undersized lobsters, and a degradable panel, a mechanism to allow nearly all legal-sized lobsters as well as most other captured fish and crustaceans to escape from lost pots, are federally mandated on these coated-wire traps. However, a functional problem may exist with certain molded plastic escape panels that also contain the escape vent. The panel opening may be sufficient to allow the release of lobsters but may not be large enough to release larger fish that enter the trap. Although we do not have definitive information on the continued fishing impact of this gear, sea sampling programs and diving observations suggest low but consistent, by-catch of Atlantic cod and other finfish (Brad Chase, Massachusetts Division of Marine Fisheries, personal communication).

The trapping of lobsters in New England, both inshore and along the continental slope, has intensified during the past 20 years (Krouse 1994). In Massachusetts and Maine, 1,440,276 traps were set in 1971 and 2,500,924 were used in 1991. This increase, along with the previously mentioned changes in gear construction and storage, has placed the lobster fishery and the mobile trawl fishery in considerable conflict.

Traps may be lost through interactions with mobile fishing gear, storm damage, cut buoy lines, theft and vandalism, and senescence (Smolowitz 1978; Breen 1990; Kruse and Kimker 1993). During the past 3 years the New England area suffered several major storms that caused devastating gear losses to the lobster fishery. Although the storms were unexpected, the industry's change to fishing more gear and to storing gear at sea exacerbated the situation. After one of the storms, financial assistance under the Federal Emergency Management Act was provided to local fishers who lost large proportions of their gear. This aid has led fishermen to expect compensation for income and gear loss caused by severe weather. As one lobsterman recently said to a state extension agent: "This winter has been hard and I haven't been able to go out and tend the traps I left out. I not only lost income, but gear. Is there government money available to reimburse my loss?" (Charles Connors, Massachusetts Division of Marine Fisheries, personal communication).

The Massachusetts Division of Marine Fisheries (MDMF) surveyed a particularly large loss of lobster gear in and around Massachusetts Bay caused by a storm in October 1991. Traps, set in trawls of 10–25, were tangled together. Up to 5 lost trawls were observed by MDMF divers at one discrete site; entanglements involved from 2 to 24 traps along the entangled trawl lines. More than 90% of the traps surveyed were wire traps, and about 80% were in excellent condition. These traps were still ghost-fishing 6 months later. The species observed in the traps were primarily Atlantic cod and lobster. Although government financial assistance was available to help retrieve the traps, lobstermen made no attempts to recover the gear, fearing it was too severely tangled to be retrievable (unpublished report, Massachusetts Coastal Zone Management).

Quality and availability of the information necessary for estimating annual trap loss rates varies considerably. Anecdotal reports of trap loss rates run as high as 20%–30% per year (Smolowitz 1978). Along costal Maine, an estimated 5%–10% of 2 million lobster traps fished in 1992 were lost and never retrieved (Jay Krouse, Maine Department of Marine Resources, personal communication), which means that from 1 year alone (1992) there were 100,000–200,000 lost traps in Maine waters. Proposed regulations in the next amendment to the federal lobster management plan may create comprehensive maximum trap limits that will vary by region, and these restrictions may help reduce the incidence of trap loss by eliminating excess gear from the area. For example, one lobstermen reduced his loss rate to only 2% of his gear in a year by fishing 300 instead of his normal 800 traps (P. Caruso, Massachusetts Division of Marine Fisheries, personal communication).

Another problem relates to the disposition of old, unusable traps. Anecdotal reports from reputable lobstermen suggest that where landfills charge for disposal of old gear, some lobstermen respond by dumping their worn-out traps at sea. No definitive data are available on the frequency of this illegal dumping.

Pacific Crab Traps

Traps are used in the Pacific Northwest for shrimp (*Pandalus platyceros*), Pacific cod (*Gadus macrocephalus*), and crabs. Dungeness crabs (*Cancer magister*) are caught with two types of traps, circular steel traps covered with stainless steel wire mesh and rectangular steel traps coated with polypropylene mesh (Breen 1987). Both trap types have shallow entrance ramps equipped with metal gates or triggers that prevent crabs from escaping (High 1976). Crab trap losses are caused by the same forces that cause lobster gear losses, with the additional problem that float lines may be cut by wind-driven ice floes (Kruse and Kimker 1993). Rates of trap loss are normally between 10% and 25% (Kruse and Kimker 1993; High and Worlund 1979). One of the authors has observed derelict crab traps while surveying areas near Juneau, Alaska, with side-scanning sonar. These two Alaskan sites, which are not necessarily representative of other fishing areas, had a trap density between 6 and 10 traps/km^2. The traps appear to have a long life, remaining intact and potentially ghost-fishing for years (Breen 1990; High and Worlund 1979).

Current regulations in Alaska require some form of degradable escape mechanism. For all trap fisheries except for Dungeness crab, an 18-in. escape slit is cut horizontally into the side of the trap within 6 in. of the bottom and sewn up with untreated (degradable) cotton twine. In some fisheries use of a galvanic time-release device may be required (Kimker, personal communication). Dungeness crab traps, which are constructed of stainless steel mesh, have lids that are sewn on with untreated cotton twine. As in the eastern U.S. trap fishery, this will aid in decreasing the mortality of the target species caused by lost gear but does nothing to alleviate the accumulation of debris in the marine environment. The accumulation of long-lived gear in areas of high marine productivity has been shown to interfere with other fisheries and to adversely affect marine mammals, seabirds, and some species of fish (Kuzen 1990; Day et al. 1985).

Bottom Trawls

Although a number of bottom trawling techniques are in use around the world (i.e., otter trawl, beam trawl, twin trawl, pair trawl, and sometimes even midwater trawl), the focus of this review is on otter trawls. Most other trawl types are modifications of the otter trawl. For example, the beam trawl uses a horizontal beam or pipe to spread the mouth of the towed net, rather than otter boards or "trawl doors."

There is a great range of scale in otter trawl fisheries around the world. Within the United States, fishing vessels rigged for otter trawling range from small 10-m boats with one owner or operator to the huge (> 100-m) factory trawlers of the North Pacific (Low et al. 1985) whose nets could catch objects the size of the inshore vessels. The size of the trawler is usually dependent on the proximity of the fishing grounds to the home port, the sea state usually encountered, and the depth of the water to be fished.

Bottom trawl operations are basically the same regardless of the size of the vessel. Conical nets are dragged along the ocean bottom, the horizontal openings of the nets are maintained by doors, and vertical openings are maintained by floats on the tops of the nets and weighted chains and ropes on the bottoms of the nets. Plastic and steel are used to construct doors and floats, and hardened steel is used for cable and chain. Several primary materials are used to construct the webbing of the nets, nearly all of which are synthetic (e.g., nylon, polyethylene, polyamide, and Dacron). The twine diameter will vary from 2 mm to more than 5 mm, and the

twine may be doubled in the body of the net. The size of the meshes used in the body and cod-end of the trawls varies from fishery to fishery and may be as small as 1.75 in. (44 mm) for the shrimp fishery to 6.0 in. (152 mm) for the New England groundfish fishery to the 128-m meshes of the Gloria midwater trawl (Anonymous 1994). The aft end of the net is the cod-end, where the catch accumulates. The cod-end is usually constructed of a similar type of twine but frequently is protected with chaffing gear. The chaffing gear is often constructed of unraveled synthetic twine laced into the cod-end.

Trawling differs in the length of the individual tows, both within and among fisheries. Also, the duration and towing speed over ground (SOG) vary considerably. Tows for certain inshore species, such as shrimp or squid, may be between 0.5 and 1.5 h long. For groundfish, tow durations vary from 5 to 10 min for pollock off Alaska, to 3–4 h for cod or flatfish in the Northwestern Atlantic, to 12–16 h for flatfish along the western coast of the United States (Dan Erickson, University of Washington, personal communication). Speed will vary from 2 to 3.5 knots for most bottom trawls to 7 knots SOG for beam trawls off Europe (Ron Fonteyne, Belgium, personal communication).

Before 1980, most trawlers operated blindly; that is, while towing their nets along the bottom, they relied on experience to know what areas of bottom were safe and what areas were obstructed. Obstructions to trawl gear consist of large rocky areas, wrecks, or anything lying on the bottom that might snag the webbing of the net, causing it to hang up or tear (Fig. 9.1). Rocky bottoms often cause sections of a net to rip out, usually along the front end of the net in the wings, mouth, or belly. Trawlers usually carry extra sections of the net so that if an area is ripped it can be easily replaced. If the tear is minor, it may simply be mended. In the worst case a trawl may hang on a wreck. If the cables that attach the doors to the net are caught or if the heavy-duty hanging lines of the net are caught, the vessel may be forced to abandon the entire net. Very soft bottoms can also cause problems. For example, if a trawl digs in and fills up with mud it may become too heavy to lift and the entire trawl may be lost. Based on data collected by observers in 1983, Low et al. (1985) estimated that between 1954 and 1982, 668 nets or large portions of nets were lost in the trawl fisheries off Alaska.

In recent years technology has enabled trawlers to better delineate the bottom through sonar imaging. Navigation systems such as Loran and the satellite-based Global Positioning System (GPS) allow exact locations of inhospitable bottoms to be noted, and computers allow these locations to be compiled in databases and displayed on navigational plotters. Improved communications between vessels using cellular telephones and fax machines has also aided in the sharing of information on the locations of hangs or in changes in the known areas of fishable bottom. This increase in technology, along with other factors such as overcapitalization, has contributed to the continued decline of the most valuable fish species. As a result, commercial concentrations of fish have become harder to find and trawlers are taking greater risks fishing in areas with higher potential for rips, hangs, and gear losses. These rocky areas have many crevices in which groundfish, cod, haddock, and flatfish may hide. In New England, offshore trawlers, 20–30 m in length, are fishing inshore on very rough bottom that they never had fished before (David Dutra, personal communication).

This increase in the need to tow on rougher bottom has led to the development of more sophisticated sweeps at the front bottom portion of the trawl. Fishermen have changed this part of the trawl—the part that makes first contact with the bottom—to include large rollers or "rockhoppers." The latter, as the name implies, makes it possible to tow over very irregular bottom. Rockhopper sweep gear can pass over 1- to 2-m boulders without injuring the trawl net. Other sweep gear available permits fishing in areas with boulders even larger than 1–2 m.

The trawl gear will impact irregular bottoms, turning over cobble and larger rocks,

thus disrupting habitat that may serve as nursery areas for postpelagic fish larvae, juvenile fish, shellfish, and crustaceans (Lough et al. 1989). Fishing near or on very rough bottom or near or on old shipwrecks will result in gear loss. Trawlermen do this purposely as they realize that the risk results in greater catches. They design their operation so that at least two nets are aboard and ready to fish. They also have crews that are skilled in repairing torn nets. The object, as in most fisheries, is to keep actively fishing at all times, even when one net is under repair.

Advances in technology are also creating stronger, lighter materials, including ultrahigh molecular weight polyethylene, with which to construct otter trawls (Fiorentini and Hansen 1994). Spectra is one such material that is also very persistent when lost. The manufacturer of this material touts "If there is a negative side to Spectra it would be for the manufacturer. It lasts so long that the repeat business is very small" (Anonymous 1994). These materials permit trawl construction with much less drag, as much as 45% less in groundfish bottom trawls. What effect will this have? Much larger trawls can be towed with little or no change in vessel horsepower or wench power. More bottom can be covered per unit time. The trawls are expensive so it is unlikely, at least at present, that trawler captains would put their trawls on rough bottom or fish on or immediately next to serious bottom obstructions or hangs.

Once lost on the bottom, trawl net may have low ghost-fishing potential. The traditional twine of the trawl has a larger diameter, is more visible, and is more readily sensed by fish than is the monofilament line used in gillnets. If a lost net were suspended upward by associated flotation, to form a curtain rising from the bottom, it could present significant ghost-fishing potential. Typically, the net losses are small and the lost materials appear to form additional habitat for such organisms as ocean pout, wolfish, and cod as well as "substrate" for sessile invertebrates such as hydrozoans and sea anemones (Fig. 9.2). Underwater observations by the author, using submersibles and ROVs (remotely oper-

FIGURE **9.1.** Trawl net caught on wreck. (Photograph by Brian Sherry.)

FIGURE **9.2.** Trawl net caught on wreck—a habitat for hydrozoans (*Tubularia* spp.). (Photograph by Brad Sheard.)

ated vehicles), have shown that in deep water where sea floor currents are minimal, derelict trawl gear usually has an overburden of silt and is quite visible, posing little threat to marine life.

Trawl netting is also often found floating at, or near, the surface of the sea. Many of the synthetic twines are buoyant, and sometimes the twine buoyancy is augmented by the trawl floats, which may remain attached to major pieces of trawl webbing. It is well known that floating debris can attract a variety of pelagic species of fish and invertebrates (Hunter and Mitchell 1966). This webbing, although visible, may attract other marine species (e.g., sea turtles and seals) that can become entangled (see Laist, Chapter 8, this volume).

The small pieces of floating trawl webbing can be the result of accidental loss or of deliberate discarding. Too few programs exist that encourage or allow for dockside disposal of any fishing gear or pieces of fishing gear (R. Bruce, Massachusetts Division of Marine Fisheries, unpublished report, 1991). Even after the ratification of Annex V MARPOL 73/78 and the passage of the Marine Plastic Pollution Research and Control Act of 1987, organizational efforts to assist commercial fishing—at least on the U.S. Atlantic Coast—have been limited.

The lack of an organized effort to recycle fishing gear along the U.S. Atlantic Coast has had implications this year (1995). Recent changes in management regulations in New England waters stipulated an increase in cod-end mesh size from 5.5 in. (140 mm) to 6.0 in. (152 mm). Consequently, fishermen must buy new gear and get rid of the old. Some landfills refuse to take the old gear and some fishermen report that the old, now useless, cod-ends and nets are simply being dumped into the sea. According to MDMF, there are approximately 600 mobile gear licenses in the state. If each trawler has two cod-ends (a common backup practice) and each cod-end is constructed of 50 by 60 meshes of 5.5-mm double polypropylene twine weighing 75 lb, that would mean 90,000 lb of plastic will have been illegally dumped. While many captains will not toss cod-ends overboard, if 25% of all licensed fishermen did so, it would still result in approximately 23,000 lb of derelict trawl netting.

Gillnets and Enmeshing Gear

In the Northwest Atlantic, gillnets are commonly used to fish for demersal species in offshore waters year round and in inshore waters seasonally. The gillnet fishery usually involves vessels of more than 20 m fishing on a day-trip basis. This means that as weather permits, they set and haul their nets on consecutive days, leaving the nets to fish unattended overnight. In these fisheries, the gillnet is made of monofilament webbing with a 2-m vertical profile (between the floatline and the leadline) and is 91 m in length. In the Gulf of Maine, demersal gillnets traditionally are set in continuous strings or sets of 10–12 nets that total 914–1097 m. A single vessel generally sets five to six strings that total between 50 and 72 nets. On each end of the sets are 5- to 15-kg anchors attached to surface floats and reflectors. The mandated minimum mesh size for groundfish is about 15 cm (6 in.).

Our observations show that the number of nets used by many of the gillnetters has sharply increased during the past 10 years. A recent letter from the Massachusetts Gillnetters' Association to the Massachusetts Marine Fisheries Commission stated that many boats were fishing 150–200 nets. This represents a threefold increase in gear fished per unit vessel over this 10-year period. The increase is probably related to the precipitous decline of groundfish and the gillnetters' need to earn a living: to survive, they set out more gillnets.

In Atlantic Canada, gillnets are of two types, pelagic and demersal. They are used to target cod, turbot, herring, and mackerel (Brothers 1992). The demersal nets are made of monofilament webbing; the pelagic gillnets are made of nylon. The Canadian demersal gillnetter usually uses a gillnet that is 100 m long by 3 m deep. Between 3 to 20 gillnets are tied together to form a "fleet" (the U.S. equivalent of a "set"). The gillnet fishery in some locations is expanding, but this is in deeper water, to about 1500 m for Greenland turbot (Gerald Brothers, Canada Department of Fisheries and Oceans, personal communication). This expansion resulted in some initial heavy losses of gillnets because the fishermen were unfamiliar with the depths, currents, and bottom types.

In northern European waters, use of gillnets, or more specifically, enmeshing gears, has been expanding (Fahy et al. 1992). Presently, more than 50% of the total catch of cod in the North Sea is taken by gillnet (David Wileman, Danish Institute for Fisheries Technology, personal communication). There are several reasons for this increasing use of enmeshing gears; foremost is the need to increase effort to compensate for a decrease in the availability of groundfish. The smaller and less sophisticated vessels have been forced to seek fish on bottom not worked by the modern, efficient mobile trawl fleet. The two most common nets in use by the smaller fleet are gillnets and tangle nets. Both nets are single layers of webbing; however, the tangle nets are more loosely hung so they tend to entangle rather than gill the target species. While gillnets target salmon, spurdog, and gadoids, tangle nets target crustaceans, turbot, anglers, and rays.

Gillnets and tangle nets are used widely in the Northeast Atlantic and Mediterranean. Wreck fishing is widespread throughout the Atlantic regions of Europe. In this style of fishing, gillnets are set directly in the vicinity of wrecks, which act as artificial reefs, increasing the danger of snags, hangs, and net losses. Fahy (1993) reported that in France vessels engaged in enmeshing fisheries are artisanal types ranging between 6 and 24 m in length. These vessels primarily exploit inshore coastal areas, using strings of gill nets as long as 25 km. Greece has a large coastal gillnet fishery of both recreational and commercial gillnetters. Gillnets and trammel nets (triple-walled gillnets that entangle rather than gill) are used by more than 20,000 vessels along 15,000 km of coastline to catch many finfish and shellfish species. It is not unusual for these vessels to use sets as long as 4 km in the commercial fleet (the recreational fleet is limited to 200-m sets). The frequency of net loss and degree of its impact has not been investigated for this area; however, the shear quantity of gear being used must create a fair quantity of ghost gear.

Derelict Gillnets and Enmeshing Gear

Most of the directed research on the impact of demersal gillnets has taken place on the U.S. Atlantic Coast. The National Marine Fisheries Service (NMFS), the National Undersea Research Program (NURP), and MDMF undertook a 3-year study that investigated the magnitude and impact of "ghost" gillnets in two traditional gillnet areas located in the Gulf of Maine (H. A. Carr et al., 1985; H. A. Carr and Cooper 1987). This study involved gear that was located more than 12 miles offshore and in waters between 50 and 120 m deep. This study found a very low density of derelict gillnets that fished at about 15% efficiency.

One derelict net that was found in 1984 was surveyed in the winter and spring of 1985 with a submersible and again in 1986 with the use of a ROV (H. A. Carr 1988). This study enabled scientists to observe the net when many important groundfish and other species were present. The results of these surveys indicate that a lost net maintains its vertical profile as a result of the positive buoyancy of the floatline and floats. The rate of continued fishing of the ghost net is primarily a function of the maintained vertical profile, visibility to the fish, and abundance of finfish species.

A recent study investigated experimental nets set in shallow inshore waters ($\approx$ 20 m) (Carr et al. 1992) (Fig. 9.3). The catch in this study differed considerably from the catch in the nets found offshore in previous studies. In inshore areas species diversity and seasonal abundance are typically higher, and this difference was reflected in our observational data. The study also provided more information on the fate of certain entangled species, such as predation by cunners (*Tautogolabrus adspersus*) on lobsters (Fig. 9.4).

The impacts of increasing gear use per vessel and the targeting or concentration of gear on wrecks are not specifically known. However, several incidents have suggested that concern is warranted. The first example is off southern New England where gillnet gear has been left unattended and preempting the use of other gear, such as longlines, lobster trawls, or jigs (Wayne Iayno, commercial fisherman, personal communication). One 30-m piece of gillnet retrieved inadvertently with some lobster traps had about 90 kg of Atlantic cod entwined, all in a state of decay. A second incident off the coast of Ireland resulted in the near-loss of expensive, remote underwater observation equipment and jeopardized the lives of several scientists. This incident occured during an expedition to survey the *Lusitania* (Robert Ballard, Woods Hole Oceanographic Institute, personal communication). The shipwreck is covered with nets. These are mostly enmeshing nets, but some trawl netting was also present. The enmeshing gear is reportedly deliberately placed on the wreck as local fishermen risk their older, worn gear in attempts to catch the concentrations of pollack and other fish around the wreck (R. Slayter, Delta Oceanographic, Oxnard, CA, April 1994, personal communication).

Demersal Longlines

Longline gear has one main line stretched on the sea bottom between two anchored buoys. Longline gear used in the northwest Atlantic

FIGURE 9.3. An Atlantic cod caught in experimental gill net. (Photograph by Peter Auster.)

FIGURE 9.4. Moribund lobster enmeshed in gillnet. Fish, probably cunners (*Tautogolabrus adspersus)*, have eaten through soft segments of the tail. (Photograph by Arnold Carr.)

typically has about 1000 single hooks attached to the mainline spaced 6 ft apart. Traditionally these hooks are baited, but in some fisheries artificial baits or lures are either in use or being investigated. The once-popular longline fisheries along the Northwest Atlantic Coast have been reduced and scattered since the 1940s. A resurgence of interest in longlining is developing as management institutions such as the New England Fishery Management Council consider alternatives to trawling for groundfish. Longlining is also being considered for use in deepwater fisheries off Canada (Brothers, personal communication).

Demersal longlining is generally considered selective in most groundfish fisheries and is believed to have little ghost-fishing potential. The first reason for this is that typical longlines are well tended. Gear is set out for only a few hours at a time with the vessels nearby; thus, gear loss is minimal. Second, longline design and operation are straightforward, limiting potential losses from mechanical failure, accidents, or mistakes. Lastly, a longline loses its fishing power rapidly as baits are lost and hooks settle onto the sea floor. One longline was observed during the submersible survey for ghost gillnets off New England (Carr and Cooper 1987); it was overgrown by benthic life and did not have any hooks showing or any fish or evidence of fish associated with it.

Jigging

Jigging is an ancient fishing method that, until the past 10–20 years, has been undertaken

TABLE **9.1.** Summary and description of gear loss problems for each major type of gear.

Gear type	Location	Major cause(s) of loss	Degree of impact on fish and shellfish	Susceptibility to user conflict	Source
Trawls	United States (coasts) Canada Australia U.S. Atlantic Coast	1, 4	a, 1	Medium to high	High 1985; Low et al. 1985
Traps	U.S. Atlantic and Pacific Coasts Caribbean Canada (both coasts)	1, 2, 3, 4	b, 2	High	Breen 1990; Breen 1987; High and Worlund 1979; Kruse and Kimker 1993; Muir et al. 1984; Smolowitz 1978
Gill nets	U.S. (all coasts) Australia East coast—Great Britian, Irish coast, North and Baltic Seas Atlantic Canada	1, 2, 3	Inshore: c, 3 Offshore: b, 2	High	Breen 1990; Brothers 1992; Carr 1988; Carr et al. 1985 ;High 1985; Wileman (personal communication)
Longlines	U.S. (Northwest Atlantic and Northeast Pacific) Canada	2, 3 [degree of operational loss is unknown]	a, 1	Low	High 1985
Weirs	New England	3	?	Low	

Major causes of loss:
  1. Operational
  2. Gear conflict
  3. Weather and storms
  4. Senescence
Impacts on fish and shellfish:
  a. Low for targets   1. Low for nontargets
  b. Medium for targets   2. Medium for nontargets
  c. High for targets   3. High for nontargets

with a single handheld line. Advances in the fishery, aside from better navigation and bottom sounders, include the use of monofilament line and jigging machines. There are two types of jigging machines. The first type has a single line with a jig or jigs on the end, and the machine is used to keep it active and facilitate easier hauling back (up) of the fish. The second device has a loop of continuous line with secondary lines with hooks or jigs spaced periodically on the loop; the machine is programmed to pay out and retrieve the loop such that the jigs are fished in a specific depth range.

Jigging is often used to catch target species located on the bottom of the sea. Productive sites for jig fishing are becoming more difficult to find as fish stocks decrease. Conflicts between jig fisheries and other gear types (especially gillnets) have increased as they are forced to fish the same grounds or wrecks. Lost monofilament line and jigs have been

caught on other types of fishing gear and on rough bottom. This lost gear has not been observed to continue fishing. Near shore, especially in very shallow waters, monofilament line from all sources has been a hazard to birds, turtles, and divers.

Weirs

Fish weirs are structures made of poles or floats that shape a long horizontal leader of netting that usually ends with a collecting net that can be collapsed (pursed up) to concentrate the accumulated catch. The structures are close to shore. The overall length of the net will differ by area but ranges from 30 to 300 m long or longer. The principal length is represented by the horizontal leader that guides fish into the collecting net.

Most weirs are maintained frequently,

sometimes daily. They are also very accessible to those fishing and to public observation. Little impact has been reported relative to incidental fishing or ghost-fishing. In New England, in the past 15 years, there has been one incident of a marine mammal becoming entangled in the leader of a fish weir.

Conclusion

In summary, the six gear types (fish traps or pots, mobile trawling, demersal gillnets, demersal longlines, jigs, and fish weirs) vary as far as ghost-fishing impact (Table 9.1) The rate of loss, cause of loss, and degree of impact are all subject primarily to the characteristics of the gear and the fishing operations which use that gear.

Changing patterns of fisheries operations, dwindling target shocks, advances in equipment handling and materials technology, solid waste disposal limitations, and loss reimbursement programs all contribute to the potential for loss or discard of fishing gear in the fisheries of New England. Pots, traps, and gillnets clearly have the greatest potential for ghost-fishing. While longlines, trawls, jigging systems, and weirs may entangle individual animals of a variety of species, they have much lower ghost-fishing potential than lost or discarded traps and gillnets. Generally, the changes in modern fisheries tend to increase gear conflicts, promote greater risking of gear, and expand fishing grounds. Each of these trends contributes to the ghost-fishing problem. To answer the question posed in our title—No, fishing practices in recent years have not reduced the potential impacts of ghost-fishing.

PART II
The Sources and Solutions to the Marine Debris Dilemma

SECTION III
The Socioeconomics of Marine Debris

SECTION III
The Socioeconomics of Marine Debris

Introduction

Rational development of solution strategies for large-scale environmental problems involves decision making based on reliable social and economic information and analysis. This section includes some of the latest thinking in economics and sociology relevant to crafting and justifying realistic marine debris solutions. The following introductory observations are presented to assist noneconomists in appreciating the importance of this socioeconomic research and analysis.

Ultimately, all decisions are value judgments. Individuals assign "values" to the outcomes of any decision or event, however important or trivial, on the basis of their personal knowledge, interests, and obligations. This may be done through conscious calculation or it may be a matter of "feelings" or "hunches." Comparison of the values assigned to outcomes permits a rational decision to be made. The outcome with the highest perceived valuation generally will be pursued.

Government policy, at least in theory, is determined by the aggregation of individual valuations across society. If the sum of valuations is positive for a particular decision or policy, then it will be a net benefit to society. Ideally, while recognizing that not all individuals will benefit, the outcome with the highest net benefit becomes the government's policy objective.

In their study of these phenomena, economists seek a common currency in which to measure and compare the "value" of all choices. Typically, these choices are to buy, sell, or trade goods or services and are made in the context of a market using money as a standard medium of exchange. The assignment of monetary value through markets greatly aids the economist in analyzing the relative importance of choices, both by providing a universal scale of value and by assigning that value through actual transactions.

If a monetary value can be assigned to a particular decision or event (e.g., a marine debris impact), then it may be relatively straightforward to determine if a particular outcome will be a cost or a benefit to an individual. For example, a fisherman discards his wornout nets at sea for years, avoiding the costs of storing, handling, transporting, and transferring this material to the port or landfill for proper disposal. One night he runs over some waste net at sea, causing the bearings on his propeller shaft to seize—he loses a number of fishing days, pays for a marine tug service as well as new bearings and shipyard time. For this individual, the costs avoided (benefits) and incurred in this trade-off are known or can be estimated. The primary cost–benefit analysis subtracts the out-of-pocket expenses and lost income caused by the vessel entanglement from the total costs avoided by dumping at sea. If the result is positive, he continues dumping his waste nets at sea. If it is negative, he ceases dumping at sea.

Implicit in this simplified decision model is the assumption that the outcome of the individual cost–benefit analysis is the same for all fishermen who dump their nets at sea (the society of fishermen). In truth, every conceivable outcome will be experienced, leaving the individual fisherman to ponder whether giving up the benefit (cost avoidance) of dumping at sea will actually reduce the probability of future damage to *his* boat. The cost–benefit calculations for those fishermen who have never been disabled by debris will suggest that their best interests are served by continued dumping. Should the aggregate of all fishermen's experiences in this example be negative (i.e., the costs of vessel entanglement exceeds the benefits of net dumping), a rationale role for government policy and regulation is identified. To protect the society of fishermen from the greater aggregate costs of vessel disablement by debris, government would need to use its authority to ensure that the members of that society incur the lesser aggregate costs associated with not dumping their wastes at sea. Unless all are bound to incur these costs, few will, and the common good will suffer. This is a simple example of Hardin's (1968) tragedy of the commons.

If we consider our example beyond the interests of fishermen, to the interests of society in general, a suite of other costs come into play. Some costs are measurable through markets, such as the impact of ghost-fishing on the long-term availability and price of seafood, and some are not, such as trashed beaches or the injury and death of entangled seals and turtles. Where there is no market operating to reflect monetary valuations, the valuation of decisions or events is not an obvious process. Factors influencing individual and societal perceptions of value may be many and complex. Economists are developing techniques to assess the value of nonmarket entities to compare the costs and benefits of, for example, government policy choices regarding public expenditures to control or reduce marine debris. At present, the economic impacts of marine debris and the public's willingness to pay to control them are very poorly measured.

Sociologists and economists are particularly interested in the factors influencing perceptions of value and how they manifest themselves in the formation of social norms. When a segment of society adopts a certain world view or belief, it makes decisions (behaves) reflecting that view (or belief), occasionally in contradiction to the expectations of sociologists or economists. These expectations are based on their current understanding of how decisions are made (their paradigms). It is commonly observed that disciplinary paradigms are subject to shifts, usually driven by contradictions between observations and theoretical expectations. Existing economic theory predicts people will litter and will spend little to control marine debris until they personally incur some cost. At the same time, emerging public behavior (social norms) shows people increasingly willing to incur costs to prevent marine debris while receiving little or no direct benefit. This contradiction has generated significant academic interest in the socioeconomics of the marine debris issue. It seems likely that the cost–benefit calculus associated with marine debris impacts and solutions involves poorly understood social and behavioral phenomena. The papers in this section pursue improved techniques to help resolve this contradiction and sharpen the focus of public marine debris policy.

This section provides thoughtful treatment of the factors influencing valuation, the measurement of value, and the relevance of both to solving marine debris problems. In the first paper, Sutinen develops a new socioeconomic model that explains apparent successes in the use of "moral suasion" to influence public response to environmental issues. In explaining the economic role of moral suasion, Sutinen contributes to resolving our contradiction between theory and observation. Kirkley and McConnell give us a primer on the economic way of thinking about marine debris so we are able to appreciate the relevance of cost–benefit considerations. Smith, Zhang, and Palmquist report the results of a preliminary assessment of people's willingness to pay for clean beaches (a nonmarket valuation), which suggests that willingness to pay is positive and perhaps substantial. Their results also provide further clarification of the marine debris cost–benefit calculus. Finally, Laska offers a conceptual framework for considering the development and effective application of actions to control marine debris.

10.
A Socioeconomic Theory for Controlling Marine Debris: Is Moral Suasion a Reliable Policy Tool?

Jon G. Sutinen

Introduction

In 1987, marine debris gained national prominence in the United States when 30 senators sent a letter to President Reagan calling for action to address the problem of plastics in the marine environment. The senators' letter, signed by Democrats and Republicans, liberals and conservatives, identified three specific policy measures for mitigating marine debris. They endorsed policies that included use of (1) degradable materials, (2) bounties and incentives, and (3) public education campaigns. Seven years later, the actual policy measures primarily consist of public education[1] and dumping prohibitions.[2]

Current policy does not include bounties or other incentive measures to encourage the retention and retrieval of marine debris, as requested in the senators' letter.[3] This may be

a serious oversight. Congress has demonstrated a preference for incentive-based policies. For example, the Clean Air and Water Acts make extensive use of tradable pollution rights. Incentive systems also are based on sound economic theory and supported by a good amount of evidence. Tradable rights to pollute are proven mechanisms for controlling pollution and minimizing the economic burden on society. Deposit programs are cost-effective ways to control littering.

Is moral suasion, such as public education campaigns, an effective policy tool for achieving compliance with MARPOL Annex V? Or, should moral suasion methods be discarded and replaced with other approaches? If kept as a policy element, should moral suasion be a major or minor part of the policy package? These are the questions this paper addresses.

In the next section we review the conventional economic theory of regulatory compliance and derive implications for marine debris policy. Following this review, the basic theory is extended to account for the influence of moral suasion on individual behavior. The enriched theoretical framework is used in the last section to identify ways to strengthen existing programs and determine what policy approaches are needed to significantly reduce marine debris.

---

[1]Public education, as a form of moral suasion, attempts to persuade people to use proper forms of disposal (Bruner 1990; Debenham 1990; Friday 1990; Humphreys 1990; Wallace 1990).

[2]The dumping prohibitions under U.S. and international law forbid disposing plastics and other materials in specific areas of the marine environment. MARPOL Annex V sets specific discharge prohibitions for plastics and other materials in and outside special areas.

Use of degradable materials is required for some fishing gear, and an increasing number of the items that end up as marine debris are being made from degradable materials.

[3]Incentive schemes for controlling marine debris are discussed by Augerot (1988), Jarman (1988),

---

and Sutinen (1988) in papers presented at the Workshop on Fisheries-Generated Marine Debris and Derelict Fishing Gear (Portland, OR 1988).

 162 Jon G. Sutinen

Table 10.1. Discharge violations under MARPOL Annex V.

Year	Number of violations	Number of civil penalties	Violations without penalties	Percent violations without penalties	Total value of civil penalties
1990	92	—	92	100%	—
1991	121	12	109	90%	$ 39,400
1992	179	55	124	69%	$595,000
1993	183	29	154	84%	$301,300
Totals	575	96	479	83%	$935,700

Basic Theory

Conventional economic theory assumes that individuals base their trash disposal choice on the costs of their disposal options. Specifically, potential dischargers of marine debris are assumed to chose the option that minimizes the cost of discharging their waste (Kirkley and McConnell, Chapter 11, this volume). To keep the discussion simple and to focus on the essentials of the regulatory compliance problem, we consider two options only: legal and illegal disposal of waste generated at sea.[4] Legal disposal conforms to Annex V provisions, and illegal disposal does not.

In this simple model of compliance with Annex V, an individual is assumed to dispose of trash legally if

Legal disposal cost < illegal disposal
cost + expected penalty   (1)

Otherwise, the individual disposes illegally at sea.[5] The model considers only direct, tangible costs associated with the two disposal

[4]Although this analysis concerns waste generated at sea, the same framework applies to terrestrial-generated waste such as from combined sewer overflows.
[5]This model is based on Becker (1968), the first formal theoretical framework for explaining illegal activity. Following Bentham (1789), Becker argues that criminals behave basically like all other individuals in that they act in their own best interest. In Becker's model, an individual commits an illegal act if the expected gain from acting illegally exceeds the gain from engaging in legitimate activity. Gains and losses are measured in terms of the direct and tangible consequences of a given action. See Heineke (1978) and Pyle (1983) for surveys of the theoretical models used in the economic literature on criminal behavior.

options. Disposal cost includes not only the out-of-pocket costs for a bag or other container, but also the inconvenience of storing trash at sea and carrying it to shore. The expected penalty is the product of the probability of detection and sanction and the monetary value of the sanction.

Under the Civil Penalty Procedures used by the U.S. Coast Guard to enforce MARPOL, recommended penalties for the first offense range from $10,000 to $15,000 per violation per day; the maximum penalty is $25,000 per violation per day. From 1990 through 1993, fewer than 600 violations of MARPOL V were detected by the U.S. Coast Guard (Table 10.1). Civil penalties were assessed in fewer than 20% of the cases. The average civil penalty was just under $9800. On average during this 4-year period, the Coast Guard detected 144 violations and issued 32 penalty assessments per year.

Because no estimate of the actual number of MARPOL Annex V violations exists, we cannot derive from these data what fraction of actual violations is being detected. We have determined neither the type of MARPOL Annex V violations nor the categories of vessels involved in the infractions. The most we can conclude from these data is that, on average, if one is cited for a violation, there is at least an 80% chance of not having to pay a civil penalty. If cited and a penalty is assessed, the expected penalty is approximately $10,000.

The framework developed in this section has a straightforward prescription for policy. If the typical cost of illegal disposal (of just throwing trash overboard) is less than the cost of legal disposal (of bagging the trash and carrying it to shore), then the expected penalty will have to be large enough to offset the difference between the costs of legal and illegal disposal to deter dischargers from illegally disposing their trash at sea. Because the probability of detection and sanction is very small in practice, the policy should be to apply very severe sanctions, sanctions large enough for the product of their monetary value with the low probability to be larger than the difference between legal and illegal disposal costs.[6] However, this is generally not feasible. The courts are not willing to mete out sanctions perceived as excessively severe. Rather, sanctions that fit the crime, as measured by the illegal gains realized or the social harm caused, are preferred. Hence, the sanctions for violations of marine debris law will generally be modest and not act as an adequate deterrent to illegal disposal at sea unless the probability of detection and sanction is significantly large. Producing a significantly large probability of detection and sanction requires extensive monitoring and surveillance, a very costly and unlikely prospect for controlling marine debris.

The model can be modified to include incentives. For example, a deposit system would provide a monetary reward for disposing of waste legally. The behavioral rule, revised to include the reward for proper disposal, becomes dispose of trash legally if

Legal disposal cost − incentive reward <
illegal disposal cost + expected penalty   (2)

Otherwise, the individual disposes illegally at sea. An incentive policy will be effective in securing compliance with Annex V only if the incentive reward for proper disposal is greater than or equal to the difference between legal and illegal disposal costs (in-

cluding the expected penalty). Even if incentives were an element of marine debris policy, it is unclear whether feasible levels of incentives would be large enough to make a significant contribution to controlling marine debris. This is an empirical question for which there is yet no evidence.

In summary, basic compliance theory has policy prescriptions very different from current marine debris policy. Because severe sanctions or extensive monitoring and surveillance are not likely feasible, deterrence from enforcement programs is not expected to make large-scale contributions to controlling marine debris. We investigate next the theoretical basis for moral suasion policies.

A Model with Moral Suasion

Michael Crichton (1988) tells the story of his search for egg-laying leatherback turtles along the east coast of peninsular Malaysia. After several nights waiting, he discovers one of the giant turtles laying her eggs near his hotel. While Crichton watches the turtle's slow and clumsy effort to dig a pit for her eggs, a vacationing family arrives. The parents take flash photographs of the children each taking a turn sitting atop the back of the turtle. After several minutes of the family's taunting the creature, a local Malay boy arrives and explains to the family what the turtle is doing, how she had laboriously struggled up the beach, turned around to face the ocean, how long it had taken her to dig her pit, the effort it would take to lay her eggs, and the many hours she would lie there, exhausted, trying to find the strength to struggle down the beach and return to the surf by daybreak. The family listened in silence, their young son got off the turtle's back. Children were encouraged to touch the turtle's shell and make peace with the great creature. With growing respect, the family stepped back from the pit. Once they understood what was happening, they became sympathetic and understanding, stopped harassing the turtle, and quietly drifted away.

---

[6]The Coast Guard's Civil Penalty Procedures recognize this in part: "Because of the special difficulty in detecting and enforcing MARPOL regulations at sea, stiffer penalty levels are warranted for discharges in violation of MARPOL. MARPOL discharges within the waters of the U.S. Exclusive Economic Zone, and discharges within Special Areas, have higher recommended penalty levels. In general, 'Maximum Level' penalties should be recommended for any of these discharges, unless strong mitigating factors exist, in which case lower level penalties may be considered." (p. 3)

In this example, the Malay boy's explanation persuaded the family to stop their environmentally undesirable behavior. Simple education functioned as moral suasion, i.e., inducing socially desirable behavior by shaping or appealing to, or both, people's personal values. Education, as a moral suasion method, provides people with information which, if successful, shapes their attitudes and values and induces socially desirable behavior. The general operating hypothesis is that information influences personal values and personal values influence behavior (Fishbein and Ajzen 1975).

The paradigm commonly used in economics to explain and predict behavior (especially the theory used for policy analysis) makes little allowance for moral suasion and personal moral values. Most contemporary economic theory typically either ignores the influence of moral considerations or, in the extreme, denies moral factors have an influence on economic behavior (Hausman and McPherson 1993). This raises the question whether moral suasion methods are ill founded, or whether the theory is incomplete.

Morality and moral norms clearly influence economic outcomes. The evidence supporting this proposition is substantial.[7] For example, a large number of experiments have shown that people do not act as free riders when given the opportunity. Instead, many people persist in investing substantial proportions of their resources into public goods despite conditions designed to maximize free riding. In experiments of repeated prisoner dilemma games, more than half of the subjects cooperate without being coerced or paid. Several experiments have shown that many people return lost wallets to their owners with all the money inside. We also witness anonymous contributions to charity—above and beyond what tax incentives can explain. And, it is customary for people to leave tips in restaurants in distant cities which they never expect to visit again.

The available evidence suggests moral

norms may be a significant factor influencing compliance with marine debris regulations. It seems reasonable to assume that for a recreational boater, in most circumstances at sea, the probability of being detected discharging trash overboard is near or at zero: the boater can wait until the authorities are out of sight before discharging at sea. There appears to be no significant deterrent to illegal discharge, and disposing trash at sea is less costly and more convenient, providing boaters with a real incentive to discharge trash illegally. Yet the evidence indicates discharging trash at sea is not very common. The Recreational Boat Owners Survey asked boaters how frequently trash is discharged into the water for a variety of reasons (including safety, heavy seas, high winds, accidents, or children playing). Of 540 total responses, 38% admitted trash was discharged into the water at least once during the previous boating season. Also, 43% of the respondents claimed seeing other boaters discharge trash into the water at least once during the season. These statistics indicate that about 60% of recreational boaters never allow trash to be discharged into the water, intentionally or unintentionally. Because deterrence is minimal and perhaps nonexistent in many cases, compliance with MARPOL Annex V regulations is likely based on moral considerations.

Morality does play a role in economic analysis; it just has not received much attention by contemporary economists. In contrast, our economist forefathers gave morality due attention. Adam Smith (1759), the father of economics, explicitly portrays human economic motivation to be multidimensional, arguing that psychic well-being is based on acting morally and receiving the approval of others, as well as enhancing wealth. For the intrinsic motivation influencing behavior, he imagined an "impartial spectator" within each of us, with which we "scrutinize the propriety of our own conduct." Nineteenth-century economists commonly account for moral sentiments in their writings, and contemporary economists do not completely ignore morality. Some prominent figures in the profession, including Akerloff, Harsanyi, Hirshman, Margolis, Schelling, Sen, and Sti-

---

[7]For reviews of the evidence see Etzioni (1988, Chap. 4), Frank (1988), Mansbridge (1990), and Thaler (1991).

gler, have examined the consequences for economic outcomes of morality (Hausman and McPherson 1993). The fact remains, however, that morality is a foreign element to most contemporary economic analysis.

Outside economics, recent research in psychology and sociology emphasizes the importance of socialization processes in affecting behavior. Compliance with rules and regulations is hypothesized to be related to both the internal capacities of the individual and external influences of the environment, where the socialization process is the linkage between the individual and society. There are two leading psychological theories to explain how socialization processes work with respect to compliance behavior: cognitive and social learning. Cognitive theory focuses primarily on the individual and stages of development (Kohlberg 1969, 1984; Levine and Tapp 1977; Tapp and Kohlberg 1977). According to cognitive theory, the key variables determining compliance are the individual's personal morality and level of moral development.

Social learning theory, on the other hand, focuses primarily on the conditioning effects of the environment (Aronfreed 1968, 1969; Bandura 1969; Mischel and Mischel 1976; Akers et al. 1979; Akers 1985). According to social learning theory, the key variables determining compliance include peers' opinions and the extent of social influence on individual encounters.

The sociology literature contains two basic perspectives on compliance: instrumental and normative (Tyler 1990). Similar to Becker (1968), the instrumental perspective assumes individuals are driven purely by self-interest and respond to changes in the tangible, immediate incentives and penalties associated with an act. The key variables determining compliance are the severity and certainty of sanctions.

The normative perspective emphasizes what individuals consider just and moral, instead of what is in their self-interest. Individuals tend to comply with the law to the extent that they perceive the law as appropriate and consistent with their internalized norms. The key variables determining compliance in the normative perspective are individuals' perceptions of the fairness and appropriateness of the law and its institutions.

In summary, the literature identifies the following factors determining compliance: (1) potential illegal gain, (2) severity and certainty of sanctions, (3) individuals' moral development and their standards of personal morality, (4) individuals' perceptions of how just and moral are rules being enforced, and (5) social environmental influences.[8] Our current task is to develop a theoretical model consistent both with this literature and with basic principles of economics.

Our theoretical development adopts the perspective of Adam Smith on individual behavior. To incorporate intrinsic motivation in the model, we imagine Smith's "impartial spectator" operating within the individual to scrutinize and evaluate the propriety of his or her conduct.[9] Because proper conduct depends on whether an individual acts morally, we must characterize the nature of moral acts.

While there is no universal definition of moral acts, Etzioni (1988) identified several characteristics that are generally agreed. One of the primary characteristics of moral acts is that they are *intrinsically motivated*, involving nonmaterial rewards internal to oneself. That is, internal satisfaction is realized independent of extrinsic consequences, such as whether others know about such behavior. A second characteristic of moral acts is that *sacrifice* and the denial of pleasure (e.g., doing penance, fasting) in the name of moral principle are often involved. An implication of this is that individuals will sacrifice income or incur costs to carry out a moral act. A third characteristic is that moral acts often concern *intentions and processes,* not outcomes. Unlike consumptive pleasure, moral satisfaction can be the result of taking proper measures, regardless of the outcome. To the extent that

---

[8]Which of these variables are significant determinants of compliance with marine debris regulations is ultimately an empirical issue.
[9]How morality should be modeled is not without controversy. In our view, how it should be modeled is ultimately an empirical issue.

moral acts are concerned with the end results, how the result was attained is significant. For example, stealing a candle to light in church would not be morally satisfying to most people. A fourth characteristic is *symmetry*: the standard defining morality is applied equally to all people under comparable circumstances; otherwise, the moral dictum is arbitrary. Each of these characteristics has significant implications for our model of compliance behavior and, it turns out, for compliance policy.

In addition to wealth enhancement, Smith recognized social influence as another form of extrinsic motivation affecting behavior.[10] Social influence plays a significant role in everyday social exchange, often taking the subtle forms of ostracism or withholding of favors.[11] Like enforcement authorities, peer groups can reward and punish their members, either by withholding or conferring signs of groups status and respect or, more directly, by channeling material resources toward or away from a member of the group. The available evidence supports the hypothesis, showing that a given individual is more noncompliant the more his community and peer groups are noncompliant (Vogel 1974; Geerken and Gove 1975; Witte and Woodbury 1985; Kuperan and Sutinen 1994).

Social influence and morality are closely linked. The symmetry characteristic of moral acts implies that the standards used to judge one's own behavior are used to judge others' behavior. Therefore, the moral principles on which individuals base their own behavior are also the basis for the social influence they exercise. The more widespread a common moral obligation is in the population, the stronger the social influence to conform with it is expected to be.

To account for moral sentiments and social influence, we assume that disposing of trash at sea in violation of Annex V is perceived by the individual and his peers to be contrary to their personal moral values. This causes the individual to experience guilt and other negative intrinsic feelings we call psychic cost. If his peers know of his illegal disposal, his social reputation will suffer, potentially subjecting him to ostracism or pressure to dispose of his trash legally. We call this consequence social influence cost. Combined with the factors in the basic model (all of which relate to wealth enhancement), the behavioral rule for disposal becomes dispose legally if

Legal disposal cost < illegal disposal cost + expected penalty + psychic cost + social influence cost   (3)

Otherwise, dispose the trash illegally at sea. The nature of the context in which individual decision makers operate will affect these costs. Recreational boaters, commercial fishermen, and freighter operators, for example, will face different disposal costs and likely have different values with respect to pollution in the marine environment. Individuals responsible for terrestrial-generated waste that is discharged into the marine environment, e.g., from combined sewer overflows, may not appreciate the impact they are having on the marine environment and therefore face no psychic and social influence costs with their decision.

We now have a rudimentary framework to explain how an individual's and the community's norms of behavior influence compliance. As it stands, the framework is not very useful for policy analysis because the norms of behavior seem to be exogenously determined and given. There is evidence to suggest, however, that individuals' moral and social standing may be closely linked to regulatory policy and practices.

The compliance literature recognizes two types of intrinsic motivation or obligation (Tyler 1990). One type is related to the individual's desire to behave according to his sense of personal morality, i.e., an intrinsic obligation to follow one's own sense of what is right or wrong; the other type is related to

---

[10]Concern for one's social reputation is recognized in modern economic analysis as a motivation important to compliance behavior (Allingham and Sandmo 1972).
[11]For a rigorous and comprehensive analysis of social exchange, social norms, and social influence, see Coleman (1990) [also see Kelman (1961); Fishbein and Ajzen (1975); Hoffman (1977); Muller (1979); Ajzen and Fishbein (1980); Wrong (1980)].

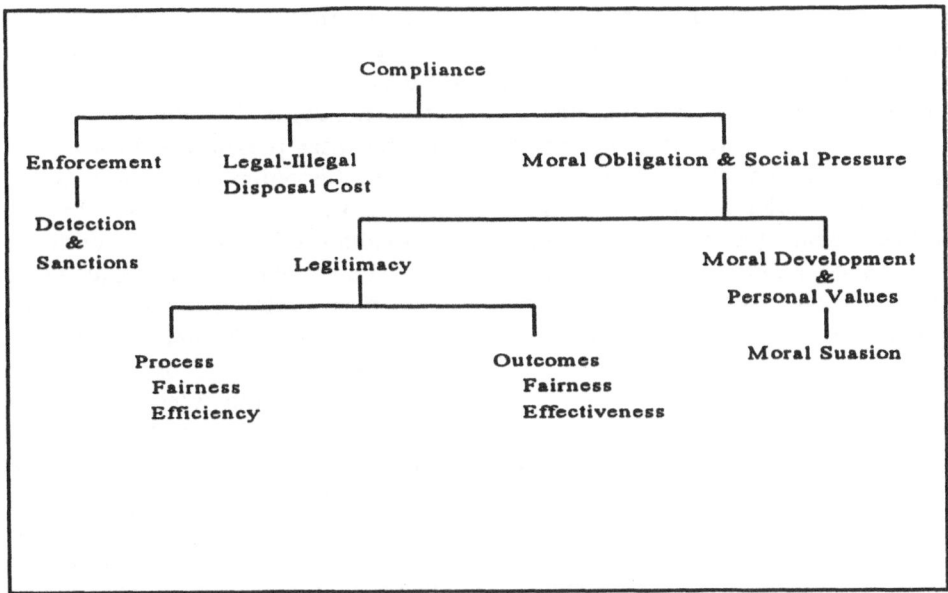

FIGURE 10.1. Determinants of compliance.

the intrinsic obligation to follow the dictates of a "legitimate" authority (such as the police, one's boss, or other authority figure). Legitimacy effectively functions as a stock of loyalty on which leaders can draw. Those who accept an authority's legitimacy are expected to comply with its dictates even when the dictates are contrary to an individual's self-interest.[12]

Legitimacy

Theories in the compliance literature identify four sets of an authority's characteristics that relate to legitimacy. Two involve outcomes (or ends), and two involve processes (or means) of the authority; and two involve issues of justice, and two do not (Fig. 10.1).

[12]Tyler argues that it is better for a rule-making body to base compliance on legitimacy rather than on personal or group morality because the scope of legitimate authority is more flexible (in that leaders usually have a wide range of discretionary authority). Personal morality, on the other hand, is double-edged, for it may or may not accord with the dictates of the authority, leading to resistance to the law and legal authority, instead of compliance with its dictates.

The first of these, effectiveness of the outcome, may involve the extent to which an unpolluted marine environment is realized and an individual is made better off. The second, distributive justice of the outcome, involves the perceived fairness of how the benefits or sacrifices are shared among the affected parties. The third, efficiency of the process, involves the speed and efficiency with which people perceive the authority responding to problems within the scope of the authority's jurisdiction. The fourth, procedural justice, involves how fairly the authority treats people and the concerns of those affected by the process.

Public choice theory in general, and some psychological theories of leadership, argue that legitimacy depends in large part on the authority's ability to provide favorable outcomes. That is, people perceive as legitimate and obey the institutions that produce positive outcomes for them. However, there is evidence that people place greater importance on procedural issues. The principal result of Tyler's (1990) study is that the perception of legitimacy is closely linked to people's views of the fairness of the procedures used by the authorities. More strongly, he demonstrates that the people he studied comply more with the law if the procedures employed by the legal or political authority are perceived by them to be fair.[13] Other research cited by Tyler indicates that distributive justice is another critical determinant of legitimacy and compliance. Tyler's overall conclusion is that justice matters most, with procedural justice being more important than distributive justice, and that the efficiency of the process and favorability of the outcome matter less in promoting legitimacy.

Moral Development and Personal Values

Social psychology emphasizes the importance of an individual's personal characteristics in determining compliance behavior. Moral development of the individual is hypothesized to be directly related to one's propensity to comply with society's rules. Kohlberg (1969, 1976, 1981, 1984) argued there are three distinct levels of moral development: preconventional, conventional, and postconventional. Each level characterizes the relationship between the individual and his social environment, based on the individual's attitude toward society's conventions. Preconventionals tend to reason in terms of fear of punishment rather than in terms of the harmfulness of their acts or the need for social order. Conventionalists tend to reason in terms of social conformity and order. Postconventionalists tend to reason in terms of moral principles that are independent of social order. Kohlberg (1969, 1976, 1984) argues that rule violations tend to decrease at higher levels of moral development. The available evidence supports this hypothesis.[14]

It is generally argued that individuals' personal values, or norms, influence their compliance behavior.[15] While there is substantial evidence that personal values influence behavior and economic outcomes (as cited previously), the correspondence between personal values and compliance with the law is mixed. Personal moral values may or may not accord with the law and dictates of an authority, leading to resistance to the law and to legal authority instead of compliance (Tyler 1990). For example, Westat (1980) reported that taxpayers are generally ambivalent about cheating on their tax return, especially when small amounts are concerned.

Attempts to induce environmentally correct behavior using moral suasion methods have not met with great success. Several years ago Baumol and Oates (1979) reviewed several voluntary, moral suasion-type programs related to recycling, use of unleaded gasoline, auto emission-control kits, car pooling, and energy conservation. They concluded:

Voluntary programs can have an effect on environmental problems but the magnitude of that effect is usually very small. There appears to be a highly reliable group of citizens (perhaps 5 percent of the population) whose consciences move them to participate voluntarily in environmental programs. However, where really substantial changes are considered essential, it appears that programs that require participation (or programs offering substantial and tangible inducements in the form of rewards or penalties) are usually necessary to carry out the job. (Baumol and Oates 1979, p 291)

Baumol and Oates found moral suasion to be helpful in brief emergencies where effective direct controls are not practical, and in cases

---

[13]Fairness, justice, and equity, like morality, can have different meanings to different people, and defy precise definition for all situations. Tyler finds, however, that for a given situation or context, people's perception of these traits are remarkably consistent across the population of people affected by the authority and its rules and laws. For example, procedural justice often is associated with having neutral, honest authorities who allow people the opportunity to state their views and authorities that treat people with dignity and respect.

[14]Kuperan and Sutinen (1994) found that individual Malaysian fishermen are more compliant with fishery regulations the higher the individual fisherman is on Kohlberg's scale of moral development.

[15]Blasi (1980) notes there are substantial definitional, conceptual, and methodological obstacles to research on this issue. A few economists who have explored the theoretical consequences of norms and moral values for economic behavior include Akerlof (1980, 1982, 1983), Frank (1985, 1987, 1988), and Scelling (1984).

in which effective monitoring or requiring procedures is not feasible (e.g., careless causing of forest fires, littering in wilderness areas). In effect, because no other methods are feasible, moral suasion is better than nothing.

We updated Baumol and Oates' review, surveying empirical studies of environmental behavior published during the last 18 years (Devitt et al. 1994). We specifically looked for evidence on the effectiveness of education and other moral suasion methods and incentives. Our review covered more than 125 peer-reviewed articles in the areas of littering, energy conservation, recycling, and transportation.[16] All the studies are based on short-term experiments; i.e., none examined the approaches used in long-term programs. Of these studies, 71 used adequate designs and statistical analyses and provided useful evidence for our review.[17]

The studies tested the efficacy of different moral suasion efforts and types of incentives to induce environmentally correct behavior. The types of moral suasion efforts include information and prompting, feedback and modeling, commitment, and social and situational factors. Some of the studies also examined the efficacy of incentives (alone and in combination with moral suasion). Unfortunately, the evidence from these studies is neither encouraging nor useful. Every method is shown to have a statistically significant effect on inducing environmentally correct behavior. However, the relative efficacy among the methods is not calibrated. Some authors concluded that incentives, while significantly influential, are less cost effective than other approaches. To the extent that the studies examined behavior after the methods were applied and removed, the evidence clearly shows that the effects on behavior disappeared. That is, the evidence indicates there is no permanent effect on behavior

from the application of any of these methods. Because none of the studies examined long-term moral suasion programs, there is no evidence on the long-term efficacy of any of these approaches.

Summary and Conclusions

This paper addresses the question whether moral suasion is a reliable policy tool for controlling marine debris. To overcome the shortcomings of the basic model, it is extended to incorporate moral obligation and social influence on individual behavior. The resulting model, a marriage of economic, psychological, and sociological theories, does account for the effects of moral suasion and explains the available evidence better than the basic model. The extended model explains how existing compliance policies are expected to work to control marine debris. Not only is the model consistent with actual policy approaches, it is consistent with received theory and supported by a wide body of evidence on human behavior. The evidence for the effectiveness of moral suasion methods to induce environmentally desirable behavior, however, is not strong. The evidence is from short-term experiments and not from studies of long-term actual moral suasion programs. About the most we can conclude from the evidence is that most moral suasion methods tested appear to work, but the extent of their impact is unknown. More importantly, the cost-effectiveness of moral suasion methods has not been calibrated. We do not know if moral suasion is superior to other approaches such as incentive programs.

From this examination of the theory and evidence we conclude that moral suasion, such as public education efforts, should be kept as an element of the marine debris policy package and carefully monitored. Careful monitoring of the effects of these efforts is essential to build up a credible body of evidence with which to determine whether, in the long run, moral suasion should be a major or a minor part of the policy package.

---

[16]We should note that studies of some prominent programs, such as the Smoky the Bear and Keep America Beautiful campaigns, have not appeared in the peer-reviewed literature.

[17]The 50-odd unusable studies generally had poor experimental designs or lacked proper statistical analysis.

The willingness to comply stemming from moral obligation and social influence is also based on the perceived legitimacy of the authorities charged with implementing the regulations. Some evidence suggests that a key determinant of perceived legitimacy is the fairness built into the procedures used to develop and implement marine debris policy. To the extent that this is valid, enforcement authorities such as the U.S. Coast Guard should determine what policies and practices are judged fair by segments of the population subject to marine debris regulations. This may mean, for example, that civil penalties and other sanctions should be comparable in value to the larger of the harm done or gains realized. This may mean that boaters and others subject to marine debris surveillance and monitoring be treated with dignity and respect. This also may mean that land-based waste disposal facilities need to be reasonably available and emptied regularly, or it may simply mean that recreational boaters have to be informed that trash receptacles are too costly and it is necessary for them to carry their trash home for proper disposal.

If a high degree of compliance can be realized via the twin forces of moral obligation and social influence, the question arises whether enforcement is necessary. We argue that it is. Enforcement is an essential element of compliance policy. This conclusion is based on the following argument. In almost any group of individuals subject to regulation there is often a core subgroup (usually small) of chronic, flagrant violators. Chronic, flagrant violators tend to be motivated only by the direct tangible consequences of their actions. Moral obligation and social influence have little or no effect on their behavior. Only changing the economic incentives, by reducing the cost of proper disposal or by increasing the expected penalty, can control the amount of marine debris contributed by this subgroup. In the absence of incentive programs, the only control mechanism for this subgroup is enforcement.

Even if the subgroup of chronic, flagrant violators is small and the amount of marine debris they cause is minor, there is a need to control their illegal activity. Eliminating enforcement would allow these chronic violators to flaunt their violation of the law. Being seemingly immune to the regulations sends two signals to the normally law-abiding public. One is that regulatory procedures are unfair, having no effect on flagrant contributors to the marine debris problem, and the other is that the regulatory program is not effectively controlling marine debris. Each of these signals weakens the moral obligation to comply and the moral basis on which social influence is exercised. As moral obligation and social influence are weakened, compliance begins to erode among those who would normally comply with the regulations. Their subsequent noncompliant behavior influences others not to comply with the regulations, and ultimately compliance breaks down.[18] Only effective enforcement can prevent this deterioration.

The issue of whether incentives ought to be part of the policy package is not a primary concern of this paper [that issue has been lightly debated elsewhere (Alaska Sea Grant 1988)]. However, the theory of disposal behavior presented here suggests that incentives could play a meaningful role in controlling marine debris. In our judgment incentive mechanisms should be fully explored as a policy element to supplement the current policy package of dumping prohibitions and public education.

---

[18]This process of deteriorating compliance is believed to have occurred in Northeast fisheries in the late 1980s (Sutinen et al. 1990).

11.

Marine Debris: Benefits, Costs, and Choices

James Kirkley and Kenneth E. McConnell

Introduction

Marine waters and the marine environment provide many services for society. In the United States, some of these services are produced and distributed as a part of the formal and organized economy that we call the market system. Large quantities of goods are shipped. Naval vessels ply the seas and offer protection to citizens of many nations. Fishermen harvest fish and shellfish. In addition, there are many services that are highly valued but less a part of the formal economy. Recreationists enjoy the amenities offered by the marine environment. The amazing variety of flora and fauna in the ocean is a source of aesthetic services to many humans. Furthermore, the ocean serves as a valuable receptor for the wastes and residuals that result from many different kinds of productive activities. This paper addresses problems and prospects of one particularly troublesome use of the marine environment, the deposition of persistent pollutants: marine debris.

Political and popular discussions produce numerous policy questions of marine debris. Who cares about marine debris and why? What are the sources and types of marine debris? Is there really a problem? What are the social, physical, biological, and economic ramifications of debris? What has been done to control marine debris? Is there a need to control debris? What is the best strategy for mitigating debris? Should zero tolerance be a target goal for controlling marine debris? Should plastics be banned from the ocean?

Questions of policy about marine debris are fundamentally about economics because they are about the allocation of resources. We explore the issues of marine debris, particularly plastic and other similar persistent pollutants, from the perspective of economics. Economics offers a framework for understanding the current levels of marine debris and analyzing potential solutions to problems of marine debris. The economic way of thinking provides the means for assessing the damages to society caused by marine debris and the need for public intervention. In cases in which the polluter is known, economics may be the only method that allows damages to be assessed against and collected from the polluter. In the United States, we have exemplary legislation in the Comprehensive Environmental Response, Compensation and Liability Act (CERCLA) and the Oil Pollution Act that enables government trustees to sue private parties for the recovery of natural resource damage.

Ultimately, the question of interest—what is the best use of the marine environment—is linked to a larger question: What is the best use of all resources? We pursue this question by trying to understand the problem of marine debris from the economist perspective.

Production Trends

Aggregate economic activity, which is the fundamental source of debris, has increased dramatically since 1950: the world population has increased by more than 240%. Concurrently, economic activity or industrial production has increased by almost 500% (Fig. 11.1). The standard of living, as measured by gross domestic product per capita, has increased by 76%. Plastics production has grown even more rapidly than aggregate economic activity. Between 1965 and 1990, U.S. plastic resin production increased by 407%. In a study prepared by Franklin Assoc. Ltd. for the U.S. Environmental Protection Agency (EPA) (1994), a simple analysis of the relationship between U.S. population and plastic waste generated between 1960 and 1993 indicates a strong correlation (.86) (Table 11.1). More important, however, is the strong upward trend in plastic waste generation per person per day. In 1960, per capita plastic waste per day was 0.01 lb; by 1993, it had risen to 0.43 lb (>4000%).

Marine activities directly connected with debris have also grown. For example, U.S. Coast Guard boater registrations increased from about 4.1 million in 1965 to 11.0 million in 1990 (Fig. 11.2). This trend continues in the 1990s. Additionally, data from the U.S.

National Marine Fisheries Service show that the number of commercial fishing vessels increased from 87,161 in 1970 to 94,540 in 1990, but peaked at 129,800 in 1985 (Fig. 11.3).

Property Rights and Scarcity for Marine Debris

Simply having more production of persistent waste does not automatically cause a greater problem. In addition to the increased production of potential debris, two other factors are essential in creating the problem we know as marine debris. The first concerns the common property nature of the ocean. The marine environment is used by many public and private enterprises for many purposes, and each user typically exploits the services of the ocean without regard for the impact of their use on the services that others receive. For example, a cruise ship dumps food debris mixed with Styrofoam packaging and serving materials. This reduces the costs of the vessel's handling of the debris but lowers the

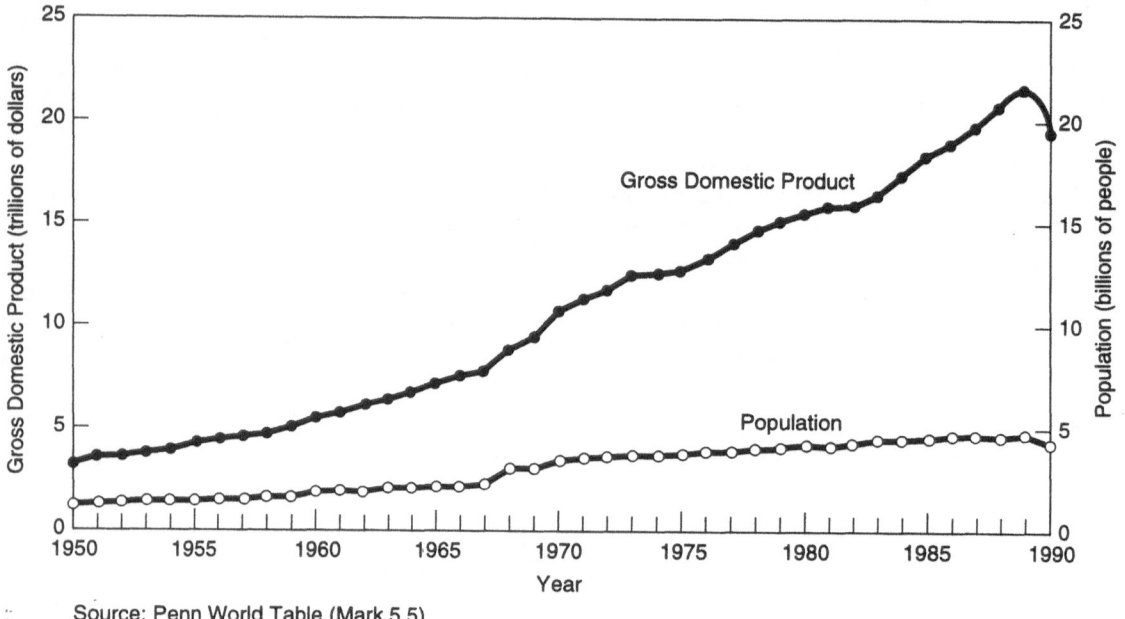

Source: Penn World Table (Mark 5.5)

Figure 11.1 Gross domestic product (GDP) and population for 150 nations (GDP in 1985 international dollars). (From Penn World Table, Summers and Heston, 1991.)

TABLE 11.1. Municipal solid waste generation, recovery, and disposal, 1960–1993

Item and material	1960	1965	1970	1975	1980	1984	1990	1993
Gross waste generated[a]	82.30	96.30	118.30	122.70	139.10	148.10	198.00	206.90
Net waste disposed[a]	76.40	91.90	109.90	112.80	123.00	126.50	165.20	161.95
Percent distribution of net waste:								
Paper and paperboard	31.10	35.00	33.10	30.40	33.60	37.10	36.80	37.60
Glass	8.40	9.20	11.30	11.60	11.30	9.70	6.70	6.60
Metals	13.70	11.60	12.20	11.80	10.30	9.60	8.30	8.20
Plastics	0.50	1.50	2.70	3.90	6.00	7.20	9.00	9.80
Rubber and leather	2.20	2.40	2.70	3.30	3.30	2.50	3.00	3.00
Textiles	2.60	2.40	2.00	2.20	2.30	2.10	3.00	2.50
Wood	3.90	3.80	3.60	3.80	3.90	3.80	6.20	6.60
Food wastes	14.60	13.10	11.50	11.80	9.20	8.10	6.70	6.60
Yard wastes	20.30	19.20	19.00	19.50	18.20	17.90	17.70	15.90
Other wastes	1.70	1.70	1.70	1.90	1.90	1.90	1.60	1.40

[a]Expressed in tons (millions).
Source: Franklin Associates, Ltd. (1994, p. 162) [see also Cavaliere 1989].

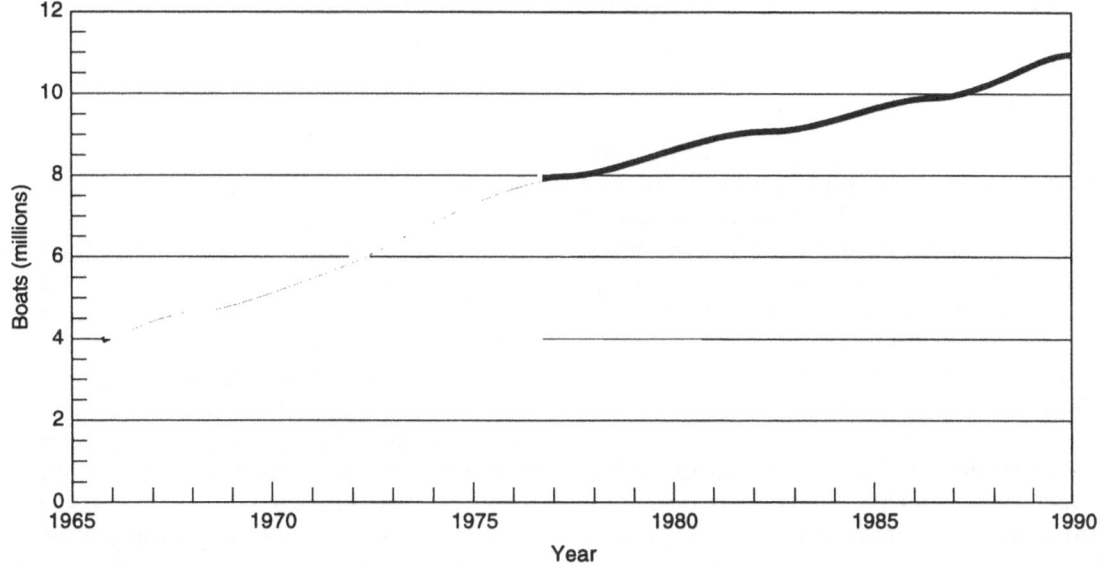

Source: United States Coast Guard

FIGURE 11.2. Millions of recreational boats licensed, 1965–1990. (From U.S. Coast Guard.)

enjoyment of recreational fishermen in the area. This element of common property is inherent in all uses of the marine environment.

Second, over the period of great increase in marine debris, the public perception of the value of a clean environment and the risks from pollution has grown significantly. This is partly the result of growth in income, increase in leisure time, and better education about the environment. It is fueled by our scientific abilities to measure what is happening in the environment. And it has been enhanced by a growth in the appreciation of all living things. For example, the killing of whales is now viewed in many places as abhorrent, but was taken for granted as a good thing in the nineteenth century. The net

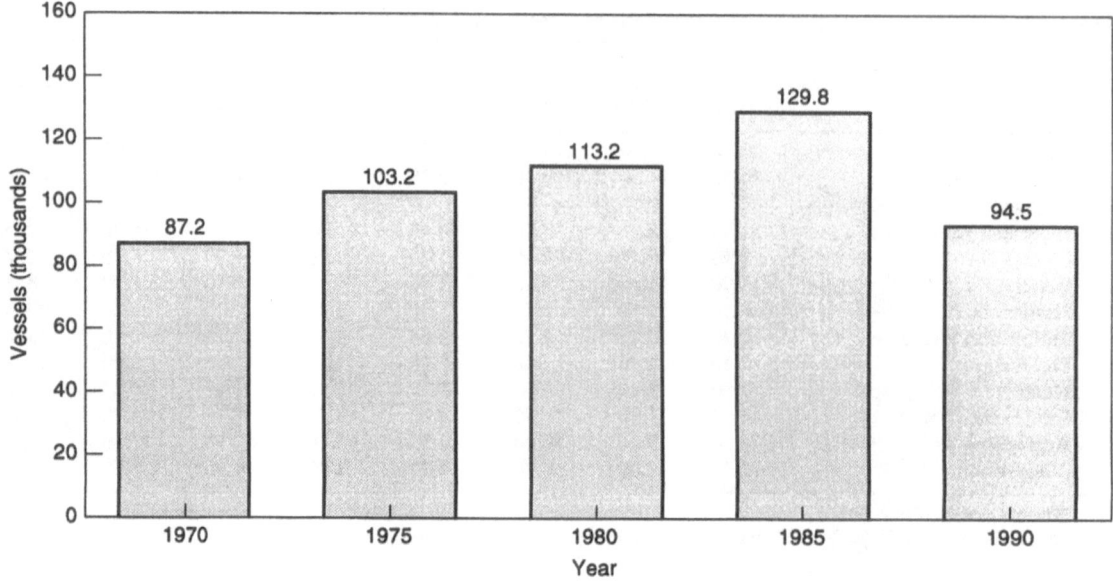

FIGURE 11.3. Thousands of U.S. commercial fishing vessels, 1970–1990.

impact of growth in appreciation of the environment, which is explored more fully here, enhances the economic value of the unpolluted natural environment. The two forces together, the common property nature of the ocean and the increasing economic value of the pristine qualities of the ocean, give rise to the problem of marine debris.

The connection between the absence of property rights and problems of resource misallocation is not unique to marine debris. In fact, property rights are poorly defined or nonexistent for almost all resources at early stages of economic development. Without property rights, resources are poorly used; that is, they are used as if they were not scarce. Property rights emerge when resources become valuable enough to warrant the use of extra resources to establish and enforce the rights. Natural resource development in the United States bears this out. In the early stages of European colonization, land on the fringes of settled and cultivated areas was still plentiful enough to be free. With the gradual establishing of property rights, colonists would move further away and settle on land that was free. But growth in agricultural techniques enhanced the land's value and

made property rights worth establishing. The same was true of water in the western United States. Water is now scarce enough so that when property rights are technically feasible, they exist, such as on the Colorado River.

In some situations, property rights are well defined for environmental resources. According to CERCLA and the Oil Pollution Act, trustees of the nation's marine resources have a right to the services from uncontaminated resources and may sue for damages if these resources are injured by certain toxic substances, including oil. An example of an emerging property right in fisheries is individual transferable quotas, which grant rights to harvest fish; these rights, like personal property, may be bought, sold, rented, or traded. When property rights are well defined, such as for agricultural land and mineral ores, ownership works toward efficient resource utilization, and usually in the best interest of society. When property rights are well defined, the role of government in the allocation of resources may reasonably be limited to enforcing laws relating to property rights. In those cases, the government does not need to know about benefits and costs to ensure efficient resource allocation.

Marine resources have one of the characteristics that make property rights worth establishing: they are steadily growing in value. In some instances, the establishment of property rights may not be feasible. This is the case for some but not all dimensions of air and water pollution. It may also be true for marine debris. Thus, even though the ocean resources are no longer without limit in their bounty, the technical aspects of ownership obviously negate any attempt at individual ownership. In such cases, efficient resource use requires direct government intervention. To understand what is required for efficient government actions, we must turn to the economic costs and benefits of marine debris.

Choosing the Level of Debris: Trade-Offs Between the Desirable and the Possible

Government involvement in helping rid the ocean of debris is taken by all (including economists) as a "good thing." But how much reduction in debris should be pursued? Is it possible to have too much of a good thing? Would outright prohibition of marine plastics be an optimal strategy for public policy? Social choices about the degree of cleanliness of the marine environment implicitly or explicitly choose between the benefits of a cleaner environment and the costs of achieving it. Policies advocated for environmental improvements are more likely to be adopted if they can be shown to be worth pursuing. That is, benefits must exceed costs, and policies must be able to realize environmental improvements in a cost-effective manner.

Two aspects of economic benefits and costs are critical to an understanding of how they are used. First, one must have an idea of the definition of benefits and costs that is intricately connected with the opportunity cost of using resources. Second, it is useful to understand how and why benefits and costs change with the level of debris.

The following analysis describes the nature of benefits and costs of programs to control marine debris. Let us first focus on the intuition of benefits and costs that is frequently at odds with public perception about economic costs and benefits. Economic resources are scarce because they have many uses. Benefits and costs are both based on the notion of economic value. The economic value of an action for an individual is the maximum amount of money a person would pay for the consequences of the action rather than go without it. Because money is exchangeable for resources, benefit is also synonymous with the amount of resources willingly sacrificed. Aggregate economic value is simply the sum of individual economic values. Benefits and economic value can be used interchangeably. Benefits are then the economic value of a change in resource allocation. Costs are defined in terms of opportunities foregone. The economic costs of an action are the economic benefits foregone. Whether something is a cost or benefit depends on the points of reference.

Some simple examples help illustrate these definitions. Consider the economic costs (benefits foregone) of closing a beach. For an individual, it is the maximum amount of money he would pay rather than have the beach closed. This depends not on what he spends on getting to the beach and while at the beach, but rather on what the alternatives are. If an equally suitable beach is available a mile away, the cost of closing the beach may simply be the cost of getting to the substitute beach. On the other hand, if the beach to be closed is unique and the poor substitutes are far away, the willingness to pay could be quite substantial, certainly more than the cost of getting to the nearest substitute.

As another simple example, consider the cost of establishing trash reception facilities at a port. This is just the opportunity cost of the facilities, that is, the economic value of the next best use of the resources that go into the facilities. When the economy functions well, this opportunity cost will be simply the market price.

These definitions of benefits and costs, based on the notion of economic value, have been hammered out by economists with the goal of using them to determine good or bad resource allocations. As we demonstrate, an improvement in resource allocation is one for which benefits exceed costs. It is good, because if benefits exceed costs, then as a consequence of the improved resource allocation in principle those who benefit can compensate those who absorb costs and have some benefits left over. This compensation criterion is the basis for an efficient resource allocation.

The definition of benefits and costs is at odds with popular perception, which typically equate benefits with spending or economic impact. Only under unusual and well-defined circumstances will the amount of money spent, or economic impact, be a good measure of economic value foregone (costs) or economic value gained (benefits).

The distinction between economic value and economic impact can be illustrated simply by referring back to the beach example. A household spends $50 a trip on gas, food, etc. to visit beach A. A perfect substitute, beach B, is 5 miles further on the same highway and costs $5 more per trip. The economic impact as measured by the amount or money spent on beach A is $50 per trip. But note what happens when beach A is closed. First, the loss to the household is $5, the amount of money it takes to get to beach B. Second, the spending is continued but goes elsewhere. The initial level of spending is not relevant to measuring the economic costs, although it certainly matters to those who care only about beach A.

This summary of the distinction between economic benefit and economic impact shows that impact is not a good measure of value. Economic impact, and the spending that generates it, is most often simply reallocated from one locality to another. For example, if a beach is improved, drawing new visitors and new spending, some part of the increased spending, probably substantial, has been reallocated from another locale. Thus, an increase in spending in one place means a reduction in another. It is unproductive to measure the worth of national policies by the economic impact in one location without

subtracting the loss from other locations. This is the rationale for the measures of benefits and costs discussed here. They show the net increase in economic value, not one region's gain over another's loss.

The same argument about the unsuitability of economic impact as a criterion for judging projects and policies also apply to jobs or employment. If we were to use the number of jobs created as a measure of the benefit of the project, we would have to determine the value of lost production that is given up to find additional workers. Creating jobs is not difficult. We could build pyramids, dig holes, or even spread more marine debris and hire workers to clean it up. The more difficult task is to design projects and policies that improve real incomes, and when successfully applied, this is what comparing benefits with costs helps to accomplish.

There are occasions when economic impact is important. The criteria that must be satisfied, to include economic impacts in a measure of value, are (1) that the impact must employ previously idle or underutilized resources (such as labor and other inputs), or it must bring about changes in prices of goods and services, which in turn change the employment of resources; and (2) measurement of the impact must extend throughout the political jurisdiction of the government considering the change. In the absence of these conditions, an impact in one location simply measures the loss from another location. To illustrate, consider a couple of examples involving reductions of debris in the Caribbean. Suppose Jamaica embarks on a policy of reducing debris on its beaches, leading to greater tourism. Tourists spend money, hiring workers away from other sources of employment. Only the extra income is a net gain, not the total spending. However, if workers are unemployed, the additional income will be a gain. In the absence of these conditions, an impact in one location simply measures the loss from another location.

To illustrate the second criterion, suppose reduced debris, which increases tourism in a wider Caribbean initiative, leads to increased tourism in Barbados at the expense of tourism in Jamaica. From the perspective of Barbados, the impact in Barbados may be a net gain. But from the view of the wider Caribbean, Jamaica's loss must be subtracted from Barbados' gain.

Economics is anthropocentric about costs and benefits. For a project or policy to have benefits, humans must gain, directly or indirectly. While this may appear unduly restrictive, some examples of benefits demonstrate that many superficially, unproductive resources do indeed benefit humans. Costs are to be measured not only as out-of-pocket costs, which most people recognize, but also in terms of opportunity costs—that is, the economic value of the best alternative use of the resource. We describe economic costs and benefits as they pertain to marine debris.

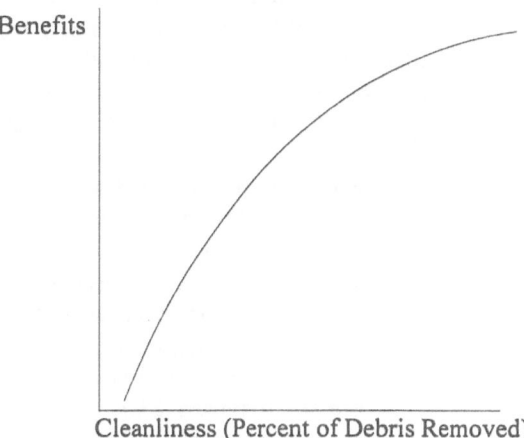

FIGURE 11.4. Benefits of reduced debris.

The Benefits of Controlling Marine Debris

Benefits of controlling marine debris can be defined as the economic value of goods and services that households and firms would give up to reduce debris for a given period of time in a given location. An excellent reference on this topic is the book by Freeman (1993b). We typically measure the economic benefits as willingness to pay, and so we might discuss household willingness to pay to obtain a debris-free beach for a year. The unit can be larger—say the Florida Bay, and the time period longer—say for 20 years. But benefits measure what other goods and services would be given up to reduce debris.

We use willingness to pay for two reasons. First, it is denominated in dollars (money) and hence we have a common unit of exchange. But more important, willingness to pay means willingness to give up resources. Because it takes resources to reduce debris, by measuring willingness to pay we can compare the resources that people are willing to give up (benefits) with the resources that are needed (costs) to achieve the desired reduction.

The benefits from improved cleanliness are graphed in Fig. 11.4, with benefits or willingness to pay on the vertical axis and degree of cleanliness on the horizontal axis. As the degree of cleanliness goes up, benefits go up, but not as fast. This gives rise to the marginal benefit curve, where the degree of cleanliness is on the horizontal axis and willingness to pay for additional cleanliness on the vertical axis (Fig. 11.5). As the household enjoys increasing levels of cleanliness or less debris, it gets fewer benefits for each incremental reduction in marine debris; the more cleanliness, the less willingness to pay for *addi-*

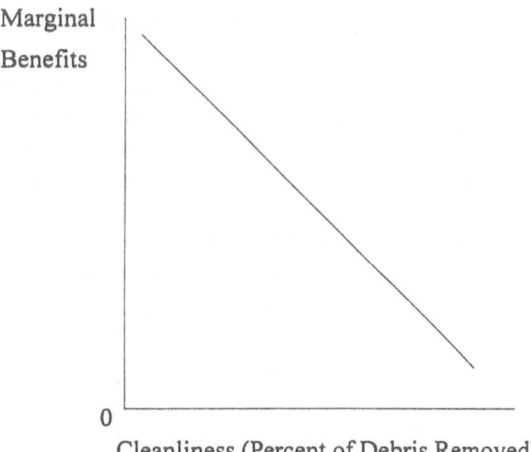

FIGURE 11.5. Marginal benefits of cleanliness.

tional cleanliness. At low levels of cleanliness, households will pay a lot for reducing debris, but as there is less and less debris in the marine environment, households will be willing to pay less and less for reducing debris.

The concave nature of the willingness to pay for cleanliness applies in several dimensions. A household might pay $100 for one beach to be completely free of debris, but it would pay $150 for two beaches and $180 for three beaches. The marginal benefit for the first beach is $100, for the second beach $50, and for the third beach $30. Not only will it not pay society to clean up a given beach completely, it will generally not be worthwhile to clean up all beaches, because the marginal benefits of cleaning additional beaches goes to zero. This concavity of preferences is a fundamental aspect of all preferences and must be accommodated when choosing policies to reduce debris. It applies, for example, to household willingness to pay to preserve species or to prevent ingestion of plastics in many species. As we add additional species to be protected from ingestion of plastics, household willingness to pay to protect these new species will be less than for the first species protected.

While benefits can accrue only to humans, all kinds of services from the environment can yield benefits. A brief catalog would be as follows.

Enhanced Recreational Value

The marine environment provides many kinds of recreational services. Sportfishing, boating, beachgoing, surfing, and many related activities occur in marine waters. For most people, the economic benefit of these activities will diminish with increases in debris. For beaches, this is obvious. Beaches covered with debris are less enjoyable than clean beaches. What people are willing to pay for clean beaches measures the value of resources that could be allocated to prevent or remove debris. Note that the declining marginal value of cleaning up works for recre-

ational beaches. When a beach is completely covered with debris, beachgoers may avoid it completely. When it is 75% cleaned up, most users would return. Going from 75% to 90% removal would enhance the value to beachgoers, but improvement from 90% to 100% might not be noticed, bringing no additional benefits.

Debris reduces the value of boating in two ways. It threatens propulsion and steering systems, and it spoils the natural beauty of the marine environment. Debris both interferes with and poses an aesthetic affront to recreational fishing. For all of these activities—beachgoing, boating, recreational fishing—there is good evidence of economic value. There are a few studies that cover all activities. For example, Bockstael et al. (1988) used demand models to estimate the economic value of sportfishing, boating, and beachgoing for the Chesapeake Bay. Most of the research on the economic value of marine recreation has been done on marine sportsfishing, which may be less susceptible to damage from debris than beachgoing activities. Freeman (1993a) has surveyed studies that estimate the economic value of marine recreation.

Research on the value of beaches typically deals with the value of access, not the value of environmental improvement. Some representative studies include those by Bell and Leeworthy (1990) on Florida beaches and McConnell (1977) on Rhode Island beaches. These studies estimate what it is worth to go to a beach, not what pollution does to beach use.

There is good evidence that reduction in pollution of various types increases the benefits of recreational activities. For example, in 1988, Bockstael et al. estimated that a 20% improvement in water quality in the Chesapeake Bay (measured by the product of nitrogen and phosphorus) would increase recreational benefits by about $40 million per year. These figures illustrate that the loss in benefits from debris can be substantial. But there is, as yet, little direct evidence on the magnitude of these benefits for marine debris.

Passive Use or Existence Value

Some people may be willing to give up control over resources just to ensure that some beaches are clean, yet knowing that they will never use the beach. This value, which economists call passive use or existence value, is measured in research by Smith et al. (Chapter 11, this volume) for debris-free beaches. The concept of passive use values, when applied carelessly, leads to controversial results. It has nevertheless been endorsed by the U.S. Department of Commerce, National Oceanic and Atmospheric Administration (Arrow et al. 1993), and given legal standing in an Ohio decision (Cummings and Harrison 1994). Existence value is especially appropriate to the biological impacts of marine debris. Humans have preferences not only for the populations of living resources but also for the means of their mortality. In the *Exxon Valdez* oil spill, households revealed a willingness to pay to reduce the oil-induced mortality of certain seabirds, even when assured that the populations were not threatened with extinction (Carson et al. 1992). While economists have not worked on marine debris in the same measure, there is little doubt that households would be willing to pay to protect various forms of marine life from ingestion of plastic and mortality caused by plastic straps, rings, lines, nets, etc.

There are many cases of natural resource or living resource damage for which economists have assessed the economic losses. A quick list includes the spotted owl, the red cockheaded woodpecker, the whooping crane, harbor porpoises, and elephants as well as unique wetlands. Based on such experiences, it seems clear that people would give up control over resources to protect sea turtles, whose populations are threatened by ingestion of plastics. These resources have been studied for the value of the services they provide to humans in their natural state. Here then is the connection between biological research on impact and economic valuation: studies of the willingness to pay to prevent these impacts from debris provide evidence that society can find the resources to prevent

the impacts. Mitchell and Carson (1989) surveyed the approaches used in estimating passive use value.

Vessel Entanglement

When the risk of entanglement to ships is reduced, there are two kinds of benefits. First, the direct cost of damage to propellers, engines, and gear is reduced. Second, the opportunity cost to the owners of vessels, in terms of downtime, is lessened. Vessel owners or operators would be willing to pay an amount up to savings in cost to reduce debris. There is no systematic study of the economic costs of vessel entanglement, although anecdotal reports are common.

Ghost-Fishing

When fishing gear is lost at sea and continues to capture fish, ghost-fishing occurs. The economic benefits of reduced debris (in this case fishing gear) are derived from the increase in resource availability and abundance. The economic benefit is variable because it is captured economically or recreationally. Less is known about the economic benefits of reduced ghost-fishing than of other services impaired by marine debris. The ability to benefit from a reduction in ghost-fishing depends on whether the fish stocks themselves are widely used. In the typical case, when fish are available on a first-come, first-served basis, fish are not managed as a scarce resource and the costs of ghost-fishing are negligible. The reason for this involves the role of property rights. When there are no property rights, individuals and firms treat resources as if they have no value: they use them up as fast as possible. Hence, adding more of these resources, as would happen when ghost-fishing was reduced marginally, would not increase social well-being.

The Costs of Controlling Marine Debris

The costs of controlling marine debris are defined as the economic value of goods and services that must be devoted to reducing debris for a given period of time and geographic area. Costs include both out-of-pocket and opportunity costs. For example, suppose a fishing vessel installs a trash compactor for storing plastics. The out-of-pocket costs include the cost of the compactor plus the operating and maintenance costs. But there may also be opportunity costs. For example, if the vessel must give up part of fish-holding capacity, the foregone capacity is an opportunity cost.

In considering the costs of controlling marine debris, costs incurred at the source where the debris is generated and costs incurred to control or clean up the debris in the environment are additive. For example, suppose we are concerned with monofilament fishing line. Reductions in the amount of fishing line in the environment can be achieved through some combination of reducing it at the source—recreational fishing boats—and cleaning it up when it reaches the ocean. The sum of the costs of reducing the monofilament is the sum over all recreational anglers who make the source-reduction efforts plus the cost of removing the monofilament from the environment.

Regardless of whether it is added up over sources or simply cleaned up from the marine environment, the cost of controlling debris typically goes up as the degree of cleanliness increases (Fig. 11.6). Furthermore, the cost increases at an increasing rate so that when incremental or marginal cost, the addition to cost caused by additional cleanliness, is graphed against the degree of cleanliness, it increases (compared to marginal benefits, which decrease with the degree of cleanliness) (Fig. 11.7). If we measure the horizontal axis as percent of debris removed, then the marginal cost grows quite large as the zero-debris point is approached at M. This point is easily illustrated by the beach cleanup example. The big pieces of debris are easy to get, probably by hand. Medium-sized pieces can be raked up. But to remove all the debris would require a mechanical device, such as a tractor pulling a wheeled rake. Obviously the costs grow dramatically.

Control costs have many subtle compo-

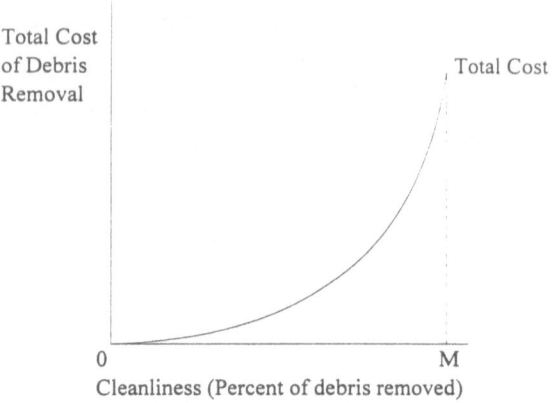

FIGURE 11.6. Total cost of controlling debris.

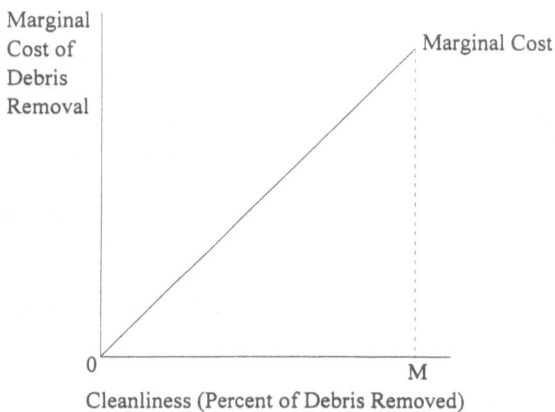

FIGURE 11.7. Marginal cost of controlling debris.

nents because they range from source substitution to cleanup costs. For example, the marginal cost of additional landfill space would be one dimension of control costs that increased with the amount of debris to be disposed of on land. The following categories suggest the nature of control costs.

Source Reductions

Source reduction can mean several different things. For example, it could be the costs incurred when shifting from plastic to paper-derived products or from plastic to paper grocery bags. Combined sewer overflows (CSOs) present a different sort of source reduction. The research by EPA (1992c) suggested that CSOs are a significant source of debris. The cost of reducing debris by solving the combined sewer overflow problem could be quite large. It would require either building complete new systems to manage and remove debris from stormwater, a prohibitively costly undertaking in many situations, or radically improving terrestrial disposal practices to keep debris out of the stormwater systems. Recycling is also a way of implementing source reduction that can reduce the cost, but only when recycling works as a market mechanism.

Cleanup

The most obvious cost of reducing debris is the simple task of using labor and equipment to remove debris from a beach, estuary, harbor, drainage ditch, or sewage treatment facility. A not-so-obvious cost of this cleanup is the implied cost of landfill space or other method of disposing the collected debris.

Enforcement and Monitoring Costs

When government regulations are enacted, they are not automatically obeyed. Compliance typically requires monitoring and enforcement. This is likely to be a substantial cost for controlling marine debris because of the widely dispersed sources, compared to other pollutants. These costs include administrative personnel at the federal, state, and local levels including a maritime force such as the U.S. Coast Guard for handling MARPOL Annex V enforcement.

Individual Adjustment Costs

Regulations that require individual vessels to dispose their debris properly have small and widely dispersed costs that may be substantial in the aggregate. For example, complying with the provisions of MARPOL Annex V requires vessels to provide onboard storage of certain plastics. Because space on many vessels is scarce, this requirement imposes costs on the individual vessels. In addition, ports must provide extra space for disposal and temporary storage of debris. Storage and disposal costs are particularly important for controlling debris from recreational boating. The following example illustrates a study of the individual adjustment costs.

An Example of the Costs of Controlling Marine Debris: Adjustment Costs for Recreational Boating.

Most calculations of the costs of controlling marine debris, whether actual or conceptual, relate to the costs of actions taken to remove debris. These are typically out-of-pocket costs. But other costs, including the opportunity costs of recreational boaters, may be equally important. Here we demonstrate one approach to determining the economic costs of controlling marine debris that includes the full costs of the effort from the perspective of recreational boaters.

This empirical analysis deals with recreational boaters who, according to responses obtained from a survey of Maryland boaters, typically dispense with their debris by carrying their wastes away from the marina when the marina does not provide adequate waste-handling facilities. This disposal activity imposes costs on the boaters. They must pack up their debris and transport it to an offsite disposal facility. Local storage facilities would reduce the implicit adjustment costs. We estimate these costs for recreational boaters in Maryland.

The analysis was based on a mail survey of recreational boaters conducted in the summer of 1993. The survey was administered to 1000 registered boat owners in Maryland counties that border the Chesapeake Bay; there were 641 respondents. The focus of the survey was recreational boaters who primarily use marine waters (i.e., the Chesapeake Bay, its tributaries, and Maryland coastal waters). The purpose of the analysis is to estimate boaters' willingness to pay to have adequate trash facilities located at their marinas. The willingness to pay is part of the estimated cost of disposing of marine debris. In an attempt to obtain recreational boaters' willingness to pay (WTP) to assure adequate trash collection facilities at their dock of origin, the following contingent valuation questions were included:

Suppose that by paying a fee you could ensure trash collection facilities. Would you be willing to pay $X per season to be sure you always had a place for your trash at the dock or marina?

The amounts ($X) for the question were varied randomly across the sample from $5.00 to $200.00. A random utility or discrete choice model was used to estimate mean willingness to pay for trash collection facilities. The dichotomous response to the contingent valuation (CV) question was assumed to be a linear function of the dollar cost from the CV question, X, and the total number of recreational boating trips taken during the past boating season (May 31, 1993 through September 6, 1993). There were 241 responses to the CV question. The remaining 400 respondents skipped the CV question by responding "no" to the prior question:

Do you ever have difficulty finding a place to dispose of trash at the dock or marina you return to after boating?

These respondents who are skipped are assumed to have a zero willingness to pay for trash facilities because they had no difficulty in disposing of debris. An estimate of the mean willingness to pay for all respondents was found by summing the estimated WTP across all individuals from the random utility model who had inadequate facilities and dividing by the number of such respondents (241). Mean willingness to pay for adequate trash collection facilities at the dock or marina was estimated to be $9 per season per recreational boater. Dividing each respondent's estimated WTP by the total number of

boating trips taken last season yielded an estimate of the mean WTP per trip of $1.10 per trip. This measure is an estimate of the cost per trip of disposing of debris for those who currently do not have adequate facilities. While this number is not large by itself, it suggests two things. First, some effort is required to dispose of debris when facilities are not adequate. A small amount of discouragement, provided by the weather or other untoward events, might be sufficient to induce the boater to get rid of the waste overboard. Second, when the implications for aggregate costs are examined, the $1.10 per trip suggests that the potential costs of disposing of debris are not negligible. For example, when expanded to the relevant population of boaters in Maryland, it yields an estimate of $270,000 for the adjustment costs of boaters in Maryland waters.

Analyzing Policies with Costs and Benefits

The schedules of incremental costs and benefits aid in evaluating policy questions. In Fig. 11.8, the vertical axis measures the incremental (marginal) cost and benefits of con-

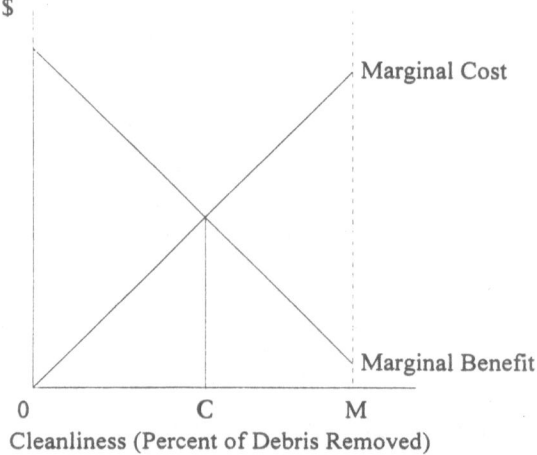

FIGURE 11.8. Marginal costs of benefits of controlling debris.

trolling debris at a beach, and the horizontal axis measures percent of debris removed. At zero level, no debris is removed; at M, all debris is removed. A policy such as a complete ban on debris would imply a situation such as M. The optimal quantity of cleanup occurs at C. At M, costs of additional cleanliness far exceed the benefit. For example, it is possible that beachgoers would not perceive cleanliness increasing beyond 95%, while the costs of such cleanup are quite high. This is the case of a ban on debris. A ban would eliminate introduction of debris into the environment. But if it could actually be enforced, it would be quite costly. It would require containing all debris from littering, combined sewer overflows, incidental loss of plastic pellets, and much greater care with synthetic fishing nets, as well as other more extreme measures. The additional benefits at the point of the ban, however, might be quite low. A complete ban on debris is an expensive proposition and likely to cost far more than the benefits that it produces.

The absence of any policy toward marine debris is equivalent to point 0. At zero cleanup, point 0, the benefits of more cleanup exceed the costs. Households and firms are willing to give up far more resources than are needed to reduce debris. At point C, increases in cleanliness have costs that exceed benefits, while for decreases, benefit costs exceed costs saved.

Figure 11.8 illustrates the use of incremental costs and benefits to help formulate policy. Frequently, economics is used to evaluate policies and regulations. For example, in a current research project unrelated to marine debris, the willingness to pay for two different strategies involving the harbor porpoise is being estimated. The research project considers elimination of mortality to the harbor porpoise caused by gill nets. This research provides evidence that Massachusetts households would pay more than $100 to protect the harbor porpoise from mortality induced by gill nets. In many ways, this is similar to projects to reduce marine debris. It illustrates that humans do, in fact, value aspects of the marine environment regardless of whether these aspects can be purchased

from a conventional market for goods and services, or even whether they have a direct-use component. People value the environment in its pristine state, even when they do not use it or even see it.

The evidence from a variety of studies that households will pay to prevent the mortality of fauna, especially marine mammals, must be taken in conjunction with biological impact studies. Such studies seem to indicate that while plastics are widely dispersed in the ocean and frequently ingested, the threat to species is more limited. For example, sea turtles appear to be especially vulnerable to plastics. Consequently, the benefits of reducing marine debris will likely be greater around resources which people use, beaches, harbors, and the like.

In sum, the economics of marine debris, like the economics of any other pollutant, involves a trade-off between the costs and gains of reducing debris. To apply this thinking for specific projects and proposals, one learns the benefits from the fate and impact of debris. The costs depend on the sources of debris. Thus, knowledge of the flow of debris is essential in making policy choices.

Some Practical Implications

It is natural to ask: why bother with economics, especially comparing benefits and costs? The simplest and best answer is that policies that reduce marine debris take real resources. In a fundamental resource accounting sense, doing the economic analysis means finding out whether those who gain from the policies will relinquish control over resources sufficient to achieve the policies. In aggregate, it is impossible for society to achieve goals beyond the constraint of the resource endowment. This constraint is well known. Economics introduces the constraint in applications. The value of the economic principles lies in their applicability to many specific settings. Practical implications of the economics of marine debris include (1) monitoring, enforcement, and compliance with laws; (2) the need for incentives; (3) complete bans on plastics in the marine environment; (4) the importance of terrestrial sources; (5) economic damages from lost ecological functions; and (6) progress in the economics of marine debris.

Monitoring, Enforcement, and Compliance with Laws

The principal regulatory device for controlling marine debris from ships is Annex V of MARPOL. There is no empirical evidence that the Annex V rules are having an impact on marine debris. The absence of discernible effect from Annex V makes economic sense. The extent to which individuals obey laws is in part a question of economics. As Sutinen (Chapter 10, this volume) shows, the probability of detection and the penalties for detection play an important role in compliance. The costs of monitoring and enforcement are roughly proportional to the area monitored, which in the case of marine debris, is enormous. At-sea enforcement is not likely to play a major role in discouraging at-sea disposal; it simply is not practical for the U.S. Coast Guard to monitor every vessel. Further, the fines necessary to induce compliance would be substantial, and courts have shown reluctance to impose fines far in excess of the damage from the legal violation. One is then left with the notion that individuals will comply through a sense of social obligation. When it is cheap to comply voluntarily, individuals will do so. But when the costs are high, they will dispose over the side. The levels that define cheap and high, however, are not easy to discern, being largely a matter of individual perceptions.

The Need for Incentives

MARPOL Annex V is written as if at-sea enforcement and compliance will be sufficient to reduce debris. There is reason to doubt that this will be true, particularly as MARPOL Annex V does not specify how compliance is to be achieved. We must look for alternatives or complementary regulations to Annex V. Specifically, we need to consider methods

that mimic the effect of property rights. Incentive systems, for example, might pay bounties for the return of plastics and Styrofoam. This is not the same as a recycling program. One need not expect that returned plastic is worth something. There must simply be a judgment that the cost of paying people to bring in plastic exceeds the damage if the plastic were left in the environment. Progress on reducing other pollutants has also reached the point where incentives are playing a more important role, as, for example, in the revised Clean Air Act in which polluters trade the rights to emit sulfur dioxide. In addition, commercial fishermen in states with a bottle bill typically retain and return for a refund not only their cans and bottles but also cans and bottles caught during fishing.

There remains the issue, however, of whether incentive systems would actually reduce the overall levels of marine plastics. Alternatively, what shoreside facilities would have to be available for a plastic recovery system to be successful? Based on survey results for Maryland, shoreside facilities appear poorly equipped to handle increased debris. This information suggests this is the rule rather than the exception in other states. Many coastal landfills have regulations against depositing plastics. Thus, incentive and property rights systems may not adequately control marine debris unless other programs are also implemented (e.g., shoreside facilities).

Complete Bans on Plastic in the Marine Environment

A total ban on the use of plastic in the marine environment raises the issue of whether benefits would exceed costs. That is, would society benefit more from a complete ban on plastic debris than it would cost society to impose the ban? There are numerous uses of plastic in the marine environment for which alternative materials may not be available (e.g., monofilament fishing line, commercial fishing nets or parts of fishing gear, and recreational fishing lures). The costs imposed on society of using alternative materials may far exceed the benefits society would realize (e.g., $75 fishing line that must be replaced every 4 weeks rather than a $12 spool of line that lasts an entire fishing season).

In examining the need for a complete ban, it is essential that benefits be evaluated relative to costs. Figure 11.8 depicts a framework for considering the issue of a complete ban. At the point of 100% reduction in debris, the incremental costs are extremely high, but the marginal value is quite low. In simple economics, cleanup costs would be reduced by imposing a less stringent policy (e.g., 75% reduction rather than complete ban). It might not even be feasible to perceive the difference between 95% and 100% reduction. Such a policy will not simply be difficult to attain; it will not be worthwhile.

The Importance of Terrestrial Sources

It seems unlikely that we will have really good data on trends, sources, and other aspects of the dispersion of marine debris. It is likely, however, that controlling terrestrial sources of debris will have a high payoff in terms of benefits and costs. The benefits of reducing terrestrial sources of debris are likely to be high because such sources contribute much to beach, harbor, and coastal water debris, where most of the damages from debris are incurred. Damages are visibly high in areas of high use such as a beach near a large city. The costs of reducing terrestrial resources can be relatively low, especially if investments like storm drainage runoff controls also help reduce other problems (e.g., flooding, coliform, and chemical pollution).

Economic Damages from Lost Ecological Functions

Except for certain types and amounts of debris, it is difficult to assess the economic damages associated with ecological functions and debris. The wide dispersion of injury to marine life from marine debris suggests that the economic damages are substantial. It is not known with precision, however, whether the damages are widespread (e.g., how many fish are no longer available to recreational anglers because of marine debris, or what do 50 dead seabirds mean for the Northwest Atlantic herring stocks?). The public and scientific community are aware of the obvious injuries (e.g., sea lions, seals, dolphin, whales, and turtles); they do not have adequate information about marine debris and ecological processes.

It is very important to design a monitoring and analytical strategy for estimating the relationship between economic losses, ecological functions, and potential levels of debris. Risk assessment would help evaluate the potential economic damages that might result from the loss of ecological functions caused by varying levels of debris. For example, if it was determined that a particular type and quantity of plastic debris, if allowed to remain in the marine environment for a given period of time, had a 95% chance of killing 100 sea turtles, it would be possible to assess the expected economic damages and develop necessary control policies.

Progress in the Economics of Marine Debris

In many ways, the homilies of economists never change. And there is growth in the economics of marine debris. Smith et al. (Chapter 12, this volume) treat the economic costs of marine debris the way they should be treated, as valuing the impairment to marine resources. It is important that work like this be continued and joined together with the biological impact work. Sutinen (Chapter 10, this volume) addresses the important issue of compliance and enforcement. Enforcement mechanisms will also grow in importance.

12.
The Economic Value of Controlling Marine Debris

V. Kerry Smith, Xiaolong Zhang, and Raymond B. Palmquist

Introduction

Marine debris pollution became a front-page news story in the late 1980s.[1] With this media attention, there has been increasing regulatory activity and some efforts to monitor the problem. In late 1987, Annex V of the international protocol for the prevention of pollution from ships (MARPOL 73/78) was ratified by the United States. Data in a National Park Service report indicate (based on 3 years of sampling at National Seashores around the United States) that amounts of debris found at five of the eight sites studies was remaining approximately constant or increasing (Cole et al. 1992).[2] Thus, debris in coastal areas remains a problem.

Little is known about how people react to debris on beaches and on resources providing habitat for protected marine species. To our knowledge, this paper reports the first attempt to measure the importance of controlling marine debris as one means to maintain or enhance what would seem to be an aesthetic characteristic of the beaches and coastal areas. Our findings are based on a contingent valuation (CV) survey designed to estimate the economic value of controlling marine debris on recreational beaches in New Jersey and North Carolina.

Overall, our research suggests the following: (1) people do consistently distinguish situations with differing amounts of debris when they are described using color photographs; (2) the data from a telephone-mail-telephone CV survey indicate that the measures of people's willingness to pay (WTP) for debris control are consistent with our a priori expectations that reducing debris enhances the quality dimensions of coastal resources; and (3) the local conditions for New Jersey and North Carolina beaches seem to have affected how respondents interpreted the situations describing beach conditions in each area.

In the next section, we describe the methods used to measure economic values for debris control. Following that discussion, we summarize the development of a survey instrument and the nature of our tests for the consistency of people's evaluations of conditions involving debris. The fourth section describes our data and results and the last discusses their implications for future research in estimating the economic values

---

[1] For example, a cover story of *Time* Magazine in 1988 (August 1, "The Filthy Seas") featured marine debris and its impact on beaches in New Jersey as well as elsewhere in the United States. About the same time comparable front-page stories appeared in *Newsweek* Magazine (August 1, "Don't Go Near the Water") and *Business Week* Magazine (October 12, 1987, "Troubled Waters").

[2] This summary is based on quarterly measures for each collection site reported by Cole et al. (1992, Fig. 16). One site (Assateague Island National Seashore) seems to have experienced a slight decline in the last year of the survey available (1990 and 1991) in the report. Cole et al. explain this reduction at Assateague as caused by a significant overwash on its beaches during March 1991.

people have for debris control as one quality dimension of coastal resources.

Modeling Marine Debris as a Quality Attribute of Coastal Resources

Most descriptions of the conceptual foundations for people's measures of economic values begin with a specification of a preference relation. A consumer's choices are defined by using this relation along with a budget constraint. In this setting, the influence of any commodity (or characteristic of a commodity) on choices will depend on how it is specified to enter the preference relation, the budget constraint, or both. For many environmental resources there are few clues available to guide these decisions.

To estimate the economic value a person places on some commodity (or some characteristic of a commodity), we must observe a choice along with a clearly defined set of circumstances that describe the consequences of that choice for the individual who made it. These choice elements serve to isolate a lower bound for an individual's willingness to pay (WTP). Choices involve some object of choice and a set of circumstances for the choice. The object of choice can be anything: it can be an apple or a beach that is free of debris. The circumstances of a choice relate to all the factors that describe the assignment of rights, timing, degree of certainty, and consequences of a choice. When both the object of choice and the circumstances are clearly defined, the choice implies the object selected is at least as good (or as valuable) as what was given up to obtain it. Thus, when a choice is made in favor of the object, the consequence related to the most important foregone alternative defines a lower bound for the economic value. When a choice is made not to select the object, we know that its economic value to the individual must be less than this most important foregone alternative.

Contingent valuation analysis of the eco-

nomic value of debris control applies this basic logic by describing to each surveyed individual the characteristics of coastal resources with special emphasis on the amount of debris that is or will be present. The interview then describes a plan to modify these conditions by controlling or removing the debris. These descriptions define the object of choice—the program to control or remove marine debris. They include a characterization of how the program will be financed and the cost imposed on each individual. This serves to specify the circumstances of choice. The cost provides a monetary measure of what will be foregone if the individual chooses the program.[3]

More formally, this evaluation can be presented within the preference function framework using an indirect utility function, $V(\cdot)$, expressed as a function of income, y, the prices of the marketed goods an individual consumes, represented here by a vector, P, and the coastal resource indexed by its quality, q, designated as $C(q)$. Equation 1 uses the model to consider how a possible choice would be evaluated with a one-time payment, t, to improve the quality of the coastal resources from q_0 to q_1.

$$V(y - t, P, C(q_1)) < > V(y, P, C(q_0)) \qquad (1)$$

If the left side of Eq. 1 exceeds the right, then t provides a lower bound on the economic

---

[3]The logic underlying this formulation is that of a simple trade-off where a lower bound for the economic value of a trade is what must be given up to obtain the desired object. At the individual level it can be expressed in anything. Comparison and aggregation across individuals require monetary measures. Kreps (1990) presented this logic in exploring the prospects for a choice function as the basic unit of analysis for modeling consumer behavior.

The important insight furnished by a choice function for this paper is the recognition that Hicksian measures of willingness to pay (WTP) are economists' constructions, not monetary measures people retain. A decision about the question model in CV surveys—namely, whether people can better report their maximum WPT or state a choice is *not* about their ability to recall a monetary value from memory. Rather, it is about eliciting preferences. See Carson et al. (1994) for further discussion of this logic.

value of the improvement in the quality of coastal conditions from q_0 to q_1. The reverse inequality implies t is an upper bound for that economic value.

It is important to acknowledge that, unlike the format implicitly maintained in Eq. 1, we do not have to specify units of measurement for q, such as pounds of apples, to be able to define an object of choice consistent with the theory underlying the definition of economic values. Nonetheless, we do have to understand the characteristics of the resource (or quality characteristics) that are important to people. This understanding is essential to presenting the proposed program to change the resource (or characteristics) so that most people interpret the program in the way the analyst intended. If this correspondence between an analyst's intentions and people's understanding is not maintained, then the choices will be misinterpreted and the measured values may not be for the intended objects of choice.

There is another important aspect of using CV surveys to elicit choices for estimating economic values. The choices do not require the financial commitment that is proposed to respondents. As a result, they are describing programs that could be offered or could be on a referendum. When an individual states a choice that would be made there are inevitable questions as to how likely these stated intentions correspond to what would actually be done if the program were offered and the implied financial obligation capable of being imposed. Debate over the reliability of stated intentions versus observable actions has been significant in this literature.[4] There is no clear-cut consensus as to what the available evidence implies. Nonetheless an expert panel, co-chaired by two Nobel laureates, assembled to evaluate whether CV could be used for natural resources damage assessment, concluded that it did offer a starting point for these evaluations.[5]

The panel's evaluation held that the burden of proof in establishing the reliability of a CV survey rested with the analysts. They proposed a set of guidelines for CV practice. Four aspects of the Panel's recommendations for CV design are important to evaluating our study. These include (a) probability sampling with sufficiently high response rates to avoid selection effects, (b) detailed information presenting the object of choice clearly, (c) use of a question format that presents a cost along with the proposed object of choice and asks for a choice, and (d) evaluation of the responses to ensure those interviewed understood questions and stated choices that recognize the consequences of their decisions. As the next two sections describe, our survey methods attempted (within the research resources available) to maintain high response rates and to correct for selection effects; developed and pretested information materials to describe the object of choice; offered a clear choice question; and finally, used a research design that allows the properties of the economic values to be evaluated because the amount of debris in the base (no plan condition) varied across respondents.

Designing a Contingent Valuation Questionnaire for Marine Debris

Presenting the control and cleanup of marine debris as a meaningful object of choice requires an understanding of how people think about debris on coastal beaches, as well as some perspective on the aspects of a program to control it and clean it up when needed. Because past research on measuring the importance of the aesthetic dimensions of environmental resources has suggested that visual image would likely be important to people's

---

[4]For an overview of the issues see the special section in the 1994 *Journal of Economic Perspectives* with articles by Portney (1994), Diamond and Hausman (1994), and Hanemann (1994).
[5]The NOAA Panel report (Arrow et al. 1993) in-

cluded a set of guidelines for CV studies used in litigation, recommendations for research, and a general conclusion that ". . . CV studies can produce estimates reliable enough to be the starting point of a judicial process of damage assessment, including lost passive use values" (p. 4610).

perceptions of debris as an influence to beach quality, we sought photographs or visual images of beaches with differing amounts of debris.

Ideally, one would like to be able to relate the photographs depicting the variations in conditions to changes in users' disposal practices and to any differences in cleanup policies but unfortunately the highly stochastic nature of the concentration of debris in coastal areas makes this type of control infeasible. Thus, we sought records about the conditions of the *same* beaches at different times reflecting variations in the amount of marine debris that could arise within the current regulatory regime. We were able to obtain assistance from Dr. Tony Amos of the University of Texas, who has maintained a photographic record of debris conditions over more than 10 years for one section of Mustang Island, Texas. Dr. Amos provided complete access to his photographic inventory. Our development process involved three sets of activities—focus groups, a pretest, and a review of the proposed survey instruments by other economists with experience in the design of contingent valuation surveys.

Focus Groups

Three focus groups were conducted over a 7-month development period. Table 12.1 describes the locations, dates, number of participants, and tasks considered in each group meeting. The objective of the first of these group meetings was to evaluate how people would describe debris on beaches and whether photographs would be effective in explaining different debris conditions.

To investigate these issues, two sets of 12 photos each were presented to each participant, who was asked to rank them independently. The first set of photos did not have people or structures featured prominently in beach scenes, but the amounts and types of debris varied; the second group did include people and structures. We hypothesized that the second group might elicit questions from respondents about who was responsible for

the debris and its cleanup. The photos were selected from those available in Dr. Amos' inventory according to attributes we hypothesized might influence people's perceptions of beach quality, including the amounts of natural debris (such as kelp) and man-made debris; unique items, such as components of ships or anchors that might attract interest; material likely to be associated with severe storms; dead fish and other marine life; and broken glass, syringes, and other material that could be considered hazardous. The ranking tasks allowed the photos to be presented and discussed without the moderator identifying specific features of the photos and thereby prompting participants to offer specific types of responses. Participants were asked to explain the reasons why they ranked one photo higher or lower than another as a desirable beach scene. Table 12.2 provides brief descriptions for each set of 12 photos used in this first group.

The second focus group included people who used beaches more frequently. Their beach use was not limited to New Jersey; two participants indicated that they visited North and South Carolina beaches regularly and one indicated that each year included a trip to Florida beaches. Several changes were made in the tasks requested of participants. In the first group both slides of beach scenes and 5 in. × 7 in. photos were used to illustrate the beach scenes. The slides were used to provide a background photo of the beach as a reference for individual photos and some other information used in preliminary CV questions asked of each respondent. In the second focus group, all information requiring a beach scene was presented with photos. In addition, only one ranking task was used. The photos selected for this ranking task included selections from both sets of 12, identified with asterisks in Table 12.2. The last focus group sought to evaluate refined versions of the contingent valuation questions in an area less likely to have active beach users. However, the discussion began with the same ranking task used in the second focus group.

The results of an evaluation of the findings from several other qualitative summaries in-

TABLE 12.1. Description of focus groups.

Focus groups	Date	Location	Number of participants	Composition	Selection effect	Goals
1	August 27, 1991	Raleigh, NC	9	Six women (age 27–65 yr), eight married; two owned beach property	Recruited using listed phone numbers	Evaluate ranking tasks; gauge importance of beach scene; experience with debris; NC beaches versus other areas; use and passive use CV questions
2	October 28, 1991	Clark, NJ	12	Seven women (age 21–70 yr); greater use of beach than FG1; one owned beach property; eight married	Recuited by marketing research firm	Evaluate refined ranking task; refined CV questions; evaluation of payment method, and questions to be asked in booklet; considered if referendum format was reasonable
3	March 2, 1992	Asheville, NC	9	Three women (age 33–65 yr); three couples (husband and wife); nine married; none owned beach property; greater attachment to SC beaches	Recruited by church; requested random group of members, but church used Social Concerns Committee	Goals comparable to FG2 on ranking and CV; used picture board and scenes with CV to mimic insert; payment method; evaluation of referendum

CV, Contingent valuation; FG, focus group; NC, North Carolina; NJ, New Jersey.

dicate that participants in all three focus groups displayed very consistent processes in forming their perceptions about the beaches presented in the photos and the role of debris for their quality. More specifically, there is a clear indication that photographs could be used to describe alternative beach conditions and that respondents seemed capable of transferring the scenes from one location (e.g., Mustang Island) to the beaches they considered to be relevant to their local area. The characteristics of the debris conditions were consistently rated by independent groups of focus group participants.

To develop this evaluation, the photos used in each ranking task were scored from 0 to 5 (with higher scores implying higher quality levels for the positive attributes and more serious impairment for the negative attributes) by each author independently, and composite ratings were developed from the three independent scores for five characteristics: density of natural debris, density of man-made debris, clarity of the photograph, importance of dead marine life in photo, and water quality. These were combined with qualitative variables indicating the presence of people, structures, and unique debris in each photo. A subset of characteristics were then used as explanatory variables in separate rank logit models (Beggs et al. 1981) for each focus group's rankings.

The subset relevant to the first focus group is somewhat different from that for the next

TABLE 12.2. Description of photos used in ranking tasks.[a]

Scene label	Description
First ranking task (FG1)[b]	
A*	Mixture of dense black seaweed and man-made debris
B (B)	Clean beach with a hotel in the background
C (A)	Littered beach with unique metal debris in background
D (D)	Dense seaweed with birds nearby
E*	Panorama of littered beach with birds and a seal
F (F)	Lage amount of plastic on beach
G (G)	Seaweed and a yellow barrel
H (H)	Scene with what appears as outline of shipwreck
I (C)	Panorama of beach covered with seaweed and plastic
J*	Beach littered with black seaweed and debris
K (J)	Beach with prominent dead marine species
L (L)	Beach with wood from hurricane
Second ranking task (FG1)	
A	Three people with binoculars standing in seaweed
B	People on a beach strewn with seaweed
C	Person approaches ocean, trash barrel in foreground
D	Aerial view of three people and seaweed
E* (E)	A family sitting on a beach littered with debris
F	A group of people on a beach with man-made debris
G* (I)	Man-made debris on a beach near private beach
H	An automobile parked on littered beach
I	A couple looking for seashells on a beach with debris
J	A family playing on a beach near a mixture of debris
K* (K)	A mixture of man-made trash with a tent
L	Man-made debris on a beach with autos and a hotel

[a]The letters in parentheses indicate the labeling used in the second and third focus groups. For more complete discussion of each individual focus group see Zhang et al. (1992a, 1992b, 1993).
[b]Photos from the first ranking task with an asterisk were replaced with photos designated by an asterisk from the second ranking task.

two focus groups. This difference arises because the set of photographs used in the second and third groups include scenes from the two sets of photos used in the first group. This composite included photos with people and structures as more prominent parts of the images. As a result, two model specifications are applied with the second and third focus groups and only one for the first. The simpler specification includes density ratings for natural and man-made debris, a qualitative variable for the presence of unique debris in the scene, an index of the prominence of dead marine life, and an index of water quality perceptions based on each photo. The rating of clarity of the photo was not important in any of the analyses of the three focus group rankings and was not considered. The expanded models added to these characteristics separate qualitative variables for the presence of structures and of people in the photos.

Table 12.3 presents these models. The models for the individual focus groups 1 through 3 are designated FG1 through FG3. To test whether models differ across focus groups, the samples are pooled and the relevant models are identified with their respective numbers as, for example, FG12 (for pooling groups 1 and 2) and FG123 for all three. Comparisons of the sign and statistical significance of the estimated parameters across the models suggest the same factors were noticed by all participants and that they all contributed to the rankings in comparable ways.

The second and third focus groups evaluated the same set of 12 photos while the first considered two sets of 12. We tested whether their rankings could be explained by the same model. This hypothesis implies that the parameters for each characteristic will be equal for the focus groups considered. We tested

TABLE 12.3. Rank logit models for photos with marine debris.[a]

Independent Variables[b]	Individual focus group models			Expanded models		Paired models			Three-way model
	FG1	FG2	FG3	FG2	FG3	FG1,2	FG1,3	FG2,3	FG1,2,3
Density of natural debris (?)	−0.04	−0.02	−0.01	−0.04	−0.05	−0.03	−0.02	−0.05	−0.03
	(−1.95)	(−0.89)	(−0.55)	(−1.81)	(−1.93)	(−2.36)	(−1.09)	(−2.70)	(−2.36)
Density of man-made debris (−)	−0.21	−0.23	−0.24	−0.24	−0.23	−0.22	−0.23	−0.23	−0.23
	(−7.13)	(−9.99)	(−8.94)	(−8.11)	(−6.53)	(−12.50)	(−13.43)	(−10.31)	(−15.43)
Unique debris	0.30	−0.95	−0.54	−1.78	−2.96	0.05	−0.78	−2.19	−0.02
	(1.67)	(−2.81)	(−1.36)	(−3.81)	(−5.25)	(0.33)	(−3.02)	(−6.13)	(−0.10)
Prominence of dead marine life (−)	−0.19	−0.15	−0.18	−0.18	−0.21	−0.19	−0.16	−0.19	−0.19
	(−5.72)	(−5.22)	(−5.10)	(−5.71)	(−5.32)	(−8.52)	(−7.31)	(−7.80)	(−10.16)
Perceived water quality (+)	0.11	0.26	0.25	0.36	0.56	0.13	0.25	0.42	0.14
	(4.53)	(8.91)	(7.08)	(10.04)	(12.55)	(6.95)	(11.30)	(15.33)	(9.01)
Presence of structures (?)	−	−	−	1.02	0.76	−	−	0.88	−
				(2.70)	(1.71)			(3.07)	
Presence of people (−)	−	−	−	−1.14	−2.39	−	−	−1.61	−
				(−2.92)	(−5.09)			(−5.36)	
Log (L)	143.43	193.80	141.31	186.57	129.51	343.11	335.65	318.78	485.80

[a]The numbers in parentheses below the coefficients are asymptotic Z statistics for the null hypothesis of no association.
[b]The signs in parentheses after the names of each variable indicate the hypothesized direction of effect based on the interpretation of each index.

this hypothesis using likelihood ratio tests in three ways: considering only groups that ranked the same set of photos (i.e., focus groups 2 and 3) with simple and expanded specifications, and with all three focus groups. In the first set of tests the results are decisive and largely independent of the p value selected for the test. The rankings imply fairly consistent evaluations.

When we considered all three focus groups, the null hypothesis places more reliance on the model specification in capturing the salient features of the rankings. This is important because we must omit from the models the quantitative variables identifying the presence of people and structures, found to be significant determinants of the rankings in focus groups 2 and 3. These attributes did not vary in the photos used for the first group of rankings with the first focus group.[6] In this

three-way comparison we cannot consider a model that assumes they vary in all three samples. Instead, we adopt a simpler specification, and this choice could influence our test. The results for this composite test suggest that our decision on consistency depends on the probability selected for a type one error. For example, at a value of .05 we would reject the null hypothesis of equality of coefficients. However, applying a more stringent standard, a p value of .01, we would

---

[6]The structure of the model used for a rank logit is the same as the categorical logit or random utility model proposed by McFadden (1974). It relies on modeling the probability of a choice of each of K

items for those applications and a ranking in this case. It also requires the independence of irrelevant alternative (IIA) assumption to simplify the specification of probabilities for rankings. In the case of a ranking, the IIA property implies that the ordering of any pair of alternatives is independent of the other possible alternatives that might be considered. The probability of orderings are specified to be functions of characteristics of the alternatives. If the characteristic does not change across alternatives, then it will not be possible to estimate its effect on the ordering. In the first focus group, the first set of photos to be ranked did not include people or structures in a prominent role.

not.[7] This difference between the comparison of the rankings for the second and third groups versus comparisons across all three indicates the importance of our ability to account for features in the different sets of photos.

Finally, the focus groups suggested that the typical household knew very little about the sources and composition of marine debris. To meet this need, together with requirements of selecting a random sample of households, we adopted a telephone-mail-telephone survey format. The first step involves a telephone interview with an adult decision maker from a randomly selected household. After completing a short interview collecting attitudinal and demographic information, the respondent is asked to participate in a second interview about coastal issues. For those who agree, an information booklet about marine debris is sent, along with the photos describing different beach conditions. After that material is received the second interview is conducted with the same individual. The CV questions rely on the photos and information booklet to help frame the program offered to respondents.

Pretest Evaluation and CV Review

Our pretest consisted of eight cognitive interviews. Using current telephone directories, four respondents were recruited from Raleigh, NC and four from Asheville, NC. These interviews consisted of 45-min discussions of a draft of the booklet and the CV question. They were conducted by phone to correspond to the conditions of the survey (after the text of these questions had been revised in response to reactions in the focus groups as well as to the suggestions of several other economists involved in CV surveys).[8] Several

changes were made to the booklet to enhance readability. In addition, the CV questions were shortened and some wording changes adopted to respond to areas of confusion identified by the pretest respondents.

Form of Final Survey

The research design for the survey sought to distinguish four factors that could influence people's evaluations of programs to control debris, including local conditions, type of coastal resources, amount and character of debris (relevant to the resource type), and the means of payment. Some of these, such as local beach conditions, amount and type of debris, and the means used to pay for the control program, were identified as important by participants in the focus groups. Others, such as distinguishing the type of coastal resource experiencing impacts from debris, were identified based on another research objective not directly relevant to this paper, attempting to distinguish use and passive use (or nonuse) values for quality dimensions of coastal resources.[9] Table 12.4 outlines the experimental design and the variations in each factor that were incorporated in the overall survey design. Because this research was intended to be exploratory, the results are best treated as comparable to a pilot study. The final sample sizes in each of our design cells are too small to develop estimates of population parameters. Our two-stage (telephone-mail-telephone) design contributed to this reduction. However, our design is also very complex and would need to be simplified in any final survey

---

[7]These results are based on likelihood ratio tests, derived using the log likelihood statistics in Table 12.3.

[8]The economists involved in providing comments on the survey instrument included Richard Carson

of the University of California at San Diego, William Desvousges of Triangle Economics Research, Robert Mitchell of Clark University, and the professional staff of the survey firm KCA, Inc. These individuals focused their attention on aspects of the CV questions.

[9]Passive use value is the term used by the Court of Appeals decision in the Ohio case to refer to nonuse values. For a discussion of the conceptual definitions for use and nonuse values see Smith (1993) and Freeman (1993b).

TABLE 12.4. Design elements for marine debris survey.

Design factor	Control method
1. Local coastal conditions affecting beaches used for recreation	Sample design included two different locations: New Jersey (N.J.) and North Carolina (N.C.). Two independent samples were drawn, one for N.J. and one for N.C. All materials adjusted to reflect the beach location identified in text of questions.
2. Impact of debris on different coastal resources—beaches used for recreation versus coastal resources providing marine habitat	Two different CV questions were asked in random order of each respondent—one relevant for local beaches (i.e., N.J. or N.C.) and one for N.C. Estuarine Research Reserves.
3. Amount and type of debris	Four photos depicting different mixes of debris used to describe conditions at recreational beaches (see Fig. 12.2A–D). Two photos without people or structures depicting undeveloped beaches were used to describe conditions for N.C. Estuarine Research Reserves.
4. Payment mechanism	CV questions for debris control and cleanup program intended for recreational beaches explained payment as annual surcharge to state income tax (independent of use of beach) and a per visit beach use fee for all beaches in state. These variations were randomly assigned to each respondent.

to maintain a sufficiently large sample size in each design cell.

Because our focus is on the estimates of respondents' willingness to pay for control programs to control and clean up debris on recreational beaches, we focus our attention on the survey questions and related information used to develop those estimates. As Table 12.4 suggests, respondents were asked about debris control on two types of coastal resources—beaches available for recreational use and North Carolina Estuarine Research Reserves, largely available as marine habitat and for marine research uses rather than recreation. In both cases photos were used to describe the impacts of different amounts of debris. For recreational beaches the photos were derived from the scenes considered in the focus group analyses. For the impacts of debris, these photos were selected to depict different types of areas without evidence of recreationists.

As Table 12.4 suggests, variations in both sets of photos were incorporated in the design: four different levels of debris conditions on the same beach relative to a cleaned alternative scene for the same area in the case of recreational beaches, and two different levels of debris for the estuarine research reserves.

The eight combinations were randomly assigned to respondents. The order of the two cleanup programs was also randomized.

Figure 12.1 provides the photo used to depict the beach condition provided the program to control and clean up debris is undertaken. Figure 12.2A–D provides the photos used to describe beach conditions if the program was *not* adopted. Each respondent received only one of these four alternatives along with the same photo (see Fig. 12.1) describing the beach conditions with the control and cleanup program (along with one of the two for the research reserves).

Figure 12.3 reproduces the text of the question used to ask about the debris control and cleanup program for recreational beaches. This was asked of adult decision makers[10] who had already been interviewed to collect attitudinal, demographic, and economic information (i.e., during the first telephone survey labeled the phase I survey). If the individual agreed to participate, he or she received an eight-page booklet that described

---

[10]For this survey, an adult decision maker was defined as an individual over 18 years of age who makes decisions about the household budget.

FIGURE 12.1. Proposed status of beach *with* cleanup and control program.

marine debris, explained where debris comes from, outlined its effects, discussed laws to control debris, and explained (for the second CV question) the national estuarine research reserves and the specific North Carolina sites considered as part of this question. The package also included an explanatory letter and short mail-back questionnaire about their recreational use of beaches.[11]

Before asking the questions about the control and cleanup program, the second telephone interview assured the interview was with the same individual (the name had been collected in the first interview), confirmed the mailing had been received and the booklet read (if not, another interview time was scheduled), and asked about experience with conditions like those described by the

photos. Following the CV question, follow-up questions were asked for open-ended responses about each individual's choices.

Results

Sample

Our research design considered the samples relevant for New Jersey and North Carolina beaches independently. Within each sample, a further distinction was drawn in defining the initial selections for the random digit-dialed (RDD) samples. About half of each sample was composed of RDD numbers selected to be representative of each state. The remaining observations were drawn from regions identified as providing recreationists for each state's beaches, based on the onsite surveys conducted through NOAA's Public

---

[11]To enhance response rates each respondent was told that they had a chance of winning one of four $50 awards if they completed the second-phase interview and mailed back the questionnaire about recreational use of beaches.

Figure 12.2A–D. Scenes (A–D) of status of beaches *without* cleanup and control program.

C

D

BLOCK A (Questions 7)

In the next part of this interview, consider the beaches in the page labeled "Alternative Coastal Conditions" and how you would feel about conditions similar to these for beaches in North Carolina or New Jersey?

7.   I would like to ask you about a proposal for controlling debris on North Carolina and New Jersey beaches, involving greater enforcement of the laws prohibiting waste disposal in the ocean and periodic clean-ups of the beaches.

Without these efforts, we can expect that ocean disposal of waste would make conditions on North Carolina and New Jersey beaches resemble **Photo A** in the time.

This program involves **both enforcement and cleanup activities** financed by those states whose citizens are likely to use North Carolina or New Jersey beaches. It would improve beach conditions to those depicted in **Photo B**.

The proposal would involve:

1. Activities financed by an annual surcharge added to each household's state income tax. This means each household's annual taxes would increase by $ _____ per year.

North Carolina Sample
The following (dollar) values were randomly selected by interviewers for respondents in North Caroline, Virginia, and Maryland: 5, 10, 25, 40, 60, 100, 150, 350, 450, 750, 1,000
For respondents in the Midwest, Pennsylvania, and New Jersey the values were: 5, 10, 25, 40, 60, 100, 150, 350, 450

New Jersey Sample
The following (dollar) values were randomly selected by interviewers for the entire sample: 5, 10, 25, 40, 60, 100, 150, 350, 450, 750, 1,000

2. Activities financed by a statewide beach use fee. The beach use fee would be $_____ per **person** and would be for **each time** a person visited a beach anywhere in North Carolina or New Jersey. It would be in addition to any other local charges or state park fees.

The following (dollar) values were randomly selected by interviewers for the entire sample: 0.25, 0.50, 0.75, 1.50, 1.75, 2.00, 2.50, 4.50, 6.00, 7.50, 10.00, 12.50, 15.00

The program's funds would be used exclusively for controlling debris on beaches so conditions would be improved to those depicted in **Photo B**.

Keeping in mind your household income and current expenditures, together with these new costs, if this proposal was placed on the ballot this fall as a referendum, would you vote for it?

01 Yes—Go to question 8
02 No—Go to question 9
03 Don't Know—Go to question 10

[The interviewers' instructions were removed from figure 12.3]

Figure 12.3. Question used to ask about the debris control and cleanup program.

TABLE 12.5. Cross–tabulations for choices of plans to control and clean up marine debris.[a]

	Photo of debris conditions			
Sample	A	B	C	D
Full sample		$\chi^2_{(3)} = 3.38$		(p value = .337)
Against control and cleanup program	63	47	60	47
	(56.8)	(58.8)	(60.0)	(70.2)
For control and cleanup program	48	33	40	20
	(43.2)	(41.2)	(40.0)	(29.8)
North Carolina sample		$\chi^2_{(3)} = 0.22$		(p value = .975)
Against control and cleanup program	41	26	34	21
	(62.1)	(59.1)	(63.0)	(63.6)
For control and cleanup program	25	18	20	12
	(37.9)	(40.9)	(37.0)	(36.4)
New Jersey sample		$\chi^2_{(3)} = 6.32$		(p value = .097)
Against control and cleanup program	22	21	26	26
	(48.9)	(58.3)	(56.5)	(76.5)
For control and cleanup program	23	15	20	8
	(51.1)	(41.7)	(43.5)	(23.5)

[a]The number in parentheses below the sample count in each contingency table is the column percentage (i.e., between against and for the program).

Area Recreation Visitors Surveys.[12] The overall response rate for the phase I sample was 52.6%. These interviews were completed in September 1992. Of the 1773 adults interviewed, 66.4% agreed to a second interview. The callbacks to complete these interviews and second-stage interviewing ended November 16, 1993, yielding a sample of 693 completions for both interviews and an overall response rate of 39.1%.

Clearly, this low response rate raises the possibility of significant selection effects and would, in the absence of some adjustment, limit the relevance of our findings for gauging population estimates of the value of debris control and cleanup programs. It does not invalidate the use of the surveys for tests of specific hypotheses. They can also serve as an approximate gauge of the economic value people may hold for improving quality conditions at beaches. This latter role is the way one might interpret results from a pilot survey. Nonetheless, the findings will be conditional to any model used to take account of selection effects. We focused our summary of the survey results on the questions involving debris conditions on recreational beaches with a state income tax used for the payment vehicle.

Choice Analysis

Table 12.5 summarizes the votes for and against the control program for each of the subgroups receiving one of the four photographs in Fig. 12.2A–D. The letters in the columns identify the subsamples according to the photo received. Three contingency tables are reported, one for the overall sample and two derived by splitting this group into the North Carolina and New Jersey subsamples. The sample sizes are smaller than the number completing the phase II survey because the remaining group in each sample was randomly assigned to the beach access fee version of the CV questions (i.e., in these questions the payment was described as an increase in a beach assess fee). With these small sample sizes, choices alone do not offer much ability to detect differences in perceptions of beach quality conditions based on the photos. These results stand in fairly sharp contrast with the ranking results from the focus groups.

---

[12]These judgments were based on the Public Area Recreation Visitors' surveys undertaken for NOAA for Island Beach State Park in New Jersey and Cape Hatteras National Seashore in North Carolina for one recreational season.

There are, however, important differences in the question addressed with this table and that considered by the models with the rankings of beach scenes. Here we consider stated choices with a specific financial consequence explained to each respondent. There are also a number of other factors changing in the responses presented in these four columns. The most important of these is the different tax amounts assigned to respondents. Because this was random (see the text of the questions reported in Fig. 12.3), we would expect an approximately uniform spread of values across the four subsamples. Nonetheless, as the sample size in each cell drops, this uniform pattern is less likely to be present in any single realization of the random assignment process to different photos of the base debris condition. Finally, as we just noted, these comparisons do not take account of any selection effects induced by the telephone-mail-telephone format used to collect these data.

In spite of these limitations, the record for discriminating among conditions is a little better for the New Jersey sample, where arguably users may have experienced debris conditions that more closely parallel the photos. Nonetheless, it remains a weak one that does not provide results as strong as the focus groups seemed to have suggested would be present in a survey that used photos to describe beach conditions.

Statistical Model for WTP

Responding to the limitations imposed by the sample size requires that we introduce a statistical model which is consistent with the economic framework for describing choice behavior to explain respondents' decisions. The Hicksian definition for willingness to pay (WTP) for a change from q_0 to q_1 measures the t^* that will make both sides of this expression equal as in Eq. 2:

$$V(y - t^*, P, C(q_1)) = V(y, P, C(q_0)) \qquad (2)$$

Solving for t^* requires that we invert the indirect utility function ($V(\cdot)$) and express t^*

as a function of y, P, q_1 and q_0.[13] Comparing t^* to the proposed t for each individual provides a model describing his or her choice (i.e., if $t^* > t$ then a respondent will vote for the program, otherwise the vote is against it). Of course, the analyst does not know the exact form for the function to be used in describing people's preferences and may have incomplete knowledge of other considerations involved in such choices. Thus, this framework generally introduces stochastic errors into its description of the choice behavior. The framework describes the probability an individual's unobservable WTP (i.e., t^*) will exceed t. This does not mean people have monetary values for everything as distinctive numbers in their consciousness. Rather, t^* is the economist's construction to "explain" what is actually known—the people's choices for specific objects of choice under a set of well-defined circumstances for those choices.

Initial research in this area used common discrete choice, maximum-likelihood methods such as probit or logit for these analyses. Because this approach has been criticized as implying some people would not support desirable programs even if they had no cost (i.e., lead to estimates of WTP that are less than zero), following the work of Carson et al. (1992) we have adopted an alternative formulation based on a Weibull survival model that constrains t^* to be a positive random variable. The survival function, $S(x)$, can be interpreted as a reduced form description of the probability that an individual's WTP is at least as great as x (i.e., $S(x) = \text{Prob}(t^* \geq x) = 1 - F(x)$, where $F(\cdot)$ = cumulative distribution function).

To implement the model with the censoring implied by our referendum model, $S(x)$ is used to describe the probability of "for" and "against" votes and the hazard rate, $\lambda(x)$ (roughly the rate of observing WPT's greater

---

[13]This framework was first proposed for describing CV responses by Hanemann (1984). More recent papers have considered whether a specification of preference functions is necessary to develop welfare measures (Cameron 1988; McConnell 1990).

than x, which can be treated as a demand measure for the program), is specified to be a function of individual characteristics. With a Weibull function describing this process, the survival and hazard rate models are given in Eqs. 3 and 4:

$$S(x) = e^{-(\lambda x)^\theta} \quad (3)$$

λ, θ are parameters

$$\lambda_i = e^{-\beta z_i} \quad (4)$$

where β is a $1 \times K$ parameter vector to be estimated and z_i is a $K \times 1$ vector of values of independent variables for the i-th respondent.

The median WTP, $\tilde{t}$, can be expressed in terms of these parameters as:

$$\tilde{t} = e^{\beta z_i}(\ln 2)^{\frac{1}{\theta}} \quad (5)$$

The use of the telephone-mail-telephone survey to provide respondents the information necessary to answer our CV question introduces the prospects for selection effects. Incorporating these effects requires that we

modify the hazard rate model to include another hazard rate (i.e., Heckman's 1979 inverse Mills ratio, IMR) that takes account of respondents' rate of participation in the second telephone interview. This relationship can be estimated from the information included in the first round, RDD telephone interviews. Following Heckman's argument for the case of linear models, this modified specification offers an approximate method to take account of selection effects and avoid biased estimates of respondents' willingness to pay for debris control and cleanup programs resulting from the low response rates.

Empirical Findings

Equation 6 provides the probit estimate for our selection model. Income, demographic, and attitudinal variables influenced respondent willingness to participate in a second telephone interview about "the quality of beaches and coastal areas."

Prob (completing phase II interview) = 0.29 + 0.004 income (in thousands)
(1.89) (3.33)
+ 0.17 environmentalist (= 1)   − 0.01 age (in years) − 0.28 below high school
(2.17)   (−6.07)   (−1.34)   education (= 1)
+ 0.29 attitude toward dumping   + 0.18 attitude toward limiting coastal
(2.94)   in ocean (= 1)   (2.58)   development (= 1)
+ 0.20 attitude toward coastal   (6)
(2.60) pollution
$n = 1671$ $R^2 = .057$

The number below each estimated parameter is the estimated asymtotic Z statistic for the null hypothesis of no association. The attitude variables are all qualitative variables defined from questions on the phase I survey. To be coded as a "1," the respondent rated reducing or restricting activities related to the identified topic as "very important." A code of "0" corresponds to those indicating it was "important," "not at all important," or were "not sure." The variable "environmentalist" is also a qualitative variable indicating whether respondents suggested that being "concerned about the environment" described them very well.

While these results suggest that the survey

method appears to have differentially attracted higher income respondents with interests in coastal issues, this does not automatically imply selection effects will be an important factor in describing respondents' choices. To investigate this issue, we estimated the Weibull survival model in two ways: (1) using separate subsamples corresponding to the groups receiving each of the four different sets of photographs without debris control and cleanup program (i.e., Fig. 12.2A–D), and (2) using a single model for the full sample. In the latter case the different photographs were treated as qualitative variables shifting the intercept for the log of the hazard rate function (in Eq. 5). Table 12.6

TABLE 12.6. Estimated willingness to pay (WTP) and survival models for program to control and clean up marine debris.[a]

Subsample	Sample size	Hazard rate		Heteroscedasticity adjustment[b]			Estimated median WTP (95% confidence interval)
		Intercept	IMR	Pseudo-R^2	Intercept	IMR	
Photo A	108	8.49	−7.06	—	−0.52	2.37	$72.18
		(8.32)	(−2.99)		(−1.08)	(2.66)	(34.20–152.32)
Photo B	75	8.05	−6.78	0.101	1.15	—	$40.97
		(3.25)	(−1.35)		(2.84)		(11.72–143.23)
Photo C	99	7.32	−4.83	0.054	0.79	—	$63.21
		(6.88)	(−2.38)		(3.61)		(29.59–135.04)
Photo D	66	7.34	−6.59	0.070	0.96	—	$21.38
		(5.09)	(−2.35)		(3.17)		(5.71–80.02)

[a]These estimates are for a Weibull survival model specified to be a function of the inverse Mills ratio (IMR), or hazard rate, derived from the probit model report in Eq. 6. Observations from the New Jersey and North Carolina samples were pooled because of the small sample. Test for the effects of pooling indicated that a qualitative variable for the sample was insignificant. The estimated medians are derived from the parameters of the Weibull models in each row.
[b]The estimates of the Weibull allow for heteroscedasticity from the estimate of the IMR. The rationale follows Heckman (1979), but implementation assumes $\sigma_i = \alpha_0 + \alpha_1 \, IMR_i$. This is incorporated when $\hat{\alpha}_i$ is judged significantly different from zero.

reports the first set of results from the analysis of the four subsamples along with the estimated median WTP derived from each subsample's model. Selection effects are important and indicate that the median WTP of respondents to the second survey is larger than those not completing the survey.

The estimates of median WTP are of particular interest given our objective of evaluating whether independent respondents could discriminate between debris conditions and whether these distinctions affected their stated choices. These findings are closer to supporting these distinctions than the contingency tables but are not statistically significant. The ordering of the estimates of median WTP by independent samples align with one's prior judgments on the severity of the debris conditions as given by the ordering of

the photos from worst to best A > C, C > < B, and B > D.

The last approach sought to improve on our ability to discriminate between each situation by pooling the four subsamples. Qualitative variables were used to identify three of the photos (allowing a relative comparison) along with the IMR effect (following White's 1982 argument for consistent estimation in the presence of specification errors) and several other variables. Equation 7a reports the model for the log of the Weibull hazard rate, λ, and Eq. 7b provides the model for the heteroscedasticity adjustment because the inverse Mills ratio is a random variable introducing the prospects for heteroscedasticity. Both models were estimated jointly (asymptotic Z statistics for the null hypothesis of no association are in parentheses below the coefficients):

$$\ln(\lambda) = \underset{(9.41)}{7.27} \; \underset{(-4.06)}{-6.49 \, IMR} \; \underset{(1.90)}{+1.00 \, \text{photo A} \, (=1)} \; \underset{(0.85)}{+0.48 \, \text{photo B} \, (=1)} \tag{7a}$$

$$\underset{(1.33)}{+0.70 \, \text{photo C} \, (=1)} \; \underset{(-1.78)}{-1.30 \, \text{own vacation home} \, (=1)} \; \underset{(1.33)}{+0.70 \, \text{composite coastal concern attitude} \, (=1)}$$

$$\ln(\sigma) = \underset{(0.87)}{0.25} \; \underset{(2.31)}{+1.24 \, IMR} \tag{7b}$$

$$n = 348$$

The results suggest that respondents do seem to discriminate between debris conditions. The model confirms that pronounced differences in the debris conditions as given in the photos are important to their stated

choices. The coefficients of photo A and photo C compare to the omitted scene (the "best" conditions of the four, photo D, with only natural debris). These estimates suggest we can estimate a significant (p value = .06) discrimination relevant to people's choices for the worst versus the best conditions with these sample sizes, after allowing for selection effects and other potential influences on these respondents' choices.

A number of other determinants of the hazard rate were considered, including income. However, these potentially important factors were not statistically significant. This finding likely arises because these variables' effects are being represented through the IMR term. Overall, these results indicate that despite a limited sample size, and the prospects for selection effects from the telephone-mail-telephone survey procedure, there is support for respondents' ability to discriminate debris conditions.

Implications

We have reported an example of how estimates of the economic value for controlling marine debris can be developed. Our approach was constrained by available resources and our ability to control the process giving rise to the problem. Nonetheless, we conducted focus groups and a large-scale pilot survey to investigate how people responded to different aspects of programs to control and cleanup marine debris and developed models to account for selection effects.

Our estimates should be considered as results from a pilot study, not "final" measures of the economic values of programs to control and clean up marine debris. The estimates do suggest independent respondents discriminate different amounts of debris consistently, at least in terms of identifying the extreme conditions. Thus, if a large-scale survey were conducted using photos with varying debris conditions, it seems likely that the resulting WTP estimates would be responsive to variation in the amount of debris controlled.

When considered together with the focus group results, these findings provide strong support for people's concern for programs to control and clean up marine debris as well as for the prospects for measuring these programs' economic values. Nonetheless, it is also important to reiterate that this survey should be treated as a pilot study, evaluating the feasibility of the task of valuing debris control and cleanup programs. To use the limited data available, we pooled respondents across areas (i.e., those relevant to NC and NJ beaches). This process assumes the differences in local beach conditions are not important to respondents' ability to consider the particular beach scenes shown in the photos (Figs. 12.2A–D). We also assumed that analysis of this question could proceed independently of the second valuation question devoted to estuarian research reserves (and not analyzed here) as well as any differences across each design point in the survey. Further research is clearly warranted, investigating whether control and cleanup programs for other areas yield comparable or greater discrimination in recognizing the effects of debris conditions on choices.

Acknowledgments. Thanks are due Tony Amos for providing access to all his photos; to Jim Coe and Don Rogers for commenting on an earlier draft; to Laura Osborne and Kurt Schwabe for research assistance; and to Drake Paul and Paula Rubio for preparing several versions of this paper. Partial support for the research was provided by NOAA (award number NA90AA-D-SG847) and the University of North Carolina Sea Grant College under project R/MRD-25.

13.
A Comprehensive Waste Management Model for Marine Debris

Shirley Laska

Introduction

The framing of the issues of how to manage solid waste in the marine environment has come a long way since 1989; clearly, the paradigm within which society perceives these issues has begun to shift. Perhaps there is no better example to demonstrate this important shift than to show its utility for reducing the waste deposited in the oceans by recreational users, a large and diverse group. This paper presents a framework for looking at, and influencing, this new perspective and provides some concrete examples related to recreational users. Before I begin, however, I want to place my comments within the question of whether we have made progress in the last 5 years by relating a brief anecdote from the Second International Conference on Marine Debris (Honolulu, 1989).

Following the presentations on environmental education at that meeting, the session where recreational user issues were most prominent in that conference, there was a general discussion among the presenters and the audience. One participant in the audience, who I believe was a member of Greenpeace from Australia, asked that the discussion be framed in terms of a waste management perspective. One of the organizers of the conference, who happened to be in attendance, turned to him and said, "We are not here to discuss such broad issues. This conference focuses only on marine debris."

Although I was taken aback by the comment, no one in the group publicly challenged the organizer. Perhaps, I reasoned, I was ignorant of how inappropriate the Greenpeacer's comment had been. Perhaps the organizer knew better. I accepted the narrower framing. The discussion ensued from that perspective.

A couple of years later, because of my participation in the second international conference, I was invited to be a member of the National Academy of Science, Marine Board's Committee on Shipborne Waste, the committee charged with addressing improvements in managing the implementation of MARPOL Annex V in the United States. The composition of the committee and its advisers was as eclectic as are the various groups who use the oceans and those charged with its health and safety.

During those meetings I found it very interesting to see the way in which each of the representatives of the various ocean user groups framed the issues. I was initially uncertain what role a social scientist should play in the discussions. In time, however, I was able to introduce a socioenvironmental model describing the creation of waste and the various opportunity points at which interventions could occur to reduce the risks posed to people and the environment. It was, in effect, a comprehensive waste management model.

Would I be told this was a committee focused only on shipborne waste, or that my proposal was too broad? Deja vu of the Honolulu conference? To the contrary, the committee has embraced the model and it has become an organizing tool for its entire report. Each chapter focuses on a particular user group and considers the ways in which the risks that come from solid waste in the marine environment can be reduced at each of the opportunity points. More importantly, other committee members have independently endorsed approaching shipborne waste from a comprehensive waste management approach. As one sponsor of the Shipborne Waste Committee defined the committee's charge: "Our primary goal of this interagency-supported research should be to identify, encourage, and enhance the use of integrated waste management policies an technologies in the implementation of the Annex."[1]

This shift in thinking, a paradigm shift from 5 years ago, represents more than the sentiments of the 20 or so people involved in that committee's activities. It reflects broad changes in thinking about marine debris and indicates serious progress in understanding the problems in a much more comprehensive fashion than was previously acceptable. The sentiments of the Shipborne Waste Committee are merely reflecting this shift. Now, let me describe this perspective on waste generation and management that we now appreciate as broadly useful.

The Technological Hazard Evolution Model

Reducing marine debris has been problematic. Evidence of the difficulties is given national media attention each fall, when the tons of debris collected by beach cleanup volunteers adorn the front pages of newspapers. Two important factors inhibiting the reduction of marine debris are (1) the ease with which people can discard waste into the

sea without detection, and (2) the difficulties experienced by mariners in managing waste at sea. While both problems could be resolved to some degree, for the former, voluntary compliance with the desired norms and regulations for disposal offers greater potential for improvement than does energetic enforcement. This judgment is based on the assumption that direct surveillance of mariners is difficult and will continue to be so, but human beings generally want to conform and can create change if the difficulties impeding voluntary compliance can be overcome.

To pinpoint opportunities to improve compliance with desired norms, a systematic approach was developed by social geographer Roger Kasperson and public administration specialist David Pijawka (1985). They considered the ways in which technological hazards pose different and more challenging problems to the public and the government than do natural hazards such as hurricanes, tornadoes, or earthquakes. They proposed that technological hazards challenge our institutions and communities because (1) these hazards are new and unfamiliar, (2) there is a lack of accumulated experience with control or coping measures, (3) there is a lack of full appreciation of the hazard chain, (4) the broad opportunities for control mechanisms make these hazards seem more controllable than natural hazards, and (5) the perception that technological hazards can be fixed by technical solutions, without regard to the social context or the significance of social costs, makes them harder to fix. In sum, technological hazards are incorrectly seen as more easily "fixable," with the result that less attention and effort are devoted to them than they warrant. For the sake of this discussion I propose that marine debris is a technological hazard: a technological hazard produced by modern technology (i.e., the manufacturing of non- or slowly degradable products and packaging materials that end up in the oceans).

To remedy this misplaced sense of certainty, Kasperson and Pijawka proposed using a hazard evolution model to clarify considerations of unfamiliar technological hazards. This comprehensive yet simple

---

[1]Presentation by Daniel Leubecker, Office of Technology Assessment, U.S. Maritime Administration to the Committee on Shipborne Wastes, National Research Council, May 7, 1992, Annapolis, MD.

model examines the ways in which a society generates technological hazards and handles the resulting impacts (Fig. 13.1). Human needs result in human wants that are satisfied by a choice of technology. The selected technology initiates a product or by-product (waste) that poses a hazard. For example, production technology may create air, water, or ground pollution that requires constant controls. In addition, the product or its packaging can create hazards after its use if disposal is not controlled. Once the material is in use or released into the environment, humans or other organisms can be exposed to the hazard (e.g., plastic six-pack rings, which entangle marine animals).

Using this model, Kasperson and Pijawka identified general types of interventions at each stage of the hazard-producing process that could prevent the hazard and its concomitant risks (Fig. 13.2). The revised model provides an organizing framework for confronting a technological hazard and permits a fuller appreciation of the intervention opportunities available, both "upstream" and "downstream" of the initiating events (e.g., putting waste into the sea). Upstream, human wants can be modified; the technology used to address the wants can be altered; an initiating event during use of the material can be prevented; and release of the materials also can be prevented. Downstream, if release or use occurs, exposure to organisms can be blocked; and finally, the negative consequences of exposure can be mitigated. As a last resort, intervention after exposure to the hazard may be able to mitigate the harm done.

In this manner, the revised model facilitates discussion of the changes needed and how they may be accomplished. Important concerns are (a) the location within the social organization of the change required, (b) the costs of the change to society and how those costs are spread or concentrated, (c) the array of segments of the society involved, and (d) ways to facilitate the change. Consideration of these issues will help determine how far the benefit or benefits of a particular change or effort would go toward mitigating the targeted hazard.

The Hazard Intervention Model and Recreational Sources of Marine Debris

The Kasperson and Pijawka model has considerable application to the problem of garbage and solid waste discharged into the marine environment and to exploration of ways to facilitate the implementation of MARPOL Annex V. In working with the original model, the Committee on Shipborne Waste found it appropriate to make two modifications. First, the "human needs" column, as shown in Figs. 13.1 and 13.2, was omitted. While modification of human needs may be possible, it is very difficult to accomplish and probably not an intervention that user groups currently have the capacity to accomplish. It is important, however, to recognize the human needs that place users in the marine environment: the needs to earn a living, to engage in recreation, and to transport resources and materials. It is also important to recognize that human needs might be modified, especially in terms of whether they are regarded as appropriate for the marine environment.

As for their second modification, the committee decided to replace the "human wants" column (of Figs. 13.1 and 13.2) with "behaviors." The approach of Annex V has been to establish new performance standards. It is more appropriate, therefore, to focus on the changes in behavior that must be engaged to accomplish this new performance. Accomplishing this new regime for garbage handling need not compel a change in human "wants." Hence, the first column of the original model should be changed to "behavior that encourages generating garbage."

Dimensions of Interventions

In addition to modifying and relabeling the original hazard evolution model to focus it on "marine vessel garbage," the committee chose to refine the intervention model by suggesting five likely arenas from which successful interventions may emerge. Based on

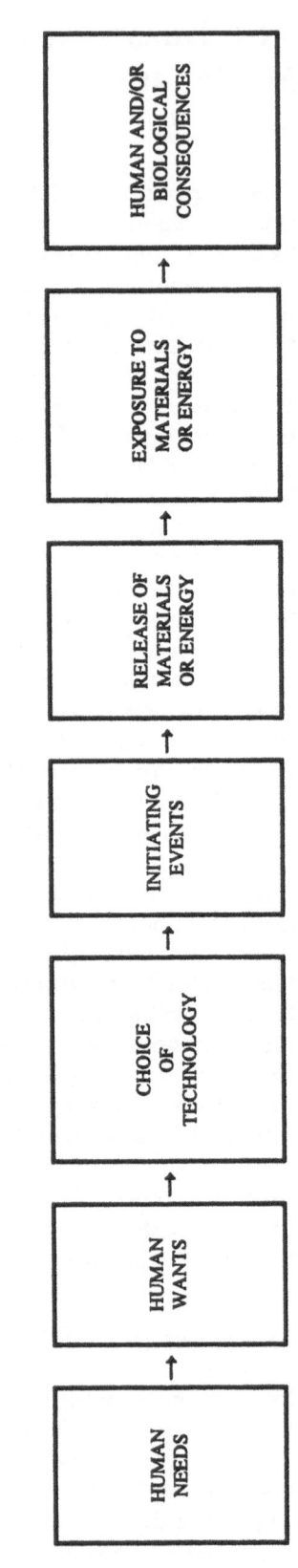

FIGURE 13.1. The chain of technological hazard evolution (the technological hazard evolution model) (Kasperson and Pijawka 1985).

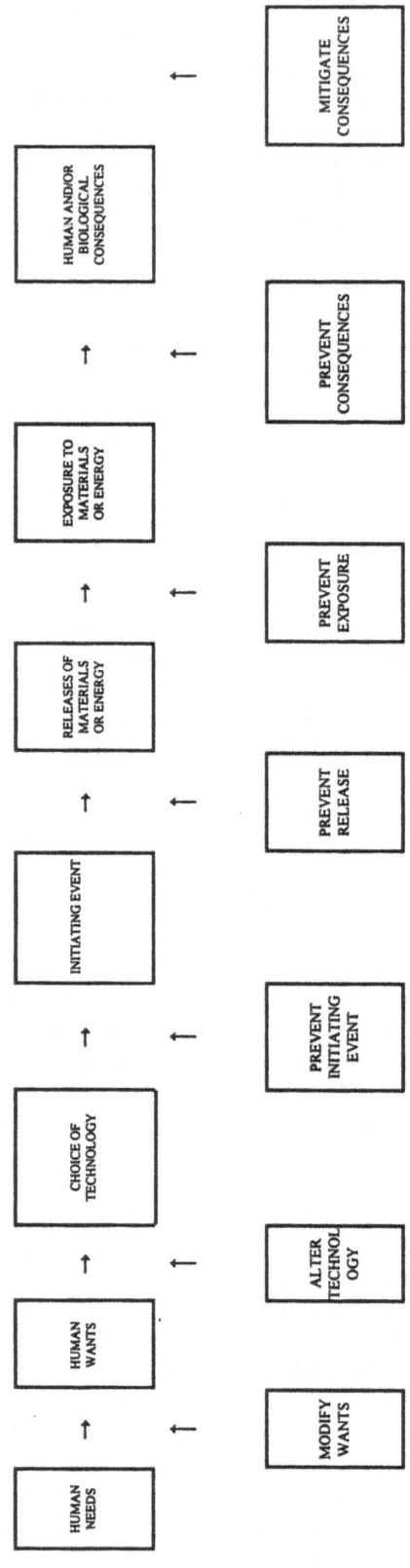

FIGURE 13.2. Intervention opportunities in hazard management (revised technological hazard evolution model) (Kasperson and Pijawka 1985).

the durable nature of the hazard created by garbage thrown from boats and ships, and the fact that the objective is to permanently end long-standing practices that have been in use across all sectors of the maritime community, the committee proposed that strong contributions to the successful implementation of Annex V can come from (1) technological innovations, (2) organizational and operational changes, (3) educational communications, (4) government and private regulation and enforcement, and (5) economic incentives.

It is important to keep in mind that considering possible interventions in all these five areas is especially important when examining recreational uses because many of those involved in recreational use of the marine environment are not involved in formal organizations. Thus, to rely on each individual recreational user changing their behavior simply because they come to recognize that their behavior is having a negative effect on the environment is asking a tremendous amount. Changing the behavior of the "freestanding" individual is very difficult to accomplish. Combine this with the generally held view that the ocean is so vast, one person's behavior makes no difference, and you can appreciate the magnitude of the educational challenge.

For there to be a significant reduction in marine debris generated by recreational users, both as individuals and as members of organizations, interventions in all five areas must be considered within each of the technological risk-production phases. Educational efforts alone will not solve the problem. Table 13.1 shows examples of using the hazard evolution model to identify interventions to reduce recreational source marine debris and its impacts.

Technological Innovations

Technological innovations offer some potential for reducing marine debris and its effects. The technology used in several phases of pollution generation could be modified and improved prospectively. While most technology to date has been designed to block the release of waste into the environment, technology also can assist in reducing both the amount of waste generated and exposure to the waste once it has been introduced into the environment.

Technologies to reduce the likelihood of garbage being taken aboard recreational vessels should focus on food- and bait-handling equipment that would permit bulk items to be stored safely and efficiently. Onboard sightseeing and fishing boats, soft drink dispensers that would not leak or pose insect problems would enhance the likelihood of their use as opposed to individually packaged soft drinks. The cups used could be reusable and thus less likely to be discarded at sea.

Technologies for marine debris intervention, once ships are at sea, are being developed by the U.S. Navy. For instance, progress is being made in developing waste-processing equipment for shipboard use. Engineers seek to ensure that these efforts result in shipboard systems that (1) are appropriate to handle the waste generated by boats and ships with various waste-generation needs, (2) are sized to the space available on boats and ships, (3) are reliable and cost effective, and (4) produce processed waste in appropriate form for safe disposal at sea or return to land. All of these qualities are important for the needs of recreational users onboard their boats. New technology tailored for use in recreational, port reception facilities presents another potential opportunity for intervention.

Organizational and Operational Changes

Organizational and operational changes may be implemented through policy changes and with new or existing technology. The purpose of such changes is to focus the human resources of corporations and organizations more effectively on the problems of marine debris. In the recreational realm, that can mean changing the various behaviors and operations of formal and informal organiza-

TABLE 13.1. Examples of possible interventions over hazard evolution: marine recreation users.

Hazard evolution model → Intervention model	Behavior that encourages generating garbage — Modify behavior that encourages generating garbage	Shipboard generation of garbage — Reduce garbage generation during voyage	Breakdown in compliance — Prevent breakdown in compliance	Discharge of garbage into sea — Block discharge of garbage into sea	Exposure to discharged garbage — Block exposure to discharged garbage	Consequence of discharged garbage — Diminish consequences of discharged garbage
Technological	Create products that need less packaging	Develop food and bait-handling equipment for bulk dispensing	Construct boats with garbage storage areas	Develop waste-processing equipment appropriate for small boats	Use collectors, retainers, or skimmers at common release or accumulation sites	Use degradable packaging where feasible
Organizational and operational	Encourage acceptance of bulk food and drinks	Remove as much packaging as possible before marine recreation	Organize waste-handling work so that compliance is easier	Organize waste-handling operations at marinas and parks for easier usage	Hold beach cleanups and debris retrieval contests	
Educational communication	Educate public to less waste-producing ways that satisfy their recreational needs; prescribe "eco-friendly" marine behaviors in eco-tourism lectures	Give specific instructions on how to reduce waste taken into marine and coastal environments	Describe MARPOL Annex V regulations; use group dynamics to reinforce desired behavior	Encourage boaters and beachgoers to report violations. Use visuals of entangled, marine life to educate		
Regulation and enforcement	For commercial recreation vessels, develop company and family waste reduction and management rules with clear rationale	Prohibit use of particular materials in recreational, maritime activities	Require marinas and parks to comply with recycling ordinances	Require filing of waste management plans with permit and license applications; limit access and use in sensitive areas	Give authority to state and municipal boating and maritime units to levy fines	
Economic	Create economical recyclable, compactable packaging and storage products	Offer such packaging and storage products near marinas and coastal parks	Publicize cost of fines; create economical compactors and other waste-handling equipment for boats	Charge significant deposits at recreation area stores for items at risk to marine discard	Offer substantial reward for recovering debris	

tions: owners and employees of recreation-based businesses (large and small) and their members and customers. For example, to change garbage-generating behavior, individuals can alter their food and beverage operations in response to marketing or organizational policy, choosing to bring bulk liquids and beverages onboard with reusable cups, choosing food with few by-products (preparing these foods on shore), choosing recyclable, compactable containers, and repackaging condiments, etc. in small reusable containers.

At the next phase, reducing garbage generation during voyages, all items that have packaging (food, equipment, replacement parts) may be removed from some packaging ashore. To prevent the breakdown in compliance, service organizations and suppliers involved in the marine recreation industry can organize the work of their employees so that it is easier for all involved to comply with sound environmental principles. Sufficient training on how to use compactors, and sufficient time to collect and store waste plus company rules for handling waste, are simple steps that prevent a breakdown in compliance and set a good example for others. Similarly, at the next phase, blocking the discharge of garbage into the sea, a port or marina can reorganize its waste management operations so that the services are integrated and easier to use, thus prompting mariners to return their waste to land. Finally, recreational users can adopt boat operating procedures that include retrieving old debris they observe in the marine environment so that it will not impact wildlife or be an eyesore to other users.

Similarly, beach cleanups are readily identifiable as interventions, however modest, against the consequences of a technological hazard. Each volunteer who bags a piece of marine debris is helping to reduce the hazard to wildlife, end the aesthetic degradation it causes, and reroute the debris back into the solid waste management system. For materials still in the water, retrieval is clearly an intervention that serves the same purpose. Recreational fishermen who retrieve marine debris from the water and bring it back to shore for disposal are helping to mitigate the consequences of someone else's illegal dumping. Rewards could be offered to encourage seafarers to retrieve marine debris; this approach has been employed in "fishing rodeos" (recreational fishing competitions) in the Gulf of Mexico.

It is important to remember that integrated waste management is an organizational and operational concept that uses technology; it is not simply the technology or, in the case of recreational users, not primarily the technology. Organizational and operational changes are possible for recreational users in each phase of waste generation identified in the hazard evolution model.

Educational Communication

Educational communication has very strong potential for effective intervention of the problems of marine debris. Because education is, from a political standpoint, one of the most accepted intervention methods for handling environmental hazards, it needs to be exploited fully. To achieve its potential, education must target specific users. Within each "community" of users, different members of the community will have distinct requirements for education, training, and information exchange.

Educational messages must be relevant to the unique situation and experiences of the particular user. They should provide information about requirements and implementation methods and persuade the user to comply: a mariner must know how to comply to do so successfully. The messages also should be structured to enhance the self-image of the mariner as someone who conforms to environmentally sound behavior voluntarily rather than seeking conformity imposed by an external control (Laska 1990). When external control is the only means used to persuade, the desired behavior will stop as soon as the external control is stopped (Arbuthnot et al. 1977). Also, the educational process should encourage the reinforcement of the message through group dynamics. The desire to conform to group norms and the power of group pressure to accomplish conformity should not be overlooked, for they are very powerful tools that can be employed to support implementation (Laska 1990).

Educational interventions are possible at every phase of the hazard evolution model. Mariners can be made aware of the way in which their human needs become articulated into wants. They also can be made aware of alternative ways to satisfy their needs or even persuaded to constrain these needs in the marine environment because of the impacts of marine debris. It may be possible to modify outdated attitudes among those who still view the ocean as perfect for dumping waste. The educational message could be that the ocean is a very inappropriate place to conduct waste-producing activities because of the difficulty of safe garbage disposal.

Specifically, educational messages can be directed toward recreational marine users to instill disgust for a polluted environment and respect for a clean environment. It should also help them know how to act on these attitudes (i.e., by reducing the amount of potential waste taken to the beach or to sea). The promotion of eco-tourism can contribute to this appreciation because the user is able to learn more quickly how such waste affects the marine organisms.

There are also promising possibilities for educational intervention at the waste-generation and waste-release stages. Objectives could include stimulating the recognition and modification of behaviors that result in waste being carried or generated onboard marine vessels and of inappropriate disposal of that waste. For people to modify their behavior, they must come to know how to do the new behavior. The easier it is for them to learn the new behavior, the greater will be the likelihood that they will do it. The desire to do something is not separate from the process one has to go through to learn it. Dieting programs that promise simple ways of learning healthy means of preparing meals are an example. Such ads recognize that the knowledge of how to do it differently is necessary to successfully change the dieter's behavior.

Compliance breakdown can be reduced through the normal reinforcing educational programs at which volunteer organizations and the media excel. To the extent that the marine user is an active member of a volunteer boaters' association, this reinforcement can extend to the normal pressures of group participation. The best learning process is reinforced by other members of the group; learning in a vacuum is much less effective. To ignore the power of the group in influencing individual behavior is to lose an important opportunity to alter behavior. Information about the management of solid waste associated with coastal and marine use should also be included in eco-tourism lectures. It is likely that individuals who take such tours will be receptive to learning more about how to protect the environment.

Educating boating auxiliaries to report violations when observed can aid in blocking the discharge of wastes into the marine environment. If more and more marine users believe that others are observing their behavior and will take action to report it, fewer will be likely to commit violations. Finally, by making marine users aware of the ways in which marine organisms are exposed to garbage and its impacts, some will act to minimize the hazard by cleaning beaches, rescuing animals, and becoming educators themselves.

Government and Private Regulation and Enforcement

Government and private regulation and enforcement have the potential to reduce hazards generated by recreational users at each phase of the waste-production process. Activities in the first phase (behavior that encourages generation of garbage) are more likely regulated by company rules. Thus, the commercial sector of the recreation industry can insist that employees act responsibly toward the environment. Such regulations come, however, only when the supervisors believe these changes as important. It is difficult to legislate such changes in behaviors of owners or supervisors or both.

At the stage of generating waste during voyages, it is possible to enact legislation that would regulate the recreational maritime use of materials that are likely to cause marine debris problems if improperly disposed. Plastics would be such materials. At the compliance stage, requiring recycling in the zoning laws near marinas would remind mariners of the importance of sound waste management practices and this would help in influencing their behavior while in the marine environment. At the stage of actually blocking discharge of the garbage, boats could be required to file waste management plans when the owners obtained permits or licenses or both. And, when the illegal discharge has actually occurred, authority to levy fines could be extended to state and municipal boating and marine authorities.

Economic Incentives

Finally, several types of interventions encourage compliance by offering an economic benefit or opportunity. Whenever possible, it is important to facilitate compliance in a cost-effective manner. Making recyclable, compactable, and reusable products for marine use that are not overly expensive should help to reduce the garbage generated during voyages. Stores located near marinas and boat launches should be encouraged to stock such items.

Publicizing possible and actual fines levied against violators can motivate mariners to comply. Moreover, the development of onboard garbage-processing equipment that is reasonably priced, yet reliable and effective, will provide motivated marine users with alternatives to discharge at sea. For those boats too small to have such equipment, deposits could be charged [probably requiring some type of incentive for the charger] on packaging items that could not be removed before use, thereby encouraging users to return their garbage to land.

Finally, economic incentives for debris retrieval become evident when nets get entangled in boat propellers and when marine animals are maimed and killed. Recreational users can also be encouraged to retrieve debris by the offering of substantial rewards for its recovery. Some fishing tournaments have already implemented this type of program.

Conclusion

Imagine a world in which recreational users did not litter. If we were to investigate how such a nonlittering situation were to prevail, we would probably discover its basis in an array of widely accepted behavioral norms, educational messages, and organizational practices. In our current society, we have such an interwoven array, but it is geared *toward* littering: wants we believe are absolutely necessary (i.e., equivalent to needs); technology that produces products which cannot degrade; norms that permit and encourage their use in the marine environment; vessels without adequate provisions for storing waste once brought onboard; inadequate waste-handling systems on shore to receive recreational user waste; and wide acceptance of the belief that the litter impact of a single recreational user is inconsequential to the marine environment.

The power of such an interwoven array of beliefs, values, and behavioral norms, whether it be to prevent or to "facilitate" marine littering, is irrefutable. To offer any prospect of truly shifting from the current paradigm to the nonlittering paradigm, the entire interwoven array must be addressed. Our example application of the revised technological hazard evolution model to address recreational sources of marine debris demonstrates a useful framework for rationalizing efforts to achieve this paradigm shift.

SECTION IV
Considering the Maritime Sources of Debris

SECTION IV
Considering the Maritime Sources of Debris

Introduction

The theories discussed in Section Three come to light in people's struggles to implement controls over the sources, pathways, and deposition of marine debris. The following account illustrates some of the primary challenges facing members of the maritime community as they decide how to address their contributions to the marine debris problem. The players are a fishing industry association (an information provider attempting to foster changes in behavior), the captain and crew of a commercial fishing vessel (information receivers implementing changes in behavior), and a fish processor operating an unloading dock.

In 1988, the fishing industry association was making their members aware of their contribution to the marine debris problem, and the impending requirements of MARPOL Annex V. In response, the captain and seven-member crew of an Alaskan fishing vessel sorted and saved all the plastic wastes generated during a weeklong trip in the Bering Sea. The crew was surprised by the volume of waste created and the work required to sort, package, and store it. At the end of its trip, the fishing vessel arrived at the fish processor's dock on Akutan Island in the eastern Aleutians on an early-morning flood tide. The crew took some pride in handing their trash ashore.

Pursuing a quick turnaround, the fishing vessel departed Akutan Island on the ebb tide that same afternoon. As the mooring lines were thrown off, the processor's trash for the day, including the trash that had been painstakingly gathered and delivered by the departing fishermen, was hauled to the pier's end and dumped into the outgoing tide. Without considering that the processor had limited options for disposing of their trash, the fishing vessel's captain and crew concluded that attempts to comply with MARPOL Annex V were a waste of their time. Word of their experience spread rapidly throughout the fleet.

Implicit in this story is the need for a broad legal and regulatory system that not only engages the three essential components of maritime-source debris control (i.e., vessels, ports and landside waste management), but also provides equal consideration to each component's unique set of social and economic challenges. Throughout history, the maritime sector has been easily defined and regulated for navigation, trade, and security purposes. Thus, the promulgation of international pollution control standards has been relatively straightforward, at least by comparison with efforts to address pollution sources on land. For ship-generated garbage, Annex V of MARPOL (73/78) and the national statutes that implement its provisions are the broadly accepted model for controlling maritime-source marine debris. Essential to its effectiveness, MARPOL Annex V is unique in that (1) it features a total prohibition on the discharge of plastics; (2) it applies to all ships, boats, and vessels, including fixed and floating platforms; and (3) ports and terminals are obligated to provide adequate reception facilities for ships' wastes. Unfortunately, Annex V does not address the landside waste management component. Implementing these features presents a different set of challenges to each participant: large and small ports, militaries, commercial fishing fleets, recreational boaters, and the tourist industry, to name a few. Also, MARPOL Annex V contains no best practice or technology requirements, which both simplifies and complicates its successful implementation.

The majority of marine debris is persistent synthetic materials, typically classed as plastics. The full prohibition of their discharge from vessels is a primary step in reducing debris. Because Annex V totally prohibits the discharge of plastics, onboard problems are encountered (e.g., space limitations, contamination, odor control, and vermin management) when handling, sorting, and storing wastes at sea. Options for shipboard waste management vary greatly between types of ships and their itineraries, each posing further challenges.

MARPOL Annex V applies to all ships, boats, and vessels, including fixed and floating platforms. Before the entry into force of Annex V, international experience with pollution control laws for ships was limited to large cargo carriers and tankers. These large vessels are tractable for regulatory purposes, making them quite different from the vast majority of vessels that do not participate in large-scale international trade. Effective extension of Annex V requirements to the mass of fishing, recreational, military, and domestic work boats of all types is one of the biggest challenges for controlling vessel-source marine debris. Typically, national capacities to enforce Annex V implementing regulations are grossly inadequate.

Ships cannot comply with the discharge requirements of MARPOL Annex V unless they have a place to deliver their wastes. Annex V requires that ports and terminals provide adequate reception facilities for ships' wastes. When ratifying Annex V, states agree to ensure the provision of adequate port

reception facilities for ships' garbage. In fulfilling this obligation, the ships and ports complete the circuit by which maritime-
source marine debris is eliminated. However, ships are not
obligated to offload garbage; ports are free to charge fees or set
other requirements (sorting, packaging, certifications, etc.) in
receiving garbage; there is competition among ports for traffic
that affects fees and services; national policies regarding entry of
foreign garbage vary; and landside disposal capacity may be
inadequate. Time and experience are working to slowly move
toward local and regional solutions to these challenges.

Unlike many other pollution prevention laws, Annex V contains no shipboard or portside best management practices or
technologies requirements. It prescribes no formal processes for
achieving its stated requirements, leaving the methods entirely
to the regulated parties. In general, the flexibility inherent in this
approach is expected to lead, after a time of confusion, to the
most economically efficient solutions to the Annex V challenges. To aid in initial implementation, the International Maritime Organization (IMO) prepared and published voluntary
guidelines. The most appropriate methods and technologies for
meeting Annex V requirements under a wide variety of circumstances are under continuing development. Along with the
Annex V Special Areas provision (see Barnett, Chapter 14, Table
14.3, this volume), which is not unique to Annex V, and, in the
case of the United States, the extension of Annex V requirements
to military vessels, these features create significant challenges
for nations, vessels, and ports as they endeavor to control the
maritime sources of marine debris.

This section includes case studies and commentaries through
which these elements of Annex V implementation are woven.
The challenges to island states and regionwide considerations
are presented from the Caribbean perspective in papers by
Barnett and Wade. Issues affecting the provision of port recep-
tion facilities are treated from an international perspective by
Ninaber and through comparative assessment by Hollin and
Shaw. We conclude the maritime-source section with case
studies on fishing vessels by Topping et al., military vessels by
Koss, and recreational boats by Ellis and Podlich, finishing with
Wallace's framework of strategies for implementing Annex V.

14.

Shipping and Marine Debris in the Wider Caribbean: Answering a Difficult Challenge

F. G. (Jerry) Barnett

Introduction

Marine debris, a global pollution problem, is especially serious for the Wider Caribbean, a region of more than three dozen diverse states and territories renowned for its fragile, natural beauty. In this developing region, already beset with a dense resident population (Table 14.1), beauty makes tourism the number-one source of foreign exchange. Fueled by the burgeoning coastal populations and exacerbated by booming tourism, the marine debris problem is magnified in the Wider Caribbean, with more and more people generating more and more garbage, much of it finding its way to the sea.

The diversity of marine debris problems in the Wider Caribbean is reflected in the number of initiatives and forums arrayed against it. As part of the problem-solving effort envisioned by the Third International Conference on Marine Debris, some of these efforts are worth exploring. To set the stage, this paper summarizes progress of the governments of the Wider Caribbean's 20 island and 15 continental states and territories in ratifying a number of important legal instruments addressing marine debris. The paper then highlights some of the important actions being taken at the international, regional, and national levels to bring marine debris under control in the region.

Conventions to Prevent Marine Debris

The International Maritime Organization (IMO) is the United Nations' agency responsible for administration of global maritime treaties. Since its creation in 1948, IMO has successfully administered improved safety at sea and protection of the marine environment through its many international conventions. There are more than 150 IMO member nations worldwide. In the Wider Caribbean Region, 26 of the 28 countries are IMO members (only Grenada and St. Kitts and Nevis are not), reflecting a broad regional interest in improving marine safety and pollution standards.

Table 14.2 displays the status (through July 1995) in the Caribbean of two IMO- and one United Nations Environment Programme- (UNEP) administered, marine protection conventions that address marine debris. The following are brief descriptions of the conventions listed in Table 14.2.

MARPOL (73/78) is the 1973 International Convention for the Prevention of Pollution from Ships as modified by the protocol of 1978. It regulates the operational discharges of wastes from ships. As shown in Table 14.2, 54% of the region's political

TABLE 14.1. Independent Caribbean Island populations and area data, 1992.

Island(s)	Stayover tourist population (×1000)	Percent of change (from 1991) in stayover tourists	Resident population (×1000)	Tourist-to-resident ratio	Island size (km²)	Total population (tourist and residents)/km²
Antigua and Barbuda	194	+6	81	2.4	441	624
Bahamas	1,399	−2	263	5.3	13,935	119
Barbados	385	−2	259	1.5	431	1,494
Dominica	47	+1	72	0.7	751	158
Grenada	88	−	91	1.0	344	520
Jamaica	909	+8	2,394	0.4	10,630	311
St. Kitts and Nevis	78	+7	42	1.9	269	446
St. Lucia	177	+12	156	1.1	616	54
St. Vincent and Grenadines	53	+3	109	0.5	388	418
Trinidad and Tobago	235	+7	1,268	0.2	5,130	293

Source: Caribbean Tourism Organization and World Bank.

entities[1] have ratified MARPOL Annexes I and II (oil and chemical pollution abatement); 49% of the region's political entities have ratified Annex III (prevention of pollution by hazardous substances carried in packages, freight containers, or portable tanks), 26% have ratified Annex IV (pollution by sewage); and 49% have ratified Annex V (pollution by garbage). Annexes I and II are mandatory; Annexes III, IV, and V are optional. Before a state can ratify Annex V, it must ratify Annexes I and II. With the exception of Annex IV, all Annexes have entered into force. [Annexes enter into force only when the combined merchant fleets of states party to the Annex exceed 50% of the gross tonnage of the world's merchant shipping. Annex IV currently stands just under 40%]. The provisions of Annexes I, II, III, and V are binding on contracting parties and can be enforced consistent with international law.

The 1972 Convention on the Prevention of Marine Pollution by Dumping Wastes and Other Matter (the London Dumping Convention or LDC) focuses on preventing or limiting the dumping at sea of land-produced wastes and is relatively successful in the Wider Caribbean, with a 57% ratification rate.

The 1983 Cartagena Convention for the Protection and Development of the Marine Environment of the Wider Caribbean Region is a comprehensive, umbrella agreement aiming to protect the marine environment of the region from pollution from ships, dumping, land-based sources, seabed activities, and airborne discharges. It also identifies environmental management issues for which cooperative efforts are to be made: specially protected areas, cooperation in cases of emergency, environmental impact assessment, and scientific and technical cooperation. There is also an article on liability and compensation. By ratifying the protocol concerning Cooperation in Combating Oil Spills in the Wider Caribbean Region, a Party accepts more specific obligations to control pollution from discrete sources and to cooperate in specific aspects of environmental management. More than two-thirds (24 of 35) of the Wider Caribbean's political entities have ratified or acceded to the convention and the protocol.

Do these conventions work? For marine debris, it is too soon to tell. The regional

[1]The 35 governments (i.e., countries, states, and territories) that are competent to ratify and implement, and are counted separately (with equal weight) in these calculations.

TABLE 14.2. Status of convention ratification related to marine debris in the Wider Caribbean Region as of July 1995.

States and territories	IMO	MARPOL Annexes I and II	MARPOL Annex III	MARPOL Annex V	MARPOL annex IV	London Dumping Convention	Cartagena Convention and Oil Spill Protocol
Antigua and Barbuda	X	X	X	X	X	X	X
Bahamas	X	X	X	X			
Barbados	X	X	X	X		X	X
Belize	X	X	X	X	X		
Colombia	X	X	X	X	X		
Costa Rica	X					X	X
Cuba	X	X				X	X
Dominica	X						X
Dominican Republic	X					X	
France[a]	X	X	X	X	X	X	X
Grenada							X
Guatemala	X					X	X
Guyana	X						
Haiti	X					X	
Honduras	X					X	
Jamaica	X	X	X	X	X	X	X
Mexico	X	X				X	X
Netherlands[b]	X	X	X	X		X	X
Aruba		X	X	X		X	X
Netherlands Antilles[a]		X	X	X		X	X
Nicaragua	X						
Panama	X	X	X	X	X	X	X
St. Kitts and Nevis							
St. Lucia	X					X	X
St. Vincent/Grenadines	X	X	X	X	X		X
Suriname	X	X	X	X	X	X	
Trinidad and Tobago	X						X
United Kingdom[c]	X	X	X	X		X	X
Anguilla							
British Virgin Islands							X
Cayman Islands		X	X	X		X	X
Montserrat							
Turks and Caicos Islands							X
United States[a]	X	X	X	X		X	X
Venezuela	X	X	X	X	X		X
Total not ratified	—	16	18	18	26	15	11
Total ratified	26[d]	19	17	17	9	20	24
Percentage ratified	—	54	49	49	26	57	69

Source: International Maritime Organization (IMO) (1995): U.S. State Department; U.S. Coast Guard.

[a]Ratification is automatically extended to French Departments (French Guiana, Guadeloupe, and Martinique) and U.S. Territories (Puerto Rico, and the U.S. Virgin Islands). Ratification is also automatically extended to the five islands of the Netherlands Antilles. These departments, territories, and islands are not included in the total count or percentage of ratification.

[b]After ratification by the Netherlands, implementation requires independent action by Aruba and the Netherlands Antilles; therefore, Aruba and the Netherlands Antilles are included in the total count and percentage of ratification.

[c]Ratification requires independent action by British Territories (Anguilla, British Virgin Islands, Cayman Islands, Montserrat, and the Turks and Caicos Islands). These territories are included in the total count and percentage of ratification.

[d]Number of IMO members in the Wider Caribbean Region.

experience in controlling oil pollution, as described by the Joint Group of Experts on the Scientific Aspects of Marine Pollution (GESAMP) (IMO 1993), suggest that these conventions will work, especially once the MARPOL Annex V Special Area rules become enforceable.

Wider Caribbean Initiative on Ship-Generated Waste (WCISW)

Ship-generated wastes properly discharged to port facilities may find their way back into the marine environment when land disposal systems are inadequate. In areas where sanitary landfill facilities are limited or nonexistent, garbage from land is sometimes collected and taken to sea for dumping. An integrated approach to waste disposal in coastal areas is therefore essential if marine debris is to be controlled.

MARPOL 73/78 allows for the designation of specific geographic areas in which the discharge of certain substances is prohibited. In these "special areas," rules can apply to any or all the pollutants covered by Annexes

I, II, or V of the convention. As highlighted in Table 14.3, the disposal of plastics, dunnage, packing materials, paper, rags, glass, metal, bottles, or crockery is prohibited within Annex V Special Areas. In fact, only food wastes can be disposed in special areas, and then only within the constraints detailed in Table 14.3.

Following the recommendation of an IMO and UNEP workshop in Caracas during October 1990, IMO's Marine Environment Protection Committee approved designation of the Wider Caribbean as a special area under MARPOL Annex V, effective April 18, 1993. Before special area rules can be enforced, states of the region must notify IMO that adequate port reception facilities are available to meet the needs of the ships operating within the area. On receipt of sufficient notifications, IMO will inform (within 12 months) all contracting parties to MARPOL that special area rules will apply to the Wider Caribbean.

Solid groundwork thus has been laid for the protection of the marine environment of the Wider Caribbean from the dangers of improperly disposed ship-generated wastes. However, much work remains to be done before states of the region can reap the benefits of special area status. Three interrelated

TABLE 14.3. Summary of at-sea garbage disposal regulations (IMO 1992).

Garbage type	All ships except platforms[a]		Offshore platforms[a]
	Outside special areas	Inside special areas[b]	
Plastics (e.g., synthetic ropes, fishing nets, and garbage bags)	Disposal prohibited	Disposal prohibited	Disposal prohibited
Floating dunnage, lining, and packing materials	>25 miles offshore	Disposal prohibited	Disposal prohibited
Paper, rags, glass, metal, bottles, crockery, and similar refuse	>12 miles	Disposal prohibited	Disposal prohibited
All other garbage comminuted or ground (e.g., paper, rags, and glass)	>3 miles	Disposal prohibited	Disposal prohibited
Food waste not comminuted or ground	>12 miles	>12 miles	Disposal prohibited
Food waste comminuted or ground[c]	>3 miles	>12 miles	>12 miles

[a]Offshore platforms and associated ships include all fixed or floating platforms engaged in exploration or exploitation of seabed mineral resources, all ships alongside or within 500 m of such platforms.
[b]Garbage disposal regulations for special areas shall take effect in accordance with regulation 5(4)(b) of Annex V.
[c]Comminuted or ground garbage must be able to pass through a screen with a mesh size no larger than 25 mm.
[d]When garbage is mixed with other harmful substances having different disposal or discharge requirements, the more stringent disposal requirements shall apply.

efforts must be undertaken. First, national legislation adequately addressing environmental matters must be enacted. Many states of the region currently lack laws concerning environmental pollution (Wade, Chapter 15, this volume). Where laws exist, they are often outdated, carryovers of laws imposed by colonial powers; they do not reflect current issues and do not encompass the control of polluting substances covered by MARPOL. Although MARPOL has the effect of international law, it can only be enforced under the domestic laws of contracting parties; neither IMO nor any other international organization can enforce it. In the absence of adequate national laws, no enforcement is possible. Technically, the ratification of MARPOL by a state cannot take place until the state has in place adequate laws to enforce the convention.

Second, MARPOL must be ratified. As noted earlier, barely 50% of the states and territories of the Wider Caribbean have ratified it. Many factors contribute to this situation, including the lack of adequate national legislation; a dearth of national institutions to conduct the required convention-related inspections, a shortage of adequate port reception facilities; and perceived higher priority issues within governments.

Third, adequate port reception facilities for ship-generated waste must be provided. In many cases this implies significant capital investments by governments, port and terminal operators, and waste management companies serving ports. Governments must be encouraged to evaluate means within their authority to establish incentive systems to manage the economic burden so that garbage and other MARPOL-controlled substances delivered to port are actually received and disposed of properly at reasonable cost. Options for this encouragement include tax incentives, loan guarantees, government subsidies, and preparation of coordinated project proposals to secure financing from international organizations.

In October 1993, an IMO–World Bank Workshop in London initiated the US$5.5 million WCISW, addressing these issues and aiming at putting in place the infrastructure necessary to fully realize the special area status of the Wider Caribbean. This groundbreaking workshop was attended by delegates from every independent country in the Wider Caribbean—government officials responsible for port activities and national legal affairs relative to the implementation of MARPOL. WCISW was launched through the Global Environment Facility and, at no cost to any government, recently has been upgraded to include fielding a project coordinator, a technical consultant, and a legal consultant in the region for 3 years to assist in the development of the critically needed pollution legislation and port reception facilities. Additionally, there is a project backstopping officer at IMO headquarters in London.

Anticipated to run through the summer of 1997, the WCISW will sponsor five workshops in the region. Three of them will be technical workshops held 12, 24, and 30 months from inception of the field program. Legal workshops are planned 12 and 24 months after the legal consultants begins work. A final workshop concluding the WCISW will be held at World Bank headquarters in Washington, D.C.

Organization of Eastern Caribbean States (OECS) Solid Waste Management Project

The OECS was established in 1981 to promote political and economic cooperation among Antigua and Barbuda, Dominica, Grenada, Montserrat, St. Kitts, and Nevis, St. Lucia, and St. Vincent and the Grenadines, with the British Virgin Islands an associate member. Population figures vary from 140,000 in St. Lucia to 14,000 in Montserrat, and land masses from 290 square miles in Dominica to 36 in Montserrat. The two British dependent territories of Montserrat and the British Virgins are not included in this project, the preparatory phase of which is almost complete.

This project, begun in early 1993, is important because of its sharp focus on island needs and its inclusion of both ship- and shore-generated wastes. Without activities such as this, impediments to achieving the MARPOL Annex V Special Area status for the Wider Caribbean will not be overcome. This initiative aims to (1) upgrade existing landfills and construct some new ones; (2) set up a system of waste collection and disposal for ship-generated, MARPOL Annex V wastes; (3) enhance waste collection by providing new equipment and, where appropriate, developing transfer stations; (4) minimize waste through governmental policy changes and recycling; and (5) strengthen appropriate institutions and develop legislation for improving solid waste management, environmental health, and public education programs for better overall environmental management.

To handle the growing amount of ship-generated waste landed annually in OECS states, the project is looking to locate adequate reception facilities in international ports of entry as well as in smaller harbors used by domestic craft. Participating states will have the opportunity to collaborate in (a) developing and employing regional environmental standards for solid and liquid waste disposal, (b) coordinating efforts to harmonize environmental legislation, (c) developing upgraded facilities for monitoring water quality, and (d) developing standards for waste management at ports, harbors, and anchorages.

The OECS-recommended system for waste produced aboard ships of 100 gross tons or more includes the following:

The comminution and disposal at sea beyond 12 miles of all food wastes

The volumetric reduction of nonfood wastes by a factor of at least 12 by compacting, developing an average density of 39 lb/ft^3, or incineration

The sterilization (e.g., incineration and cooking) aboard ship of all food-contaminated waste before compacting and transfer ashore

Transfer ashore to rugged bins compatible with transportation systems.

It is estimated the cruise ships without incinerators, but with compactors, would need about 10 bins; cargo vessels would need only 1 or 2. For smaller vessels under 100 tons coming in from international voyages, Annex V wastes would be stored in plastic bags and deposited into one of the bins on landing. Garbage generated by small vessels sailing domestically would be treated as domestic waste, disposed as it is now by marina operators.

To ensure only ship-generated waste is taken ashore, a set of safeguards will be observed. First, large vessels will submit a waste management plan to the Ministry of Health or similar authority at least 90 days in advance of the port call. This plan would specify onboard waste management, food-sterilization, and waste-reduction practices, and would estimate the volume of waste to be landed; the national authorities would have 90 days to approve the plan or suggest modifications. Second, ship operators will certify only wastes generated since the previous port call are being landed, and on arrival, the ship's captain will certify that the management and landing of the ship's waste is per the approved plan. Third, national authorities will have the right to inspect the vessel or its waste at any time while in national waters. Finally, noncompliance would result in the requiring of a waste management performance bond by the culpable shipping line, to be exercised in the event of future noncompliance.

Estimated costs of the project run to around US$50 million, being financed through a consortium of organizations including the World Bank, with another US$5–$10 million coming from the participating states. The recovery of recurring costs for waste collection and disposal is vital to the small island economies maintaining the updated waste systems planned. Under the OECS Project, fees for the management of ship-generated waste are to be harmonized through common guidelines, although some variation is anticipated. Plans call for the charges either to be incorporated into existing port charges or levied separately as a special environmental fee.

Regional Program for the Assessment and Control of Marine Pollution (CEPPOL)

CEPPOL is one of five regional programs forming the Action Plan for the Caribbean Environment Programme (CEP) under the Cartagena Convention. The Caribbean Environment Programme was established to provide a mechanism whereby the diverse states and territories of the Wider Caribbean could collectively protect their marine and coastal resources, the basis for much of their economic development. The Regional Coordinating Unit (RCU) in Kingston, Jamaica, coordinates and implements the program, operating under the joint authority of the UNEP and the governments of the region. Support comes from (1) the Caribbean Trust Fund (from governments participating in the Action Plan), (2) the environmental fund of UNEP, and (3) national, bilateral, and multilateral contributions.

The RCU plays an important role coordinating and consolidating the relevant work of a host of partner organizations, including the Advisory Committee on Protection of the Sea, the Caribbean Coastal Marine Productivity Programme, the Food and Agricultural Organization, Greenpeace, the Intergovernmental Oceanographic Commission (IOC), the IMO, the Marine Environmental Laboratory of the International Atomic Energy Agency, the Pan American Health Organization, the United Nations Development Programme, and the World Bank.

CEPPOL itself is a jointly sponsored IOC–UNEP activity. One of 10 (approximately) CEPPOL program activities is the monitoring and control of pollution by marine debris. The presence of marine debris, especially plastic wastes on beaches and in coastal waters, is a major concern in the Wider Caribbean. CEPPOL is in the process of assessing the extent and impact of marine debris in the region. This assessment will enable the design and evaluation of the effectiveness of measures aimed at controlling marine debris.

Completion of data collection on sources, fluxes, and levels of marine debris is scheduled for 1995. The RCU will then prepare a summary of the findings, including a layman's version for broadening public education and awareness about marine debris-related problems in the Caribbean. The RCU is continuing its efforts to coordinate and plan regionwide activities for educational and monitoring activities.

The Barbados Port Authority Incinerator

The Barbados Port Authority's incinerator project is an outstanding example of initiative and resourcefulness in handling the marine debris problem. Begun as an idea in the Barbados Port Authority's Engineering Department in 1986, the incinerator project started in earnest in 1991 when government permission to proceed was granted. At the same time, an IMO and World Bank study of port reception and disposal facilities for garbage in the Wider Caribbean recommended financing the construction of a 16-ton daily-capacity, municipal waste incinerator in Bridgetown.

Landfill capacity in the Caribbean Islands is being rapidly consumed. This is particularly true in Barbados, an island country with one of the highest population densities in the world, more than 600 people to the square kilometer. Bajan government regulations require sterilization of ship-generated wastes before landfilling. Before construction of the incinerator, wastes received at the port were burned in the open and the ashes were landfilled. While having an obvious negative effect on ambient air quality, the open-burning method caused other problems. The smoke sometimes entered the ventilation systems of cruise ships in the harbor, making life uncomfortable for some passengers and miserable for asthmatics. The billowing smoke from the shoreside fires set off shipboard fire alarms, further changing the desired image as an idyllic cruise destination. To stop these problems, the Port Authority embarked on an international search for an inexpensive, smokeless, easy-to-maintain incinerator.

Finance

Original construction estimates were around US$1 million for the entire incinerator project. The Port Authority's Engineering Department was able to cut this cost to about $800,000 by preparing the site themselves, digging trenches for drainage, roughing-in plumbing for seawater cooling, working with the Barbados Natural Gas Company setting up the specially constructed gas line to provide the right gas pressure, doing electrical work, and pouring the extensive concrete slab.

The Port Authority made a 10% down payment in February 1992 to Long Island's (NY) Tanner Management Corporation for the Hoskinson Pyrolytic Smokeless Incineration System, intending to spread the balance of the cost over 2 or 3 years. Fabrication in New Jersey was completed and the incinerator was shipped in two 40-ft containers in October 1992. The incinerator was commissioned, and the outstanding balance paid in cash, on May 6, 1993, during a period of rising tourism in Barbados. Bridgetown was voted the Caribbean Shipping Association's 1993 Port of the Year. Port Authority officials anticipate the incinerator paying for itself in 7 or 8 years and having an operating life of 20 years. Ships are charged US$75 for three 1.5 m-capacity bins of garbage. The Port levies a separate US$3 head charge for visiting passengers.

Not only does the incinerator generate revenues per garbage bin for the Port Authority, the port's clean, efficient handling of cruise ship wastes serves as a magnet for more cruise ship port calls, which generates even more revenues for the port and the country. As a further benefit, moving from open-hearth burning to incineration freed 5000 ft^2 of valuable space in the port for more productive uses; the new incinerator site required only 4000 ft^2 versus the open-hearth site's 9000.

Operation

To handle the load of more than 6000 tons of waste brought into Bridgetown annually by cruise ships, the incinerator runs Monday through Friday in two 8-h shifts daily with a 1-h overlap; one shift is scheduled Saturday and Sunday if ships are in. Occasionally, the burning day runs shorter than 15 h, and maintenance is performed.

Ship-generated waste in 1.5-m^3 bins is fed into the incinerator at 22.5 m^3 h^{-1}. However, a growing segment of the input is coming from on-island sources—government agencies and financial institutions—to destroy shredded documents. The present operation reduces volume to the landfill by factor of roughly 19, converting about 340 m^3 of garbage to about 18 m^3 of sterile ash daily. The incinerator ash is removed first thing each morning before starting the day's burn cycle.

Environment

Two emissions tests done in October and December 1993, just before the high tourist season, easily surpassed the stringent U.S. carbon monoxide and particulate standards. As was done in the open-hearth days, the untreated incinerator ash is deposited directly into the landfill.

Expansion

Because of its great success from almost every perspective, a US$450,000 expansion is planned. The planned expansion will include:

A 43%-enlarged bottom chamber that boosts capacity from 35 to 50 tons per day

A scrubber that provides a membrane between the bottom and top chambers which slows the smoke, ensuring a more complete burn and ultimately releasing less smoke to the atmosphere

A magnetic separator that facilitates the removal of metals from the ash (ships are already required to sort their garbage via color coding)

A compactor that crushes and recycles sterile metal cans.

The Port Authority also plans to convert some of the incinerator's energy into hot water for use in the port authority complex.

The Nassau (Bahamas) Waste Conversion Project

Only in the proposal stage, this US$12 million project located in Nassau would include a dockside facility. It would utilize "low technology" similar to that used in a 350-tons-a-day plant operated successfully in Jamaica for 12 years before it was closed by the government in 1974. This endeavor aims to convert 95% of Nassau's solid, nonhazardous waste into usable agricultural and horticultural material using a clean, safe, vermin- and odor-free method of accelerated decomposition. Modern engineering and materials-handling technology, involving no separation of wastes, make this process feasible and affordable. This project has important implications for solid waste management throughout the region.

Caribbean Marine Debris Workshops

To date, there have been four Caribbean Marine Debris Workshops: in April 1991 in Puerto Rico, August 1992 in Mexico, January 1994 in the Bahamas, and in August 1995 in the Dominican Republic. A fifth workshop is planned for July 1996. These workshops are conducted for, and by, scientists, policy makers, educators, regulators, port and terminal operators, industry, marine user groups, conservation organizations, and solid waste managers to preserve and strengthen the marine debris network, assess progress toward predetermined goals, and exchange information on recent events and emerging issues. Support is lent from the IOC, IMO, the U.S. National Oceanic and Atmospheric Administration, UNEP, the U.S. Environmental Protection Agency, the U.S. Coast Guard, the U.S. Sea Grant College Programs, and the World Bank.

A primary product of the workshop series is a 15-point Marine Debris Waste Management Action Plan for the Caribbean (IOC 1994) that specifies the main activities needed to reduce marine debris in the Wider Caribbean Region. The Action Plan strengthens regional cooperation between national and municipal agencies and organizations as well as the tourism and waste management industries. The Plan identifies key participants and their potential roles in marine debris and waste management programs.

Conclusion

All the challenges to achieving the objectives of MARPOL Annex V are manifested in the Wider Caribbean Region: legislation, port reception, waste management and disposal technologies, enforcement, intraregional competition, and special area requirements. The diversity of views on the best solutions to marine debris problems in the Wider Caribbean Region, and elsewhere, is not a problem. It is a product of the simplicity of MARPOL Annex V and the diversity of needs and capacities in the Caribbean community. Innovative solutions born of this diversity can be shared to the benefit of other regions and communities facing similar challenges.

15.

The Challenges of Ship-Generated Garbage in the Caribbean

Barry A. Wade

Introduction

The ratification of Annex V of the MARPOL Convention marks an important step forward in the protection of the oceans. However, many countries have failed to fully understand its significance, and the leaders of Caribbean States do not seem to have paid sufficient attention to this development. One reason may be that in the area of marine pollution, ship-generated garbage is considered less significant than oil spills or spills of hazardous substances. At the local level, ship-generated garbage has not evoked society-wide responses. Nonetheless, if left unaddressed ship-generated garbage could have damaging environmental and public health consequences.

Ship-generated garbage has a leprous character; it is unwanted by both ships and ports. Before the development of international obligations, Caribbean States had full justification for refusing to receive this unwanted waste. Ships were left with one of two choices: retain garbage onboard or dispose of it at sea. Many were indifferent and unconcerned about which option was exercised. Now, Annex V is challenging Caribbean States not only to meet their new obligations, but to do so while maintaining and improving the competitiveness of their ports.

The fact is that the opportunity cost of inaction under Annex V is too great for Caribbean States to ignore for two simple rea-

sons. First, although ship-generated garbage has been found to contribute only 8.5% (by weight) of the litter found on Jamaican beaches, the persistent character of its principal constituents (plastics and glass) make it a source of long-term pollution (Wade et al. 1991). Additionally, this percentage is likely to grow with the projected rapid increase in merchant and cruise ship activity in the Caribbean.

Shipping Trends in the Caribbean

Massive investments have been undertaken as Caribbean countries strive to establish and maintain market shares as transshipment hubs and home ports for cruise ships. Every state including Cuba, Haiti, the Dominican Republic and Venezuela have undertaken port expansion or enhancement projects in recent years (Table 15.1). Where cruise shipping is concerned, almost all the ports in the Caribbean have embarked on projects to develop or extend berths to accommodate larger vessels. Many countries have significantly upgraded and improved shore facilities to take full advantage of the cruise trade. Competition between Puerto Rico, Jamaica, and now Trinidad and Tobago to become transshipment hubs for the major shipping lines has led, in the case of Jamaica, to the investment of U.S.$58 million in the develop-

TABLE 15.1. Cruise port developments in the Caribbean (1992 to present).

Country	Planned expansion
Bahamas	Shopping complex (U.S.$50 million)
Barbados	Passenger terminal with shopping complex
British Virgin Islands	Cruise ship pier and passenger terminal
Cayman Islands	Cruise ship pier (U.S.$16 million)
Dominica	Cruise port
Grenada	Welcome center
Jamaica	Cruise ship terminal (U.S.$50 million)[a]
Martinique	Cruise ship terminal
Puerto Rico	Passenger terminal[b]
St. Croix	Cruise ship pier (U.S.$15 milion)
St. Maarten	Passenger terminal, shopping mall, and cruise berth (U.S.$60 million)
Trinidad and Tobago	Cruise ship terminal

Source: *Cruise Industry News Quarterly* (spring 1992).
[a]*Daily Gleaner* [Jamaica] (May 1994).
[b]*Seatrade Review* (March 1992).

ment of new berths at Gordon Cay. In Barbados a duty-free shopping center was recently opened in conjunction with the upgrading of the port.

These projects are a reaction or adjustment to the developments in the international shipping industry. Shipping lines are ordering larger ships to take advantage of economies of scale and reduce operational costs. The "round-the-world service" provided by the major shipping lines is a distinct feature of the container trade and a prime example of the trend toward the utilization of larger vessels. The cruise trade, especially in the Caribbean, is characterized by the onslaught of major new ship construction. These new megaliners with passenger capacities of 2000 and more will dominate the world market, 70% of which operates in the Caribbean (Cruise Industry News 1993/94). Further, competition among cruise lines is rife in the Caribbean market and has led to discounting wars to attract passengers in a market that is said to be suffering from "overtonnaging."

The foregoing trend in the market for shipping services has significance for Caribbean states in that (1) larger ships with larger passenger populations mean larger amounts of garbage will be generated; (2) competition in the market also means that as ships look for ideal ports of call, the economic cost of using the port facilities will weigh heavily in the decision; (3) ship owners, whether cruise, tanker, container, or fishing, are motivated to maximize profits and when it is economically expedient to do so, they may illegally dump their garbage; and (4) the International Maritime Organization (IMO) estimates that more than 50% of cargo transported by sea today can be regarded as dangerous, hazardous, or harmful under the IMO classification, designation, or identification criteria, respectively.

The projected increase in transshipment and ship tonnage in general will mean that the potential health and environmental risks posed by cargo-associated, maintenance, and operational wastes will be great. Caribbean States are therefore placed in a dilemma. Serious steps will have to be taken to manage increasing volumes of shipborne garbage while not adversely affecting the costs of using port facilities.

Caribbean island states have large coastlines and proportionally large Exclusive Ecomomic Zones that cannot, on the resources available, be effectively policed to prevent marine debris. Furthermore, even if offenders are caught, outdated marine legislation such as the Jamaican Harbours Act of 1897 that provides for a paltry fine of $2000.00 (U.S.$65.00) for marine pollution are not effective deterrents. If a suitable legal and institutional framework is not established, increasing amounts of garbage will be dumped at sea, some portion of which will eventually arrive on the beaches. Based on these points, we conclude that the Caribbean States cannot afford to cling to the status quo, but must face the challenges thrust upon them by the growing maritime industry and the international laws that govern it.

The Legal Setting: International Law

The challenges facing Caribbean States arise primarily from the obligations ensuing from international law and its incorporation into domestic law. The primary source of law governing ship-generated garbage is the International Convention for the Prevention of Pollution from Ships (MARPOL 73/78), optional Annex V. The convention addresses pollutants in five Annexes: (I) oil, (II) noxious liquid substances in bulk, (III) harmful substances carried by sea in packaged forms, (IV) sewage, and (V) garbage.

Garbage under Annex V is composed of several types of waste including food waste, domestic waste, operational waste, cargo-associated waste, maintenance waste, and plastic. Plastic is a notable component of garbage because of its broad utility, including packaging (bottles, containers, liners), disposable eating utensils and cups, bags, sheeting, fish nets, strapping bands, rope, and line (IMO 1988). Plastics along with cargo-associated wastes (dunnage and pallets, packaging material, plywood, cardboard, wire, and steel strapping) are the most persistent forms of garbage and pose the most problems in their recovery and disposal. As mentioned, the high percentage of dangerous goods being transported by sea means that operational and cargo-associated wastes may pose safety hazards and require special handling not normally provided by garbage reception facilities.

It should be noted that the ratification of Annexes I and II is compulsory on ratification of the convention. This means that the obligations of Annex V need to be addressed alongside the obligations under the compulsory Annexes that include and are not limited to the provision of reception facilities for oil, oily wastewater, and noxious liquids.

The Amendments to MARPOL 73/78 designating the Wider Caribbean as an Annex V Special Area were formally adopted in April 1993, giving Caribbean States added protection against marine pollution from ship-generated garbage. In the Special Area only food waste may be discharged under certain specified conditions. However, IMO regulations specify that the special area requirements will not apply unless all the port states have notified IMO that they can provide adequate port reception facilities. The added

protection has therefore brought with it added responsibilities and challenges, especially in the provision of adequate reception facilities for the ships that will be forced to retain most of their garbage on board for discharge ashore.

Ratification: The First Challenge

The new rights granted under MARPOL Annex V can only be exercised and enforced by Caribbean States if they have, in fact, ratified the convention and incorporated its provisions into domestic legislation. However, only 49% of Caribbean States have ratified the convention (see Barnett, Chapter 14, Table 14.2, this volume). The reasons for this may be unassociated with the perceived costs of implementing the MARPOL obligations.

As explained here, the majority of the Caribbean countries are shipper-dominated, not ship-owning, countries; therefore, pressures for the passage of maritime legislation in general are weak. In Jamaica, for instance, the Merchant Shipping Act and the Harbours Act both date back to the nineteenth century and have been in the process of being amended for more than 10 years. The first challenge therefore involves the creation of enough pressure to encourage the various governments to ratify the convention. However, even in those cases in which the convention is ratified, other legal considerations, particularly the nature of the local legislation, will play a critical role in ensuring its successful administration. [The IMO and World Bank's Wider Caribbean Initiative on Ship-Generated Waste, (see Barnett, Chapter 14, this volume), addresses this legislative challenge.]

Domestic Law

A State will not be able to enforce its right under the convention unless the provisions the convention are incorporated into domestic law. Inconsistencies between existing domestic laws and international laws dealing with the same subjects tend to weaken the validity of the international laws. Although no local laws in Jamaica, for example, specifically address shipborne garbage, there exists a number of laws that may pose problems in the implementation of MARPOL Annex V, especially regarding the duty to receive shipborne garbage. The same is probably true for all the Caribbean states that operate under English Common Law (Table 15.2). Because shipborne garbage is a relatively new issue, it is likely that other Caribbean States will experience similar legal conflicts in implementing Annex V. A number of laws under the Jamaica jurisdiction can be examined to highlight these problems.

The Natural Resources Conservation Authority (NRCA) Act, 1991

The NRCA Act is the primary piece of environmental legislation in Jamaica and as such has implications for the management of wastes. Section 12 (1)(a) of the Act prohibits the discharge "on the ground or into the ground" of any polluting matter except in accordance with a license granted by the Authority. Section 12(1)(b) also prohibits the construction of any works for the discharge of any polluting matter unless under a license granted by the Authority. By virtue of Section 38(1) of the Act, the relevant Minister of Government may make regulations governing *inter alia* (d) "the importation, collection, storage, recycling, recovery or disposal of substances that may be hazardous to the environment" and (g) "the design, construction, operation, maintenance and monitoring of facilities for the control of pollution and waste disposal."

The powers granted to the NRCA under this Act can easily be used to govern the management of shipborne waste. The importation of waste under this Act and the construction of reception facilities as considered

TABLE **15.2.** National legislation affecting shipborne waste management.

Country	Legislation
St. Christopher and Nevis	Litter (Abatement) Act 1989
	Public Health Act 1969
	Quarantine Act (Chap. 227)
Trinidad and Tobago	Customs Act (Chap. 78:0)
	Public Health Ordinance (Chap. 12/14)
	Quarantine Act (Chap. 28:05)
	Trade Ordinance 1958
St. Lucia	Public Health Act 1975
	Quarantine Ordinance (Chap. 158)
	Customs Act 1967
	Litter Act 1983
Turks and Caicos	Customs Ordinance 1973
	Public Health Ordinance 1976
	Quarantine Ordinance (Chap. 47)
Montserrat	Importation Protection Ordinance (Chap. 47)
	Public Health Ordinance 1981
	Quarantine Act (Chap. 210)
St. Vincent and Grenadines	Customs Act (Chap. 183)
	Litter Act 1991
	Public Health Act 1977
Belize	Customs Excise Duties Act (Chap. 38)
	Public Health Act (Chap. 31)
	Quarantine Act (Chap. 32)
Grenada	Customs Act 1960
	Public Health Act (Chap. 237)
	Quarantine Act (Chap. 247)
Jamaica	Customs Act 1941
	Public Health Act 1974
	National Resources Conservation Authority Act 1991
	Quarantine Act 1951
	Litter Act 1985
Barbados	Customs Act (Chap. 66)
	Quarantine Act (Chap. 53)

under Section 9 of the Act are subject to a permit or license, the terms of which are dictated by the Authority. Additionally, Section 32(1) of the Act gives the NRCA broad powers to guard against any condition that may tend to endanger the environment where no laws specifically address the condition. Consequently, any domestic legislation developed to implement Annex V would have to recognize that the duty to receive shipborne waste could be severely curtailed or overridden by the NRCA, should that receipt present additional threats to the environment.

The Quarantine Act, 1951

The Quarantine Act governs the control of the spread of infectious diseases in cargo or persons arriving or departing through Jamaican airports and seaports. Although there exists no explicit prohibition on the importation of garbage, the Act allows a quarantine officer to refuse the granting of pratique (clearance for berthing for health purposes) and subsequently detain a ship if he considers it to be in an unsanitary condition. Regulation 29 of the Quarantine (Maritime) Regulations provides that no merchandise or other article can be removed from a ship that is not granted pratique. Much discretion is given to the relevant quarantine officer to determine whether or not a ship has onboard substances likely to cause the spread of infectious diseases. The relevant officer can therefore determine whether shipborne garbage can be discharged or not. Section 7(1) of the Act also gives the relevant Minister of government powers to make regulations for preventing "danger to health from ships . . . or things therein arriving at any place. . . ." The Quarantine Authority can therefore be given specific powers under the Act that can be used to restrict the importation of garbage. It should be noted that the equivalent regulations governing aircraft specifically prohibit the discharge of matter capable of producing an outbreak of infectious disease.

The Customs Act, 1941

Although this Act concerns the financial as opposed to environmental consequences of the importation of goods, Section 76 provides that no goods may be unloaded from any ship without the written permission of the Commissioner of Customs. As garbage is defined as ship's cargo under Section 70 of the Act, problems may arise when the customs authorities exercise their powers under the Act over cargo for which there is no bill of lading or equivalent document and is generally of no value to anyone.

The Litter Act, 1985

Under this Act, litter includes garbage. Section 3 of the Act makes it an offense to deposit litter in a public place, which by definition includes a wharf, jetty, quay, or any other place to which the public has access. The importation of shipborne garbage can therefore be considered a criminal offense under the Act. Hence, the existing laws in Jamaica inadvertently restrict and complicate ports' reception of garbage, and major amendments will have to be made before the provisions of MARPOL V can be implemented. The same can be said for other English-speaking Caribbean countries that adopted similar legislation from the United Kingdom when they achieved independence.

Implementing Flag State Obligations

Once incorporated into national legislation, the implementation of the legislation governing the management of shipborne garbage poses substantial challenges. These challenges will be posed to Caribbean countries in their capacity as Flag States and Port States. The obligations of Flag States under MARPOL surround the implementation of waste minimization practices and shipboard garbage-handling and storage procedures. Major investments will include the installation of incinerators, compactors, comminuters, and associated hardware. Caribbean States, however, apart from the few fledgling open ship registries in Antigua, St. Vincent and the Grenadines, Barbados, and Grand Cayman, are not shipowning or flag states in the traditional sense.

Implementing Port State Obligations

As Port States, Caribbean countries will be responsible for the provision of adequate reception facilities as well as the transport and final disposal of the garbage received.

Reception Facilities

The provision of reception facilities will be determined by a number of factors including the type of garbage received, the ship type and design, ship operating routes, number of persons on board, duration of voyage, and the time spent in areas where discharge into the sea is restricted.

Research has been carried out by the World Bank on the considerations involved in providing reception facilities in the Caribbean (World Bank 1991). The study showed that the various port authorities were aware of the need for reception facilities, but after weighing the costs and benefits, they found themselves faced with costs associated with (1) potential import of disease vectors, (2) additional burden on municipal waste management, and (3) risk to workers from handling uncontrolled wastes. The benefits identified can include both the attraction of port traffic and the increase in port revenue from fees charged for reception facilities. In Costa Rica, after weighing the costs and benefits, legislation was passed to prohibit the importation of Annex V wastes (World Bank 1991). However, in view of the influence of tourism on Caribbean economies and the part that

cruise shipping plays in that industry, one has to assume that all the Caribbean Island States will ratify the convention and seek to satisfy the obligations therein. The convention provides that reception facilities must be adequate and that undue delay must not be caused to ships. It has been said in similar fora that "no reception facilities means no delays," but that position is not acceptable because it promotes violation of the law and continued pollution of public resources.

Adequacy

What is considered adequate will vary from port to port, but such facilities must be efficient, reliable, and able to fulfill the statutory obligations without regard to flag, ship size, or waste type. Transshipment ports such as in Puerto Rico and Jamaica will have to consider the collection of a wide variety of garbage from other regions that are part of the round-the-world service of the major shipping lines. Container ships will increasingly demand shorter turnaround times, and therefore reliability and availability of reception facilities must be ensured. While cruise ships may spend longer times at each port of call, ports will have to accommodate the greater volume of waste produced, which is estimated at 3.5 kg person^{-1} day^{-1} compared to 1 kg person^{-1} day^{-1} for the local population (Table 15.3).

While the obligations with respect to adequacy are being met, the facility must be compatible with the other stages in the waste management process, ending in disposal. A very large reception facility may meet the needs of all ships but may be incompatible

TABLE 15.3. Quantities of Annex V wastes produced at sea in the Caribbean.

Vessel type	Average number of passengers and crew	Rate of generated Annex V wastes (kg/person/day)	Quantity of Annex V wastes (kg/day) produced on board vessels
Cargo ship	30	2.0	60
Cruise ship	1200	3.5	4200
Yacht	4	2.0	8
Fishing boat	5	2.0	10
Miscellaneous	20	2.0	40

with the transport system, leading to long storage times in the port before the waste is cleared. In tropical climates, long storage times produce hot, moist conditions, a recipe for the breeding of disease-carrying bacteria and vectors, and an unnecessary increase in the risk of health hazards.

Location

Reception facilities will, of necessity, be located within the vicinity of the port to prevent undue delays. The port, however, must be careful that the waste reception facilities do not negatively affect their patrons. This is especially true for cruise ship terminals that are visited during the day and are the first features of the island that greet the passengers. Careful planning is therefore necessary to ensure that the aesthetics of the port are not compromised by the presence of efficient but unsightly reception facilities.

Cost Recovery

The financial costs of providing these facilities and the waste management programs that will govern their use will have to be justified against environmental benefits that cannot readily be given a dollar value. The recovery of most, if not all, of the costs of the investment will be one of the main aims of government. Several cost-recovery strategies have been outlined as (1) direct charge to the waste producer, (2) generic increase in port charges, and (3) sale of recyclables (World Bank 1991). Charging the waste producer is a delicate issue for many reasons. Ports are built in response to shipping needs and not the reverse. Charging ships for receiving garbage may make ports more expensive to use and, therefore, less competitive in relation to neighboring ports.

In the Caribbean, cost recovery has to be considered in the context of the intense competition among cruise lines and ports of call. Most Caribbean ports can ill afford to provide reception facilities free of charge, as is the case in some European ports. They must

identify ways of making it "pay" for ships to use the facility. Ships should in some way be attracted to (or at least not repelled by) the port because of the presence and ease of use of the facilities. Schemes for reimbursing ships for the costs incurred in using the facilities might be considered.

Transportation

Where the legal obligations under MARPOL V end, the most serious of the challenges begin. Once the garbage is received it should be transported on a timely basis to its final resting place. Transportation will, in most cases, involve the physical movement of the garbage by truck. Where it has been decided (based on ship traffic and land space considerations) that a neighboring state will accommodate the disposal, transport by barge or rail may be used.

In many countries, safety and equipment employed in the transport of municipal garbage are already casualties of inadequate budgets. The capital cities of most Caribbean States are centered around ports, and the transport of garbage will frequently involve transit through densely populated areas. When one considers that the international garbage being transported may include disease-spreading food wastes and perhaps dangerous chemicals in cargo-associated wastes, careful and effective management of transport will be crucial. This garbage cannot be carried in open trucks like municipal garbage, more so because of the public perception that receipt of international waste is a form of "garbage imperialism." The adequacy of containers, unloading and loading procedures, packaging, and emergency response measures will have to be addressed. Further, to minimize health hazards, accurate and up-to-date information must be provided to garbage handlers.

Disposal

The disposal of solid waste is already a serious problem for the governments and private sectors of all Caribbean countries. At present the majority of the countries use dumpsites as the final resting place for municipal garbage. Typically, the same places and methods are used for ship-generated garbage. This is an unsatisfactory state of affairs and is wholly inconsistent with the need to prevent and minimize risks to public health.

Many of the Caribbean states solid waste dumps are poorly located (Table 15.4). Most dumps contaminate groundwater in the surrounding environment. In Barbados, Antigua, and St. Kitts, dumpsites are located in mangrove swamps (World Bank 1991). The landfill at Plata, Dominican Republic, actually has a stream running through it. Similarly, the Duhaney River bisects the Riverton City dump in Kingston, Jamaica. Transport of floating debris and leachates are therefore serious problems that may be worsened if Annex V wastes are deposited at existing sites.

Caribbean dumpsites and landfills support human and animal scavengers. Human scavengers carry on recycling and subsistence activities at, or near, these dumpsites. Adding to this dilemma, direct human contact with garbage allows for the easy contraction and spread of disease. Furthermore, the nature and content of foreign garbage, especially cruise ship garbage, is likely to attract even more scavengers.

Spontaneous combustion can easily occur at poorly managed dumpsites, raising concern over the chemical composition and toxicity of smoke, ashes, and partially oxidized residues. A comprehensive waste management policy governing disposal at well-sited,

well-monitored, and generally well-managed landfills is essential to accommodate the disposal of both municipal and shipborne garbage.

The creation of further landfills, however, is impossible without incurring significant costs. Land space is already a scarce commodity in many islands, leading to the poor siting of existing dumpsites. Proposals for new sites will have to compete with those for housing, industry, and agriculture. Once lands are identified for use as landfills, strict management guidelines will have to be developed that take into account internationally accepted standards. Site monitoring and evaluation will have to continue long after the landfill has been closed to ensure that materials are not released and that hazardous conditions do not develop.

The Approach

The proper management of shipborne garbage involves several stages over which various government departments exercise control. With the minimization of health and environmental hazards as the major objectives, a comprehensive plan must be formulated that identifies the role of each agency and also clearly sets out their priorities and responsibilities. The effectiveness of a comprehensive waste management policy will depend on the ability to measure the performance at each stage, namely reception, transport and disposal, and to formulate and take necessary corrective action. Collabora-

TABLE 15.4. Description of disposal sites in selected Caribbean States.

State	Location	Problems
St. John's, Antigua	Mangrove swamp	Scavengers and animals
Freeport, Bahamas	Near ocean (50 m)	Scavengers
Bridgetown, Barbados	Mangrove swamp	Full; new site being developed
Cartagena, Columbia	Near dwellings	Scavengers
Limon, Costa Rica		Scavengers and contaminated groundwater
Kingston, Jamaica	A river on site, dwellings are near	Scavengers and contaminated surface water
Basseterre, St. Kitts	Mangrove swamp, near dwellings	Scavengers
Castries, St. Lucia	Near dwellings (600 m)	Scavengers
Port of Spain, Trinidad and Tobago	Mangrove swamp, near dwellings (300 m)	Scavengers

Source: World Bank (1991).

tion within an institutional framework and the careful utilization of scarce resources will be key factors in achieving the stated objectives.

Acceptance of the risk of unknown diseases to meet new international obligations is not an easy decision for a government to make. Development of political will is the first step toward taking this decision, and governments and citizens alike must be made fully aware of the costs and benefits of proper waste management practices. National policies of collaboration and education to effect the necessary awareness must also be recognized in any regional solutions to the challenges of shipborne garbage.

Cooperation Amidst Competition

As mentioned earlier, all the Caribbean States have undertaken some form of port expansion to be able to cater to the needs of the shipping industry. However, ports, like bridges, cannot easily be sold or converted to other uses and the large amounts of money invested have to be amortized by the increased earnings of the port. Consequently, every effort must be made to attract ships to the port and to maintain their business.

This aim may be successful in much of the shipping trade, but the cruise trade poses its own set of difficulties. Cruise lines, for example, are known to change their itineraries from time to time and with short notice. The cruise trade has been described as volatile in the Windward Islands of the Caribbean where passenger statistics fluctuate widely because of frequent changes in itineraries

(Economic Intelligence Unit 1989). Caribbean countries are therefore very reluctant to make their ports more expensive than their neighbors for fear that they may lose their share of the cruise trade. The decision to provide reception facilities and embark on a new waste management program will therefore have to be weighed against the possible loss of cruise business.

Conclusion

The Caribbean region is in a strong position in the cruise and other markets because of its proximity to the largest single market in the world. Collaboration of effort is therefore essential to ensure that the obligations of Annex V are met with no undue hardship to each island's trade. Every effort should be made to develop an environmental ethic in the citizens of each state who may have to indirectly subsidize the waste management programs that need to be developed. This ethic can only be created through public education and the free flow of information about effective waste management practices and the range of benefits that result.

All shipping interests in the Caribbean should therefore join in promoting public education and awareness of pollution prevention in general, and of the specific need for good and proper management of solid wastes, including those generated at sea. The MARPOL Convention provides the Caribbean with not only a challenge to be among the world leaders in this endeavor, but also an opportunity to improve its entire range of port- and landside services to preserve the sustainable value of the national, regional, and local environment.

16.

MARPOL Annex V, Commercial Ships, and Port Reception Facilities: Making It Work

Ellen Ninaber

Introduction

Annex V of MARPOL (73/78) is the key international authority for controlling ship sources of marine debris. States representing nearly 70% of the world's registered shipping tonnage have ratified Annex V and presumably are striving toward its full implementation. Annex V requires states "to ensure the provision of facilities at ports and terminals for the reception of garbage, without causing undue delay to ships, and according to the needs of the ships using them." Interestingly, while Annex V prohibits the discharge of plastics, ships are generally allowed to discharge all their other garbage at sea, at specified minimum distances from shore. In Annex V Special Areas, the discharge norm is far more strict: only food wastes may be discharged outside 12 nautical miles from shore; all other disposal into the sea is prohibited. However, the International Maritime Organization (IMO) Guidelines for the Implementation of Annex V (1988) recommend that, regardless of location, ships should endeavor to use port reception facilities as the primary means for disposal of their wastes. In the context of commercial shipping, this paper briefly examines some institutional factors relevant to the ability of ships and ports to meet these challenges. In particular, we consider whether it is possible to use port reception facilities as the primary means to dispose of ships' wastes.

The MARPOL Challenge

MARPOL Annex V regulations contribute to the difficulty of controlling pollution by garbage from ships. It can be said that the regulations of the Annexes of MARPOL are very difficult to enforce. There is no mandatory port discharge requirement for garbage. This means that ships are free to choose when and where they will deliver their wastes. This freedom makes it difficult to limit ships' opportunities and incentives for illegal disposal without an extensive enforcement and surveillance program. Under the MARPOL system, a ship may leave port with a full load of garbage. To police illegal discharge, authorities must either follow the ship or inform the next port of call that the ship should be inspected on arrival. These actions involve costly enforcement systems and international coordination that most countries are reluctant or unable to support.

Landside Waste Management Constraints

There are three basic problems encountered by ports receiving ships' wastes. First, the types and quantities of wastes delivered by ships may be atypical for the port. This problem is usually encountered by small communities in remote places and small islands that are visited by cruise ships. Second, national disease control laws may effectively prohibit the landing of certain ships' wastes. Many countries prohibit or restrict the landing of foreign-source garbage to protect against entry of human, agricultural, and livestock diseases. Third, waste management infrastructure and strategy may be absent or inadequate for handling the wastes received. The majority of ports should not have any difficulty providing reception facilities for disposal of ships' wastes provided they have an adequate waste management system in place for the wastes generated on land. Unfortunately, for much of the world, basic solid waste disposal infrastructure is inadequate.

Ship-generated waste is usually only a minor part of a port's total waste stream. Once received on shore, ship-generated wastes become part of the total waste stream of a port. Both ship-generated wastes and land-based wastes should be handled in an environmentally sound manner. If this is not the case, port receipt of ships' wastes merely transfers the disposal problem to land. For the developed countries it may be difficult to understand what it really means not to have an infrastructure for solid waste management. To illustrate, in 1993 the IMO financed a marine debris awareness campaign in Indonesia. One of the activities involved a beach cleanup. It was a very successful event, with whole villages enthusiastically participating. When all the litter was collected a truck arrived to pick it up. When they asked where the rubbish would be taken, the volunteers were told that it would be dumped back into the sea! Similar reports abound regarding the fate of garbage delivered by ships in ports worldwide. It is not uncommon for ships' wastes collected in a port to be dumped directly or indirectly (via beach or riverside dumps) back into the sea. These stories create major disincentives for the delivery of ships' waste to ports and for the control of littering.

For these situations (ships' waste directly or indirectly dumped back into the sea), the problem is not the ship-generated waste stream but the entire waste management system. Thus, the MARPOL obligation to provide adequate port reception facilities for ships' wastes may be added to the list of pressing reasons why the international community should give higher priority to development of a solid waste management infrastructure on land.

The problems of remote ports and islands may only be solved with a regional network of port reception facilities that can receive all types of waste, together with a sound waste management strategy for the ships. It may be worthwhile for the IMO and regional organizations to consider making a distinction between type A and type B facilities: type A would receive all ships' wastes (e.g., solid wastes, oil, chemical residues, etc.); and type B would receive wastes needing frequent disposal (e.g., food and food-contaminated wastes). If a region is faced with insurmountable problems in providing adequate port reception facilities in most ports, this two-level approach might be viable solution provided that (1) ships are not required to make additional port calls to dispose of their wastes, and (2) there is true regional agreement and cooperation in the provision of a sufficient number of adequate facilities.

The Packaging Dilemma

In considering the primary waste control strategies—reduction, reuse, and recycling—the most important one for marine debris control, but also the most neglected one, has been reduction. The preponderance of plastics in the postconsumer waste stream arises from product packaging. In general, packaging design is driven first by product protection and handling and second by marketing considerations. Little thought is given to how packaging designs will contribute to postconsumer waste management problems, much less to maritime industry waste management problems. This is not surprising because for most consumer products the maritime market is a minuscule fraction of the total market. Hence, ships in the international or domestic trades are not in a position to influence the packaging of the stores they order in every port (major purchasers such as cruise line companies and the U.S. Navy may be exceptions to this).

With regard to managing pollution from ships, packaging wastes from cargo and stores fall into two categories: (1) packaging for hazardous substances and (2) packaging for normal cargo and stores. The primary consideration in selecting the packaging for hazardous substances is its "seaworthiness" should the cargo be lost at sea, limiting options for its reduction [this important element of marine environmental protection is addressed under Annex III of MARPOL (73/78)]. For cargo-related packaging, considerable reduction has already been achieved through containerization, but for ships' stores there has been little progress. To promote the objectives of Annex V, packaging of nonhazardous cargo and stores should be drastically reduced. Without specific changes in packaging industry priorities, ships will remain at the mercy of market and technological forces that dictate packaging designs for general land-based usage.

Ship and Shipping Industry Problems

There are three major obstacles to port delivery of ships' wastes. First, most shipping companies and their crews are not organized to implement a good waste management strategy. The importance of this cannot be overestimated. Due to fierce competition, most ships are operated by the absolute minimum number of crew, who may be poorly trained, overworked, and underpaid. A company that is willing to skimp on the safety of crews and ships is not likely to invest much time and effort in waste management strategies. Second, many ships lack the space and facilities to store much waste. An environmentally conscious captain of a modern container vessel reported that the company had stopped building ships with good storage facilities for waste and that the swimming pool on deck was now lined with litter bins. Unfortunately, Annex V is silent on the issue of onboard storage facilities. A third problem

is caused by the odor and vermin that are inevitable when food wastes are stored for a number of days or weeks. The use of pesticides in enclosed spaces also poses health and safety problems.

It may be said that such problems are easily solved. The fact is that there is a paramount lack of knowledge among ship operators about waste management strategies and technologies for their implementation. Information on, and experiences with modern waste treatment technology, vary enormously and are often in conflict. This confusion is aggravated when shipowners purchase equipment without providing extra labor and money for proper operation and maintenance.

A consequence of all of this is that purchases of waste-handling equipment are often inappropriate for a particular ship or ship-port interaction. For example, compactors make it impossible to segregate waste (after compaction). In ports where rules for segregation of wastes apply, the use of compactors may lead to additional disposal cost. Another example, incinerators are often regarded as a technological option to dispose of all types wastes, including plastics and fuel sludge. This misuse of incinerators can generate a variety of potentially harmful by-products in the form of air pollutants and contaminated ashes and residues. Improper use of incinerators may also create fire hazards onboard ships. Ironically, in the majority of cases, incinerators are not creating environmental (or environment health) problems on ships because most incinerators are inoperable because of misuse, yielding instead additional disposal costs for owners.

Ship and Port Conflicts

The waste management strategy that is adopted by the shipowner will be determined by the types and quantities of waste which are generated on the ship and by the discharge possibilities in the ports of call. For the commercial fleet this means that all kinds of situations are encountered. The ship may call on ports with no facilities at all, or with inadequate facilities, or with expensive facilities, or with bad service. But the most frustrating situation will be to find reception facilities with different, even conflicting, rules for segregation and delivery of waste types.

Communities have different capacities and techniques for the treatment of wastes and therefore they may demand waste to be separated in different ways. There are perfectly legitimate reasons for some countries to have strict laws controlling foreign wastes, but many countries do not have these regulations. Such countries may put greater emphasis on, for example, recycling opportunities. Some countries, especially in Africa, have public health laws from the last century that make it impossible for ports to accept ships' wastes under any circumstances. Some developing countries have concluded that providing port reception facilities for ships wastes makes them a dumping ground and have adopted regulations prohibiting landing of ships' wastes.

The international community must strive to reach agreement on practices for garbage separation onboard ships and the way ships' wastes should be received in ports. Without some convergence on these standard practices, the problem of a lack of reception facilities will be replaced by the problem of ships not being able to comply with conflicting or arbitrary regulations. To dispose of the waste that has not been separated according to the rules in a particular port will probably involve very high costs. High costs, or any costs for that matter, are the greatest incentive for illegal behavior.

Capital, Fees, and Incentives

Disposal fees are a delicate matter. The provision of waste reception and treatment facilities requires initial capital investment, for which sufficient funds must be available. A distinction can be made between waste reception facilities and waste treatment facilities. Usually the investment for treatment facilities will be higher than for reception facilities. Depending on the situation in each individual port and country, the investment for treatment facilities may be of such magnitude that governmental participation or international assistance, or both, would be required. In most cases it will be most efficient to set up facilities for the combined treatment of land- and ship-generated wastes.

Once the initial capital investments have been made, a mechanism must be established to recover the operating and maintenance costs for the facilities. When making a decision on cost-recovery mechanisms, the following questions should be considered:

Will the fee system stimulate the delivery of wastes in port?
Does the fee system interfere with interport competition?
Is an extensive enforcement system needed to prevent illegal discharges?
Is an extensive administration needed to operate the fee system?
How do you collect the fees?

The "polluter pays principle" is widely adhered to for land-based polluters; however, at sea, the real polluter pays nothing—he simply discharges at sea, knowing all too well that the risk of being caught is small and the risk of being fined almost beyond the limits of probability. Thus, when choosing a particular fee system, unless it is developed in cooperation with the other ports in the region, an extensive enforcement system may be necessary to provide ships with an incentive to pay. If ships find some facilities too expensive to use, they will either dump at sea or at cheaper ports. For such ports, fewer customers means relatively higher reception facilities costs. At the least, a relatively uniform regional fee system will minimize the apparent benefits of illegal dumping.

An increasing number of ports provide facilities "free of charge." Because there is no such thing as free of charge, this usually means that costs of disposal are subsumed in the port dues. This option may affect interport competition by making the port look more expensive than its neighboring ports. Also, the revenues may be used by the port for other than waste reception costs. The best option seems to add an explicit new component to the tariff system or an explicit surcharge on the port dues so that the ship knows what is charged for waste reception and disposal. The charges should be differentiated for particular types of ships. The crew numbers may be used as a general measure of waste volume, hence costs, for bulk carriers or containerships. For passenger ships the number of crew and passengers should be used. In case of cargo ships, it should be taken into account that wastes may be cargo related. The revenues of the surcharges are subsequently applied to operate waste reception and treatment facilities. The system assures a relatively stable income for the port to finance these facilities.

If the port charges are reasonable and unavoidable, the costs of discharges will be less of a disincentive for legal disposal and illegal discharges will be less likely to occur. However, if the service is not adequate and discharge procedures are long and troublesome, this will provide incentive for illegal disposal. Ships visiting ports at short intervals may pay relatively more than other ships for the disposal of their wastes. In many cases, the remedy is to exempt the ship from paying for the rest of the year once it has called at the port a specified number of times.

Whether a fee system interferes with interport competition depends mainly on the size of the surcharge relative to the standard port fees. A large surcharge may affect a ship's choice of ports. Experience shows that waste disposal is only a fraction of the total costs for a ship to visit a port. However, in all cases it would be best to have regional agreements on what fee system is to be applied.

It has been surmised that so-called free-of-charge port reception systems will attract wastes, especially when neighboring port facilities are absent or expensive. In such cases, national laws and port bylaws specify that the wastes received should have arisen during the normal operation of the ship. The IMO's Guidelines for the Implementation of Annex V (1988) provide a definition of "cargo residues" and their treatment that supports this interpretation. Thus a ship that arrives with a cargo of rotten oranges, for example, demanding free-of-charge disposal, will have to pay as this is not normal operational waste.

Conclusions

Many speakers at the Third International Conference on Marine Debris gave us fine examples as to how to tackle the problem of ship-generated waste. We must not forget that shipping is an international business, involving many different parties. I do not believe that the problem of ships' waste can be solved by only looking at the ships. Effective prevention of ship-source marine debris requires tackling the problem in an integrated manner. We need to address the waste management deficiencies on land and the problems of packaging just as urgently as we need to provide waste management strategies onboard ships.

So, is it possible for ships to use port reception facilities as a primary means to dispose of their wastes? In this author's opinion the answer is, over the long term, yes—provided the international community is prepared to (1) develop mandatory MARPOL rules based on environmental considerations for packaging and storing materials taken aboard ships; (2) develop MARPOL rules for onboard waste storage space and facilities; (3) develop MARPOL rules for onboard waste management and delivery plans; (4) develop IMO standards for crew size and training that recognize pollution prevention duties; (5) emphasize MARPOL Article 17 (promotion of technical cooperation); (6) work toward the development of international norms for packaging to facilitate reduction, reuse, and recycling, worldwide; and finally (7) work to raise the priority of land-based waste management infrastructure development for national and international funding institutions.

17 ■

Comparison of MARPOL Annex V Port Reception Facilities for Garbage in the U.S. Gulf of Mexico and the United Kingdom

Dewayne Hollin and Duncan F. Shaw

Introduction

Annex V of the International Convention for the Prevention of Pollution from Ships, as modified by the protocol of 1978 (MARPOL 73/78), entered into force on December 31, 1988, changing the way ships' wastes are managed. Ports, terminals, and marinas are required to provide adequate waste reception facilities for ships to offload waste generated onboard. Under Annex V, ship-generated garbage includes "all kinds of victual, domestic and operational waste excluding fresh fish and parts thereof, generated during the normal operation of the ship."

Both the United Kingdom and the United States have implemented MARPOL Annex V through legislation. The U.K. implemented MARPOL Annex V requirements through the Merchant Shipping (Reception Facilities for Garbage) Regulations 1988 (SI 1933 22 93). Because the North Sea is designated as an Annex V Special Area, the discharge of all waste within that sector of U.K. waters, except for food waste, is prohibited. The U.K. Department of Transport was interested in how U.K. port reception facilities were being utilized and commissioned the Centre for Marine and Coastal Studies at the University of Liverpool in 1991 to survey eight U.K. port areas.

The U.S. implemented provisions of MARPOL Annex V through passage of the Marine Plastics Pollution Research and Control Act (MPPRCA) of 1987. MPPRCA amended the Act to Prevent Pollution from Ships to prohibit the discharge of any plastics into waters of the United States, set limitations on other discharges, and require that every port, terminal, and marina provide adequate waste reception facilities. The Annex V Special Area designation for the Gulf of Mexico (GoM) has yet to enter into force, so nonplastic waste may still be discharged at sea under basic Annex V restrictions. The U.S. Environmental Protection Agency and the Texas General Land Office were both interested in how the MARPOL Annex V requirements were being met in GoM ports. The Sea Grant College Programs at Texas A&M University and Louisiana State University were contracted to survey the garbage- and waste-handling practices at port reception facilities, including costs, adequacy of services, and levels of utilization. A survey by Hollin and Liffman (1991) looked at reception facility operations, and in 1993 Hollin and Liffman surveyed how vessel operators used the reception facilities.

This report is based on the findings of these three studies and primarily focuses on a comparison of the findings of the surveys. While the objectives of the U.K. and U.S. studies were slightly different, comparison of the survey results yields a number of conclusions about MARPOL Annex V waste reception facility operations that are of interest to other areas. The surveys found a great deal of similarity in port reception facilities in the

TABLE 17.1. Types of Gulf of Mexico ports, private terminals, marinas, and offshore staging areas surveyed.

Type of facility	Number of responses	Approximate number of vessels using facilties annually	Approximate percentage range of vessels offloading garbage[a]
International and coastwise shipping ports	22	18,000[b]	0%–20%
Private terminals	7	1,500	0%–20%
Commercial fishing ports	7	2,350	50%–100%
Recreational boating facilities	6	2,170	50%–100%
Offshore staging operations	4	6,800	20%–100%
Totals	46	31,620	

[a]For international and coastwise shipping ports and private terinals, percentages ranges are for restricted garbage only. No estimates were reported for general waste offloading. Commercial fishing ports, recreational boating facilities, and offshore staging operations handle only general or unrestricted waste.
[b]This does not include barge tows, only ships calling at port from international destinations or engaged in coastwise trade.

U.K. and U.S. Gulf of Mexico, notwithstanding the fact that Annex V Special Area requirements for the North Sea area around the U.K. (including six of the eight ports surveyed) entered into force on April 18, 1991, while the U.S. GoM was not yet designated (Annex V was formally amended to designate the Wider Caribbean, including the GoM, as a Special Area on April 4, 1993). This similarity is what makes the report's application to other areas interesting.

Comparison of Research Methods and Study Areas

The U.K. study involved a survey of eight port areas covering U.K. ports and terminals of different types and sizes. The port areas included Forth, Tees, Flexistowe, and Dover on the North Sea, Southampton and Poole on the English Channel (but included in the North Sea Special Area), and Mersey and Clyde on the Irish Sea. The survey was required to (1) determine if reported reception facilities did exist, (2) assess the level of utilization of facilities and identify problems of accessibility, (3) compare the views of ship operators using facilities, (4) compare disposal costs to ships, and (5) compare charges imposed by ports providing their own facilities with charges in ports using subcontractor facilities. Personal interviews were conducted with port authorities and terminal operators as well as ships' agents and masters.

The GoM study was conducted in 1991 by Texas and Louisiana Sea Grant personnel. Sixty-one people representing 46 landing sites (public ports, private terminals, marinas, commercial fish houses, and offshore oil and gas staging areas) along the Texas, Louisiana, Mississippi, Alabama, and Florida coastline were interviewed. Also included in the survey were waste management system operators, ships' agents, and U.S. Department of Agriculture (Animal and Plant Health Inspection Services, APHIS) officers. The primary objectives of the study were to determine the garbage- and other waste-handling practices, the nature, costs, and adequacy of existing facilities, and levels of facilities utilization. Table 17.1 shows the types of ports, private terminals, marinas, and offshore staging areas surveyed and the number of vessel arrivals annually.

Comparison of Survey Results from Both Study Areas

Types of Ports and Waste Reception Facilities

Based on the level of service and the fee system employed, U.K. ports were categorized as (a) ports that rely on ships' agents and waste management contractors; (b) ports that provide dumpsters (referred to as a skip in the U.K.) accessible from the ship's berthing area and include this service in port charges; (c) ports that provide dumpsters as in (b) but charge all ships for service; (d) terminals that provide a dumpster at the berth, monitor its use, and charge whether used or not, and (e) same as (d), but include the cost in port charges. Most U.K. port operators seem to prefer leaving garbage disposal arrangements to ships' agents and independent contractors.

U.S. GoM ports and private terminals also preferred to use third-party contractors to handle garbage offloaded at their facilities. Of the 22 ports contacted, 18 used third-party contractors exclusively while the remaining 4 ports had joint arrangements with third parties. All ports provided some dumpsters for general waste disposal, but these were not always near the ships' berthing areas. Any vessel needing a dumpster for large-volume disposal would have to contract for this service. Private terminals also preferred to use third-party contractors, but some of these terminals secured any dumpsters on their property and closely monitored the use of the facilities. Both commercial fishing ports and recreational boating facilities provided general waste disposal dumpsters without direct charges to patrons.

Pest and disease control laws in the United States require that foreign vessels have their garbage inspected by the U.S. Department of Agriculture (APHIS) before being offloaded in special containers for sterilization or incineration. This "restricted" waste disposal has always been paid for by the vessel through a third-party waste management company. GoM data showed that about 10%–14% of the foreign vessels offloaded "restricted" waste of this type.

The types of waste reception facilities at larger U.S. GoM ports and terminals varied considerably by type of installation and the extent of vessel use (Table 17.2). The most common type of waste-handling receptacles are 10–40 yd^3 dumpsters with regularly scheduled pickup by municipal or commercial haulers. Public ports customarily provide some dumpsters at no charge for general waste, but temporary dumpsters contracted by the vessel for "excess" garbage disposal are paid for by the vessel. These temporary dumpsters are normally provided by the same waste management companies under contract with the ports to provide permanent dumpsters.

Both the U.K. and U.S. GoM studies indicate the preference for third-party contractors to handle garbage disposal with merchant ships' agents making arrangements for waste disposal services. Both areas also provide some type of disposal containers near the berthing area for general waste disposal and include the cost for this in port charges. The two regions differ in how they charge for the service. Some U.K. ports charge the vessel for the service whether they use it or not,

TABLE 17.2. Type of reception facilities available at U.S. Gulf of Mexico ports, private terminals, marinas, and offshore staging areas.

Type of facility	General waste dumpsters		APHIS incinerator		APHIS commercial autoclave		APHIS refrigerator and incinerator system[a]	
	Yes	No	Yes	No	Yes	No	Yes	No
International and coastwise shipping ports	19	3	20	2	3	19	3	19
Private terminals	4	3	7	0	0	7	1	6
Commercial fishing ports	7	0	N/A[b]		N/A		N/A	
Recreational boating facilities	6	0	N/A		N/A		N/A	
Offshore staging operations	4	0	N/A		N/A		N/A	
Totals	40	6	27	2	3	26	4	25

[a]Wastes are refrigerated at port facilities until picked up by waste management companies, usually once per week.
[b]Not applicable. Commercial fishing ports, recreational boating facilities, and offshore staging operations do not normally handle Animal and Plant Health Inspection Service (APHIS) waste.

while U.S. GoM ports provide some disposal facilities without charge but require that vessels pay third-party contractors to handle the disposal of "excess," unrestricted garbage. All restricted waste disposal in U.S. GoM ports is handled by third-party waste contractors and paid for by the vessels.

U.S. GoM smaller fishing ports and marinas that do not handle restricted wastes generally provide garbage reception facilities without charge to the vessels doing business with the port or marina. The costs of these services are however increasing, and small ports and marina operators are looking at ways to recover these costs. Most of the small ports and marinas utilize dumpsters of various sizes with regular pickup from commercial waste management firms; this was also true of small ports in the U.K. study.

Utilization of Waste Reception Facilities

Utilization patterns for U.K. port reception facilities are difficult, if not impossible, to accurately determine without a better record-keeping system. With few exceptions, no reliable records of utilization of waste disposal facilities are kept by U.K. ports. This is true primarily because ports pass the responsibility for waste disposal to the ships' agents and local waste management contractors. The vessel master is required to keep records on the volume of garbage landed, incinerated onboard, or dumped at sea, but port and terminal operators using contractors rely heavily on estimates based on the number of skips used, which sometimes contain considerable amounts of waste not originating from the ship. Only about 15% of the ships ask for waste skips to be provided.

Use of third-party contractors for waste disposal has also made it difficult for U.S. GoM ports to determine the utilization patterns of their reception facilities. Records for restricted waste disposal (APHIS) are more accurate because the number of disposal entries are reported to the U.S. Department of Agriculture (USDA) along with the disposal weights. Table 17.3 shows the fiscal year 1993 activity levels for 10 of the largest GoM ports. This report shows that about 14% of the foreign vessels arriving in these ports requested disposal of restricted (APHIS) garbage. Contributing to this low utilization of GoM shoreside disposal facilities is the apparent lack of flexibility in scheduling pick up for restricted garbage; most contractors require 24-h notice before pick up.

Information for unrestricted garbage is not available except by number and size of dumpsters used for general waste and by the number requested by ships' agents for excess waste disposal. Currently, 10–15% of vessels docking at both U.S. GoM and U.K. ports actually offload garbage; however, most GoM ports are reporting a slight increase in the utilization rate. Most GoM waste manage-

TABLE 17.3. Fiscal year 1993 (10-1-92 to 9-30-93): U.S. Department of Agriculture APHIS activity reports.

Gulf of Mexico port	APHIS garbage disposal entries	Garbage disposal weight (kg)	Foreign ships inspected
Tampa, FL	399	648,052	991
Mobile, AL	132	66,518	396
Gulfport, MS	83	41,886	369
New Orleans, LA	439	219,500	2,209
Galveston, TX	153	72,732	762
Port Arthur, TX	25	23,299	226
Corpus Christi, TX	68	17,040	415
Baton Rouge, LA	68	5,440	91
Houston, TX	359	127,205	2,207
Brownsville, TX	0	0	907
Totals	1,726[a]	1,221,672	8,573

Source: U.S. Department of Agriculture.
[a]Percentage of total foreign arrivals, 14.1%.

ment firms report that they have additional waste-handling capacity. With port discharge utilization rates as low as 10%, garbage is being disposed of at sea, retained for disposal at other ports, or incinerated aboard the vessel.

U.K. and U.S. GoM ports share the problem of a lack of reliable figures on utilization of their waste disposal facilities, which results from the multiplicity of agents and contractors involved and the failure of ports to generate statistics on waste disposal. Because the master keeps these records for his ship, the ports could request this information, but its assembly into a useful database would be a major task. Another problem is the lack of monitoring of the disposal facilities and the apparent mixing of ship-generated waste with other materials not originating from ships. In both regions, the actual volume of ship-delivered waste can only be estimated. There should be greater port monitoring and control, but not so much that ships are deterred from using these facilities.

Costs and Arrangements for Waste Disposal

In the U.K. garbage disposal costs vary from one part of the country to another, by size and type of port and by distance to the landfill site. Also, requirements for special treatment of food wastes, because of possible health hazards from foreign sources, may add to disposal costs. In some instances, the cost of disposing a small quantity of garbage can be excessive when a vessel is charged for a skip containing only a few bags of garbage. There were few differences found in the cost of disposal service to a ship and service to a land-based facility. The only differences found were for delivery and collection of the skip and special treatment of the ship's garbage when required.

In the U.S. GoM, as expected, there were cost differences between restricted (APHIS) garbage and unrestricted garbage. These costs vary from port to port depending on factors such as transportation, special handling, type

of disposal, and treatment. Where unrestricted garbage disposal may cost \$8–\$10 per cubic yard, restricted disposal may cost up to \$500 per cubic yard because of special treatment and disposal methods. The standard 1-yd^3, plastic-lined, corrugated box used for APHIS waste in the U.S. averages about \$25 per box for disposal alone, but sterilization and disposal of a cubic yard of APHIS waste may cost from \$100 to \$500 depending on the distance to the sterilization facilities. Another important cost for processing restricted ship-generated waste is labor to sort garbage and place it in special containers.

Costs for processing restricted ship-generated garbage are always going to be higher than those for land-generated waste. However, there should be no major difference in processing unrestricted ship-generated waste and land-based waste because the basic collection and disposal methods are the same. Some savings may be realized when the ports provide permanent dumpsters through a contractor. Dumpster service is more expensive when requested and paid for by the vessel rather than as a regular service negotiated by the port.

Common Complaints About Waste Disposal Services

Both the U.K. and U.S. GoM ports receive the same complaints about waste disposal—high cost and inconvenience for ship agents to arrange for garbage disposal, particularly if additional skips or dumpsters are requested. Having to make these arrangements 24 h in advance also creates some problems for agents and ship masters. Some private terminals also place restrictions on the use of their disposal facilities and closely monitor their use.

All ports in these studies provide basic garbage disposal facilities near the ship's berthing area or could provide them if requested and given reasonable notice. However, the arrangements in the majority of ports do not provide any positive incentives to ship operators and masters to offload their garbage.

Port Users' and Port Managers' Opinions on Waste Reception Services

U.K. port users would like to see garbage disposal facilities provided at each berth by the port authority and have the cost included in port charges. The port authorities are generally unwilling to do this because of the difficulty of policing the way the skips are used. Smaller, general-cargo ports can handle these types of arrangements more easily.

Most U.S. GoM port users think that the ports do provide adequate reception facilities for unrestricted garbage, but would like to see more flexibility in scheduling pickups of restricted garbage. Although they have accepted the current arrangements of placing ship-generated waste into 1-yd^3 boxes for disposal by incineration and using 4-yd^3 containers for sterilization, they do not like this inconvenience. Approximately 60% of the maritime vessels operating in the GoM have some type of onboard disposal capability, most frequently an incinerator. Most maritime vessels also continue to discharge garbage at sea under full compliance with MARPOL Annex V. How these vessels will respond to the entry into force of the Wider Caribbean Annex V Special Area is uncertain.

Conclusions

Our results suggest the following conclusions for U.K. and GoM port reception facilities for ship's garbage:

Larger ports generally rely on vessel agents to make arrangements for disposal of vessel garbage, utilizing third-party waste management contractors to collect and dispose of garbage and assess charges for this service.

Smaller ports and marinas handling large numbers of small vessels that do not employ agents provide unrestricted garbage disposal facilities without charge. These smaller ports and marinas utilize both municipal and commercial waste-handling contractors to collect and dispose of waste, and they absorb the cost of the service without direct charges to the vessels.

The principal complaints about port garbage disposal services are (1) the high cost of handling small quantities of garbage, (2) the lack of convenience in handling restricted waste in larger ports, and (3) the restricted use of disposal facilities at some private terminals.

Utilization of port reception facilities is low at U.K. and U.S. GoM ports, with 10%–15% of all arriving vessels using their facilities. This indicates garbage is being disposed of at sea, retained for disposal at homeports or other ports of call, or incinerated aboard the ship. The utilization rate is increasing in some U.S. GoM ports.

Larger ships calling in the GoM and U.K. are generally *more* aware of garbage disposal requirements and the availability of port facilities. Ships are installing incinerators for onboard garbage disposal and establishing better source-reduction and recycling programs.

Better recordkeeping measures are needed to provide more reliable information on requests for disposal service and the volume and types of garbage received.

Recommendations

Ports and private commercial terminals need to take more responsibility in ensuring that garbage reception facilities are adequate. The cost of providing these facilities should be covered by standard, mandatory port charges and not separate optional fees. Ports could continue to use contractors, if convenient, but should cover the cost outs of their normal port charges. This allows ships to benefit from the ability of the ports to negotiate long-term, competitive-rate service contracts, which in turn makes the port more appealing to ships. This procedure would provide a positive incentive to ship masters to land their garbage, thus avoiding the temptation to discharge at sea. This could include "berthside" service on arrival and before departure with the cost covered by berthing charges.

For small ports and marinas, the emphases needed to promote use of reception facilities shifts away from cost management and toward education. The users of small ports and terminals and marinas (e.g., commercial fishermen and recreational boaters) do not generally handle restricted wastes and have less exposure to maritime authorities than the cargo vessels frequenting the larger ports. As such, their motivation to use port reception facilities may be raised through education programs describing legal disposal requirements and penalties and stressing the damage to the environment created by discharging garbage at sea.

The results of these studies show, for both small and large ports, that no apparent combination of fee systems and levels of services seems to elicit higher disposal rates (i.e., greater than 10%–15%). Further information and data for determining the best management practices for waste disposal facilities, under a variety of conditions, should be developed and provided to port managers and marina operators. Government authorities could take the initiative in developing and publishing best practices guidelines.

18.

Waste Disposal Practices of Fishing Vessels: Canada's East Coast, 1990–1991

Paul Topping, David Morantz, and Glen Lang

Introduction

There is a growing awareness that persistent plastic debris in the marine environment threatens marine life and reduces economic potential. Data from beach surveys and high seas observations collected over the years demonstrate a long-term problem and identify a common source of debris: offshore fishing fleets (Buxton 1990; Lucas 1992). An extensive fleet of both foreign and domestic vessels operates in the Atlantic Ocean within the 200-mile economic zone off Canada's east coast. Before this project, formal data concerning waste disposal practices of fishing vessels or of other vessels at sea were scarce.

The discharge of garbage from the operations of ships is controlled by the Canada Shipping Act, as administered by the Canadian Coast Guard. The Act specifically prohibits the discharge of garbage containing more than 4% plastics (by weight); furthermore it requires garbage to be comminuted. Enforcement of the Act's requirements poses problems for the Coast Guard because of its limited resources and the long-standing acceptability of discharging garbage into the sea. Another problem is the availability of facilities to receive ships' waste. While major ports and harbors have adequate facilities, smaller fishing harbors often lack them. Efforts to improve facilities for managing wastes at small craft harbors are ongoing, but the task remains daunting; in 1990 there were 2282 small craft harbors and 535 cargo terminals in Canada (Department of Fisheries and Oceans 1993; Transport Canada 1990). A key challenge continues to be long-term operation and maintenance of waste reception facilities. Maintaining new facilities often required fishermen and other harbor users to pay fees to cover these costs. Given the costs, it was expected that ocean discharge would remain the preferred method to dispose of ships' wastes.

Certain types of debris, such as net fragments and floats, have been clearly attributed to the fishing fleet, but no data on their actual waste management practices were available (Department of Fisheries and Oceans 1993; Hargreaves and Carter 1989; Lucas 1992). The Fisheries Observer Program, operated by the Department of Fisheries and Oceans (DFO) to collect data for fisheries management purposes, presented a unique opportunity to collect this important information. Fisheries observers are mandated to monitor the catch of fishing vessels operating within Canada's economic zone to ensure compliance with Canadian fisheries regulations, namely catch limits for various species. The observers board fishing vessels and remain onboard while the vessel operates in the Canadian Exclusive Economic Zone (EEZ). Thus, observers were in position to record the waste management practices for each fishing trip. These data, collected over the duration of the study, characterize waste management practices and trends among Canadian and eight foreign fishing fleets operating in Canada's Atlantic EEZ.

Methods

DFO observers, while aboard foreign and domestic fishing vessels, were asked to record the vessels' waste disposal methods and the types of waste disposed, including the types of plastics. Unfortunately, circumstances did not permit quantities of waste to be estimated by the 97 observers who took part in gathering these data as part of their normal observer duties. A marine plastic debris occurrence report form was created to encourage consistency. The form provided standardized categories for four basic disposal methods and 18 types of waste, 9 plastic and 9 other types. When waste was disposed, the observers marked the appropriate categories on the form. Additional space was provided for comments, and on each trip, the vessel's name, nationality, and the trip dates were recorded.

Data were entered into a computer database during the 2-year study period as trips were completed. Each record represented one trip and comprised the trip details and tallies under the appropriate categories. The data were transferred to a spreadsheet and checked for consistency and completeness of records for disposal practices. A total of 772 trips were recorded, and after editing, the records of 739 trips were included in this study. Sorting routines were used to tabulate the data into trends, and counting functions were used to score the number of entries under a category.

Results

Fisheries observers reported the disposal practices of 181 domestic and 95 foreign fishing vessels during the 739 trips they were aboard. All trips took place in 1990 and 1991 within 200 km of the shore of Nova Scotia. The length of the trips varied from 1 to 83 days. Details on the composition of the domestic and foreign fleets observed are given in Table 18.1 The number of foreign vessels observed represented 100% of their fleet operating within Canadian waters during the study. The number of domestic vessels observed represent only about 3% of the entire Canadian fleet. Most Canadian vessels are small, operating for about 24 h or less per trip. Fisheries observers tended to board larger vessels, and about 15%–20% of vessels exceeding 30.5 m in length were sampled.

The sizes of the observed vessels varied considerably. Although precise figures were not recorded, domestic vessels tended to be smaller with crews of 6–12, but factory vessels of 300 dead weight tons (DWT) and more, with crews of more than 30 people, were also in use. Foreign vessels were gener-

TABLE 18.1. Composition of the offshore fishing fleet observed off the coast of Nova Scotia, 1990–1991.

Country	Number of vessels observed	Percentage of fleet observed	Number of trips observed	Mean (trips per vessel)
Canada	181	65.58	532	2.94
Japan	19	6.88	23	1.21
USSR [former]	49	17.75	112	2.29
Cuba	17	6.16	46	2.71
France	3	1.09	6	2.00
France (St. Pierre)	2	0.72	3	1.50
Faroe Islands	2	0.72	10	5.00
Bulgaria	1	0.36	3	3.00
Denmark	1	0.36	3	3.00
Norway	1	0.36	1	1.00
Totals:	276		730	

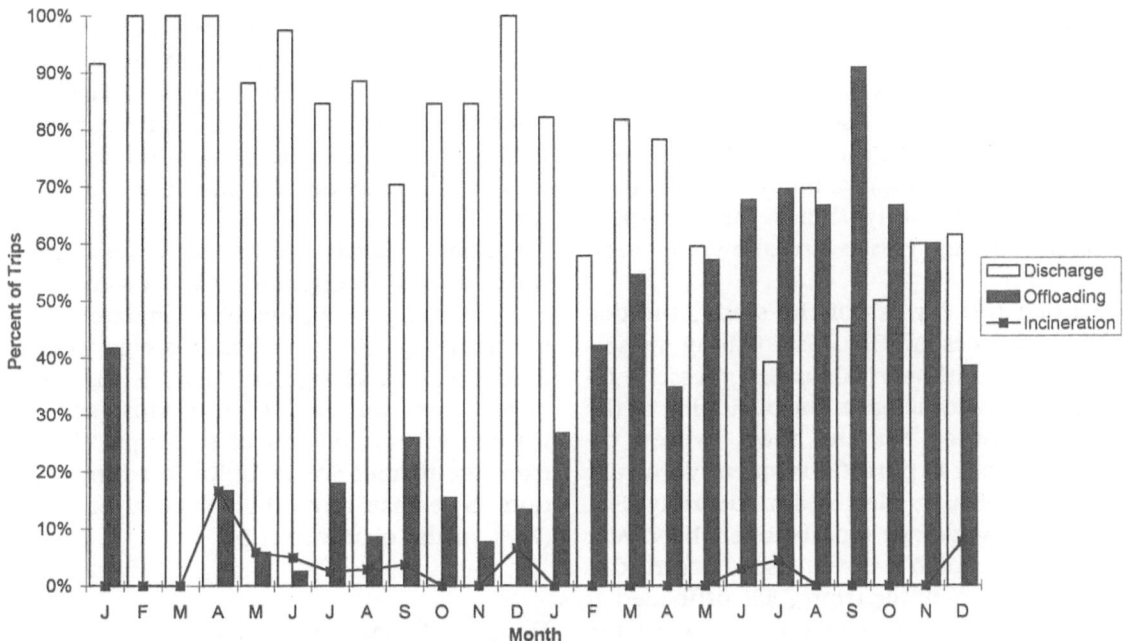

Figure 18.1. Monthly trends for the disposal practices of Canadian fishing vessels operating off the coast of Nova Scotia, 1990–1991 (percent of observed vessel trips using each method, by month). *White bars,* discharge; *black bars,* offloading; *squares,* incineration.

ally large, ocean-going factory trawlers over 300 DWT.

Disposal Methods

The waste disposal methods observed were (1) discharge over the side of the fishing vessel, (2) discharge through a galley chute, (3) incineration aboard the vessel, and (4) offloading at port. Discharge into the sea, either over the side or through a galley chute, was the most common waste disposal method observed for both foreign and domestic fishing vessels. Disposing of waste in this way has always been the most convenient and practical method for mariners because of its simplicity. On about 18% of the trips, however, fishing vessels used more than one disposal method. A vessel would generally use more than one disposal method when cost effective or convenient. Incinerators, typically used by foreign vessels, also provided heating for the crew and when their waste exceeded the incinerator's capacity it

would be discharged into the sea. A domestic vessel close to shore may prefer to offload at port rather than discharge their waste close to shore and their homes. A detailed breakdown of the frequencies in which these disposal methods were used is presented in Table 18.2.

Waste and discharge into the ocean, either over the side or by a galley chute, during 76%

Table 18.2. Usage of waste disposal methods onboard all observed vessels, 1990–1991.

Method	Incidence (trips)	Percentage
Offloaded at port only	134	18.1
Incinerated at sea only	41	5.5
Discharged into the sea only	434	58.7
Incinerated at sea and discharged into the sea	60	8.1
Incinerated at sea and offloaded at port	4	0.6
Offloaded at port and discharged into the sea	62	8.4
All methods	4	0.6
Totals:	739	100

of all trips; it was offloaded at port during only 26% of all trips and incinerated on 14% of all trips. Note that because more than one disposal method was used during single trips on the same vessels, the percentages of trips utilizing specific disposal methods total more than 100%. Monthly trends for Canadian, foreign, and all vessels are shown in Figs. 18.1–18.3.

During the 2 years of the study, the disposal practices of Canadian fishing vessels improved. Although Canadian vessels favored ocean discharge at the beginning of the project, toward the end they favored offloading waste in port. Consequently, the use of ocean discharge declined moderately (Fig. 18.1). In an average month, over the 2-year period, Canadian vessels used ocean discharge on 76% of the trips, 35% offloaded at port, and 1% used incineration. During the first 5 months, 96% of the observed trips (on Canadian vessels) discharged waste into the sea. In the last 5 months, 57% of the trips

discharged waste into the sea and 65% offloaded waste at port. This improvement was likely because of efforts by the Maritime Fishermen's Union and several Canadian fishing companies who prohibited their factory vessels from discharging their garbage at sea.

Foreign vessels continually discharged waste at sea throughout the project; however, most also continually incinerated their wastes (Fig. 18.2). Overall, foreign vessels used ocean discharge on 72% of the trips, incineration on 46% of the trips, and offloaded in port on 3% of the trips. Vessels from the former Soviet Union used incineration more than any other disposal practices (75% of the observed trips).

Waste Types Disposed

Domestic and foreign fishing vessels in Nova Scotia's offshore waters dispose of a variety

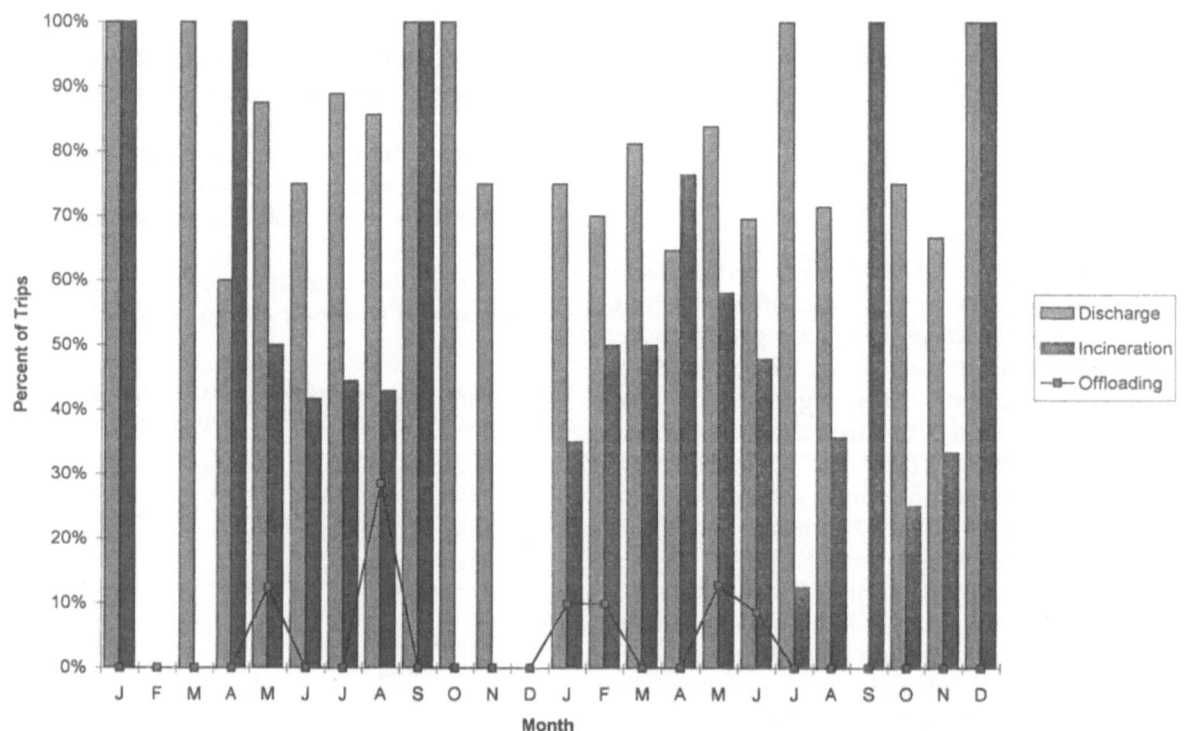

FIGURE **18.2.** Monthly trends for the disposal practices of foreign fishing vessels operating off the coast of Nova Scotia, 1990–1991 (percent of observed vessel trips using each method, by month). *Shaded bars,* discharge; *black bars,* incineration; *squares,* offloading.

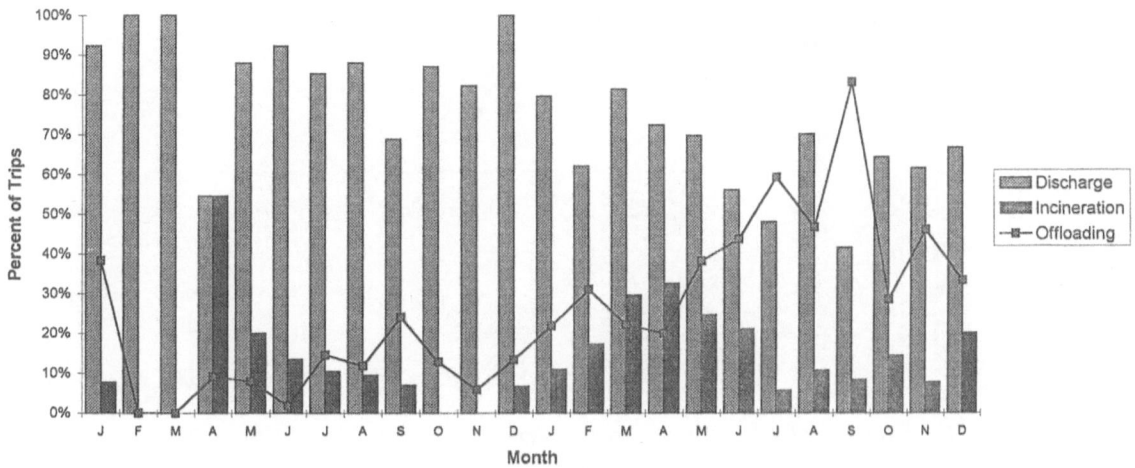

FIGURE 18.3. Monthly trends for the disposal practices of all fishing vessels operating off the coast of Nova Scotia, 1990–1991 (percent of observed vessel trips using each method, by month). *Shaded bars,* discharge; *black bars,* incineration; *squares,* offloading.

of different plastic and other waste types, including bottles, bags, polystyrene, packing bands, rope, fishing nets, food wastes and food packaging, metals, paper, cardboard, glass, and other plastic products. Categories for nine plastic waste types and nine other waste types were recorded during the study. Observers could not always record waste types during their observations; hence, the collected data did not permit waste types to be linked to their disposal methods. The data did, however, allow estimation of the per-

centages of trips that disposed of specific waste types for all vessels combined and for vessels from four countries with adequate sample sizes. Figures 18.4 and 18.5 present details on the estimated total percentages trips that disposed of plastics and other waste types. Figure 18.6 and 18.7 compare disposal practices between Canada, the former Soviet Union, Japan, Cuba, and the combined observations from fishing vessels of the five other nations listed in Table 18.1.

Plastics were disposed on all trips. Of the

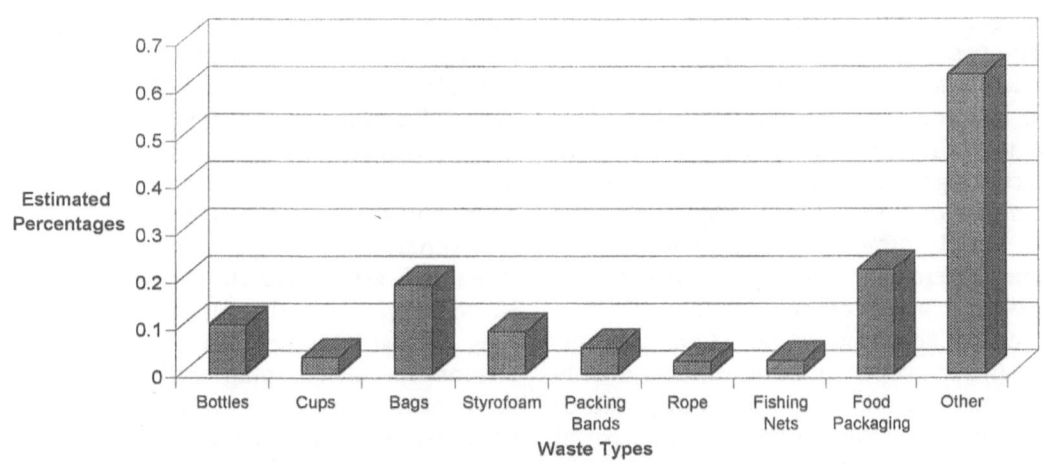

FIGURE 18.4. Estimated percentage of observed vessel trips, off the coast of Nova Scotia (1990–1991), that disposed of various plastic waste types.

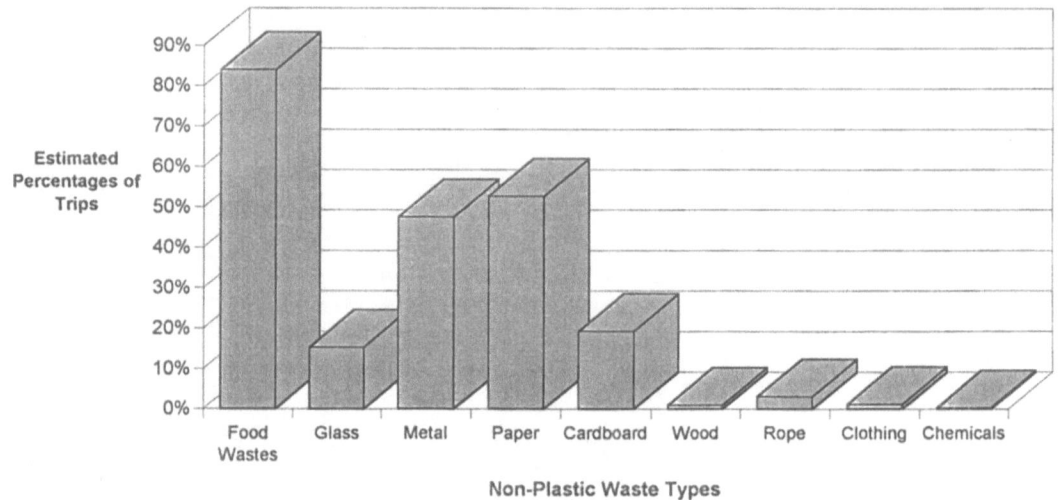

Figure 18.5. Estimated percentage of observed vessel trips, off the coast of Nova Scotia (1990–1991), that disposed of nonplastic waste types.

nine plastic waste types recorded, the most common category disposed by vessels were items classified as "other" by the observers. These items, which did not fit into one of the prespecified plastic waste type categories, included broken equipment such as fishing buoys, electrical conduits, or plumbing fixtures. The next most common waste type disposed was food and galley wastes, which occurred on 85% of all trips. Some trips that lasted less than 1 day did not produce food wastes.

Of the four most frequently observed nationalities, vessels from the former Soviet Union generally had the lowest estimated percentages of trips disposing each of the recorded waste types. Observers generally noted the crews were very careful to conserve anything of use. Cuban crews were similarly frugal, but their disposal frequencies by waste type were generally higher than the Soviet vessels' crews (Figs. 18.6 and 18.7).

Some differences were observed between the disposal practices of Canadian and foreign vessels. For example, only Canadian vessels disposed of plastic cups, albeit on only 3.5% of all trips. Foreign vessels disposed of plastic rope and strapping more often, about 7% of the trips, while Canadian vessels disposed of such waste on only 2% of the trips (Fig. 18.5). Japanese vessels appeared to dis-

pose of all waste types more frequently than all other countries. Although appearing to dispose of wastes more frequently, the size of their fleet (19 vessels) and number of trips (23) during the study reduced the relative impact of such disposal activity in Canadian waters.

Discussion

In interpreting these data, the effect of the observers' presence aboard a fishing vessel should be considered. An observer is an official of the Government of Canada, and it is possible that some crews will use different disposal options out in the open ocean when no government official is observing them. All disposal practices observed, however, have been considered normal seafaring practice for generations; many crews likely feel "tossing trash" over the side is only good seamanship, even though the Canada Shipping Act has specified conditions and prohibitions for discharging garbage. As well, many of the vessels return regularly to fish in Canadian waters and are familiar with the observers. The overall data for this study suggest that the effect of the observer's presence was minimal on foreign vessels but al-

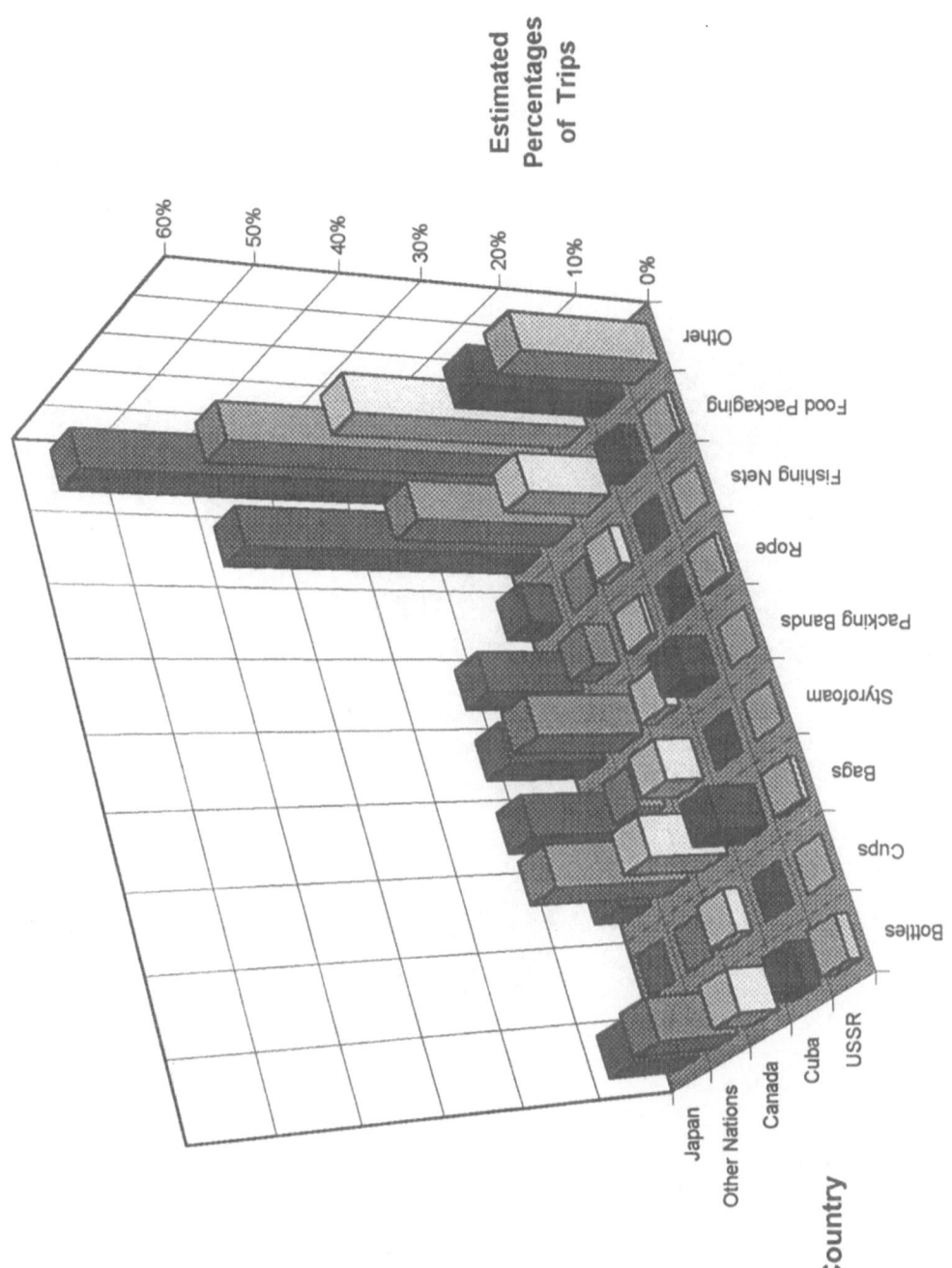

Figure 18.6. Comparison of the estimated percentages of observed trips by fishing vessels from Canada, the former Soviet Union (USSR), Japan, Cuba, and the other nations operating off the coast of Nova Scotia (1990–1991) that disposed of various plastic waste types.

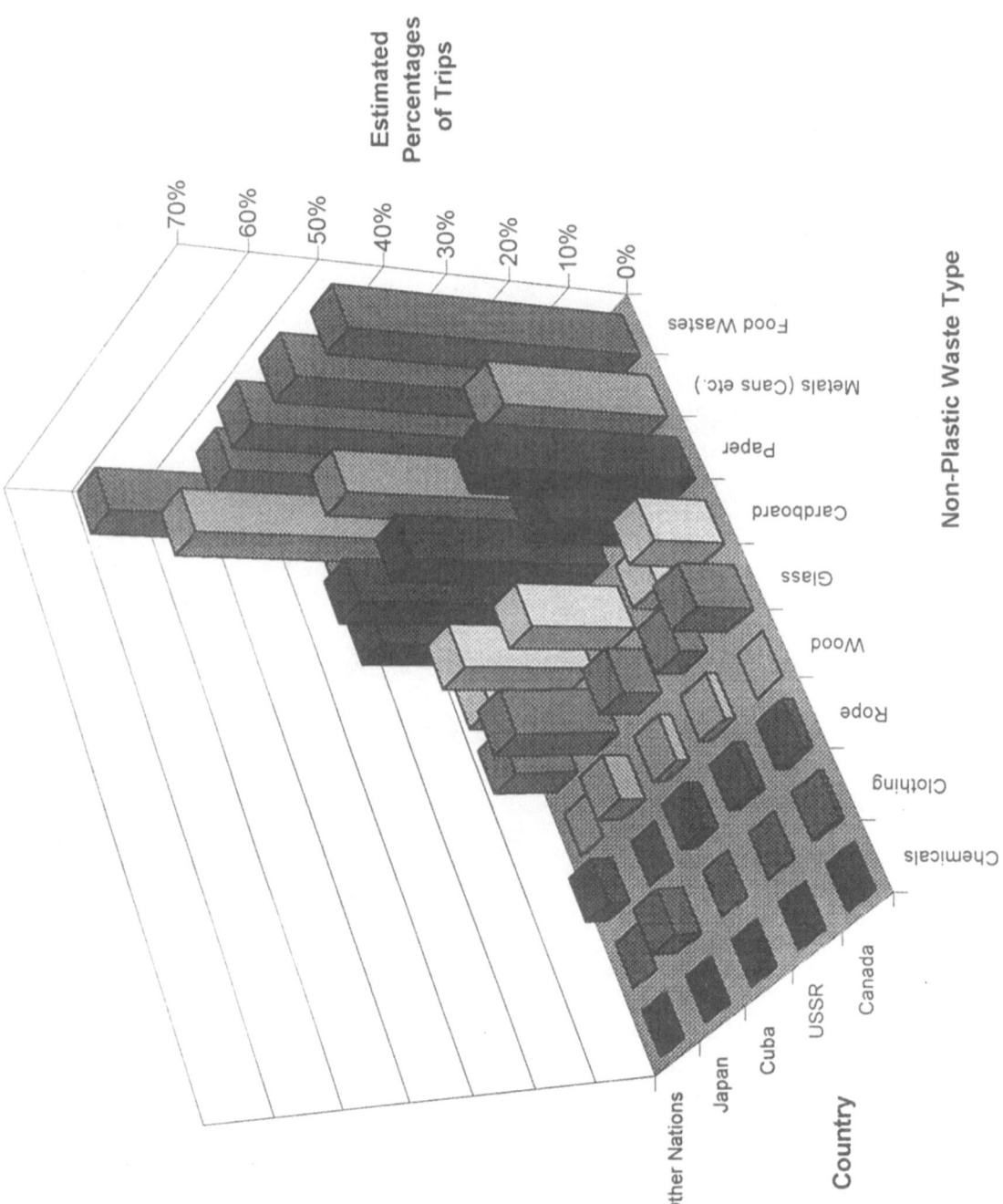

Figure 18.7. Comparison of the estimated percentages of observed trips by fishing vessels from Canada, the former Soviet Union (USSR), Japan, Cuba, and the other nations operating off the coast of Nova Scotia (1990–1991) that disposed of various other nonplastic waste types.

most certainly is responsible for the trend toward port discharge by the observed Canadian vessels. Nonetheless, on the observed vessels discharging wastes into the sea was still the most prevalent disposal method, occurring on 76% of all trips.

The results clearly show that Canadian vessels improved their disposal practices over the study period, discharging into the sea less frequently and offloading waste at port more frequently. One reason for this has been a concerted effort on the part of the Nova Scotia Maritime Fishermen's Union encouraging fishermen to bring their waste back to port. The project was supported by Environment Canada, and successfully educated the fishing community, encouraging port managers to provide facilities (e.g., barrels and dumpsters). As well, several Canadian fishing corporations adopted policies prohibiting the discharge of their vessel's wastes into the sea (J. Andrews, Seafood Processors Association of Nova Scotia, Halifax, NS, personal communication 1994). Despite this progress, Canadian vessels continued to discharge waste into the sea during an average of 57% of their trips. Likely causes for this high frequency include the limited availability of waste reception facilities and the disposal fees charged by ports. Many small ports continue to lack facilities because managing the collected wastes can be expensive.

Foreign fishing vessels generally avoided offloading their wastes in Canadian ports. Instead, they frequently used incineration and ocean discharge as their primary waste disposal methods. Two reasons for the lower usage of port facilities may be port fees and the Health of Animals Regulations that require foreign garbage to be incinerated once offloaded in Canada. Most foreign vessels were large factory trawlers of sufficient size to accommodate an incinerator. Another reason that foreign vessels favored incineration was energy recovery. Observers noted that waste incinerators also provided heating for the crew, especially on vessels from the former Soviet Union, which favored incineration over all other methods, using it on 76% of the trips. Incineration was not favored by Canadian vessels, largely because most ves-

sels were small and lacked sufficient space for an incinerator. In addition, few fishermen could afford the air pollution control equipment required under Canadian law. Many incinerators aboard foreign vessels would not meet Canadian requirements; generally, observed shipboard incinerators were less complex and expensive, but also produced higher air pollution emissions.

Another waste reduction option, reusing plastic material, was not studied in detail but was observed to be practiced by some foreign crews. Crews from the former Soviet Union were noted to carefully conserve and reuse plastics, especially containers and nets. This is illustrated in Figs. 18.6 and 18.7, showing Soviet Vessels with the lowest estimated disposal frequencies for most waste types. Future studies could examine shipboard reuse and recycling when studying waste management methods to provide better insight into these practices and how to encourage them.

The results showed that data on a broader range of categories of plastics in the waste stream would be desirable. The variety of waste types disposed of was greater than anticipated when planning the study. The categories included in the reporting were limited to packaging material, fishing equipment, and food wastes. Other items that should have been included were broken equipment, fixtures, fittings, and containers.

Figures 18.4 through 18.7 show that fishing vessels generate wastes from their own crews as well as from discarded fishing gear. Beach surveys in Atlantic Canada have often noted fishing gear debris that is obviously attributable to fishing vessels, but some researchers point out that fishing vessels certainly contribute other types of wastes (Buxton 1990; Lucas 1992). While nonfishing vessels contribute their share of debris, the combined domestic and foreign fishing fleet undoubtedly constituted a significant portion of the vessels off the Nova Scotia coast during the study period. Given this, the effects of the general domestic wastes generated by the crews themselves, and discharged into the sea, almost certainly contribute to Canada's marine debris problem.

Given that the crews living on fishing vessels generate domestic waste, it follows that similar-sized crews living aboard other types of vessels generate similar wastes. Some of the findings of this study, related to waste disposal practices of fishing vessels, might therefore be applicable to cargo vessels. Crews from the medium-sized to the larger fishing vessels are similar to crews aboard cargo vessels in that they would likely generate similar waste types and quantities. The disposal practices, however, may vary as cargo vessel crews are likely to have a greater awareness of the Canada Shipping Act. Additionally, modern cargo vessels may be better equipped to manage their wastes by incineration, comminution, and discharge or by compaction and storage. Although the disposal methods may vary, the waste types generated by the crews themselves would likely be similar. The primary waste types generated by cargo vessels, therefore, would be plastics, food wastes, paper, cardboard, metals, and glass. Food packaging, bags, bottles, and "other" items, previously discussed, would make up a significant portion of the plastics waste stream. Given that much of the waste is generated by the vessels' crews, in addition to their cargo or fishing activities, targeting education efforts toward the crews could have a positive effect.

Other types of vessels, namely passenger vessels and military vessels, have different-sized crews and generate different waste types. Passenger vessels carry a greater number of people and generate wastes typical of a tourist facility or hotel. Military vessels generate their own specialized wastes such as spent ammunition, disposable equipment, and unique packaging. Comparisons based on fishing vessel crews would not be valid.

This study has examined the fishing vessel disposal practices in terms of behavior. From this we may conclude that while there has been some improvement, vessels continue to discharge their wastes and will likely continue to be in violation of domestic and international laws. Establishing and maintaining adequate port reception facilities, coupled with education efforts directed toward vessel crews, should provide further improvement.

While additional research examining the quantities of waste discharged would provide a better assessment of the overall impact of vessel-source debris, widespread evidence indicates wildlife is impacted. During the course of this study, observers noted many instances of entanglement and ingestion among fish, seal, and seabird species. In some cases the entanglements were dramatic; one observer reported seeing more than a thousand porbeagle sharks (*Lamna nasus*) entangled by plastic strapping (R. Sciocchetti, Canada Department of Fisheries and Oceans, Halifax, Nova Scotia, personal communication, 1990).

Clearly, some actions can be taken to improve compliance with the Canada Shipping Act before overwhelming proof of the negative effects of marine debris is in hand. Environment Canada is implementing a national marine plastic debris program, which includes research to gain further understanding and education to promote better waste management practices and facilities. In addition, Canada signed the MARPOL (73/78) Convention in February 1993 and is considering ratification of Annex V.

19.
Dealing with Ship-Generated Plastics Waste on U.S. Navy Surface Ships

Lawrence J. Koss

Introduction

In the 1980s, the U.S. Navy began a long-term program to develop shipboard equipment to manage solid waste. The primary objectives were to improve the efficiency of shipboard waste handling, to reduce the security risks associated with a ship's "trash signature," and to comply with the potential ocean discharge restrictions on trash and garbage, even though the pending international agreements exempted military vessels. Plastics were considered a normal component of the solid waste stream. In 1987 when the U.S. Congress made the international ban on plastics discharge at sea applicable to Navy ships, the U.S. Navy was caught somewhat by surprise. International maritime regulations have always recognized the unique operating constraints of warships and have allowed navies to comply to the extent practicable. Nevertheless, the Navy responded to the challenge and undertook an aggressive program to achieve compliance. This paper summarizes the Navy's response to the unexpected plastics discharge limitations, the regulatory background, the Navy's program strategy, and the status of the Navy's efforts. The formal requirements and policies, adopted by the Navy for shipboard solid waste handling and disposal, are detailed in OPNAVINST 5090.1B of 1 November 1994.

Regulatory Background

The Marine Plastic Pollution Research and Control Act of 1987 (MPPRCA), signed by the President on December 29, 1987, modified the Act to Prevent Pollution from Ships (APPS) (33 U.S.C. 1902), implemented Annex V of the International Convention for the Prevention of Pollution by Ships (MARPOL 73/78) as U.S. law, and mandated certain studies of plastics pollution and compliance reports by Federal agencies. The effective date of the Act for the maritime industry was December 31, 1988, the day Annex V entered into force for the United States.

Annex V of MARPOL prohibits (subject to limited exceptions) the disposal from ships into the sea of all plastics, including but not limited to synthetic ropes, synthetic fishing nets, and plastic garbage bags. It also restricts the discharge at sea of other types of garbage to specified distances from the nearest land. Public vessels are exempt from the restrictions but are expected to comply to the extent possible. The basic requirements of Annex V are summarized in Table 14.3 (see Chapter 14 by Barnett, this volume).

Unlike Annex V of MARPOL, the MPPRCA did not exempt public vessels and directed the U.S. Navy to comply with the discharge controls beginning 5 years after Annex V enters into force (January 1, 1994). However, under the National Defense Authorization Act for Fiscal Year 1994, Congress established new dates by which Navy ships and submarines must comply with the solid and plastics waste discharge restrictions. Specifically, all surface ships must comply with the plastics discharge prohibition by December 31, 1998 and with the special area requirements by December 31, 2000; submarines must comply with plastics and special area requirements by December 31, 2008.

The Navy must submit to Congress by December 1996 a plan for compliance by all Navy ships. Congress directed the Secretary of the Navy to consult with the Secretary of State, the Secretary of Commerce, the Secretary of Transportation, and the Administrator of the Environmental Protection Agency. In addition, the Navy must provide opportunity for public participation in preparing the plan and for public review and comment before submitting the plan.

Navy Actions

Although the MPPRCA gave the Navy 5 years to comply with the plastic discharge ban, the Navy immediately took significant steps to reduce the amount taken aboard ships and discharged at sea. In November 1988, the Secretary of the Navy asked all Navy commands to take the extra effort necessary to ensure the Navy does its part in promoting a clean and safe environment. Fleet Commanders responded by instructing their ships to separate and store plastics waste onboard. In March 1989, all U.S. Navy ships began retaining all plastics waste onboard for shore disposal, if they are at sea for less than 3 days. If ships are at sea for longer than 3 days, they must retain food-contaminated plastics waste for the last 3 days at sea and nonfood-contaminated waste for at least 20 days. This Navy self-imposed 3-day/20-day rule has since been codified as U.S. law (33 U.S.C. 1902).

To help sailors understand and comply with the plastics waste discharge ban, the Office of the Chief of Naval Operations sent an educational package to each ship. The package contained guidelines, posters, brochures, and videotapes about plastics waste, the Navy's program, and the new requirements.

Navy Program Strategy

Historically, the U.S. Navy has led the maritime industry in addressing the problem of shipboard solid waste management, primarily because the Navy's equipment requirements for size, weight, safety, reliability, and maintenance generally preclude using commercially available waste management equipment. Before passage of MPPRCA, the Navy's strategy for shipboard solid waste management was to direct ships to discharge solid waste only where permitted, and to provide ships with equipment to grind up pulpable waste and compact unpulpable trash into sinkable slugs for overboard discharge where permitted.

The revised Navy strategy for shipboard plastics waste management includes five additional elements for compliance:

Source reduction (i.e., fewer plastics in the supply system)
Source separation (i.e., onboard separation of plastics and nonplastics waste)
Onboard storage of plastics waste
Education of ships' officers and crews
Onboard plastics waste-processing equipment.

Navy compliance with plastics provisions of Annex V will be achieved in two stages: near-term operational and supply system changes to reduce plastics discharges and longer term equipment installations to eliminate plastics discharges. The first stage includes source reduction, source separation, onboard storage, and educational efforts. This stage has been largely completed by implementing new Navy instructions to all ships and making changes in the Navy's supply system to reduce the amount of plastics taken aboard ships. The second stage in the revised compliance strategy includes the installation of shipboard plastics processors. This will enable Navy ships to fully comply with the plastics waste provisions of Annex V.

Ship Demonstration Project

Navy ship commanders and crews do not like throwing trash and plastics waste overboard. However, before commanders could require ships' crews to make drastic changes in shipboard solid waste management practices to reduce plastics waste discharges, the Navy had to learn what practices would be effective, feasible, and safe. The Navy needed to learn (1) the nature and quantities of plastics waste on Navy ships, (2) feasible shipboard plastics waste management practices, and (3) sailors' attitudes about plastics pollution control. In 1988, the Navy initiated the Shipboard Plastics Waste Reduction Demonstration Project to collect this information.

During the Demonstration Project, Navy researchers traveled on seven ships to quantify shipboard plastics waste and evaluate prototype equipment and plastics waste management procedures. They sorted and inventoried all plastics waste generated onboard. On two ships, they evaluated Navy-developed prototypes of a trash compactor and a waste pulper. The researchers also surveyed the sailors' knowledge about the plastics problem.

The Demonstration Project found that solid and plastics waste generation on Navy ships is similar to municipal solid waste generation in the United States. The plastics waste generation rate on Navy ships is between 0.1 and 0.2 lb per person per day and only 7% (by weight) of all solid waste generated onboard. Most of the plastics waste (55%) comes from food service activities and therefore is contaminated with food waste. The food-contaminated waste (even when stored in plastic garbage bags) developed odor and sanitation problems after 3 days' storage. The other plastics waste (generated in work and berthing areas throughout the ship) was relatively clean and could be separated and stored at the source.

Fleet Operations

As soon as results of the Demonstration Project were available, the fleets took actions to control the Navy's plastics waste discharge at sea. In March 1989, Fleet Commanders instructed ships to minimize plastics waste dumping at sea by making operational changes in the way plastics and nonplastic solid wastes are managed onboard. Specifically, all surface ships were instructed to follow the 3-day/20-day procedures. An evaluation of ship operating schedules in 1988 indicated that implementation of the 3-day/20-day policy would immediately reduce Navy-wide plastics waste discharge at sea by 70%.

Crew Education

In conjunction with the new policy of separating and storing plastics waste onboard, the Navy developed and sent to all ships an educational package to help sailors understand the reasons for the new requirements and comply with them. The education strategy focused on motivating the entire chain of command, ships' officers and crews, by providing justification for and useful information about the new requirements.

The Navy's plastics education package includes guidance material, videotapes, posters, and general literature. A Ships' Guide contains information on the problems caused by plastics in the oceans, pertinent Navy requirements, essential elements of a successful shipboard program, example approaches used on the demonstration ships, and general information about related issues. The guide also includes lists of common plastic and substitute nonplastic items, sample ship instructions to implement the program, and Navy points of contact for further information. To educate and motivate the crew members, the Navy made a 10-min videotape about plastics waste, the Navy's program, and appropriate shipboard actions. To show support for the program from the top levels of the Navy, the Vice Chief of Naval Operations made a statement on the videotape. The first educational package sent to all ships was so well received by officers and enlisted personnel that, in 1991, the Navy sent all ships an updated package with a revised Ships' Guide, new posters, and a new videotape.

Supply System

The Navy initiated a comprehensive program to reduce, to the extent practicable, the volume of plastics material going aboard its ships. The program addresses the three ways plastics materials are taken onboard ships: plastic items, plastic packaging, and plastic packing material. The focus is on providing the fleet with substitute nonplastic items and packing or packaging material where acceptable alternatives exist or can be developed. Supply system personnel have identified plastic items in ships' trash streams and are now searching for potential nonplastic substitutes. The initial effort focused on the plastic items most amenable to immediate replacement. The long-term effort addresses plastic items needing technology development before nonplastic substitutes would be acceptable.

Identifying potential substitutes is rarely simple. The substitute must be commercially available at a reasonable cost, most have minimal impact on ship operations in terms of weight and storage space requirements, and must still meet performance standards. As an example, no material except plastic currently provides an adequate barrier for items requiring moisture or electrostatic discharge protection.

After acceptable substitutes are identified or developed, the proper changes in the supply system specifications must be made. Some changes can be made by the Navy alone. Others require the consent of and coordination with non-Navy organizations, such as the De-

fense Logistics Agency (DLA) and the General Services Administration (GSA). The Navy is working closely with those agencies and industrial organizations to make changes where possible. The Navy has found that it can only exert limited influence over industry where plastic is the material of choice. For this reason, a joint working group with representatives from Army, Air Force, DLA, and GSA has been formed to coordinate efforts with the other military services and present a united position to industry. The initial focus of the joint committee is on reducing plastic packaging in Navy-used items managed by the other services. The DLA has directed its field activities to reduce plastic packaging for Navy-used items. The DLA expects to review the packaging of 635,000 active Navy-used items by mid-1996.

Internally, the Navy is making significant progress in changing packaging requirements. Since the program was initiated, the Navy has reduced or eliminated plastic packaging for more than 350,000 Navy-managed items by eliminating unnecessary plastics, using alternate materials when practicable, and packaging more in bulk. Based on annual demand projections for the items reviewed to date, an estimated 475,000 lb of plastic waste will be eliminated as a direct result of these changes.

Examples of other supply initiatives include the following:

Naval Supply Centers are reducing the amount of plastic packing and packaging materials they are using by substituting fiberboard boxes, paper cushioning and dunnage materials, paper bags and envelopes, rope, and metal strapping, and by maximum use of reusable containers.

Navy Clothing and Textile Research Facility will revise specifications affecting 500 items to substitute plastic bags with paper bags or tissue or eliminate bags totally.

Navy Publications and Printing Service will specify nonplastic packaging in all new contracts, and Navy Publications and Forms Center is experimenting with cold-seal paper packaging.

Shipboard Equipment Development

The Navy is evaluating innovative approaches for processing shipboard plastics waste through an accelerated, multiphased research and development program. The primary objective is to consolidate plastics waste and make it safe for long-term onboard storage. In 1991, the Navy designed, fabricated, and laboratory tested two breadboard-level, prototype plastics waste processors. In 1992, full-scale development models of the two concepts were designed and fabricated. In 1993 the final design was selected for further development and testing. The preproduction prototype was installed on the aircraft carrier USS *George Washington* in May 1994 for final testing. The first production unit of the plastics waste processor will be installed on an operating Navy ship in July 1996 and the last one by the end of 1998. The Navy's plastics waste processor shreds and then melts and compresses mixed plastic wastes into 10-lb disks (about 1 in thick and 20 in. round). The disks are then placed in storage until the ship returns to port.

Alternative Plastics Waste Management Practices Considered

Without proper equipment onboard to compact and sanitize food-contaminated plastics wastes, Navy ships cannot store such wastes for more than 3 days without causing unacceptable odor problems, increasing risks of fire and pestilence, and exceeding onboard storage capacity. The Navy assessed various alternatives to discharging food-contaminated plastics waste, including odor-barrier bags, washing and sterilizing wastes, and at-sea waste transfers to garbage ships. None has proved sufficiently practical or safe to adopt as interim measures until plastics processors are installed on ships.

Navy researchers developed an odor-barrier bag that can contain odors longer than 3 days if the bags are properly sealed. However, none of the potential problems of onboard storage (e.g., space limitations, fire hazards, and health risks) has yet been resolved. Sterilizing or washing the wastes to reduce odors has proved impractical as a routine procedure. At-sea waste transfer to other ships for storage and transfer to shore would be impractical and could pose unacceptable risks for the sending and receiving ships.

The difficulties and hardship associated with separating and storing plastics waste onboard ships are exacerbated on warships. Navy ships were built for combat. All spaces onboard are already being used for equipment, spare parts, or homes for the sailors. Living quarters are cramped and have very little personal storage area. Sailors already give up personal space to store groceries so ships can complete operations without resupply. Now, they are giving up more living space to store the trash they once threw over the side.

Odor-Barrier Bags

The Navy has been experimenting with different materials for making odor-barrier bags and different methods for properly sealing the bags. Navy researchers have identified a plastic resin that can be fabricated into bags capable of containing odors from decaying food wastes for 30 days; however, the bags must be sealed carefully and properly. Two researchers were needed to manually evacuate and properly seal a bag using a portable pump and a hand sealer. The researchers did identify and successfully test a commercial machine (costing approximately $8,000) that allows one person to evacuate and seal a bag.

The Navy has demonstrated that special odor-barrier bags can contain odors under experimental conditions, but several practical problems have not yet been resolved. First, the bags are not commercially available and would have to be specially fabricated for the Navy's use. Second, handling and storing large numbers of bags for up to 30 days without puncturing some bags may be difficult to achieve in practice. Third, Navy ships do not have sufficient extra space onboard to dedicate to waste storage. Last, there are potential health and fire risks associated with storing bags of food-contaminated wastes onboard for extended periods.

The Navy plans to continue investigating the issues associated with using odor-barrier bags to store food-contaminated plastics wastes onboard for more than 3 days. If suitable storage space can be found on ships and the health and safety risks are acceptable, odor-barrier bags may be an option.

Sterilizing or Washing Plastics Waste

The options of sterilizing or washing plastics waste are not practical onboard Navy ships until plastics processors are installed. The existing autoclaves onboard ships are located in the ships' infirmaries and are used daily to sterilize medical equipment, clothing, and medical wastes. Even if the units were not being used daily for medical applications, they are too small to process the large volume of food-contaminated plastics waste generated each day. Furthermore, routinely carrying food-contaminated wastes into the "clean areas" of ships' medical spaces would pose additional sanitation risks for the ships' personnel.

Washing food-contaminated plastics waste onboard using improvised equipment and facilities is impractical, ineffective, and too labor intensive to be a routine shipboard practice. During the Navy's plastics waste management demonstration projects, Navy researchers experimented with washing food-contaminated plastics waste from the galley and scullery areas. Just simple rinsing of milk bladders, meat wrappings, cottage cheese containers, yogurt cups, and other plastic food packaging was tedious and impractical. Rinsing was ineffective unless each item was fully opened and handwashed. The grease and oils on meat, fish, and shellfish wrappings were not removed by cold-water rinsing.

At-Sea Waste Transfer

The Navy assessed the feasibility and risks of transferring solid and plastics waste to other ships for storage and transfer to shore. The practice, while theoretically feasible, is impractical and would pose unacceptable risks for the following reasons:

Mobilizing the waste material topside would pose an unnecessary logistical burden, unnecessarily expose personnel to potentially unsanitary material, and cause dangerous topside clutter on the sending ship while under way

Increasing the amount of material transferred by high wire increases the risk of personnel injuries on the sending and receiving ships

Extending the "alongside" period while transferring wastes would unnecessarily endanger the sending and receiving ships

Receiving ships would be exposed to unnecessary sanitation risks because the ships do not have onboard facilities to properly store the odorous and unsanitary plastics waste.

Garbage Barges

Garbage barges are feasible for waste transferred in port and are used by the Navy in foreign ports. They are not suitable for routine at-sea transfers because towing speeds for the barges are too slow to keep up with ships under way. A fleet of special high-speed garbage ships would have to be designed, constructed, and maintained for the Navy to routinely transfer wastes at sea. Such a fleet, again while theoretically feasible, would be costly and its operation would impose the same logistical, sanitation, and safety risks as transferring wastes at sea.

Annex V Special Areas

Navy ships operating for extended periods in special areas designated by Annex V cannot fully comply with the nonplastic discharge limitations of Annex V because Navy ships simply do not have sufficient storage space for solid waste or technologies for destroying the wastes onboard. In 1993, the Navy submitted a report to the Congress that described the solid waste management equipment and the Navy's strategy for achieving maximum compliance with APPS. The Navy's compliance strategy focused on developing and installing plastics waste processors, solid waste pulpers, and metal-and-glass shredders throughout the fleet. The Navy concentrated on these low-technology approaches because they were suitable for Navy ships, affordable, and available. No advanced technologies were identified that met the Navy's needs. Therefore, the Navy requested that Congress change the APPS's requirements to prohibit discharge of plastics and "floating" debris, rather the current prohibition of all solid waste dischargers (except food wastes beyond 12 nautical miles) in special areas.

The Navy was developing solid waste pulpers and metal and glass shredders to use on ships to grind up nonplastic, pulpable solid wastes into nonfloating forms for overboard discharge. The pulpers would produce a liquid slurry from food wastes, paper, and cardboard for direct overboard discharge when the ship is beyond 3 nautical miles from shore. The shredder would shred metal and glass into small fragments that could be bagged and would sink when discharged overboard.

The solid waste pulper and shredder would not achieve, however, full compliance with the nonfood solid waste discharge restrictions in special areas. Navy ships need additional waste-destruction equipment onboard to achieve zero discharge of nonfood solid wastes. Meeting the 2000 compliance date for surface ships in special areas is not a question of new technology. The Navy does not have enough time between now and 2000 to develop a technological breakthrough for shipboard solid waste destruction and complete the required research, development, test, evaluation (RDT&E), procurement, and installation cycle for new shipboard equipment. To install new equipment on existing ships by 2000, the Navy must find an existing, mature technology that can be readily

adapted for shipboard use and installed on existing Navy warships. The Navy is now searching for such technology.

Beyond Annex V: The Twenty-First Century Ship

To protect marine environmental quality, the Navy is taking actions that go beyond MARPOL requirements. The Navy has established the goal of achieving environmentally sound ships of the twenty-first century that will be able to treat or destroy all wastes onboard. We expect naval ships operating in the twenty-first century to meet increasingly stringent environmental regulations. The Navy has a comprehensive Shipboard Pollution Abatement Program under way that will enable ships of the twenty-first century to be environmentally sound. The goal of the Navy's Shipboard Pollution Abatement Program is for ships to operate worldwide without potential for regulatory constraints, inappropriate dependence on shore facilities, or unreasonable costs imposed by environmental regulations. The basic strategy is to (1) design and operate ships to minimize waste generation and optimize waste management and (2) develop shipboard systems that will destroy or appropriately treat the waste generated onboard.

If wastes are unavoidable and cannot be destroyed or retained onboard for recycling ashore, they must be sufficiently treated to make all overboard discharges environmentally insignificant. We have not yet achieved the ultimate solution for onboard destruction for any shipboard waste stream, but we have made considerable progress in developing onboard capabilities to treat or process solid waste, oily waste, hazardous materials, and medical waste.

Summary

Since the passage of MPPRCA, the Navy has made progress toward compliance with its requirements by taking aggressive actions in the areas of shipboard operations, supply systems, equipment development, and education. By directive, Navy ships are currently in full compliance with the nonplastics waste requirements of Annex V (93% of total solid wastes, by weight), except in the Special Areas. Ships are also in full compliance with the zero-plastics discharge requirement when they are at sea for 3 days or less, and 70% in compliance overall. The Navy achieved this level of compliance beginning March 1989, when all U.S. Navy ships were directed to retain all plastics waste onboard for at least the last 3 days and nonfood-contaminated plastics waste for at least the first 20 days they are at sea. The Navy plans to completely eliminate plastics disposal from its surface ships by the end of 1998, as they are outfitted with Navy-developed plastics processors, which will compress and sanitize plastics waste for onboard storage.

In the area of supply, the Navy changed packaging specifications for 350,000 items, reducing the amount of plastics taken onboard Navy ships by an estimated 475,000 lb annually. The Navy is working with DLA, GSA, and industry on making other changes where practicable. Navy supply centers are reducing the amount of plastic overwrap and intermediate packaging on supplies sent to ships by switching to reusable containers whenever possible.

In the area of education, the Navy twice sent educational packages to all ships. These included guidelines, posters, brochures, and videotapes about plastics waste, the Navy's program, and the new requirements. Navy schools are adding formal training and awareness about plastics waste to their curricula.

The Navy is now searching for an existing, mature technology that will destroy solid wastes onboard and that can be readily adapted for shipboard use and installed on existing Navy warships by 2000.

20.

Recreational Boaters and Marine Debris: How We Can Effectively Reduce Littering

Jim Ellis and Margaret Podlich

Introduction

Historically, there has been considerable emphasis on identifying the sources of marine debris to narrow the focus of remediation. This has been a valuable exercise, principally because it has shown that every group using the water contributes to the trash problem. Recreational boaters, however, have earned the dubious distinction of being a major source of marine debris. There is no question that boaters have ample opportunity to contribute, because collectively they spend more time on the water than any other potential source group. However, the estimates of debris attributable to recreational boaters are based on assumptions about generation and dumping rates that have not been confirmed.

The commonly referenced studies of marine debris sources cite recreational boaters as second only to commercial fishermen as contributors to marine debris. In our experience, based on many years exposure to the recreational boating community, the 212,018 metric tons of overboard trash attributed to U.S. boaters by Cantin et al. (1990, see Table 6) is a gross overestimate. Although there are roughly 70 million boaters in the United States, using about 19 million boats, it is difficult to believe that every boater throws a minimum of 6.5 lb of trash overboard every year. As most recreational boating trash consists of Styrofoam or paper drinking cups, aluminum cans, and the like, it probably would take several large garbage bags full per person to make up 6.5 lb.

In reality, boaters have a disincentive to pollute the waters they use for pleasure. This was borne out in a 1990 survey of BOAT/U.S. members, which also revealed more than 80% of boaters are seriously concerned about environmental degradation. Additionally, a 1991 study by M.J. Mudar showed that boaters generate far less than the 25 lb/year attributed to them in Cantin et al. (1990). The fact is, most boaters have a vested interest in keeping the waters as clean as possible. What fun is it to go boating in dirty, littered water?

The real issue for MARPOL V today is not so much who creates the debris as how best can it be reduced? We already know that enforcement alone cannot accomplish this goal, and more study is unlikely to change the focus of reduction efforts. Better communication and realistic prevention are the best ways to get where we want to go: to solutions. Education coupled with efforts to alter the habits of recreational boaters will lead to long-term, ethically based changes in the behavior of boaters with respect to the way they manage their trash.

Know Your Audience

It is obvious that some boaters litter. Some types of recreational activities, such as fishing or food preparation, generate more trash than other activities, like water skiing. To reduce the contribution of recreational boating to marine debris, we have to reach those boaters who litter, whoever they are.

When we address marine debris and recreational boaters, the first step is to know as much as possible about the audience, which is not as simple as it first appears. What is a typical recreational boater? A swarthy sailor on a 40-ft yacht? A young speedboater? In reality, recreational boaters are as varied as the types of boats they use and the activities in which they participate. From duck hunters using 15-ft prams in the shallows to sailors in ocean-going sloops, from whitewater kayakers to multihull racers, the contrasts are stark. Clearly, profiling the specific boating subgroup you wish to address, its paraphernalia, and their habits must be the guiding principle when creating educational or enforcement programs.

An effective marine debris educational program for recreational boaters, then, must satisfy several requirements: it must identify the specific boating audiences being addressed, it must relate to the similarities that these groups have with all boaters, and it must address the activity-specific practices that generate debris. Boater profiles may be derived from the types and number of boats the groups use; what they use them for; how often and for how long they boat, the supplies they typically carry; how often and how many passengers they take; where they tie up or take their boats out of the water; whether they are local or transient; and demographics such as age, sex, owner occupation, and income.

Appealing to Boaters

For recreational boaters, like all marine user groups, new ways of thinking and acting are best introduced by their peers. Environmental groups should be challenged to identify the problems, and the boaters should formulate the solutions for themselves. Certainly, it is more acceptable to take direction from within a group than from without. In

this case, boaters educating and appealing to boaters would be ideal. If that is not a ready alternative, however, it is possible to construct educational programs so that they appeal to boaters rather than repel them.

It is an excellent idea to consult with boaters before "going public" with an idea because pollution solutions that make perfect sense ashore maybe very difficult or ridiculous aboard a boat. Experienced boaters in target groups can help refine programs so they offer realistic alternatives for boaters. This, in turn increases credibility and the boaters' willingness to consider further suggestions.

Taking the boaters' perspective can be very helpful. Telling a boating audience about sea life eating plastics may not be so good an incentive to prevent trash discharge as showing how their propellers can become tangled in discarded monofilament line. Use of the right terms is also important to the credibility of the message. In a room full of waste management professionals, phrases like "marine debris" and "inadequate reception facilities" are unlikely to be misunderstood, but to a boater these terms have limited meaning. The average boater deals with "garbage cans" and "trash in the water." Another important ingredient in a successful educational effort is understanding how boaters think about recreation. Why are they boating? What are they getting out of it? Are they boating to "get away from it all," to enjoy nature, or for some other reason?

Boaters are facing several negative factors that also should be considered. Coupled with a general decline in the amount of spare time enjoyed by our society, boaters are subject to increasing licensing and bureaucratic requirements for owning and operating a boat. Educators must remember that recreational boating is for fun, and that education efforts need to be developed with the attitude of the boater in mind.

Finally, to reach boaters, you need to know how they learn about new boating issues. Are they only involved with boating on weekends, with no weeklong exposure to boating publications or boating-related activities? Do they belong to clubs, and, if so, when do they meet or how do they communicate with each other? Are they "proactive," in that they actively seek new information, or do they wait for news to come via hearsay, then becoming "reactive?" Are they active at all, or so widely scattered that there is no umbrella organization to work through? Do school or other community organizations or agencies address or influence this boating group? Can these existing communication channels be used to get the message across?

The bottom line in producing behavior changes in any boating group comes down to taking three progressive steps. First, increase the boaters' awareness of the problems trash causes and offer solutions. If that steps succeeds, in step two boaters will internalize the information and develop an attitude about trash. Ultimately, this attitude will develop into a commitment to take specific actions that eliminate throwing trash overboard.

Innovative Solutions

Here are some success stories of programs that have made a difference in the United States and some further ideas that may help boaters generate less trash.

Recycling Monofilament Fishing Line

Recreational fishermen periodically replace the monofilament line on their fishing reels. Manufacturers generally recommend annual replacement. Typically, some fishermen have unreeled the line at sea and replaced it once they returned to shore. The results of this practice, as well as other improper disposal methods, are significant accumulations of monofilament line both in the water and on the beaches.

One industry group has responded to this problem by recycling the used line. In 1990, Berkley Tackle of Monticello, MN, started a unique recycling program at its U.S. dealers by providing cardboard collection boxes where customers drop off their used line. The filled boxes are mailed postpaid to the manufacturer, and the used monofilament reappears in the marketplace as other useful outdoor products. Since its inception, this program has recycled more than 4 million miles of old fishing line.

A Sea Grant agent in Florida took this idea one step further by enlisting the help of a local highschool club and a county task force. The club made a dozen covered wooden collection boxes (4 × 4 × 4 ft), which they placed at marinas, fishing piers, boat ramps, and tackle shops. The boxes are monitored and emptied by 4-H youth or National Park Service staff, and the collected line is returned to Berkley Tackle for recycling.

Citizen Pollution Patrol

In 1991, the Center for Marine Conservation (CMC, a nongovernment organization) received a grant from the U.S. Environmental Protection Agency (EPA) to design and test a Citizen Pollution Patrol model. CMC, in conjunction with the New Jersey Sea Grant Marine Advisory Service and Coastal Environmental Service, ran projects in Annapolis, Maryland, and the Barnegat Bay area of New Jersey. The projects were designed to achieve four objectives:

Increase awareness of the marine debris problem in the community

Test a pilot program in which citizens report suspected violations

Bring members of marine user groups into compliance with MARPOL Annex V

Educating the boating public about marine debris problems and solutions.

Opportunities to reach the boating public were as varied as the boaters themselves. Presentations were made to more than 1500 people during the 9-month boating season. Audiences ranged from the Maryland Department of Natural Resources (DNR) police captains to charter boat captains, recreational boating clubs, and junior sailing camps. Written materials were also developed to reach other boaters. In all, they distributed more than 100,000 fact sheets, placards, and other materials by mail and through cooperating marine stores, boating course instructors, and one-on-one outreach.

The project staff regularly sent out press releases to keep the marine debris issue before the people who could publicize the message. They wrote articles and provided artwork to local publications, and participated in joint radio interviews with the Maryland DNR and a television show in New Jersey. In short, the test areas were saturated with marine debris information aimed at boaters. Active boaters had many chances to receive accurate information from several reliable sources during the boating season. Also, the staff quickly answered all requests for additional materials or information. The effects of this small project were significantly amplified through use of the press, presentations, and printed materials.

Boaters for the Bay

In 1992, the Center for Marine Conservation worked in Los Angeles to create a dialogue with boaters on environmental issues. Sponsored by the local arm of the EPA National Estuary Program, the project had several objectives:

Working with local boating leaders to determine environmental interests

Gathering information and resources, locally and nationally, on those topics

Conducting a participatory workshop for boaters about environmentally sound boating

Sharing information from the workshop with the boating community to encourage boaters to become involved in developing long-term solutions.

A steering committee of 10 boating representatives provided guidance for the project and helped foster local participation and commitment to holding the workshop and pursuing long-term goals. During the workshop, boaters were introduced to local representatives from the Santa Monica Baykeeper, U.S. Coast Guard, environmental groups, and the local government. Several solutions were

explored for each environmental topic introduced and later compiled into action agendas for the Santa Monica Bay. In addition, participants wishing to continue their involvement in controlling bay pollution circulated a sign-up list to help them maintain project momentum.

The Boaters for the Bay project demonstrated the powerful impact of readily applicable information, as well as the need for local resources to be linked with those groups and individuals who want to deal with the problems. Just because environmental professionals are aware of local resources does not mean boaters know about them or how to engage them.

Information in Registration Packets

Boat owners generally receive an annual or biannual bill for registration fees. Including a simple message about the boater's role in preventing marine debris can be an effective tactic, where feasible. In some states, registration packets are already stuffed to capacity and another piece of paper could make it more expensive to mail and further dilute the recipient's attention to the contents.

In 1990, the New Jersey Department of Environmental Protection sent out in the registration packet a small "Dear Boater" sheet explaining the problems and solutions to marine debris. The New Jersey Sea Grant Marine Advisory Service helped develop the sheet. In 1993, Maryland used another approach. State officials did not want to add another piece of paper in their boating registration packet, so they rewrote their standard cover letter to focus attention on the marine environment, advocating pollution prevention ranging from using onshore sewage pump-out stations to bringing trash back home after boating.

State agencies are not the only organizations with access to lists of registered boaters. In some states, a copy of the list (on diskette or mailing labels) may be purchased and is sometimes sorted by zip code or type of boat. During the CMC Citizen Pollution Patrol project in Annapolis, some 30,000 postcards with marine debris messages and regulations were sent using mailing labels obtained from the Maryland DNR. This direct approach provided nearly all local boaters with accurate information needed to comply with the law.

An Exemplary School Program

Four years ago, the Boating Education Office of the Washington State Parks and Recreation Commission decided to initiate an environmental education program for middle-school students (ages 12–15 years) on the premise that it's easier to teach children good boating habits than to reteach adults who already have bad ones.

With the help of the State Department of Education, which provided interested teachers as advisors and writers, the Boating Education Office produced a curriculum packet that has been very effective in the classroom. The packet is a looseleaf notebook with topical chapters that contain standard classroom lesson plans, each of which can stand alone or be taught in sequence. The chapters cover environmental topics, including marine debris, safety, boat handling, navigation, and other boating subjects. In addition, there are accompanying video segments performed by Bill Nye the Science Guy, who recently signed a contract with the Walt Disney Company for a syndicated general science program.

This program also offers summer teacher-training courses and encourages teachers to copy and share the curriculum materials in their schools. According to the Boating Education Office, the program has been going strong for the past 3 years in Washington and is now spreading to other states.

The Boater's Pledge

The Boater's Pledge program was created in 1992 by the Gulf of Mexico Program Initiative. There are active pledge programs operated through university Sea Grant offices and other state offices in Alabama, Florida, Loui-

siana, Mississippi, and Texas. Each participating state is trying to get boaters and anglers to take a pledge to bring trash ashore, to encourage others to be good stewards, and to help pick up trash and bring it ashore, recycling it when ever possible.

The program works by building relationships among state agencies, marinas, marine businesses, and boaters. Participants hold cleanups at boat ramps and marinas; they put on workshops, boat show outreach events, and presentations for a variety of organizations. The objectives, of course, are to get as many boaters as possible to take the pledge and to encourage boating-related groups to be as supportive as possible. The boaters receive a decal for their boats, periodic program and other boating-related communications from the Boater's Pledge Coordination Center, and opportunities to volunteer for pledge activities in their areas. Pledge takers in Florida also have a chance to win vacations at area theme parks (e.g., Sea World and Busch Gardens). whose sponsorship has proven an invaluable asset to the program. Boats displaying the pledge decal also receive discounts or free services at some cooperating marinas. Florida Sea Grant also has worked out an arrangement whereby it can accept grant funds from industry and other sponsors to have the prison industries produce aluminum "Don't Splash Your Trash" signs for $10 apiece. These signs are placed at strategic marina locations to remind boaters about marine debris. Sponsors are identified on the signs they fund.

In Louisiana, the Department of Environmental Quality's Outbacker Program is handling the Boater's Pledge program. This program is also encouraging boaters not to throw cigarette butts overboard. The makers of Camel cigarettes provide cardboard "butt boxes" that are handed out to pledge takers and at all outreach locations. Also, Louisiana Sea Grant has installed several environmental bulletin boards in cooperation with marinas around Lake Pontchartrain. Topics, presented in professionally drawn cartoons, are changed every six weeks or so. Responses from boaters so far have been positive, with many commenting to marina owners that the bulletin boards are a great idea. The boaters on Lake Pontchartrain are bringing their trash ashore more than ever before.

Other Activities to Raise Boater Awareness

The following sections are brief descriptions of further ideas for activities to raise boater awareness of (1) marine debris, (2) boaters' obligations under the law, and (3) how to comply with the law.

Real Boaters "Use RE-tensil"

This program convinces boaters to take reusable utensils and food containers aboard instead of throwaways. The program would show boaters that use of cloth napkins, cotton dish towels, sponges, and reusable forks, knives, and spoons cost less than using disposables and minimizes marine debris. It would introduce the old Girl Scout trick of using a net bag to rinse dishes and utensils overboard before washing them ashore. The program might combine a utensil giveaway with a Boater's Pledge program if suitable donations can be arranged.

Pair Marinas and Local Agencies

This pairing could help marinas start and get recognition for recycling programs. Bureaucratic snags can discourage marina owners from getting involved in recycling, making it worth the effort to work out those problems ahead of time. Recycling incentives for boaters could include credit in the marine stores or on marina fees, per-pound reimbursement for turning in bottles and cans, or charges for delivering nonsorted trash. Making closable plastic bags available to boaters and ensuring easy access to recycling bins on the docks will also encourage boater recycling.

It's The Rule

Organizers for this program approach fishing tournament and racing organizations with suggestions to include no-dumping regulations in their contest formats. Anglers or boaters who dump trash overboard would be disqualified or penalized in some way by the organization. Information is readily available to show fishing groups the dangers of discarded monofilament line and fishhooks, as well as information to impress racing groups about the dangers of marine debris to propellers and marine life. Also, tournament and racing winners can be encouraged to become role models for good environmental stewardship.

Less Is Better

This program would impress boaters that it makes good sense to leave unnecessary packaging ashore, prepare food ahead of time, and have less trash to bring back. A good slogan, perhaps "It's a Boat, Not a Trash Barge," and humorous drawings can go a long way to convey this important message. The point is to show boaters that it is more fun to just boat without having to worry about all the trash—so don't bring it aboard to begin with.

Conclusion

The message recreational boaters of all types should internalize and act on is "don't throw trash overboard." There are many ways to avoid the need to dump trash at sea. The basics are (1) do not bring things aboard that are destined to become trash, and (2) plan to bring all your trash ashore. Either choice works. The challenge is to facilitate boaters' transition from general littering behavior to compliance with laws protecting the marine environment. There are few shortcuts to taking all the necessary steps between the awareness and action phases of education. There are, however, many ways to take those steps, and many people and groups who are willing to share their ideas, materials, and encouragement.

21.

A Strategy to Reduce, Control, and Minimize Vessel-Source Marine Debris

Barbara Wallace

Introduction

The principal focus of this paper is on commercial and publicly owned vessels, including commercial shipping, commercial fishing vessels, passenger cruise lines, military fleets, research vessels, passenger ferries, tugboats and barges, offshore oil and gas platforms, and offshore service industry vessels. There are few common elements among these categories of vessels. Some vessels are privately owned, others are public vessels. Some primarily use private ports and terminals, while others tend to use public ports and terminals. Some vessels move worldwide, while others have more regional or local movements. Some vessels carry a crew, while others carry crew and passengers. The crew on some vessels totals fewer than 5 persons, while on others the crew totals 500 persons or more. Nonetheless, every vessel generates wastes.

Sources of Marine Debris from Vessels

The sources of marine debris from vessels potentially include domestic waste (including galley or kitchen waste), operational wastes (including maintenance wastes, cargo residues, and miscellaneous wastes), and municipal and commercial wastes carried as cargo.

These wastes include materials made from plastics, paper, glass, metal, and cardboard.

Some debris is obviously from vessels. Commercial fishing is associated with debris items such as nets, salt bags (large, reinforced plastic bags used by commercial fishermen to preserve or separate their catch), bait boxes and bags, fish baskets or totes, fish and lobster tags, and gill-net or trawl floats. Debris from cruise ships includes small containers of shampoo, conditioners, and body lotion, shoe polish, and plastic cups, often labeled with the cruise line's or vessel's name. The offshore petroleum industry contributes operational waste such as plastic write-protection rings (used on computer tapes on vessels doing seismic surveys) and hard hats, often labeled with the company logo. Some marine debris such as wooden pallets, plastic sheeting, and galley wastes (such as vegetable sacks and egg cartons) are used by more than one type of industry and, while not directly identified with a particular marine user group, are typically categorized as vessel-generated waste.

The source of other debris types is less obvious. For example, plastic pieces, one of the most common debris items found during beach cleanups, could come from a variety of marine and land-based sources.

The Key Players

There are a number of key players working toward solving the problem of vessel-source marine debris. These key players can work independently or in various combinations to reduce, control, and minimize vessel-source marine debris. At a minimum, however, key players fall into at least one of the following eight categories:

International organizations (e.g., United Nations Environment Programme and the International Maritime Organization)
Agencies of national governments (e.g., environmental protection, natural resources, Coast Guard, public health, tourism, and agriculture)
Financial organizations (e.g., international economic development organizations)
Maritime industry (e.g., public and private fleet owners and operators, industry trade associations, and shipping agents)
Ports and terminals (e.g., individual ports and terminals, and industry trade associations)

Solid waste management industry (e.g., waste management companies and industry trade associations)
Nongovernmental groups
Academia (e.g., institutions and researchers).

The Strategy

There are three components to MARPOL Annex V implementation: the vessel, the port, and ultimate disposal. For vessels, MARPOL Annex V specifies restrictions on at-sea disposal of vessel-generated garbage, but it does not specify how they are to comply with those disposal requirements. Figure 21.1 shows the options available to vessels to comply with the at-sea disposal limitations. Essentially, there are three options: legal at-sea disposal of nonplastic ma-

Disposal options

Phase of Waste Management

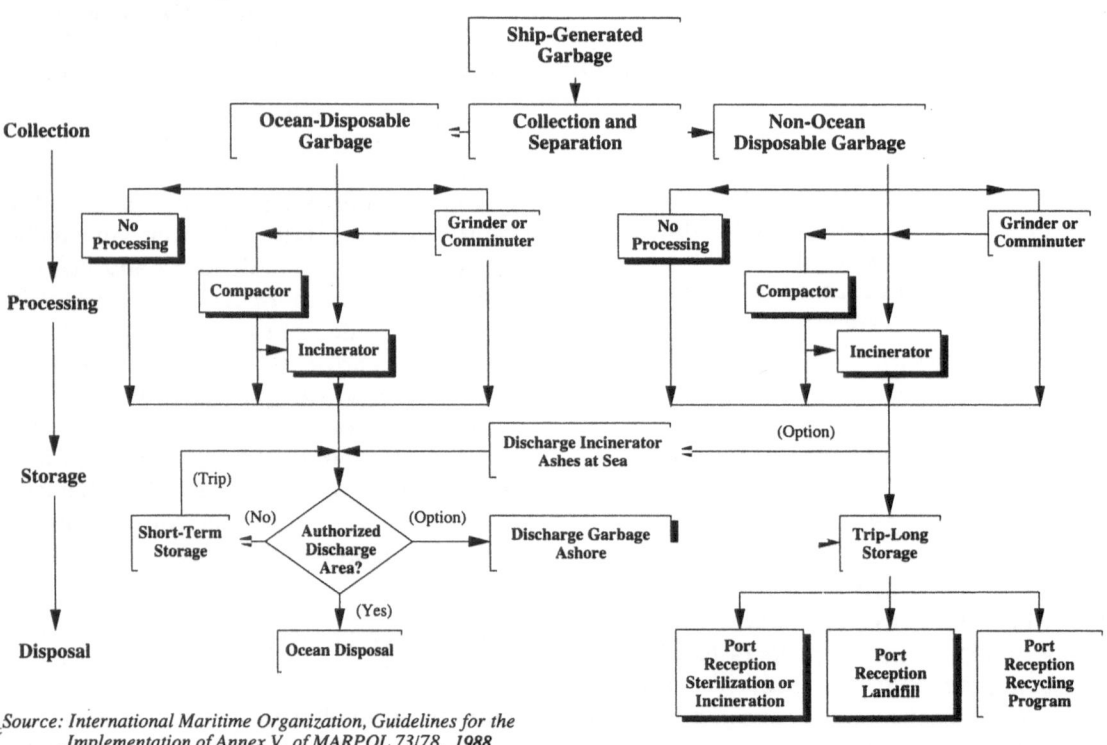

Source: International Maritime Organization, Guidelines for the Implementation of Annex V, of MARPOL 73/78, 1988

FIGURE 21.1 Options for shipboard handling and disposal of garbage. (From International Maritime Organization Guidelines for the Implementation of Annex V, MARPOL 73/78, 1988, with permission.)

terials, at-sea incineration, and offloading in port. There is, of course, a fourth option: illegal at-sea disposal. For ports, MARPOL Annex V requires that adequate reception facilities be provided in ports to accept and handle garbage retained onboard for shoreside disposal. Some countries regulate garbage coming from foreign countries, meaning that foreign or regulated garbage requires special handling and a different type of port reception facility than nonregulated garbage.

Figure 21.2 shows options for port handling and disposal of MARPOL Annex V wastes. Ultimate disposal, the third component to MARPOL Annex V implementation, is not explicitly addressed by the Convention. However, successful implementation is a series of linkages not only between the vessel and the port reception facility, but between the port reception facility and ultimate disposal. All the parts of these components must work together if the linkages are to be made.

When the links are not made or are broken, MARPOL Annex V implementation is incomplete.

The Objectives

The strategy to reduce, control, and minimize vessel-source marine debris involves meeting five objectives:

1. To promote the ratification of MARPOL Annex V, enactment of domestic legislation, and the support other programs that restrict disposal of solid wastes into the marine environment
2. To develop, promote, improve, and support MARPOL Annex V implementation efforts
3. To promote and support efforts to enhance land-based solid waste management

Options for port handling and disposal of MARPOL Annex V wastes

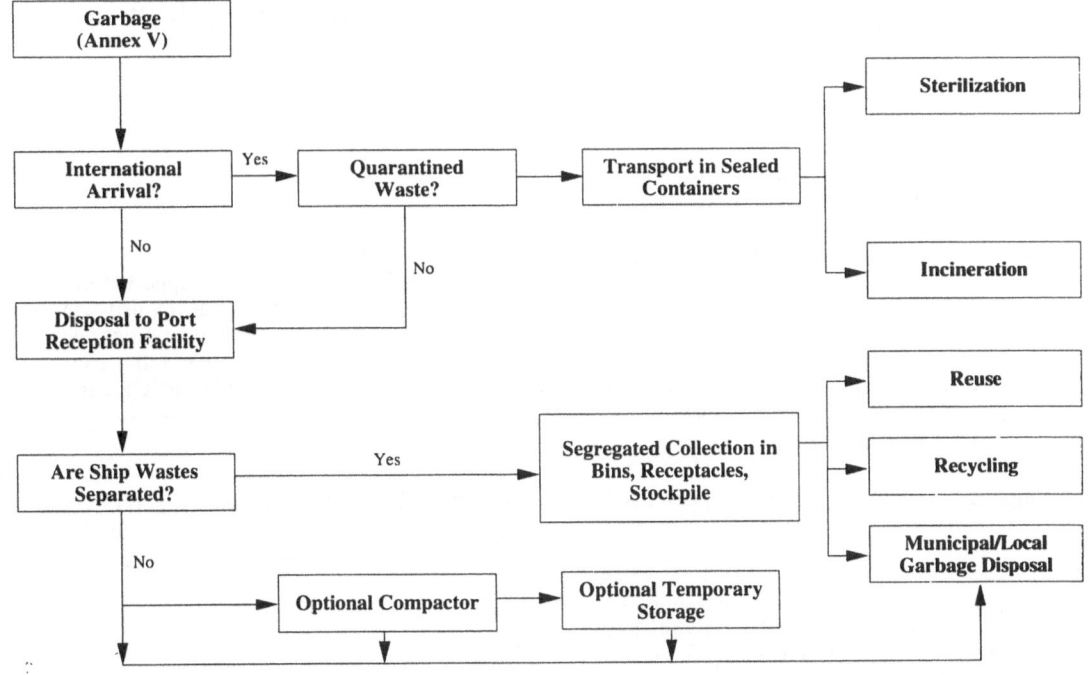

Source: International Maritime Organization, Comprehensive Manual on Port Reception Facilities, November 1995

FIGURE 21.2 Options for port handling and disposal of MARPOL Annex V wastes. (From International Maritime Organization Comprehensive Manual on Port Reception Facilities, 1995, with permission.)

4. To promote and support efforts to assist developing countries in building the institutions and infrastructure necessary to ratify and implement MARPOL Annex V

5. To develop, promote, and support strategies to achieve compliance with MARPOL Annex V and other waste discharge requirements.

Activities to Achieve the Objectives

The objectives address the need for international agreements and domestic legislation to restrict at-sea disposal of solid waste from vessels, their implementation, and compliance; they also recognize the linkage of implementation and compliance to land-based solid waste management; and they also recognize the special efforts needed to assist developing countries to build the institutions and infrastructure to handle garbage from vessels. These objectives incorporate related issues of technology, training, education, enforcement, and economics. While these objectives are interrelated (Figs. 21.1 and 21.3), activities to achieve each objective do not need to be undertaken sequentially. Activities to achieve each objective require that the key players work independently and cooperatively in various combinations. The following activities are recommended to achieve each of the five objectives. Listed within each activity is the most likely lead or responsible party(ies), and its relationship, if any, to other objectives and activities.

Strategy to reduce, control, and minimize marine debris pollution from vessels

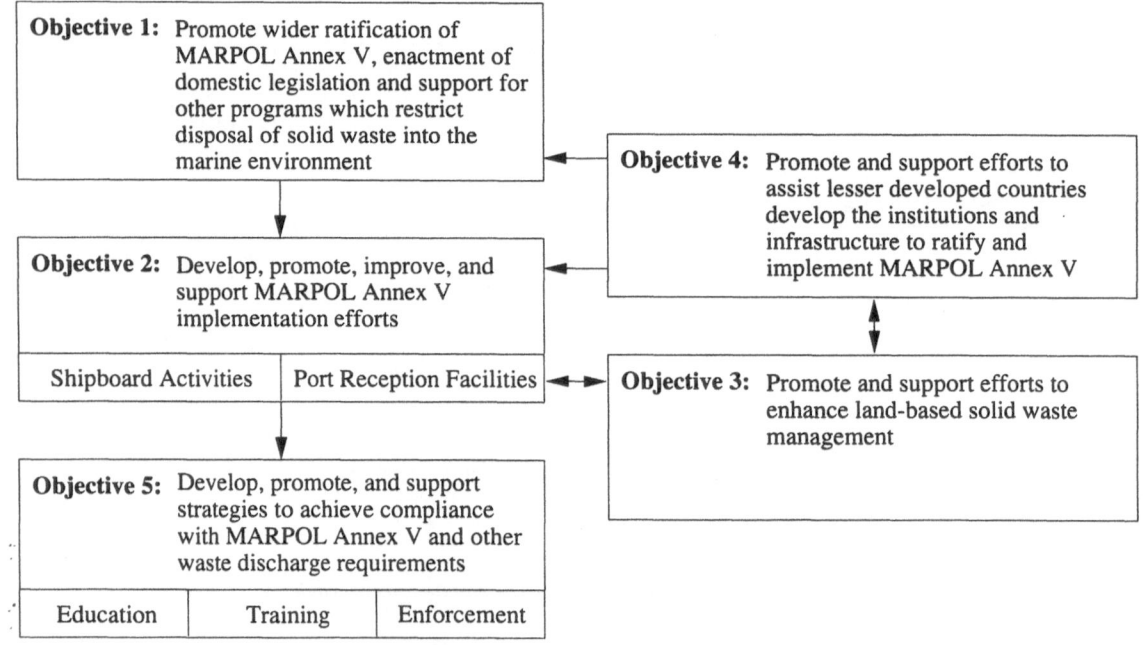

FIGURE 21.3. Strategy to reduce, control, and minimize marine debris pollution from vessels.

OBJECTIVE 1. Promote ratification of MARPOL Annex V, enactment of domestic legislation, and the support for other programs that restrict disposal of solid waste into the marine environment.

Activity 1	Encourage noncontracting parties to ratify MARPOL Annex V and enact domestic laws that restrict disposal of solid wastes into the marine environment.
	Identify, for prospective ratifying countries, specific benefits derived from improvements to solid waste management and the marine environment, and develop internal support for ratification
	Identify and systematically remove disincentives or obstacles to ratification
	When appropriate, provide technical and financial assistance
	Assist countries considering ratification to undertake public outreach programs in support of ratification
	Provide forums for discussion of regional implications of ratification
Lead and responsible party(ies)	Contracting parties, IMO, and other UN agencies, and financial organizations
Relationship to other activities	All activities within Objective 4
Activity 2	Seek to coordinate the objectives and activities of programs and action plans designed to reduce, control, and minimize vessel-source marine debris, such as:
	UNCED Agenda 21, Chapter 17B, prevention, reduction and control of degradation of the marine environment from sea-based activities (17.30a from shipping)
	Regional Seas Action Plans
	Other regional programs (e.g., Helsinki Convention on the Protection of the Baltic Sea Area)
	The Gulf of Mexico Action Plan
	The Marine Debris Action Plan for the Caribbean
Lead and responsible party(ies)	UNEP and participating nations, national governments, nongovernmental organizations
Relationship to other activities	Activity 1 of Objective 1
Activity 3	Encourage enactment of domestic laws (where they do not currently exist) that regulate the transport of garbage by vessel
Lead and responsible party(ies)	National governments, nongovernmental organizations
Relationship to other activities	
Activity 4	Develop support and amend the U.S. legislation and related regulations implementing MARPOL Annex V with the purpose of improving enforceability
Lead and responsible party(ies)	U.S. Congress, U.S. Coast Guard, maritime industry, and nongovernmental organizations
Relationship to other activities	Activity 2 of Objective 2
	Activity 5 of Objective 2

OBJECTIVE 2. Develop, promote, improve, and support MARPOL Annex V implementation efforts

Activity 1 (shipboard activity)	Encourage the reduction in wastes produced on vessels by:
	Working with suppliers to reduce packaging waste
	Providing assistance to the maritime industry in identifying reusable, recyclable, or nonpersistent materials to substitute for disposable materials
	Developing collective purchasing agreements
	Enacting domestic legislation (where it does not exist) to reduce packaging
Lead and responsible party(ies)	National governments, solid waste management industry, shipping companies, and maritime industry trade associations
Relationship to other activities	Activity 6 of Objective 2
	Activity 8 of Objective 2
Activity 2 (shipboard activity)	Develop better documentation of shipboard waste-generation and waste-handling practices
Lead and responsible party(ies)	IMO, national governments
	Activity 4 of Objective 1

Relationship to other activities	Activity 3 of Objective 2
Activity 3 (shipboard activity)	Facilitate exchange of information on technology and shipboard waste-handling practices through formal channels (e.g., existing clearinghouses such as the IMO and informal networks)
Lead and responsible party(ies)	Maritime industry, solid waste management industry, national governments
Relationship to other activities	Activity 2 of Objective 2
Activity 4 (shipboard activity)	Support development of appropriate measures to reduce air pollution from ships
Lead and responsible party(ies)	IMO, contracting parties, national governments
Relationship to other activities	
Activity 5 (shipboard activity)	Encourage universal use of vessel waste record books, waste management plans, and placards
Lead and responsible party(ies)	Contracting parties
Relationship to other activities	Activity 4 of Objective 1 Activity 2 of Objective 2
Activity 6 (shipboard activity)	Establish a "Green" Vessel Program for solid waste management by setting performance standards and formal requirements for a voluntary, self-certifying program that promotes zero discharge of solid wastes (this may need to be tied into a larger program addressing all environmental issues for vessels, such as oil, fuel, and sewage)
Lead and responsible party(ies)	National governments, maritime industry, solid waste management industry, IMO
Relationship to other activities	Activity 1 of Objective 2 Activity 8 of Objective 2
Activity 7 (shipboard activity)	Encourage vessels to report inadequate port reception facilities by facilitating the reporting process and providing a feedback mechanism on results of the report
Lead and responsible party(ies)	Contracting parties, IMO
Relationship to other activities	Activity 6 of Objective 5
Activity 8 (shipboard activity)	Develop guidance on how to reduce vessel wastes through purchasing practices
Lead and responsible party(ies)	National governments, shipping companies, maritime industry trade associations
Relationship to other activities	Activity 1 of Objective 2 Activity 6 of Objective 2
Activity 9 (shipboard activity)	Initiate company policy to honor designated Special Areas (MARPOL Annex V) immediately
Lead and responsible party(ies)	Shipping companies, maritime industry trade associations
Relationship to other activities	
Activity 10 (shipboard activity)	Aggressively encourage and support U.S. Navy compliance with MARPOL Annex V by the established deadlines
Lead and responsible party(ies)	U.S. Congress, U.S. Coast Guard, and nongovernmental organizations
Relationship to other activities	
Activity 11 (port reception facility activity)	Continue to support projects in assessing alternative approaches to provide and finance port reception facilities; conduct surveys of port reception facility users to identify their preferred payment method; and include follow-up evaluations, and distribution of results, as an integral part of projects
Lead and responsible party(ies)	National governments, international economic development organizations, ports, and terminals

Relationship to other activities	Activity 1 of Objective 4
Activity 12 (port reception facility activity)	Establish "Green" Port Program for solid waste management by setting performance standards and formal requirements for a voluntary, self-certifying program (this may need to be tied into a larger program addressing all environmental issues for vessels, such as oil, fuel, and sewage)
Lead and responsible party(ies)	National governments, IMO
Relationship to other activities	
Activity 13 (port reception facility activity)	Distribute IMO guidelines on port reception facilities to noncontracting parties
Lead and responsible party(ies)	IMO
Relationship to other activities	Activity 3 of Objective 4
Activity 14 (port reception facility activity)	Update Animal and Plant Health Inspection Service (APHIS) regulations to reduce unnecessary restrictions on regulated garbage
Lead and responsible party(ies)	U.S. Department of Agriculture
Relationship to other activities	Activity 7 of Objective 5

OBJECTIVE 3. Promote and support efforts to enhance land-based, solid waste management.

Activity 1	Develop a primer or framework, or both, for environmental regulations, particularly for islands and remote areas, and developing countries, that promotes integrated solid waste management (reduce, reuse, recycle, incinerate, and landfill)
Lead and responsible party(ies)	Appropriate UN agencies (e.g., UNEP), international economic development organizations, national governments
Relationship to other activities	Activity 2 of Objective 4
Activity 2	Support land-based recycling programs by Participating in recycling programs Supporting market development for recycled products through the purchase of goods made from postconsumer materials, and providing financial or personnel assistance or both Participating in cooperative market contracts for recyclables
Lead and responsible party(ies)	Ports and terminals, shipping companies, maritime industry trade associations, and national governments
Relationship to other activities	
Activity 3	Increase communication and working relationships among government agencies responsible for land-based solid waste management and those responsible for vessels and the marine environment
Lead and responsible party(ies)	All levels of government
Relationship to other activities	

Objective 4. Promote and support efforts to assist developing countries in building the institutions and infrastructure necessary to ratify and implement MARPOL Annex V.

Activity 1	Undertake projects designed for institution building, provision of solid waste management, port reception facility infrastructure, and MARPOL Annex V enforcement capabilities with follow-up evaluation and distribution of results
Lead and responsible party(ies)	International economic development organizations and national governments
Relationship to other activities	Activity 11 of Objective 2
Activity 2	Develop a primer or framework, or both, for environmental regulations, particularly for islands and remote areas, and developing countries, that promotes integrated solid waste management (reduce, reuse, recycle, incinerate, and landfill)
Lead and responsible party(ies)	Appropriate UN agencies (e.g., UNEP), international economic development organizations, and national governments
Relationship to other activities	Activity 1 of Objective 3
Activity 3	Distribute IMO manual on port reception facilities
Lead and responsible party(ies)	IMO
Relationship to other activities	Activity 13 of Objective 2
Activity 4	Provide forums for discussion of regional approaches to solid waste management, adequate port reception facilities, and the relationship of both to vessel-source marine debris
Lead and responsible party(ies)	IMO, international economic development organizations, and national governments
Relationship to other activities	

Objective 5. Develop, promote, and support strategies to achieve compliance with MARPOL Annex V and other waste discharge requirements.

Activity 1 (education activity)	Continue basic public and industry outreach programs that address the consequences of marine debris, but with added emphasis on identifying and tailoring programs to various industry components (e.g., management, vessel operator, onboard waste handler, linkage to land-based disposal); include formal evaluation of outreach programs as an integral part of all programs
Lead and responsible party(ies)	National governments and nongovernmental organizations
Relationship to other activities	
Activity 2 (education activity)	Enhance international network to exchange information on education programs by Including marine debris on the agendas of international conferences and meetings on the environment and solid waste management Incorporating agencies, such as UNEP, FAO, WHO, and international development banks, into information exchanges
Lead and responsible party(ies)	IMO, national governments, and nongovernmental organizations
Relationship to other activities	
Activity 3	Identify, recognize, and publicize industry success stories, particularly within that industry
Lead and responsible party(ies)	Maritime industry trade associations, shipping companies, and ports and terminals
Relationship to other activities	
Activity 4 (education activity)	Encourage the usage of placards and signs at ports and terminals outlining relevant waste discharge regulations and waste-handling procedures for that particular port or terminal

Lead and responsible party(ies)	Ports and terminals (for voluntary efforts) and national governments (for required efforts)
Relationship to other activities	
Activity 5 (training activity)	Enhance pollution prevention training and environmental literacy of maritime professionals in areas such as source reduction, reuse, recycling, available technologies for onboard waste handling, legal and policy restrictions on at-sea solid waste disposal, and the consequences of marine debris by
	Incorporating pollution prevention training and environmental literacy into academic training of maritime professions
	Conducting continuing education workshops
	Encouraging presentations of pollution prevention and environmental issues in professional forums (e.g., maritime publications and conferences)
	Developing model crew and port environmental training programs
	Using formal evaluation techniques to learn from and improve training efforts
Lead and responsible party(ies)	Academic institutions, maritime training institutions, national governments, and maritime industry
Relationship to other activities	All activities of Objective 2
	All activities of Objective 3
Activity 6 (enforcement activity)	Encourage vessels to report inadequate port reception facilities by facilitating the reporting process and providing a feedback mechanism on results of the report
Lead and responsible party(ies)	National governments and maritime industry
Relationship to other activities	Activity 7 of Objective 2
Activity 7 (enforcement activity)	Increase responsibility of APHIS in MARPOL Annex V enforcement
Lead and responsible party(ies)	U.S. Congress, U.S. Department of Agriculture, U.S. Coast Guard
Relationship to other activities	Activity 14 of Objective 2
Activity 8 (enforcement activity)	Enact legislation to ensure that only vessel-generated garbage is permitted to be offloaded to MARPOL port reception facilities for garbage
Lead and responsible party(ies)	National governments
Relationship to other activities	

Discussion

There are a number of recurring themes evident in this strategy to reduce, control, and minimize vessel-source marine debris:

Involvement of Key Players. The key players need to be actively involved in activities to reduce, control, and minimize vessel-source marine debris.

Cooperation. The key players and lead or responsible party identified for each activity need to work cooperatively to successfully complete each activity.

Coordination. There is a need for the key players to coordinate their activities with one another and with activities to reduce, control, and minimize vessel-source marine debris under authorities or action plans other than MARPOL Annex V.

Exchange of Information. There is a need for the key players to share information, particularly with those outside their own group. The formal networks for the exchange of information must be supported financially and provided with information in a timely fashion. Dissemination of the information must follow. Informal networks must also be fostered.

Evaluation. There is a need for formal evaluations of activities and programs designed to reduce, control, and minimize marine debris pollution from vessels, particularly those related to port reception facilities, technology development, training, education, and enforcement. These evaluations must then become part of the information-sharing among the key players.

Recognition of Good Performance. There is a need to not only assist vessels and ports in their efforts to improve their solid waste management but to recognize successes and good performance.

The more integrated the work on these activities becomes, the more likely that their objectives will be broadly and permanently realized. The history of international identification and regulation of the commercial maritime community, coupled with the relatively small number of well-recognized institutions championing efforts to control vessel sources of marine debris, suggest that progress toward each of the five objectives will continue to be made.

SECTION V
Considering the Land-Based Sources of Debris

SECTION V
Consideration the Land-Based Sources of Debris

Introduction

In traveling the world one quickly grows accustomed to the high densities of trash accumulated in the coastal areas surrounding cities of developing countries. The transition from the rural, agrarian, self-sustaining way of life in undeveloped regions to the urban, industrial, service-oriented way of life in the fully developed regions generally encompasses a solid waste generation and disposal evolution that is highly relevant to the marine debris problem. At the rural-agrarian level, wastes are minimal: they are mostly organic or aggressively reused; otherwise, there is virtually no solid waste disposal system. With the dramatic increase in the development and urbanization of rural areas, however, there comes an equally dramatic increase in per capita generation of persistent wastes without concomitant increases in waste collection and disposal systems (see Fig. 22.0). Meanwhile, at the fully developed end of this spectrum, per capita generation of persistent wastes is huge, but efficient collection and disposal systems typically are in place.

Real data (on per capita, persistent waste-generation and disposal rates by socioeconomic strata and population distributions) are not available to give quantitative interpretation to this solid waste generation and disposal model. However, a first approximation suggests that much of the persistent waste entering the marine environment originates from coastal- and upriver-population centers in developing countries, as much as the world's population lives in varying stages of economic development. While this hypothesis cannot be tested without a significant research effort, general demographic trends suggest that this phenomenon will worsen: Populations in undeveloped and developing areas are growing faster than in fully developed areas. There is a general worldwide trend toward urbanization, drawing people away from the rural-agrarian lifestyle into more consumer-oriented (waste-generating) occupations and modes of living. This movement is especially strong toward the coastal

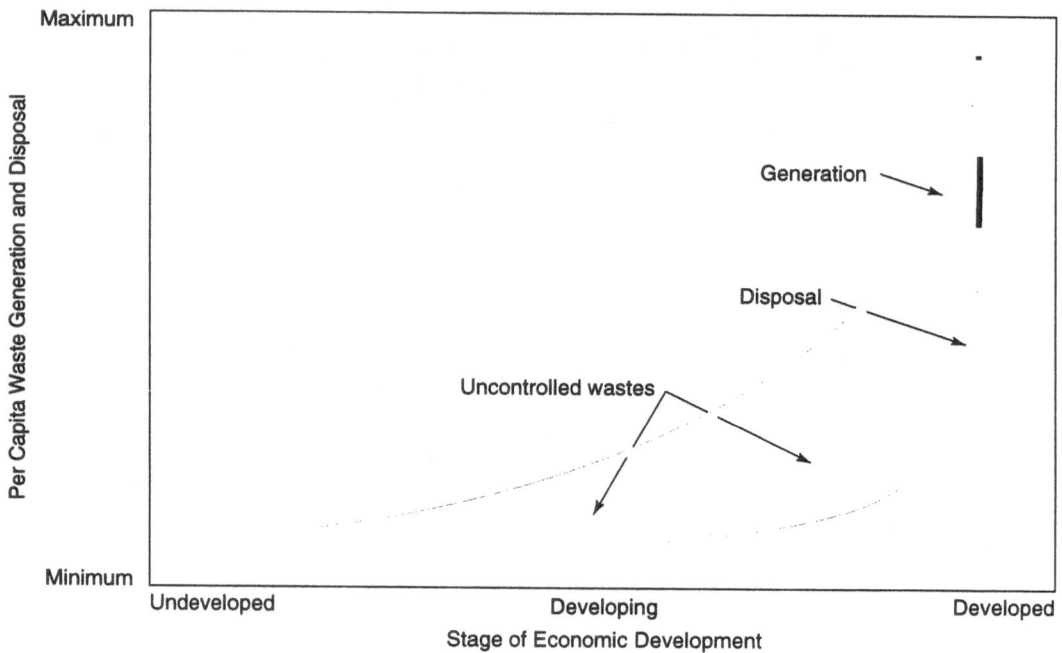

Figure 22.0. Theoretical solid waste generation and disposal model based on stage of economic development.

zones of developing countries, which further compounds the marine debris aspect of the solid waste dilemma. More broadly, worldwide observations confirm that the development of an effective waste management infrastructure is not a priority for many communities or governments.

The United Nations Environment Programme's 1994 State of the Seas Report (UNEP 1994) estimated that 80% of the pollution in the marine environment arises from land-based activities. If this estimate is accurate, and there is much evidence to suggest that it might be, controlling marine debris presents an unprecedented global-environmental challenge. Solutions to this challenge are likely linked to public health, economic development, and social and cultural change at all levels. Per capita waste generation must be pushed down through behavioral changes, facilitating source reduction, materials reuse, and recycling. In addition, waste collection and effective disposal rates must be increased through capital investment in an appropriately scaled waste management infrastructure, as well as through enforcement of regulatory mandates. Politicians will likely be compelled to advocate the needed investments and institutional commitments only through broadly increased community awareness and organizational efforts. This new awareness and shift in attitudes, of course, will have benefits far beyond the control of land-based sources of marine debris: the benefits will also be linked to higher productivity through improved public health and, where relevant, sustained natural production and tourism. In this regard, the marine debris issue often provides simple graphic lessons on the environmental consequences of individual and institutional actions. These lessons can help stimulate public support for the solutions to an array of environmental problems.

Awareness of the magnitude of marine debris coming from land-based sources is fairly recent and is increasing. This section constitutes some of the earliest writings that focus directly on this phenomenon. Of necessity, our papers draws heavily from U.S. urban and Caribbean island experiences in measuring and responding to the land-based sources of marine debris. We hope these experiences will form the basis for energetic investigations and attention to all facets of the immense challenge posed by this marine and terrestrial pollution problem.

In our opening paper, Nollkaempher assesses legal frameworks for an international response to the land-based sources of marine debris. Two suites of papers follow: the first five papers address urban-debris sources and solutions and the next three papers consider rural and Caribbean island experiences. The section's final paper is a preliminary set of strategies and recommendations for controlling land-based sources of marine debris compiled from the working group reports from the Third International Conference on Marine Debris sessions on urban, upland rural, and recreational sources of marine debris.

22.■
Legal Regulation of Upland Disharges of Marine Debris: From Local to Global Controls and Back

André Nollkaemper

Introduction

Although land-based discharges account for much of the problem of marine debris, national and local regulation of such discharges has been utterly inadequate. In marked distinction to the regime for control of sea-based pollution, international rules to induce appropriate national and local policies on upland discharges have been largely ignored. For most regions no such rules have been developed. Regional rules that do exist (mainly for the Northeast Atlantic, the Mediterranean Sea, the Baltic Sea, the Black Sea, the Persian Gulf, the Southeast Pacific, and the Arctic) rarely have addressed the specific problems of marine debris and, to the extent they have, have focused on symptoms (waste disposal) rather than on causes (waste generation).

This situation may begin to change. International environmental policy developments following the 1992 United Nations Conference on Environment and Development (UNCED) have finally curtailed the unjustified reliance on inadequate regional instruments to prevent pollution from land-based sources. The stage has been set for the reconsideration of regional programs and for the development of the Global Programme of Action to Protect the Marine Environment from Land-Based Activities (GPA) for controlling land-based sources of marine pollution.

These developments make it timely to reflect on the international regulation of upland discharges of marine debris. This paper outlines the major legal issues involved in the control of upland discharges of marine debris and assesses the main regulatory developments necessary for more effective international control of such discharges. It proceeds in three sections that (1) set forth the main arguments for current international controls of upland discharges of marine debris; (2) outline regional and global regulatory challenges and developments in regard to waste disposal, recycling, and source reduction and control; and (3) draw together the leading threads and present the case for an integrated approach.

One preliminary observation is in order. Although legal rules furthering the prevention of upland discharges of marine debris are proliferating, only rarely are such laws directly induced by concern over marine debris. The main contribution to prevention of upland discharges of marine debris is to be expected from installation of sewage treatment facilities and from recycling and solid waste source-reduction actions. These developments are only marginally influenced by concern about the marine environment. The present analysis focuses not on the various objectives of regulatory programs but rather on their contributions to effective control of upland discharges of marine debris.

Arguments for Current International Controls

Most if not all states have been wrestling with the causes of discharges of marine debris. The argument that international regulations and accompanying programs should be developed to induce and improve national and local efforts is well accepted. In summary, international regulations serve (1) to induce states to install sewage treatment works, prevent industrial discharges of plastics, and undertake other preventive actions where they may be otherwise unwilling or incapable of doing so; (2) to prevent economic disruptions that may be caused by unilateral actions such as mandatory packaging controls or recycling requirements[1]; (3) to prevent transboundary pollution by marine debris[2]; and (4) to foster economies of scale in such fields as public awareness and education and as development and application of clean production and disposal processes.

Mostly, these benefits of international regulations have been taken to support regional rather than global controls. The preference for regional regulatory actions is based on interregional variability in the extent and nature of environmental problems caused by discharges of marine debris and in states' economic capacity to address these discharges. Population density and patterns (urban, suburban, and rural), consumption patterns, and environmental awareness may be vastly different. Although states may be equally concerned about reducing discharges of marine debris, they face different risks–benefit circumstances and may make different choices. Under these circumstances, two decades of regional marine pollution control should have covered upland discharges of marine debris. Global action, on the other hand, has been limited to the hortatory provisions of the 1982 United Nations Convention on the Law of the Sea[3] and the legally

nonbinding 1986 Montreal Guidelines for the Prevention of Marine Pollution from Land-Based Sources,[4] which address the issue of marine debris only in the most general ways.

It is now clear that most existing regional programs have been misconceived and that exclusive reliance on regional approaches is inadequate. The main shortcomings of the present situation are as follows.

First, too many states have been unwilling or unable to participate in regional instruments. Some 20 years after the first regional instrument was signed only a few regions (the Northeast Atlantic,[5] the Mediterranean Sea,[6] the Baltic Sea,[7] the Black Sea,[8] and the Arctic[9]) are covered by more-or-less operational programs for land-based pollution. Programs for the Southeast Pacific[10] and the Persian Gulf[11] are as yet dormant, and other regions are devoid of such programs.

Second, existing regional regulations are inadequate. They have been unable to induce states to reset their priorities so as to install adequate sewerage controls, develop land-

[1]See John Warren Kindt, Solid Wastes and Marine Pollution, 34 *CATHOLIC UNIVERSITY LAW REVIEW*, 37(1984), at 92; Howard B. Latin, Ideal Versus Real Regulatory Efficiency: Implementation of Uniform Standards and Fine-Tuning Regulatory Reforms, 37 *STANFORD LAW REVIEW* (1985), 1267 at 12.

[2]In 1990, beach cleanups in the United States found trash from 44 countries; see 22 *BNA Environ. Rep.* (1991), at 265.

[3]United Nations Convention on the Law of the

Sea, Montego Bay, 10 December 1982, 21 ILM 1261 (1982).

[4]Adopted by UNEP Governing Council Decision 13/18 of 12 June 1985, UNEP/GC.13/16.

[5]Covered by the Convention on the Prevention of Marine Pollution from Land-Based Sources, Paris, 4 June 1974, 13 ILM 352 (1974); to be superseded by the Convention for the Protection of the Marine Environment of the North-East Atlantic, Paris, 22 September 1992, 8 IJMCL (1993), at 50.

[6]Protocol for the Protection of the Mediterranean Sea Against Pollution from Land-Based Sources, 17 May 1980, 19 ILM 869 (1980).

[7]Convention for the Protection of the Marine Environment of the Black Sea Area, Helsinki, 22 March 1974; to be superseded by the Convention on the Protection of the Marine Environment of the Black Sea Area, Helsinki, 9 April 1992.

[8]Protocol on Protection of the Black Sea Marine Environment Against Pollution from Land-Based Sources, Bucharest, 21 April 1992, 32 ILM 1122 (1993).

[9]Artic Environmental Protection Strategy, Rovaniemi, 14 June 1991, 20 ILM 1624 (1991).

[10]PRotocol for the Protection of the South-East Pacific Against Pollution from Land-Based Sources, Quito, 23 July 1983.

[11]Protocol for the Protection of the Marine Environment Against Pollution from Land-Based Sources, Kuwait, 21 February 1990.

use planning controls, and adopt demanding waste reduction policies. These regulations have focused too strongly on monitoring rather than on source control. They have concentrated on phasing out individual hazardous substances rather than on addressing industrial sectors and processes. Most important in this context, they have dealt with land-based pollution as an isolated problem rather than as an integral part of the larger problems of waste generation and management.

Third, many states are economically[12] or technologically unable, or lack the information and expertise, to implement policies to reduce generation and improve disposal of debris. Where resources are scarce, marine debris will not emerge as a priority. Lack of adequate mechanisms (global or regional) to provide necessary inducements by way of information exchange, technical cooperation, and financial assistance ensures that marine debris will not move higher on national agendas.

Fourth, it is necessary to go beyond isolated and inadequate regional approaches to prevent economic displacement caused by divergent local prescriptions for manufacturing processes aimed at reducing solid waste.[13] A certain uniformity in environmental requirements would free states from the temptation of relaxing local limitations to prevent the loss of local businesses and facilities.[14] In addition, proliferation of unilateral legislation on packaging and recycling may lead to trade barriers.[15] These arguments gain considerable weight because successful regulations on waste may encroach deeply into industrial processes and operation of markets.

Finally, because there is no valid reason to offer some species of wildlife or fish more protection against marine debris than other wildlife or fish, the existing differences between regional controls are unsatisfactory. The irrebuttable, although poorly implemented, argument that people in one area should not face health risks higher than people in other areas (recently conceived in terms of environmental justice and equity)[16] applies equally to the marine ecosystem and should be reflected in comparable levels of control in different regions.

These considerations do not necessarily call for global rather than regional regulation. The protocol on land-based pollution currently being prepared for the Wider Caribbean Region may, after all, prove the strength of a regional approach, if adequately supported by international funding mechanisms. Global mechanisms will be unable to stimulate the targeted programs that are required to induce adequate waste disposal and waste reduction practices at national and local levels without translation through (sub)regional mechanisms. However, the bleak prospects for adequate solutions in most regions has set a global process in motion, transcending the prevailing regional approaches.

Regional and Global Regulatory Challenges and Developments

The international community has undertaken a two-stage global process for addressing land-based pollution. The first stage culminated, through meetings in Halifax[17] and Nairobi, in

---

[12]André Nollkaemper, Marine Pollution from Land-Based Sources: Towards a Global Approach. *MARINE POLLUTION BULLETIN* 8 (1992).

[13]A bill recently proposed in New York State for packaging reductions explicitly took account of such effects: 24 *BNA Environ. Rep.* (1994) at 332. See also John Warren Kindt, *SOLID WASTES,* at 92.

[14]This applies similarly to a national context; see the considerations on this point in Weyerhauser Co. v. Costle, 590 F2d 1011 (D.C. Cir. 1978), at 1042.

[15]This is the main inducement of the efforts of the European Union to develop a Council directive on packaging and packaging waste; see Lucas Bergkamp and Gail N. Martiri, Take-Back Schemes,

Product Standards and Eco-Taxes, 17 *BNA Int. Environ. Rep.* (1994), at 192.

[16]See "Not in My Backyard," 20 HUMAN RIGHTS, 26 (1993).

[17]Report of the Intergovernmental Meeting of Experts on Land-Based Sources of Marine Pollution,

the adoption of Chapter 17B of Agenda 21 at the UNCED.[18] This chapter contains several substantive requirements explicitly applying to land-based pollution, including upland discharges of marine debris. In addition, several other chapters of Agenda 21, although not explicitly linked to marine debris, address the more fundamental sources: consumption patterns (Chapter 4), demographic dynamics and sustainability (Chapter 5), sustainable human settlement development (Chapter 7), and, most of all, environmentally sound management of solid waste (Chapter 21).

At the same time, Agenda 21 envisages the elaboration and implementation of its imperatives in a second stage. It recognizes that the previously dominant regional approaches to land-based pollution should be supplemented by a commanding global strategy. A global strategy should reactivate existing regional instruments and stimulate the introduction of new ones where land-based marine pollution is as yet unregulated. In particular, Agenda 21 calls on states to perform the following:

Consider developing a new set of global guidelines applying to land-based marine pollution, updating, strengthening, or extending the Montreal Guidelines.
Assess the effectiveness of existing regional instruments and initiate and promote the development of new regional agreements.
Develop means of providing guidance on technologies to handle the major forms of marine pollution from land-based sources.
Develop policy guidance for global funding mechanisms.
Identify steps requiring additional international cooperation (paragraph 17.25).

In response to Agenda 21, the Governing Council of the United Nations Environment Program (UNEP) developed a preparatory process,[19] through meetings in Nairobi (December 1993)[20] and Montreal (June 1994), that led to the adoption of the GPA by an intergovernmental conference in Washington, D.C. (November 1995).

Challenges for International Regulation of Upland Discharges of Marine Debris

The regulatory targets to be addressed in the international process can be divided into two groups: adequate waste disposal (including sewage treatment and stormwater discharges), and recycling and source reduction. Only the second track offers a true solution to the problem. Yet, even with aggressive recycling and waste reduction, a substantial quantity of solid waste still needs to be disposed of adequately to prevent it from reaching the marine environment. The following describes how regional instruments have tackled the main regulatory challenges and how developments in UNCED have provided a basis for improved international regulation.

Waste Disposal: Regulatory Challenges

The dominant responses to the major targets of regulatory action concerning upland discharges of marine debris have been the limited regional policies undertaken with respect to point-source discharges and sewage. Five other sources of marine debris that have rarely been addressed are also discussed here. The discussion of these regulatory strategies refers to experiences in the United States. Although domestic experiences obviously cannot be transplanted to the international level, they do indicate the gap between what is necessary at the national level and what is offered by international law.

Chap. II of Land-Based Sources of Marine Pollution, Report of the Secretary-General to the Third Session of the Preparatory Committee for the UNCED, A/CONF.151/PC/71 of 17 July 1991.
[18]Annex II to the Report of the United Nations Conference on Environment and Development, Rio de Janeiro, 3–14 June 1992, A/CONF.151/26 of 12 August 1992.

[19]UNEP/GC.17/L.21/Add.2.
[20]Report of the Preliminary Meeting of Experts to Assess the Effectiveness of Regional Seas Agreements, Nairobi, 6–10 December 1993, UNEP/L.S./W.G.1/1/L.3.

The single appropriate response is the prohibition of any point discharges of solids,[21] and most existing regional agreements initially sought to eliminate point discharges. Such agreements identified floatables as one of the black list substances, thereby making them subject to the general obligation to eliminate their discharges.[22] However, while agreements in the Northeast Atlantic, Baltic Sea, Mediterranean Sea, and the Arctic have developed specific prescriptions and targets for most other black list substances like mercury and cadmium, none have been developed for floatables, which *de facto* if not *de jure* means that discharges of floatables are unregulated.[23]

Inadequate sewage treatment installations also are recognized sources of debris, especially in developing states where the construction of sewage treatment plants has not kept pace with population growth.[24] This problem is also evident in some developed areas such as around the Black Sea[25] and in parts of the United States.[26] Regional fora in the North Sea,[27] the Mediterranean Sea,[28] the Black Sea,[29] the Baltic Sea,[30] the United States, and Mexico[31] have set target levels for sewage treatment (although more in response to nutrient inputs than to solid waste discharges). These regulatory targets share two crippling weaknesses. First, most of them are disturbingly general, which is what one would expect to find at the global rather than the regional level. They are not targeted, lack specification, and provide no support for national or local programs in this field. Second, they lack adequate funding. This weakness prohibits the setting of more meaningful targets (certainly in the cases of the United States–Mexico, the Black Sea, the Mediterranean Sea, and the Baltic Sea).

Practice in the United States gives some insight into the types of regulations needed to actually address the sewage-source floatables issue. All sewage treatment works have been made subject to federal construction standards and effluent standards. Federal construction standards for sewage treatment

---

[21]See for the United States s. 301 of the Clean Water Act, 33 U.S.C. § 1301; see also Michael J. Bean, LEGAL STRATEGIES FOR REDUCING PERSISTENT PLASTICS IN THE MARINE ENVIRONMENT, paper presented at the Sixth International Ocean Disposal Symposium, April 1986, at 8.

[22]See, e.g., art. 6(1) of the 1974 Baltic Sea Convention; art. 4(1)(a) of the 1974 North East Atlantic Convention. Black list substances are substances that because of their persistence, toxicity, or tendency to bioaccumulate need to be strictly controlled.

[23]See André Nollkaemper, *THE LEGAL REGIME FOR TRANSBOUNDARY WATER POLLUTION: BETWEEN DISCRETION AND CONSTRAINT,* Martinus Nijhoff, Dordrecht, 1993, at 115–116.

[24]S. 19(I) of the Conclusions and Recommendations of the 1991 Halifax Meeting, see note 19.

[25]See Ellen Hay and Laurence D. Mee, The Ministerial Declaration: An Important Step, 23 *ENVIRONMENTAL POLICY AND LAW,* 1993 at .., (noting that economic stagnation has left many coastal settlements with half-completed sewage treatment plants, or in some cases, with no sewerage system at all).

[26]A 1992 study by the Natural Resources Defense Council and the Center for Marine Conservation revealed that billions of gallons of raw waste pour into lakes, rivers, and coastal areas each year when combined sewers, which carry stormwater and wastewater in the same pipe, overflow during heavy rains; see 23 *BNA Environ. Rep.* (1992), at 13.

[27]Statement of Conclusions of the Intermediate Ministerial Meeting, 7–8 December 1993, Copenhagen, Annex II.

[28]In 1991 the Contracting Parties to the Barcelona Convention adopted a plan for the collection, treatment and disposal of sewage for each Mediterranean coastal city with a population of over 10,000 inhabitants; see UNEP(OCA)/MED IG.2/4 of 11 October 1991.

[29]S. 3 of the Ministerial Declaration on the Protection of the Black Sea, Odessa, 7 April 1993, where it is agreed to protect human health by the urgent construction of sewerage systems and sewage treatment plants in areas where there may be detrimental effects to the sustainable development of the marine environment for such activities as tourism and fisheries.

[30]See, e.g., HELCOM Recommendation 12/4 of 22 February 1991 on Industrial Connections to Municipal Sewage Networks.

[31]The Agreement on Cooperation Between the United States of America and the United Mexican States for Solution of the Border Sanitation Problem of San Diego, California–Tijuana, Baja, California, San Diego, 18 July 1985, *BNA Int. Environ. Rep.* Ref. File 31:1403, provides for funding of Mexican wastewater treatment.

plants in coastal areas require reduction of discharges of floating plastics to the maximum extent practicable.[32] Effluent standards require secondary-level sewage treatment, defined, *inter alia,* in terms of removal of a specified amount and percentage of suspended solids.[33] In addition, it has been recognized that states and municipalities will not, or cannot, implement effective treatment policies unless they are substantially supported by federal grants.[34] International programs are a long way from providing this level of guidance and support for any floatables policies. Apart from the mostly unsuccessful programs concerning point discharges and sewage treatments, international fora have ignored stormwater discharges, combined sewer overflows (CSOs), landfills, litter, and inadequate coastal zone planning as primary sources of marine debris.

Stormwater runoff from urban areas, construction areas, and landfills is a major land-based source of debris, in particular of plastics.[35] National regulation of these sources has proven troublesome. In the United States, the Environmental Protection Agency (EPA) left these sources unregulated for years and then engaged in a trial-and-error policy, issuing and frequently amending stormwater regulations that were not successfully implemented.[36] Congress has called

on the EPA to expedite permitting for discharges of stormwater.[37] The Clean Water Act reauthorization proposal of the Clinton administration[38] includes a renewed attack on these sources, requiring urban areas with populations greater than 100,000 to obtain permits for stormwater discharges. Remaining stormwater discharge sources, including smaller cities, would be addressed through enhanced state non-point-source pollution control programs. On the international level, a recent recommendation for the Baltic Sea on reduction of discharges from urban areas by proper management of stormwater is an isolated effort to bring about such rules through international inducements.[39]

Another marine debris source for which international rules have failed to contribute to effective national and local policies is CSOs. Where stormwater and sanitary sewage systems are combined, in wet weather the capacity of sewage treatment facilities can be overwhelmed, resulting in untreated discharges. Overflow mechanisms (i.e., CSOs) are built into such systems to divert the wastewater and stormwater to avoid overloading the sewer systems. These CSOs may contain considerable amounts of

---

[32]Bauer and Iudicello, *STEMMING THE TIDE, at 119.*

[33]40 CFR § 133.102(b).

[34]William H. Rodgers, 2 *ENVIRONMENTAL LAW,* St. Paul, Minn. West Publ. Co. (1986), § 4.23, at 332. The FWPCA grants the U.S. EPA the authority to make grants to local authorities for construction of treatment plants; see s. 201(g)(1), 33 U.S.C. § 1281(g)(1).

[35]20 *BNA Environ. Rep.* (1990), at 1842. See also 20 *BNA Environ. Rep.* (1992), at 2569 (U.S. EPA noting that stormwater runoff is one of the most significant remaining threats to the quality of surface water in the United States).

[36]General permits for stormwater discharges associated with industrial activities and construction were issued by U.S. EPA in 1992; see 23 *BNA Environ. Rep.* (1992), at 1307 and 1513. On November 19, 1993, the U.S. EPA proposed a general stormwater permit that would cover approximately 11,000 facilities representing 29 industrial

sectors in 12 states. 58 *Fed. Reg.* 61.146 (1993); 24 *BNA Environ. Rep.* (1994), at 1357.

[37]139 *Congr. Rec.* S 8479.

[38]24 *BNA Environ. Rep.* (1994), at 1956. For the most part this proposal conforms to proposals currently pending in Congress; see S 1114 and HR 3948. The proposed authorization embodied in S 1114 has been severely criticized by the environmental community for proposing to exempt large categories of municipalities from stormwater permits and by weakning substantially stormwater control requirements for even the largest cities; see 24 *BNA Environ. Rep.* (1993), at 343.

[39]HELCOM Recommendation 11/2 of 14 February 1990 on Reduction of Discharges from Urban Areas by Proper Management of Stormwater calls for measures at the source to prevent the deterioration of the quality of stormwater (e.g., by street-cleaning); treating contaminated stormwater as polluted wastewater in separate sewer systems, and decreasing the amount of overflow in areas with combined sewer systems, *inter alia* by appropriate design of sewerage systems and by providing retention facilities.

plastics and other floatable debris.[40] Here also, the U.S. experience is exemplary considering the lack of easy, low-cost solutions. Legal control of CSOs has been inconclusive and undirected, leaving modalities of controls to be determined on a case-by-case basis rather than by uniform standards.[41] A draft combined sewer overflow policy, imposing obligations on combined sewer systems permittees to control overflow as a result of wet weather, has only recently been published.[42] The limited efficacy of existing combined sewer overflow policy induced a bill in the U.S. Congress that would authorize further financial assistance to local governments for the construction of facilities to control CSOs.[43] This assistance may be essential, as the capital requirements are considerable.[44]

Landfills are another unregulated, major land-based source of marine debris. Runoff from landfills sited near rivers or in coastal areas is particularly problematic.[45] The U.S.

EPA Final Rule on Solid Waste Disposal Facility Criteria[46] suggests some of the regulatory options. It stipulates that solid waste facilities must (1) not be susceptible to washout by strong floods so as to pose a hazard to the environment; (2) be covered by runon and runoff systems to prevent flow onto, and to control flow from, the active portion of the facility during the peak discharge from a 25-year storm; and (3) operate in compliance with standards established under the Clean Water Act. Again, there are no international regulations supporting any such actions.

Another poorly controlled, yet significant, land-based source of marine debris is beach litter.[47] While there is a clear need for strong legal condemnation of littering,[48] littering by individuals on the beach or along rivershores and bridges has not been effectively controlled by any law.[49] Adding stringent, international prohibitions would do little to improve this situation. A seemingly more effective approach to managing the litter problem would include programs focusing on educating the public.[50] Regional efforts seeking to coordinate, stimulate, or support such programs are minimal or absent.

Finally, land-based marine pollution instruments have not handled issues of coastal zone planning. Coastal zone planning partly overlaps with previously mentioned approaches such as location of landfills and installation of

[40]Statement on Introduction of the Coastal Waters Improvement Act of 1993, 139 *Congr. Rec.,* No. 21, September 15, 1993, S 11916, at 11917. See also 20 *BNA Environ. Rep.* (1990), at 1842 (U.S. EPA noting that combined sewer outflows are one of the major land-based sources of plastics).
[41]40 CFR § 133.103(a); Rodgers, *ENVIRONMENTAL LAW,* § 4.31, at 455–457.
[42]58 *Fed. Reg.,* No. 11, January 19, 1993, 4994.
[43]Section 210 of the Coastal Waters Improvement ACt of 1993, S. 1459, 139 *Congr. Rec.,* No. 21, September 15, 1993, S 11916, at 11917. In 1981 Congress provided for special funding to address combined sewer overflow problems; FWPCA s. 201(n), 33 U.S.C. § 128(n).
[44]A recent plan to reduce overflows in Portland, Oregon, with 96% over present levels was estimated to cost $700 million by 2011; see 24 *BNA Environ. Rep.* (1994), at 1659. The U.S. EPA estimated its draft policy for controlling combined sewer overflows would cost approximately $3.45 billion per year. Under a stringent interpretation of the existing Clean Water Act (CWA), efforts to control CSOs could cost as much as $14.14 billion; see 24 *BNA Environ. Rep.* (1994), at 1956. The 1992 wastewater needs survey estimated that it will cost $41 billion over the next 20 years.
[45]20 *BNA Environ. Rep.* (1990), at 1842. Many estuaries have been affected by trash as a result of solid waste sitings near estuaries; see U.S. EPA, Second Report to Congress on the National Estuary Program; 23 *BNA Environ. Rep.* (1992), at 1428.
[46]56 *Fed. Reg.,* 50.978 (1991).
[47]20 *BNA Environ. Rep.* (1990), at 1842.
[48]In the United States, federal law has long since declared the discharge of litter by any persons unlawful; see s. 301 of the Clean Water Act, 42 U.S.C. § 1311; s. 13 of the Rivers and Harbors Appropriation Act of 1899, 33 U.S.C., § 407.
[49]See for the problems of enforcing anti-litter laws in the United Kingdom: Philip Circus, Litter Capital of Europe?, *NEW LAW JOURNAL,* January 11, 1991, at p. 23.
[50]Donald C. Bauer and Suzanne Iudicello, Stemming the Tide of Marine Debris Pollution: Putting Domestic and International Control Authorities to Work, 17 *ECOLOGY LAW QUARTERLY,* 71 (1990).

sewerage systems. It may also have some independent relevance in coastal development policies and control of nonpoint sources in coastal areas. Indeed, in the United States coastal zone management is a separate target for prevention of marine debris.[51] By defining their material scope strictly in terms of considering the introduction of substances into the marine environment, most regional agreements are prevented from addressing coastal zone management issues.

Waste Disposal: Agenda 21 and the GPA

Against this background, the UNCED had to identify and outline future strategies that would reinforce and stimulate regional, national, and local programs to improve the foregoing picture. There is little evidence that the activities of the UNCED were in any way concerned with upland discharges of marine debris. Indeed, Agenda 21 does not address such issues as point discharges of debris, runoff, littering, and landfill location. The GPA, however, did introduce some relevant objectives.

Agenda 21 contains four relevant targets. Its first and main concern is sewage treatment. It calls on states to ensure that, by the year 1995 in industrialized countries and by the year 2005 in developing countries, at least 50% of all sewage, wastewater, and solid wastes is treated and disposed of in conformity with national or international environmental and health quality guidelines. These requirements rise to 100% by the year 2025 (paragraph 21.29). Obviously, this requirement is only meaningful when sup-

---

[51]The Coastal Zone Management Act contains several programs for financial assistance, most notably coastal zone enhancement grants for reducing marine debris entering the coastal or ocean environment by "managing uses and activities that contribute to the entry of such debris." 16 U.S.C. § 1456b. The Coastal States Organization has noted that until now federal funding for coastal state non-point-source pollution control programs has been entirely inadequate; see 23 BNA Environ. Rep. (1992), at 1676.

ported by demanding national or international environmental and health quality guidelines, which are (as yet) absent in many states. Elsewhere, Agenda 21 is little more than suggestive, noting that priority actions to be considered by states may include the following:

Incorporating sewage concerns when formulating or reviewing coastal development plans, including human settlements
Promoting primary treatment of municipal sewage discharged to rivers, estuaries, and the sea (paragraph 17.27).

Second, Agenda 21 addresses the problems of coastal zone management. This again is not linked to the specific issue of upland discharges of marine debris. Of some relevance is the requirement for coastal settlements, inter alia, to improve treatment and disposal of solid wastes (s. 17.6(f)). Third, chapter 36 of Agenda 21 contains a wide variety of recommendations in the fields of public awareness and education that, although not explicitly linked to any substantive issue area, may be applied to littering. Fourth, Agenda 21 sets objectives for extending waste service coverage. Induced by the unnerving estimates that by the end of the century half the urban population in developing countries will be without adequate waste disposal services, paragraph 21.29 sets the target that by the year 2025, all urban populations should be provided with adequate waste services.

The GPA elaborates on these objectives. It repeats the objectives for sewage treatment set in paragraph 21.29 of Agenda 21 (paragraph 106) and formulates a list of national measures that should achieve those objectives (paragraph 107). In addition, the GPA lists the objective to establish controlled and environmentally sound facilities for waste collection and disposal from coastal area communities (paragraph 152(a)). A range of activities at the national level are designed to achieve that objective, including the following:

Installing garbage containers; establishment of proper operation of solid waste management facilities

Formulating and implementing awareness and education programs

Increasing local planning and management capacity to avoid locating waste dumpsites near coastlines and waterways

Establishing campaigns or permanent services, or both, for collecting solid wastes that pollute coastal and marine areas (paragraph 154).

These national actions are to be supplemented by actions at the regional level (e.g., regional agreements for solid waste management (paragraph 155)) and at the global level, including a clearing house on waste management technologies (paragraph 156).

Recycling and Source Reduction: Regulatory Challenges

Recycling and source reduction are much to be preferred over waste disposal regulation. As early as 1976, the U.S. EPA recognized that waste recycling and reduction were the preferred strategies for coping with solid waste, rather than treatment or disposal of such waste.[52] Fourteen years later this priority was reinforced when the Pollution Prevention Act of 1990 declared it to be a national policy goal to prevent waste at its source; in 1992, the U.S. EPA set the national policy goal to reduce the solid waste stream by 50% by 1997.[53] These issues are directly relevant to reduction of marine debris. For instance, it has plausibly been suggested that there are direct links between recycling programs and the amount of debris washing up on beaches.[54]

Existing regional agreements addressing land-based sources of pollution have not addressed recycling or source control at all. At the time of their conception, it was deemed appropriate to confine the objectives and mandates of these instruments to land-based pollution, *strictu sensu,* and states failed to consider land-based pollution of the marine environment as part of overall waste management and reduction policies. Such separation now can no longer be deemed acceptable. To identify the main regulatory options needing reinforcement by (sub)regional programs, consider the dominant options identified in national legislation. For recycling, the main regulatory options include these:

Setting targets for recycling to be achieved by manufacturers or in municipal waste recycling; such standards have been set at the state level in Germany,[55] France,[56] and the United States, and are currently being proposed at the federal level in the United States.[57]

Placing recycle-content mandates on packaging; this is not a popular option. An analysis of 110 amendments of state recycling laws in 1993 revealed that states are unwilling to place new recycle-content mandates on packaging, allegedly because of opposition by industry.[58]

Imposition of taxes or fees to encourage recycling, for instance, on products made with virgin materials[59] or on products that do not meet recycling standards.[60]

---

[52]41 *Fed. Reg.* 35.050 (1976).
[53]Robert V. Percival, Alan S. Miller, Christopher H. Schroeder, and James P. Leape, *ENVIRONMENTAL REGULATION, LAW, SCIENCE AND POLICY,* Little Brown and Company, Boston (1992) at 418.
[54]16 *BNA Int. Environ. Rep.* (1993), at 655 (suggesting that the relatively small amount of plastic found on Scandinavian beaches was the result of "progressive reuse and recycling" of plastic in Scandinavia").

[55]Bergkamp and Martiri, *TAKE-BACK SCHEMES,* at 194.
[56]Bergkamp and Martiri, *TAKE-BACK SCHEMES,* at 194.
[57]22 *BNA Environ. Rep.* (1991), at 2408.
[58]24 *BNA Int. Environ. Rep.* (1993), at 1601. See, however, the proposed Environmentally Sound Packaging Act of New York State, envisaging that packages contain 50% postconsumer recycled content; see 24 *BNA Environ. Rep.* (1993), at 332.
[59]Such taxes have several times been proposed in Congress; see Percival, *ENVIRONMENTAL REGULATION,* at 420.
[60]For example, the eco-tax proposed by Belgium for soft drink and beer bottles that are not recycled; see 17 *BNA Int. Environ. Rep.* (1994), at 272, and the taxes proposed in Oregon; see 23 *BNA Environ. Rep.* (1992), at 2108.

Incentives for development of new technologies furthering recycling.[61]

Creating markets; lack of markets is an endemic problem of recycling policies. States and municipalities hesitate to support recycling programs without a guarantee of a market for recovered materials. This hesitancy is paralleled by a hesitancy on the part of industry to invest in recycling without a guaranteed supply of usable materials or a secure market for recycled products.[62] Many states are undertaking efforts to improve markets by,[63] *inter alia,* enacting minimum content standards for products made with recycled goods[64] or by procurement policies with purchasing preferences for recycled products.[65]

National strategies regarding source control are at a more primitive stage. States have generally refrained from issuing mandatory regulations for reducing the waste stream, relying mainly on financial and technical assistance and providing information. The emphasis has been more on setting broad targets for voluntary source reduction, leaving it up to dischargers to decide how best to reduce the amount of pollution they generate.[66] Various regulatory approaches have been tried, including these:

Banning products that may end up in the environment; an example is Maine's ban on six-pack ringcarriers.[67]

Product standards; the best example is the standards for nondegradable, plastic ringcarriers, proven to have damaging effects on fish and wildlife. As of the end of 1991, 27 U.S. states had passed legislation prohibiting the sale of nondegradable, plastic ringcarriers.[68] On March 1, 1994, the U.S. EPA enacted its final rule on degradable ringcarriers,[69] requiring that manufacturers and importers of plastic ringcarriers test them to ensure they meet degradable performance standards.

Standards for the packaging industry, so as to reduce the amount and control the types of packaging; until now, this has been mainly pursued by voluntary reduction programs adopted by industry.[70] A key question is to what extent government should step in and impose standards.

Fees or taxes to remedy the distorted incentives that do not take into account true disposal costs and thus distort consumer and manufacturing decision making[71]; this

[61]Enhancing recycling of plastics is one of the aims of the draft strategy for the development of clean technology issued by U.S. EPA in February 1994; see 24 *BNA Int. Environ. Rep.* (1994), at 1733.
[62]Recent Developments: Federal Regulation of Solid Waste Reduction and Recycling, 29 *HARVARD JOURNAL ON LEGISLATION,* 251 (1992), at 261. See also Kindt, *SOLID WASTES,* at 89; Peter S. Menell, Beyond the Throwaway Society: An Incentive Approach to Regulating Municipal Solid Waste, 17 *ECOLOGY LAW QUARTERLY,* 655 (1990), at 669–670.
[63]24 *BNA Int. Environ. Rep.* (1994), at 1601.
[64]This was envisaged in the Amendments of the Resource Conservation and Recovery Act discussed by the Senate in 1991, S. 976, 102nd Congr., 1st Sess. sections 101–503 (1991); see also RECENT DEVELOPMENTS, at 264–265.
[65]In 1991, 40 states and the District of Columbia had passed laws to stimulate recycling markets by encouraging state agencies to purchase products with recycled content; see 22 *BNA Environ. Rep.* (1991), at 1371. In 1993, 14 states issued new laws on purchasing preferences for recycled products; 24 *BNA Environ. Rep.* (1994), at 1601; see further Percival, *ENVIRONMENTAL REGULATION,* at 421, and Menell, *BEYOND THE THROWAWAY SOCIETY,* at 678–679.

[66]Percival, *ENVIRONMENTAL REGULATION,* at 420.
[67]24 *BNA Environ. Rep.* (1994), at 336.
[68]See, for an earlier overview and discussion of this approach, Bean, LEGAL STRATEGIES, at 9–11.
[69]59 *Fed. Reg.* 9866.
[70]See 24 *BNA Environ. Rep.* (1994), at 1570 (describing the practice of voluntary programs for reduction of packaging in seven northeastern states). This emphasis is comparable to the policy of the Netherlands (in 1992 the Government signed a packaging covenant with the Foundation of Packaging and the Environment (14 *BNA Int. Environ. Rep.* 372) that called for substantial reductions in such waste by 2000; the amount of new packaging used in 2000 will be the same or 10% less than was used in 1986).
[71]Menell, BEYOND THE THROWAWAY SOCIETY, at 658–659.

may be pursued by imposing taxes on "litter-generating products",[72] packaging,[73] or waste disposal (this would create a powerful incentive for consumers to reduce the quantity of waste they generate).[74]

Incentives for development of clean technology; a draft strategy issued by the U.S. EPA in February 1994, supplemented with $36 million in funding for fiscal 1994, aims to speed development of innovative environmental technologies. The strategy aims to remove obstacles and create incentives for development of new technologies, assist development of new technologies, help small businesses adopt pollution prevention technologies, and encourage the use of new technologies.[75]

Providing information to manufacturers and policymakers, which was the main focus of the U.S. Pollution Prevention Act of 1990, intended as a first step to attain voluntary reduction of wastes created during the manufacturing process by improving the quality of relevant information available to industry and states.

Recycling and Source Reduction: Agenda 21

While not explicitly linked to marine debris,[76] recycling and source control feature prominently in Agenda 21. Agenda 21 addresses several of the previously mentioned regulatory strategies. Not surprisingly, however, the global level was found too high to agree on specific targets and timetables for recycling or source control. Instead, most of the provisions are suggestive in nature.

In regard to recycling, Agenda 21 only calls upon states to address the issue. It is hardly a demanding requirement to promote sufficient financial and technological capabilities to implement waste reuse and recycling policies and actions by the year 2000, or to have by the year 2000—in all industrialized countries, and by the year 2010, in all developing countries—a national program for (to the extent possible) efficient waste reuse and recycling (s. 21.18).

In addition, Agenda 21 offers some substantive guidance; states should do the following:

Modify existing purchasing standards to avoid discrimination against recycled materials (s. 21.19(d))

Develop public education and awareness programs to promote the use of recycled products (s. 21.19(e))

Increase funding for pilot programs to test options for reuse and recycling (s. 21.19(c))

Identify potential markets for recycled products (21.19(f))

Consider the use of incentives to stimulate recycling (s. 21.24; s. 17.22).

In regard to source reduction, most emphasis is placed on improvement of monitoring to assess waste quantity and its effects on the marine environment (21.8(b), 21.11). As to substantive issues, the provisions are undemanding. Consider the main objectives of this program area:

To stabilize or reduce the production of wastes over an agreed time frame (s. 21.8(a)).

---

[72]For instance, recent Italian legislation imposes a 10% tax on plastic film as a disincentive to plastics use; see 17 *BNA Int. Environ. Rep.* (1994), at 311. See, for this option, Bean, LEGAL STRATEGIES, at 13.

[73]In the United States, several states have proposed charges on products sold in certain types of packaging; see for examples Menell, *BEYOND THE THROWAWAY SOCIETY,* at 677–678; see for other countries; Italy (16 *BNA Int. Environ. Rep.,* at 903); Japan (16 *BNA Int. Environ. Rep.,* at 93).

[74]*RECENT DEVELOPMENTS* at 275–276; Percival, *ENVIRONMENTAL REGULATION,* at 422; Menell, *BEYOND THE THROWAWAY SOCIETY,* at 675–676.

[75]24 *BNA Environ. Rep.* (1992), at 1733.

[76]Curiously enough, although Chapter 21 is considered to be related to the chapters on freshwater resources, human settlement development, protecting human health, and changing consumption patterns, protection of the marine environment is not considered as a related program area.

By the year 2000, industrialized countries
should have policies in place to stabilize or
reduce, if practicable, production of wastes
(s. 21.9(b)).

Developing countries should work toward
that goal without jeopardizing their development prospects (s. 21.9(b)).

Initiate by the year 2000, particularly in developed countries, programs to reduce
packaging materials (s. 21.9(c)).

The GPA applies only a small part of these
ideas to land-based marine debris and certainly does not elaborate on them. At the
national level, states are to do the following:

Introduce appropriate measures "to encourage reduction in generation of solid
wastes".

Formulate and implement awareness programs on the need to reduce waste generation (paragraph 154).

At the regional level, states are called upon
to cooperate for the exchange of information
on practices and experiences regarding waste
management, recycling and reuse, and
cleaner production (paragraph 155). At the
international level, states should participate
in a clearinghouse that is not only to involve
waste management, but also recycling and
reuse, and waste-minimization technologies
(paragraph 158).

The Case for an Integrated Approach

Agenda 21 and the GPA provide a basis for
improving on preexisting international approaches to waste disposal and waste reduction. At the same time, several key issues
remain to be elaborated. The follow-up to
these instruments should include four imperatives for global and regional cooperation.
First, Agenda 21 and the GPA have insuffi-
ciently linked the basis for future action on
land-based marine pollution to the recycling
and waste-generation programs. Land-based
marine pollution, including pollution from

upland discharges of marine debris, must not
be considered an isolated set of problems.
Rather, they are an integral part of sound
environmental management and pollution
control for every waste management
activity.[77] The same policies that reduced
solid wastes on land will prevent them from
entering the environment. Consideration of
land-based pollution as a disposal issue un-
connected to source reduction will be a misconceived strategy. In this context it is dis-
concerting that the GPA has only very
partially incorporated policies with regard to
recycling and the prevention of waste generation.

In implementing the GPA at regional levels,
states should take into account these interre-
lationships and make recycling and source
reduction subject to its regulatory programs.
The first region where this opportunity
presents itself will be the Caribbean, where
preparations for a land-based pollution agreement are presently under way. This protocol
to the Cartagena Convention could provide
for a specific set of rules on waste management that would cover waste management
and disposal, including sewage and wastewater treatment, design, siting, and operation of landfills, wastewater conduits, storm
drains, and general measures for managing
wastes.[78]

Second, not all issues are adequately addressed by Agenda 21 and the GPA. In particular, they do little to induce appropriate
policies at regional, national, or local levels
for control of combined sewer overflows and
stormwater discharges, and for location and
operation of landfills in coastal areas. It is
essential that the full realm of sources of
land-based debris be covered in any future
regional programs. Third, for the most part
Agenda 21 and the GPA confine themselves
to identifying the issues, rather than setting
demanding targets. Although this is under-
standable and inherent in the cumbersome
negotiation process of these instruments, it
puts considerable pressure on regional follow-

[77]Kimball, THE PROTOCOL.
[78]Ibid.

up processes. These processes should set targets and timetables for the percentage of coastal populations to be provided with (for example) sewage treatment and stormwater discharge control.

Fourth, one of the major features of the UNCED is its focus on linkages between substantive requirements relating to the prevention of marine debris and the financial and technical cooperation and support that are necessary to make these requirements practicable. Regarding sewer-source debris, this would imply that setting target dates for the management of sewage of coastal populations shall be supplemented with strategies to achieve that objective. These strategies shall include mechanisms for inventory and transfer of low-cost systems for management and treatment of sewage and effective mechanisms for financial support, such as an international fund for planning, design, and construction of sewage treatment facilities. The key elements of an effective strategy for waste minimization are transfer and dissemination of technology to support waste recycling and waste minimization, and particularly the transfer of waste reduction technologies to industry in developing countries.[79]

There is evidence of an increasing willingness of donors to support relevant projects. Examples include support by the Global Environmental Facility for wastewater treatment in the Black Sea,[80] the funding by the European Union of Black Sea protection programs,[81] European Investment Bank lending for solid waste disposal,[82] funding for reduction of sewage problems and solid waste disposal in Rio De Janeiro by the Inter-American Development Bank and the Japanese Overseas Economic Cooperation Fund,[83] and investments in Mexican wastewater treatment facilities by the International Finance Corporation.[84] The development of substantive, internationally agreed strategies for waste disposal and source reduction should help focus the activities of funding institutions.

Elaboration of the issues identified in Agenda 21, supplemented with guidance on the issues left unconsidered, should help to induce more integrated and effective regional, national and local programs. The UNCED and its follow-up Global Program of Action for the Protection of the Marine Environment from Land-Based Activities will not fundamentally change the reliance on regional agreements as the most appropriate instruments for effective action. However, now that regional programs will have the opportunity to be integrated into, or linked with more general, global programs for waste disposal, recycling, and reduction, regional instruments of the 1990s should be much better suited to meet the challenges of the upland discharges of marine debris than those devised in the 1970s and 1980s.

---

[79]These latter two forms of support are envisaged in sections 21.23 and 21.14(f) of Agenda 21.
[80]16 *BNA Int. Environ. Rep.* (1993), at 626.
[81]16 *BNA Int. Environ. Rep.* (1993), at 170.
[82]16 *BNA Int. Environ. Rep.* (1993), at 930.
[83]16 *BNA Int. Environ. Rep.* (1993), at 912.
[84]17 *BNA Int. Environ. Rep.* (1994), at 197.

23.
Comparison of the Results of Two EPA Marine Debris Studies

Wayne R. Trulli, Roy K. Kropp, Heather K. Trulli, and David P. Redford

Introduction

Domestic and international concerns about aquatic plastic debris prompted the U.S. Congress to pass the Marine Plastic Pollution Research and Control Act (MPPRCA) of 1987. Title II of this Act required the U.S. Environmental Protection Agency (EPA) to issue a Report to the Congress on methods for reducing plastic pollution (EPA 1990a). One section of this report to Congress discusses the types and sources of marine plastic debris, the transport and fate of this debris, and its effects on the marine environment and on human health and safety. It also provides a list of aquatic debris items that are of particular concern to the EPA. These items of EPA concern are pellets, condoms, tampons, syringes and other medical items, nets and traps, line and rope, six-pack yokes (or similar beverage yokes), and plastic bags and sheeting.

The limited amount of data available before the preparation of the report to Congress prompted the EPA to conduct field studies to characterize plastic aquatic debris and identify potential sources of that debris at targeted locations along the coastal United States. In 1988, the EPA initiated two major programs, the Harbor Studies Program (HSP) and the Combined Sewer Overflow (CSO) Studies Program, to address these two issues. The focus and objectives of each program are summarized in Table 23.1.

Under the Harbor Studies Program, the EPA conducted 20 surveys to characterize aquatic debris in the harbors of 14 major metropolitan areas along the coastal United States and in Puerto Rico (PR). These areas (Fig. 23.1) include New York, Boston, Philadelphia, Baltimore, Norfolk, Miami, Houston, Tacoma, Seattle, Oakland, San Francisco, Honolulu, and San Juan and Mayagüez, PR.

Under the CSO Studies Program, field studies were conducted to identify CSOs and stormwater discharges (SWD) as possible sources of aquatic debris. These field studies were conducted in Philadelphia, PA (Fig. 23.2) and Boston, MA (Fig. 23.3), two of the cities sampled during the Harbor Studies Program. The EPA collected, processed, and characterized debris from targeted CSO and SWD outfalls to determine whether the composition of that debris was similar to debris compositions found under the Harbor Studies Program. In addition, debris was collected from selected sewage treatment plants (STP) to identify potential relationships among STP debris, CSO, and SWD outfall debris, and harbor debris. In Philadelphia, STPs are called water pollution control plants (WPCP); in Boston, they are called publicly owned treatment works (POTW).

This paper describes the field and analytical methods used to gather and process data during the conduct of each program; summarizes and compares the results of the two programs; and demonstrates some possible relationships that identify CSOs and SWDs as

TABLE 23.1. Focus and objectives of the Harbor Studies Program and the Combined Sewer Overflow (CSO) Studies Program.

Focus and objectives	Harbors Studies Program	CSO Studies Program
Focus	Examined plastic and other anthropogenic non-point-source debris in major harbors (along the Atlantic, Pacific, and Gulf Coasts) of the United States and Puerto Rico	Examined plastic and other anthropogenic debris from CSO and SWD outfalls in two U.S. cities
Objectives	Characterize types, relative amounts, and distribution of debris in representative harbors across the U.S. and Puerto Rico	Characterize CSOs and stormwater discharges (SWDs) as land-based sources of anthropogenic debris in the aquatic environment
	Identify potential sources of harbor debris	Determine the types and relative amounts of floating debris from CSOs and SWDs

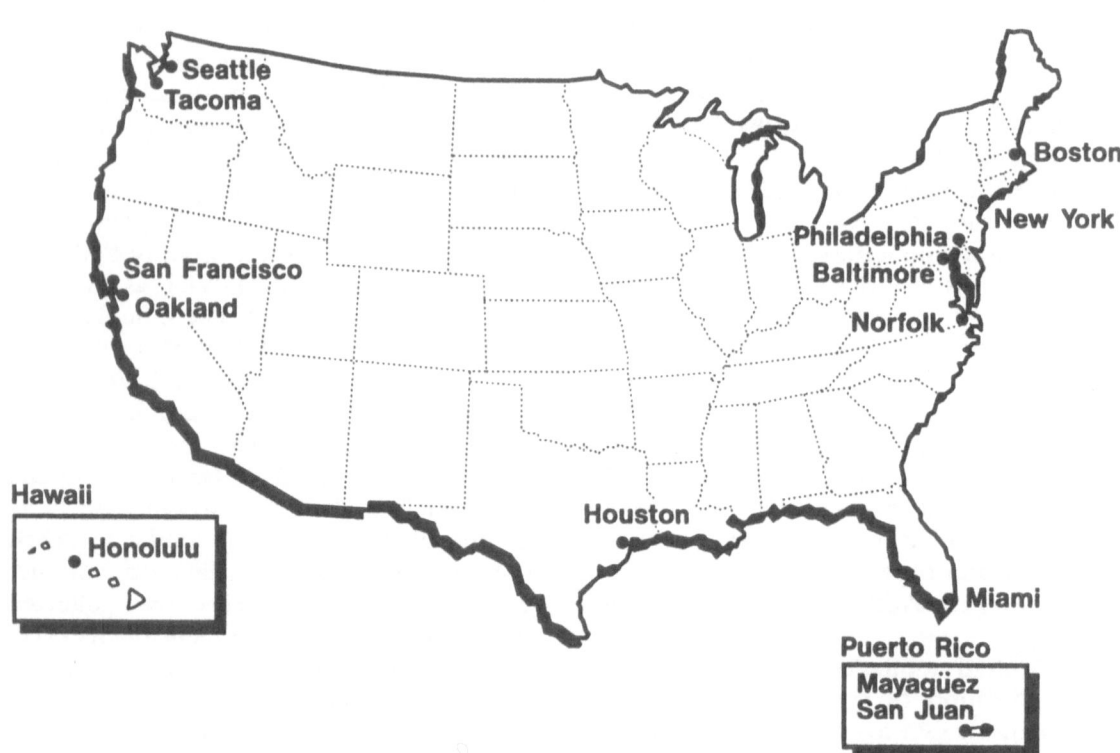

FIGURE 23.1. Locations of all harbors studied during the Harbor Studies Program (HSP), 1988–1992.

probable sources of debris in U.S. harbors. This paper attempts to answer three questions: (1) Are the relationships between the two programs evident in the raw debris composition data; (2) are those relationships also apparent when selected descriptive statistical tests (cluster analysis) are applied; and (3) do

those similarities suggest CSOs and SWDs as potential sources of harbor debris?

It is important to note that the Harbor Studies Program was designed to provide primarily qualitative information (normalized to percent composition per sample) while the CSO Studies Program was designed

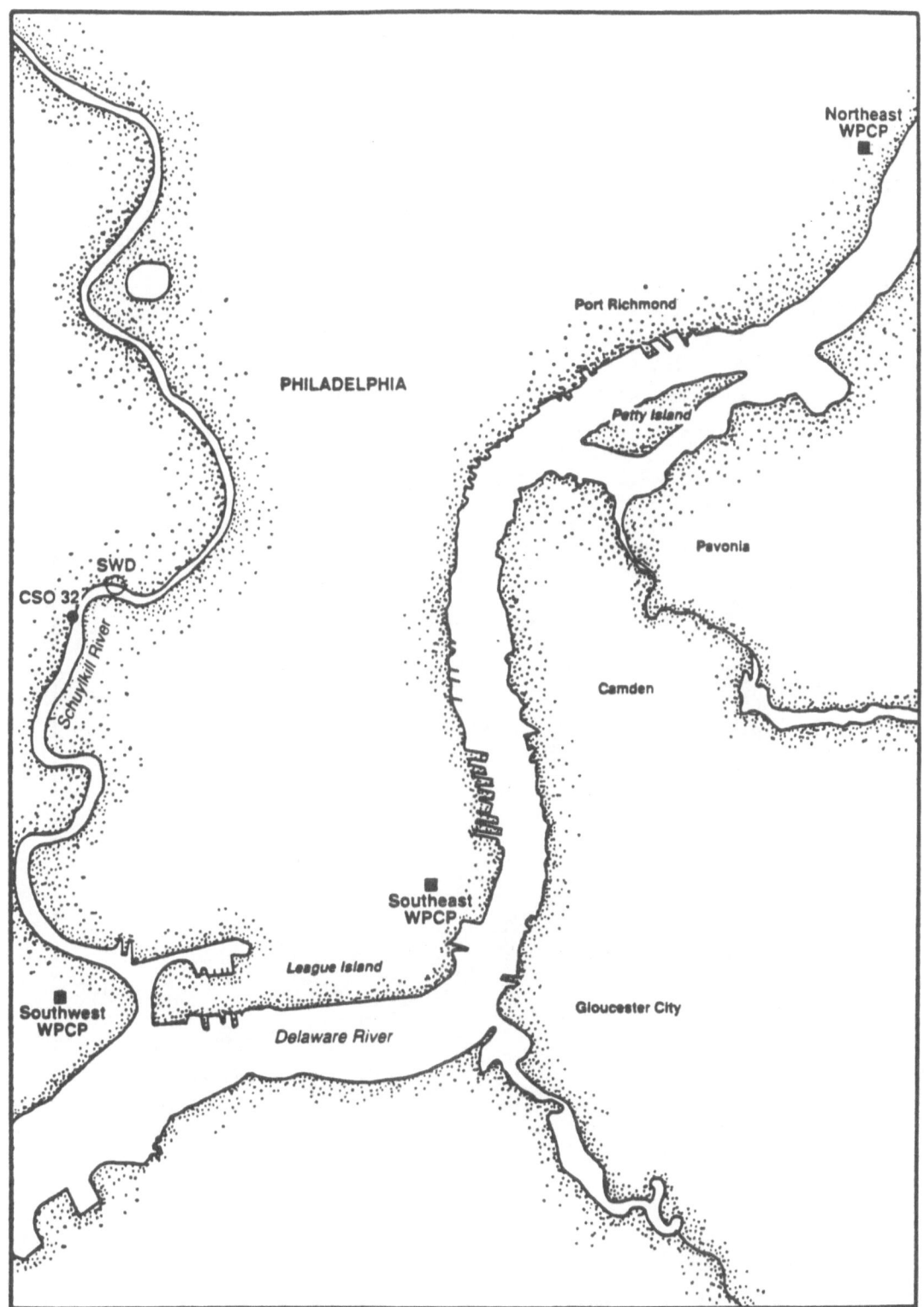

FIGURE 23.2. Locations of the stormwater discharge (SWD, *circle*) and three water pollution control plants (WPCPs, *squares*) samples in Philadelphia, PA, under the CSO Studies Program, 1989–1990.

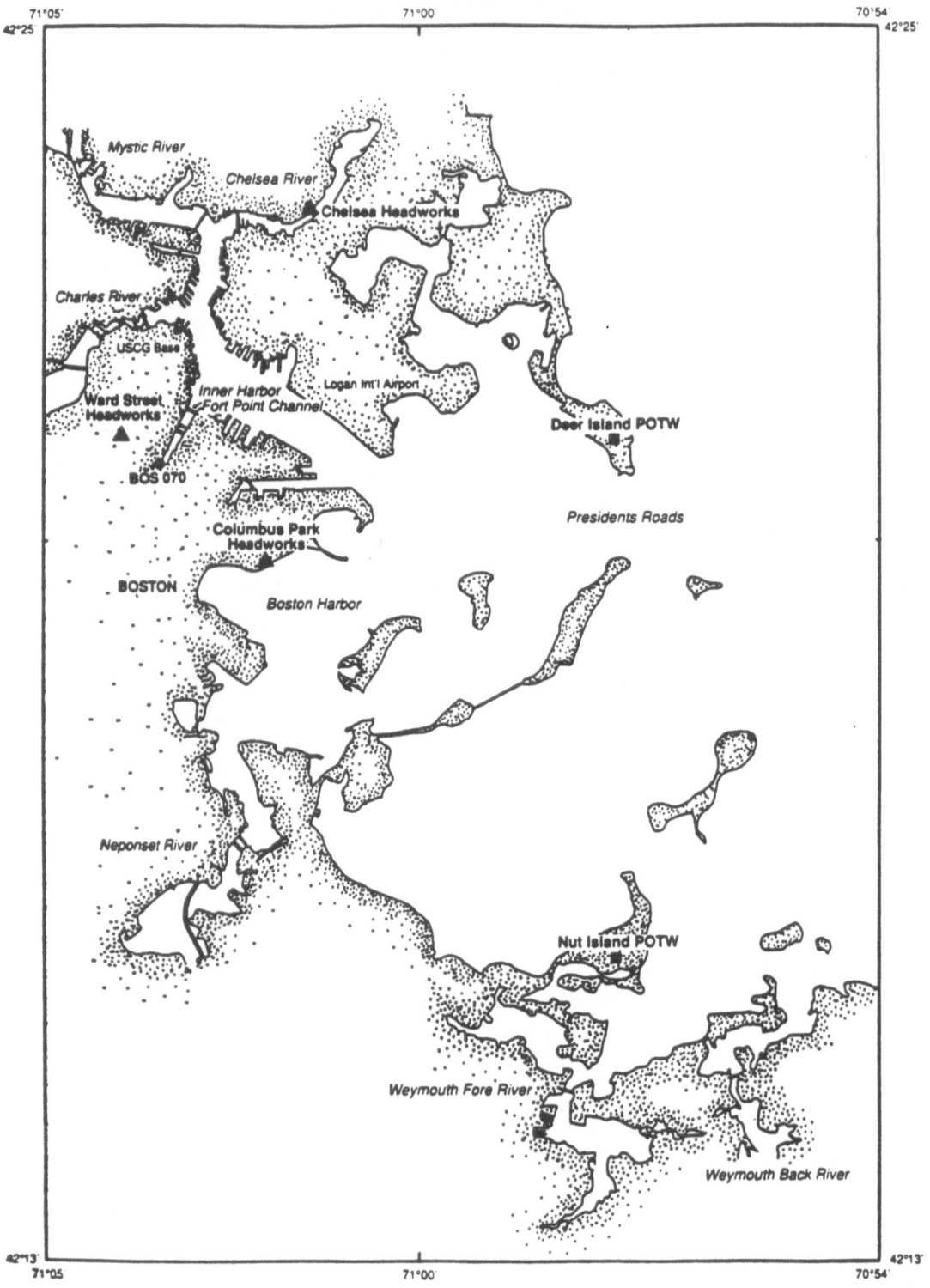

Figure 23.3. Locations of the combined sewer overflow (CSO), the two headworks (*triangles* Ward Street and Chelsea), and Deer Island publicly owned treatment works (POTWs, *squares*) sampled in Boston, MA, under the CSO Studies Program, 1989–1990.

to provide more quantitative information (percent composition per inch of rainfall for CSOs and SWDs or percent composition per day for WPCPs and POTWs). Therefore, extreme caution should be used when making assumptions about the apparent relationships existing between data from the two programs.

The Rationale for Comparing the Harbor Studies and the CSO Studies Programs

During the Harbor Studies Program, both land-based and marine-based sources were found to contribute to the aquatic debris problem and included releases from both point- and non-point-source discharges. Some of those sources of debris include SWD outfalls, CSOs outfalls, municipal landfills, industrial discharge outfalls, general population and non-point-source surface runoff (land-based sources), recreational vessels, and fishing vessels (marine sources). The main focus of this paper is to identify possible relationships among the data gathered under the Harbor Studies Program and the CSO Studies Program.

One of the objectives of the Harbor Studies Program was to identify potential sources of aquatic debris in U.S. harbors. Similarly, one of the objectives of the CSO Studies Program was to characterize the debris from CSOs and SWDs confirming their contribution as land-based sources of aquatic debris, particularly plastic debris. In the northeastern United States, CSOs and, to a lesser extent, SWDs are of major concern to the EPA because some types of debris released by these sources into the aquatic environment have been identified as items of EPA concern. The Interagency Task Force (ITF 1988) estimated that approximately 600 of 2000 STPs located in coastal communities discharge directly into estuaries and coastal waters; of those 600 systems, 135 contain one or more CSOs (ITF 1988). According to the EPA (1990a), 36 of the 100

largest STPs have associated CSO systems, 30 of which are located on the East Coast, with 12 located in New York State alone.

The potential magnitude of the harbor debris problem caused by CSOs and SWDs has prompted the EPA to more carefully examine data gathered under both the Harbor Studies and CSO Studies Programs. To confirm the EPA's belief that CSOs and SWDs have contributed significantly to the harbor debris problem, the EPA investigated possible relationships (both qualitatively and quantitatively) in the debris compositions determined for the two targeted outfalls and the two targeted STP systems under the CSO Studies Program and the debris composition determined for combined harbors under the Harbor Studies Program.

Methods

The data used for establishing the comparisons presented in this paper were obtained exclusively from the EPA-sponsored Harbor Studies Program and the CSO Studies Program. Sample collection and processing methods, analytical procedures, and database development and manipulation are briefly described for each program. The methods for making the qualitative and quantitative comparisons of data from the targeted sources are also discussed.

Sample Collection, Processing, and Analysis

Sampling methods for the Harbor Studies and CSO Studies Programs differed considerably; however, methods for processing and final analysis of the samples were identical. The following section briefly discusses all sampling, processing, and analytical methods used for each program.

Harbor Studies Program.

Floating debris (i.e., floatables) was collected in accordance with the methods detailed by

EPA (1990b, 1992a), which are briefly summarized in this section. Floatables samples were taken from two to four areas within each of the 14 harbors studied. Two or more small vessels (most often a 20-ft Boston whaler or similar-sized vessel) were used to collect the samples. In some cases, a large vessel, the EPA-owned OSV *Peter W. Anderson* (165 ft in length), was also used as a sampling platform.

Collections were made by using a neuston-type net composed of 0.33-mm-mesh Nitex®. Two different net sizes were used: (1) one for deployment from a small Boston whaler-type vessel [net dimensions, 0.5 m (opening height) × 1.0 m (opening width) × 20 m (length)] and (2) one for deployment from the OSV *Anderson* [net dimensions, 1.0 m (opening height) × 2.0 m (opening width) × 4.0 m (length)]. Tows were made from a stationary boom rigged near the bow or abeam of each vessel type.

Sampling was directed toward the densest areas within the targeted debris slicks or other areas where debris was visible. Very large or dense slicks were typically sampled more than once. When approximately 15 gal of debris (enough to fill about half of a 33-gal trash bag) was collected, towing was stopped. The debris was stored in a 33-gal trash bag until final processing, and analysis was conducted either board the OSV *Anderson* or at a shore-based laboratory. To ensure consistency throughout the conduct of the Harbor Studies Program, sample collection and analysis were conducted or directed by the same sampling and analytical team. General sampling information, including survey dates and the total number of samples collected at each location during the Harbor Studies, is listed in Table 23.2.

All man-made debris was subsequently classified and counted according to 239 different item designations (EPA 1990b, 1992a, 1992b) within nine major categories [plastic, polystyrene (foamed plastics), rubber, metal, wood, paper, textiles, glass, and miscellaneous (e.g., tar balls, fecal matter)]. Percent composition calculations were based on the total number of items per sample. Detailed descriptions of the sampling rationale and

TABLE 23.2. Locations, survey dates, and total number of samples collected during the Harbor Studies Program, 1988–1992.

Location	Survey dates	Total number of samples collected
New York, NY	November 11–13, 1988	43
	July 7–9, 1989	27
Boston, MA	December 2–4, 1988	49
	April 10–12, 1990	32
Philadelphia, PA	January 26–27, 1989	29
Baltimore, MD	January 29–30, 1989	29
	June 21–23, 1989	12
	June 3–5, 1990	17
Norfolk, VA	June 6–8, 1990	29
Miami, FL	February 3–5, 1989	31
	February 13–15, 1990	20
Houston, TX	February 6–8, 1990	15
	September 26–27, 1990	8
Tacoma, WA	February 15–17, 1989	11
Seattle, WA	February 15–17, 1988	6
San Francisco, CA	February 21–23, 1989	15
Oakland, CA	February 21–23, 1989	12
Honolulu, HI	February 18–20, 1992	11
San Juan, PR	April 22–24, 1991	8
Mayagüez, PR	April 23–25, 1991	12
Total samples		416

methods, calculations, and database management have been compiled by the EPA (1990b, 1992a, 1992b). The number of samples per city used in the results of this report are presented in Table 23.2.

CSO Studies Program.

Two types of samples, outfall samples and STP samples, were collected. Samples were collected by using two separate methods specified by EPA (1993a): one method specifically for CSO and SWD outfalls and the other for POTWs and WPCPs. Debris flowing from the CSO and SWD outfalls was captured by using a 1.8-mm-mesh Nitex net measuring 210 ft (span) × 17 ft (depth) for CSO sampling or 60 ft (span) × 15 ft (depth) for SWD sampling. These containment nets were designed to enclose the discharge pipes and surrounding area so that debris larger than 1.8 mm in diameter was retained within the net. The bottom of each net was weighted to keep the net at or near the bottom of the outfall. In addition, several 8 in. × 12 in.

floats were threaded through a sleeve at the top of the net to provide enough flotation so that the floats were buoyed above the water surface at all times and to allow the net to span the entire depth of the water column.

The samples were removed from the containment net within 12–24 h after a discharge event. Man-made debris was enumerated and classified by the methods used during the Harbor Studies Program; percent composition of each individual item was calculated on the basis of the total number of items in each sample. In addition, the average number of items collected per inch of rainfall for each outfall was also calculated for each event (for this study, events were combined and averaged).

The method used to collect debris from the three Philadelphia WPCPs (Northeast, Southeast, and Southwest) and the three components of Boston's Deer Island POTW [Ward Street and Chelsea Headworks (HW) and the Deer Island sewage treatment clarifying tank] differed from those used to collect debris at the CSO and SWD outfalls. Separate methods were necessary because STP debris is collected daily at the screening stations and clarifying tanks and is not available for discharge into the environment as is trash from the combined sewer and stormwater systems during overflow conditions. Two samples were collected, one on each of two consecutive days, from each HW (screening station) and clarifying tank of the Boston POTW and from each screening station and clarifying tank of the three Philadelphia WPCPs.

Approximately one-tenth of the material retained by the 1-in. screens was collected from each WPCP and HW. Scum samples, composed of debris and sewage residue that passes through the screening stations, were collected from the clarifying tanks at each of the STPs. Collections were made by using a device similar to a skimming net for removing floating material from a swimming pool; approximately one-tenth of the scum was removed each sampling day.

All man-made debris identified in the scum and screening samples (to an approximate size of 1 mm) was enumerated and classified by the methods used during the Harbor Studies Program. The numbers of samples collected (as well as the location, survey dates, and sample type) during the CSO Studies Program are listed in Table 23.3.

Percent Composition Calculations for Both Programs.

All man-made items in all samples were classified and enumerated as much as possible. Under the Harbor Studies Program, percent composition of each item was determined for each city sampled as well as for the entire program. Under the CSO Studies Program,

TABLE 23.3. Locations, survey dates, number of samples collected, and sample type during the CSO Studies Program, 1989–1990.

Location	Survey dates	Number of samples collected	Sample type
Philadelphia, PA	September 15, 1989	1	SWD Event 1
	September 22, 1989	1	SWD Event 2
	October 6, 1989	1	SWD Event 3
	September 25, 1989	1	SE WPCP
	September 25, 1989	1	SW WPCP
	September 25, 1989	1	NE WPCP
	September 26, 1989	1	SE WPCP
	September 26, 1989	1	SW WPCP
	September 26, 1989	1	NE WPCP
Boston, MA	May 11, 1990	1	CSO Event 1
	May 22, 1990	1	CSP Event 2
	Septmber 28, 1989	1	Boston POTW
	September 29, 1989	1	Boston POTW

SWD, stormwater discharge; WPCP, water pollution control plants (sewage treatment plants); CSO, combined sewer overflow; POTW, publicly owned treatment works (sewage treatment plants).

numerical data were normalized to reflect number of items per inch of rainfall for each outfall (CSO and SWD) and number of items per day for each STP (including the three Philadelphia WPCPs and the Deer Island POTW). Percent composition of each item was then determined for each debris source. Percent composition was determined by dividing the item enumeration by the total number of items from each sampling location and multiplying by 100. However, special consideration was given to some items due to extenuating circumstances.

Some items were found in numbers so large that enumerating them would have required excessive amounts of time or in the case of grease balls, would be very difficult to enumerate accurately, owing to their very small size (1–2 mm in diameter) and delicate structure (easily broken when handled). The presence of these items as "too numerous to count" (TNTC) was noted on the respective data inventory sheets, and these records are noted in the Harbor Studies Program report (EPA 1990b, 1992a, 1992b) and the CSO Studies Program report (EPA 1993a). No numerical estimates of these TNTC items were included in the calculated percentages reported in this report; however, TNTC reports were labeled in the database.

Resin pellets were found in high numbers in several samples; they were never reported as TNTC and were always enumerated because of their significance as a potential hazard to wildlife. The high numbers of pellets masked the relationship between the other debris items in the individual sampling locations and in both programs. Therefore, the enumerations for resin pellets are not included in the percentage calculations presented in this report unless otherwise indicated. The significance of resin pellets relative to their sources and environmental impacts are detailed in several reports (EPA 1990b, 1992a, 1992b, 1992d). As a result, pellets are discussed only briefly in this paper.

During the conduct of both programs, items were enumerated but they were not weighed or measured in any other manner. All cited percentages were calculated based on numbers of items (Harbor Studies Program) or numbers of items per inch of rainfall (CSO Studies Program). As a result, relationships within and between programs are presented in terms of item numbers and percent composition only.

Methods of Data Comparison Between Programs

In this study, the results of the Harbor Studies and the CSO Studies Programs were compared both qualitatively and statistically. As a preliminary step, a qualitative comparison of the programs was conducted to identify possible similarities in the percent composition data patterns. Evidence that obvious similarities did exist prompted the EPA to compare the programs statistically by using Bray–Curtis and Dice similarity measures. A detailed description of both qualitative and statistical methods of comparison is presented next.

Because of the qualitative design of the Harbor Studies Program and the problems associated with collecting CSO and SWD samples (EPA 1993a), interpretations based on the results of these comparisons should be viewed cautiously. In addition, although STP debris samples may reflect the types of debris available for discharge from CSOs into the harbors, they may not always provide an accurate accounting of the types of material actually discharged. Weather conditions may also influence the composition of STP debris samples. For example, during dry weather conditions, street litter may be expected to comprise a small percentage of debris in STP debris samples. Conversely, during wet weather conditions (under CSO overflow conditions), street litter might comprise a much larger percentage of STP debris. As a result, interpretations of comparative results for STPs, harbors, and outfalls should be considered preliminary.

Qualitative Method

Under this method, percentages calculated for the Harbor Studies and CSO Studies Programs were compiled under four basic data groups for comparison. These data groups are nine major debris categories, items of EPA concern, related items (sewage, medical, and drug), and street litter (caps and lids, tobacco products, and food packaging products). In addition to the four major data groups, the most common items were also compiled for each source sampled, including the combined harbors, CSOs, SWDs, and the STPs. The nine major debris categories were selected from the data card developed for the Harbor Studies Program. These categories are plastic, glass, metal, paper, rubber, polystyrene, textiles, wood, and miscellaneous (EPA 1990b, 1992a). The items of EPA concern (listed previously) were identified by the EPA as items that potentially threaten wildlife survival and human health.

Related items consist of debris associated with sewage, medical waste, and illegal drug paraphernalia. Related debris is composed of several individual items arbitrarily designated for inclusion under each of the three categories. Street litter is an artificial data group created to encompass all debris that may enter the stormwater discharge system and ultimately the aquatic environment from the street. Three categories were developed to account for caps and lids, tobacco products, and food packaging products (especially those from fast-food restaurants).

Descriptive Statistical Method

The Harbor Studies Program was designed only to provide a qualitative (not quantitative) characterization of the debris found in harbors along the coastal United States. Conversely, the CSO Studies Program was designed to provide both a qualitative and quantitative characterization of debris collected from the targeted CSO and SWD effluents and the selected STPs. Because of the similarities observed among some of the data sets (individual and combined harbors, CSOs, SWDs, and STPs) throughout the qualitative comparison of the two Programs, the EPA decided to test these qualitative findings using a descriptive statistical approach, the classification analysis. The data used for the statistical classification analyses were obtained from the Harbor Studies Program (EPA 1990b, 1992a, 1992b) and the CSO Studies Program (EPA 1993a). Modifications to the databases and methods used for measuring similarity are briefly described next.

Data Modifications.

The data sets from the two EPA programs were modified to establish a separate data set for use in the classification analyses. The separate data set (1) excluded all discrete samples (EPA 1990b, 1992a, 1992b) and included samples collected only through net tows (i.e., Harbor Studies Program); (2) excluded samples collected during an offshore Mid-Atlantic Bight survey (EPA 1992a) (i.e., Harbor Studies Program); and (3) excluded all paper items (from both programs). Mid-Atlantic Bight data were excluded to allow these analyses to focus on harbor debris. Paper items were excluded because extremely large numbers of toilet paper pieces and paper towels were found in some of the WPCP and POTW samples; these items rapidly sink and break apart when saturated with water and would not be found floating in the surface waters of a harbor. The CSO Studies Program database described in EPA (1993a) was modified for use in this study by averaging the daily debris totals to create a single value for each facility and discharge. Because sampling efforts were not consistent among stations and surveys of the Harbor Studies Program, percent composition was used as the quantitative estimate of debris abundance. The separate data set was subsequently reduced as necessary to perform specific analyses, as described here.

Analytical Approach.

The Harbor Studies Programs have yielded very large, complex data sets consisting of more than 400 samples collected from 14 cities. The goals of the present study were to summarize these data sets and to begin exploratory data analyses that would eventually lead to the development of testable hypotheses. Classification was chosen as the method of data analysis because both these goals were listed among its purposes (Gauch 1982). There was also an interest in examining the performance of biological techniques in analyzing nonbiological data. It is recognized that there may be appropriate alternative analytical methods.

Quantitative Similarity.

Normal (or Q-mode) classification analysis, which classifies sampling units (i.e., stations) by comparing the distribution of "attributes" (in this case debris items) among those units, was run using the unweighted pair-groups method (UPGMA) as the clustering method and the Bray–Curtis coefficient as the measure of similarity. The Bray–Curtis coefficient was chosen from the many similarity measures available because it has been one of the most commonly used measures in marine ecology (Boesch 1977) and still receives considerable use (Clarke and Ainsworth 1993). It frequently has been used with percent standardized data as used for these two programs. With Bray–Curtis similarity, attributes with high scores strongly influence the value of the measure; those with low scores are relatively unimportant (Boesch 1977). The implication for this study is that similarity between sites may be determined by the presence of relatively few, abundant debris items. Bray-–Curtis similarity is also useful in situations where the data matrix contains many zero scores, such as in the case with the marine debris data, that would render measures such as Euclidean distance less useful in measuring similarity (Clarke and Ainsworth 1993).

Qualitative Similarity.

The Dice similarity coefficient was used as a qualitative measure of resemblance among stations. The use of a qualitative measure allows a comparison of stations to be made on the basis of the debris items present,

regardless of abundance. The Dice coefficient is among those that have been listed as desirable for use in ecological studies because they do not consider shared absences in the estimation of similarity (Boesch 1977; Ludwig and Reynolds 1988). Other advantages of the Dice coefficient are that it places particular emphasis on shared positive attributes and it is the qualitative equivalent of the Bray–Curtis coefficient (Boesch 1977). Normal classification analysis was performed using UPGMA as the clustering method.

Results and Discussion

Qualitative Comparison

The results of the qualitative comparison are discussed in terms of the apparent similarities between the debris composition of the targeted CSO, SWD, and STPs and that of the combined harbors. In general, percent composition data gathered for each outfall type (CSO and SWD) under the CSO Studies Program compared well with the combined harbors data (excluding resin pellets) in each of the four data groups examined and relative to the most common items. Conversely, data from the STPs (POTW and WPCP) did not compare well with data from the outfalls and combined harbors. All percent composition data used in this comparison are detailed here and are displayed graphically in Fig. 23.4A–D for the combined harbors (including and excluding resin pellets), outfalls, and STPs, respectively. Table 23.4 summarizes the percent composition data for some of the more significant items and categories used for comparing each of the targeted debris sources.

Although resin pellets and their sources are of major concern to the EPA, they are discussed only briefly here. The production, transport, environmental impacts, sources, and ubiquitousness of resin pellets are detailed in other reports and publications (EPA 1990b, 1992a, 1992b, 1992d). In this paper, pellets are discussed in terms of their presence in samples from every debris source studied. When resin pellets are included in

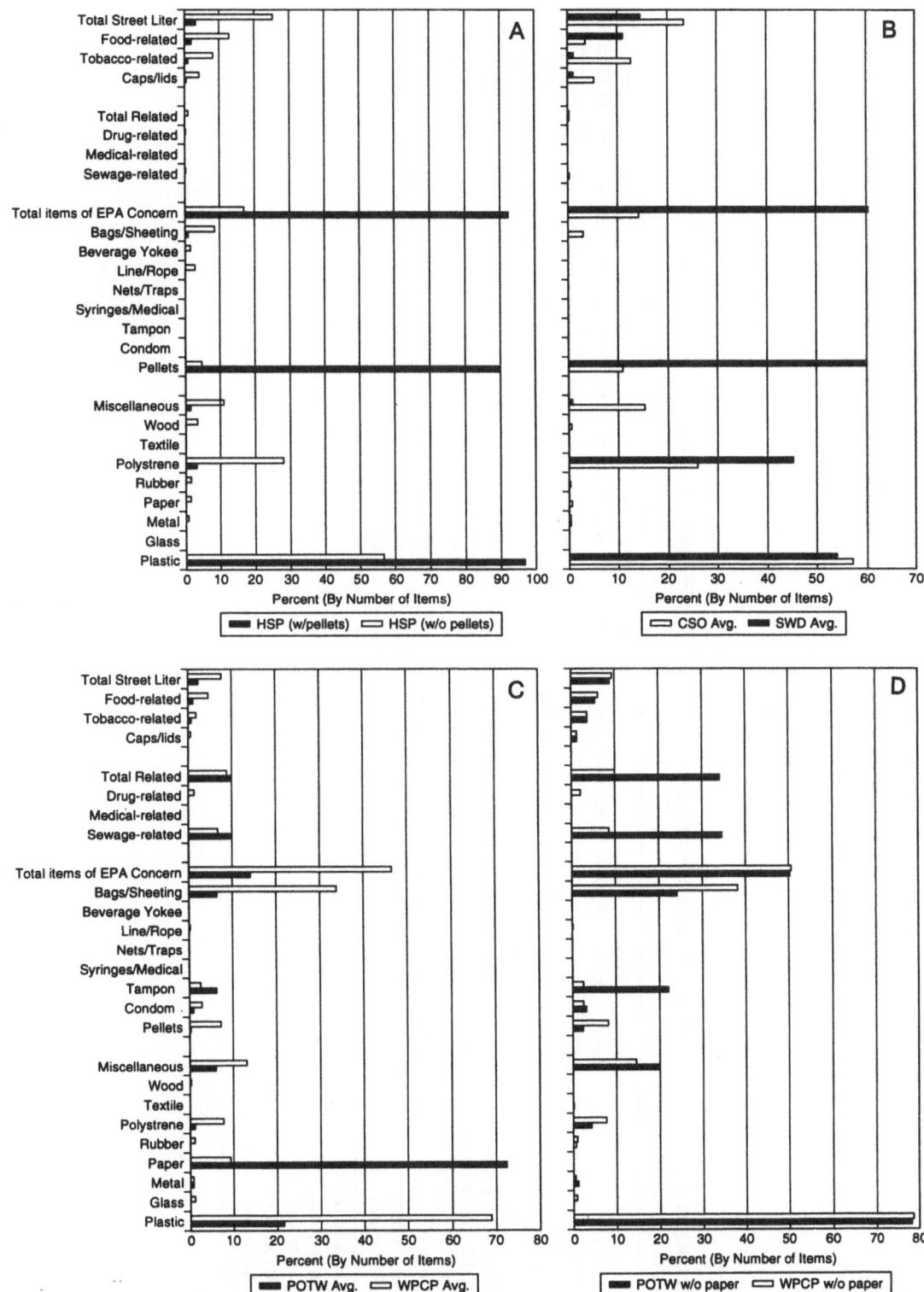

FIGURE 23.4A–D. Relative percent compositions from the four data groups examined in the qualitative comparison of (A) combined harbors [including (*shaded*) and excluding (*white*) resin pellets]; (B) CSO (*white*) and SWD (*shaded*) outfalls; (C) POTW (*black*) and WPCP (*white*) sewage treatment facilities, averages; and (D) POTW (*black*) and WPCP (*white*), without paper.

TABLE 23.4. CSO Studies Program (1989–1990) and Harbor Studies Program (1988–1992) percent composition data for significant items and categories of targeted debris sources.

Source	Total items	Total pellets	Plastic	Polystyrene	Paper	Combined street litter	EPA items of concern	Sewage-related items	Medical-related items	Drug-related items	Combined related items
Boston CSO	5,084[a]	559[a]	56.55	25.65	1.00	23.57	14.68	0.30	0.07	0.01	0.38
Philadelphia SWD	2,557[a]	1,528[a]	53.52	44.77	0.08	14.86	60.23	0.04	0.21	0.21	0.46
Boston POTW (with paper)	100,124[b]	405[b]	21.20	0.95	71.56	2.30	14.34	9.53	0.03	0.00	9.56
Boston POTW (without paper)	28,474[b]	405[b]	74.54	3.34	–	8.10	50.41	33.50	0.12	0.00	33.62
Philadelphia WPCP (with paper)	183,110[c]	13,015[c]	68.54	7.29	8.78	7.88	46.33	6.97	0.13	1.66	9.61
Philadelphia WPCP (without paper)	167,035[c]	13,015[c]	75.14	8.00	–	8.31	50.79	7.65	0.14	1.82	8.79
Philadelphia WPCP	549,330[d]	39,045[d]	68.54	7.29	8.78	7.88	46.33	1.66	0.13	6.97	8.76
Harbor Studies Program	814,407[e]	731,509[e]	95.29	2.92	0.14	2.70	91.23	0.05	0.02	0.05	0.12
Harbor Studies Program	86,290[f]	3,392[f]	55.59	27.57	1.29	25.45	17.25	0.50	0.23	0.43	1.16

[a] Average per inch of rainfall.
[b] Daily average for entire system.
[c] Daily average per WPCP.
[d] Daily average for entire system (all three WPCPs).
[e] Totals including resin pellets.
[f] Totals excluding resin pellets.

the combined harbor data, plastic dominates all other debris categories at 95% (almost 90% is attributable to resin pellets collected in the Houston Ship Channel) (Table 23.4). All eight remaining categories amount to less than 5% of the total and are of little significance. Because of their overwhelming abundance, resin pellets have been excluded from the combined harbors data that are used to make the comparisons in the following sections.

Major Categories.

Close examination of the composition data for the major debris categories under each program reveals a striking similarity between the combined harbor debris and the CSO and SWD outfall debris. Data from both outfall types compare well with the data from combined harbors (excluding resin pellets) in most of the major categories (Fig. 23.4). For the Harbor Studies Program, plastic comprised just over 55% of all harbor debris collected (ranging from less than 30% in Tacoma to 68% in Honolulu). Foamed polystyrene constituted almost 30% of the total debris (ranging from <10% in Mayagüez, PR, to almost 50% in Tacoma, WA). Plastic and foamed polystyrene accounted for more than 80% of harbor debris. Miscellaneous items were approximately 10% of all harbor debris; the remaining six categories accounted for less than 10%. Data for the Harbor Studies Program are presented in Fig. 23.4A.

Similarly, CSO and SWD outfall data (Fig. 23.4B) from the CSO Studies Program exhibit a pattern that conforms closely to data from the combined harbors. For the Boston CSO (BOS 070), plastic averaged just over 55%, ranging from approximately 20% to 80% for two events. Foamed polystyrene accounted for about 25% of all the CSO debris and ranged from approximately 15% to 40%. Plastic and foamed polystyrene accounted for more than 80% of harbor debris. Miscellaneous items constituted approximately 15% of all harbor debris; the remaining six categories accounted for less than 5%. For the Philadelphia SWD, plastic (about 55%,

ranging from more than 35% to almost 90%) and polystyrene (about 45%, ranging from about 10% to just over 60%) accounted for more than 98% of all debris collected over three events. Miscellaneous items constituted approximately 2% of all harbor debris, and the remaining six categories accounted for less than 1%.

There appears to be little relationship among debris data derived from the STPs (Fig. 23.4C) and those data from either the CSO and SWD outfalls or the combined harbors. Of the two STPs (POTW and WPCP) studied, the Boston POTW appeared to be least similar to outfall and harbor data. Paper, primarily paper towels, constituted more than 70% of the debris and was overwhelmingly the most abundant of the nine major debris categories. Plastic was about 20% of all debris collected from the POTW, while foamed polystyrene was less than 2%. Miscellaneous debris constituted just over 5% of all POTW debris and the five remaining categories less than 1%. The debris from the Philadelphia WPCPs more closely matched the harbor and outfall debris than did the Boston POTW. This phenomenon may have partially resulted from severe rainfall in Philadelphia before collecting the WPCP samples (a considerable amount of litter may have washed into the WPCPs from the street). Plastic dominated the WPCP debris at almost 70%. Miscellaneous items made up almost 15% of all WPCP debris, while paper (mostly toilet paper and paper towels) was about 9%. Foamed polystyrene contributed more than 7% of the debris. With paper excluded from the database, the POTW data more closely resemble the data from the WPCP, combined harbors, and the outfalls.

The relationship among the combined harbors, outfalls, and STPs demonstrated in the data of the major categories is clearly defined in Fig. 23.5. Four categories, plastic, polystyrene, paper and miscellaneous, account for more than 90% of all debris from each of the sources. However, the composition of those categories, as discussed previously, suggests a high degree of similarity between the combined harbors data and the outfall data (CSO and SWD) and little similarity between the

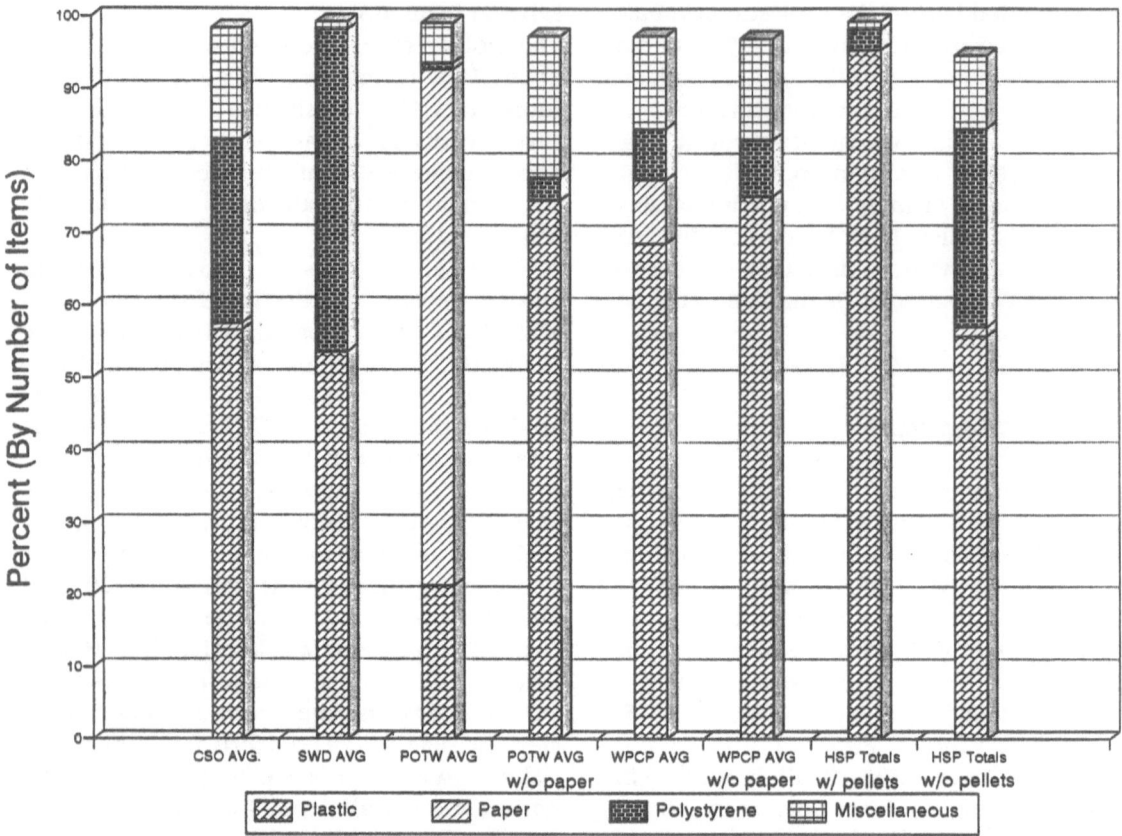

Figure 23.5. Relative percent compositions of the four most significant major debris categories for each of the sources sampled (combined data from 1988–1992). See key on figure.

combined harbors and outfalls and the STPs (WPCP and POTW).

EPA Items of Concern.

With resin pellets excluded from the combined harbors data, items of concern comprised more than 15% of all debris. Plastic bags and sheeting (just under 10%) were the most abundant of all items in this data group. Pellets (polystyrene foam only) contributed almost 5% of the total, while the remaining items of significance (line and rope, beverage yokes, and condoms) combined accounted for about 5%.

Similarities between the outfall debris and the combined harbor debris (Fig. 23.4A) are much less evident than those demonstrated for the major categories. The data pattern exhibited by the CSO debris (Fig. 23.4B) for

this data group more closely resembles the combined harbor data pattern than do the data for the Philadelphia SWD. For BOS 070, items of EPA concern accounted for almost 15% of all CSO debris. Pellets were the most abundant item, just over 10% of that debris. Plastic bags and sheeting were less than 5% and the remaining items (condoms, tampons, and beverage yokes) less than 1%. For the Philadelphia SWD (Fig. 23.4B), items of concern accounted for more than 60% of all debris collected. Pellets (plastic and polystyrene foam), comprising almost 60% of all debris, were overwhelmingly the most abundant item. Plastic bags, sheeting, syringes, and medical were the only other items of concern found, accounting for less than 1% of all debris.

A relationship among the STPs and the combined harbors and outfalls is not easily demonstrated in the debris data patterns for this data group. Items associated with sew-

Two EPA Marine Debris Studies

age, as would be expected, constitute a greater proportion of the STP debris. For the Boston POTW (Fig. 23.4C), items of EPA concern constituted almost 15% of all debris collected. Plastic bags and sheeting were the most abundant items in this category, more than 5% of all debris. Although these percentages resemble those for the combined harbors and the CSO outfall, all similarity ends there. Sewage-related items of concern (e.g., tampons and condoms) comprised almost 10% of the total debris; the remaining items (pellets, syringes and medical, and line and rope) contributed less than 1%. For the Philadelphia WPCPs (Fig. 23.4C), items of concern accounted for more than 45% of all debris. Plastic bags and sheeting, comprising more than 30% of all WPCP debris, were the most abundant items. Pellets accounted for approximately 7% of all WPCP debris and tampons and condoms combined about 5%.

The remaining items of concern (e.g., line and rope, as well as syringes and other medical waste) accounted for less than 1% of the total debris.

Figure 23.6 demonstrates the relationship between pellets and the remaining seven items of concern for each debris source. Pellets were collected from every debris source (harbors, CSO and SWD outfalls, and STPs). However, pellets were the most abundant item of concern for only the combined harbors and the two outfalls. Plastic bags and sheeting, tampons, and condoms were most abundant at the STPs.

Related Items.

For this group, the data pattern for the combined harbors is similar to that derived from

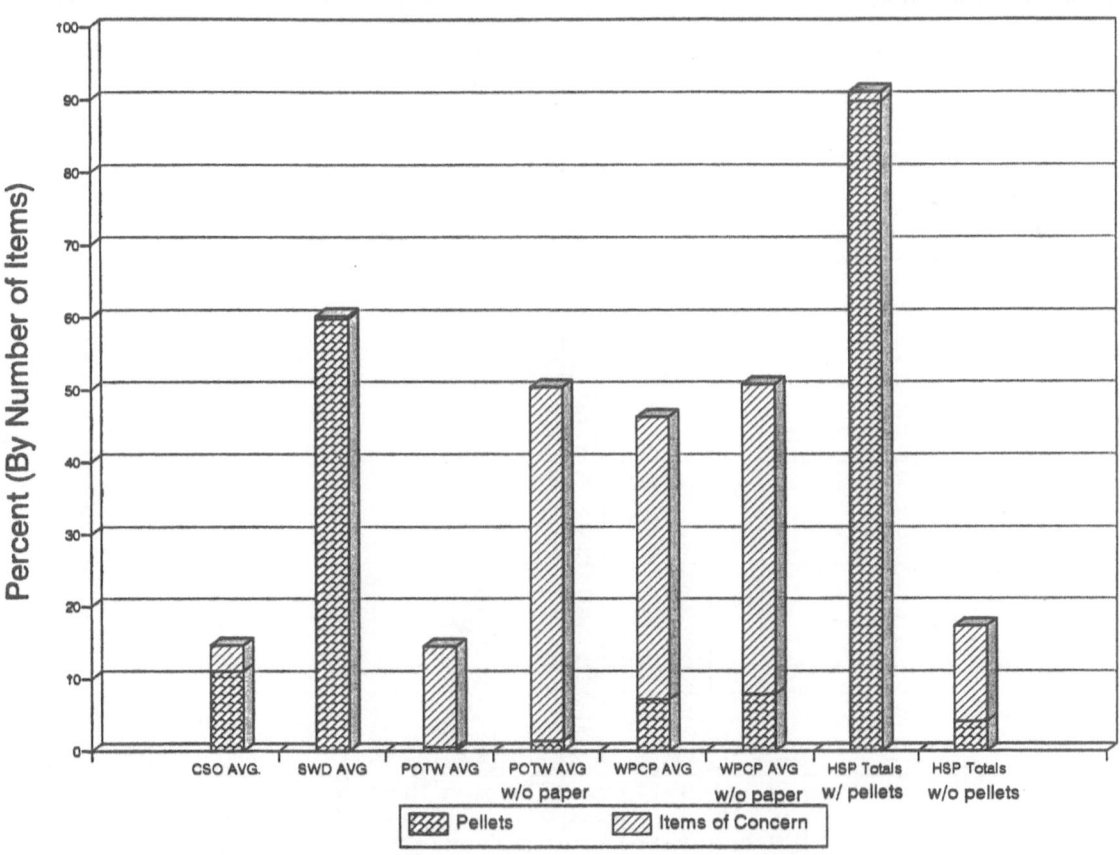

FIGURE 23.6. Relative percent compositions showing the relationship between plastic pellets (*double hatching*) and all other EPA items of concern (*shading*) for each of the sources sampled (combined data from 1988–1992).

the CSO and SWD outfall data; however, these patterns exhibit little similarity with those derived from the STPs. For the combined harbors (Fig. 23.4A), related items accounted for more than 1% of all harbor debris when pellets are excluded. Sewage- and drug-related items each accounted for almost 0.5% of the debris and medical-related items almost 0.25%.

For both outfalls (Fig. 23.4B), the data pattern was very similar to that for the combined harbors. In the Boston CSO samples, related items accounted for almost 0.4% of all debris. Sewage-related items contributed 0.3% and medical- and drug-related items combined less than 0.1%. In samples from the Philadelphia SWD, related items accounted for almost 0.5% of the debris from the SWD. Drug- and medical-related items each contributed approximately 0.2% of all the SWD debris. Sewage-related items accounted for less than 0.1% of the SWD total.

In the STP samples (Fig. 23.4C), related items comprised almost 10% of all debris collected (including paper) at both the Boston POTW and the combined Philadelphia WPCPs. At the POTW, sewage-related items were the most abundance, comprising almost the entire 10% of related debris. Drug-related items were not found, and medical-related items accounted for less than 0.1% of related debris. At the WPCPs, sewage-related items were the most abundant and accounted for almost 7% of all WPCP debris. Drug-related items were almost 2% of the total debris and medical-related items less than 0.2%.

Figure 23.7 clearly demonstrates the similarities and differences in the overall composition of related items when considered among the different debris sources. The low percentages (1% or less) of related debris derived from the outfall and combined harbors data suggest a high degree of similarity among the three sources (harbors, CSOs, and

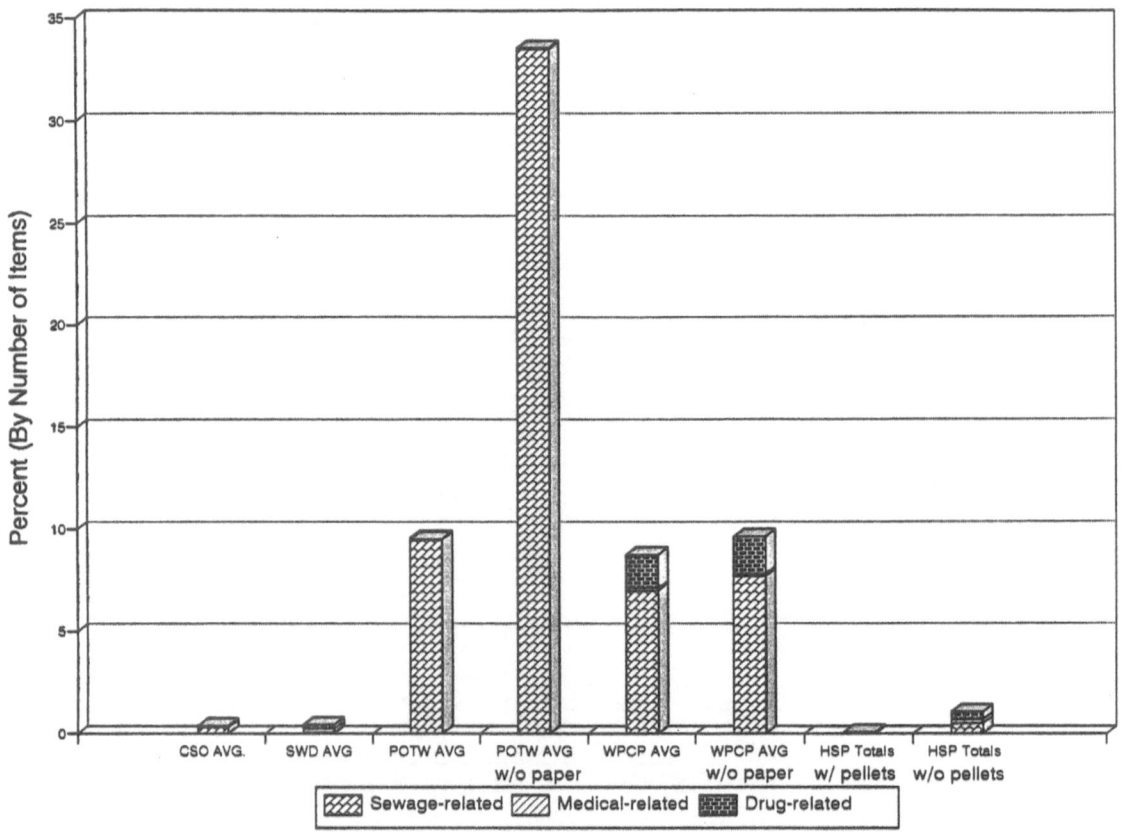

FIGURE 23.7. Relative percent compositions showing the relationship among sewage-, medical-, and drug-related items for each source samples (combined data from 1988–1992). See key on figure.

SWDs). The opposite is true when comparing the related debris data from the outfall and combined harbors with that from the STPs. The high percentages (almost 10%) collected from the STPs compared to those collected from the outfalls and combined harbors suggest that STP debris differs from CSO, SWD, and harbor debris; this may indicate that the two outfalls are a significant sources of harbor debris.

Street Litter.

Under this data group, debris data from the combined harbors correlate well not only with the CSO and SWD outfall data but also to a lesser degree with the data from the STPs. For the combined harbors (Fig. 23.4A), street litter accounted for more than 25% of all debris. Food-related items were the next most abundant, almost 15% of the total. Tobacco-related items contributed nearly 10% of all debris, while caps and lids accounted for about 5%.

For both outfalls (Fig. 23.4B), the data pattern corresponded closely with that from the combined harbors. In samples collected from the CSO, street litter accounted for almost 25% of all CSO debris collected. Tobacco-related items, almost 15% of the total, were most abundant among all street litter items. Caps and lids accounted for just over 5% of the total and food-related products almost 5%. For debris collected at the SWD, street litter comprised about 15% of the total. The most abundant of these items were food related, more than 10% of all SWD debris. Caps and lids together and tobacco-related products each accounted for almost 2% of the total.

In general, street litter percentages for the POTW and combined WPCPs appear low when compared with street litter percentages from the combined harbors and CSO and SWD outfalls. However, if the categories (caps and lids, tobacco-related, and food-related debris) within this data group are considered relative to one another for each source, then STPs correlate very well proportionally with the combined harbors and the outfalls. In debris samples collected from the

POTW, total street litter accounted for less than 3% of the total. Food-related items, more than 1% of the total debris, were the most abundant of all street litter. Tobacco-related items accounted for almost 1% of the total and caps and lids for approximately 0.25%. For the combined WPCPs, street litter comprised almost 8% of the total debris collected. Almost 5% of the total was contributed by food-related items, approximately 2% by tobacco-related items and less than 1% by caps and lids.

Street litter is the one data group that links the STP debris with the harbor and outfall debris. The fact that street litter (larger forms that cannot get into the sewage systems unless carried by stormwater runoff) is a significant component of all STP debris suggests the possibility of a common link among the five targeted sources: POTW and WPCP, CSO and SWD outfalls, and the aquatic environment into which the outfalls discharge. Figure 23.7 demonstrates the significance of the relationship among the five sources. Clearly, the street litter data from the combined harbors are most closely related with the outfall data. Although items identified as street litter were found in the debris of both STPs, the Philadelphia WPCPs contained the greatest percentage of street litter. Data from the WPCPs are more similar to the harbor and outfall data than to the Boston POTW data. (As mentioned previously, this may have resulted from significant rainfall in Philadelphia before sampling.) The presence of street litter in the debris of the STPs suggests that it is entering STPs through the CSO system and that this type of debris is available for discharge into the aquatic environment during overflow conditions.

Most Common Items.

The most common items found under each debris source are listed in Table 23.5. The combined harbors shared the greatest number of most common items with the two outfalls, nine with the Philadelphia SWD and seven with the Boston CSO. The items common to those three sources include resin pellets, miscellaneous plastic pieces, polysty-

TABLE 23.5. The most common items found during the CSO Studies Program (1989–1990) and Harbor Studies Program (1988–1992).

Items	Harbor Studies Program (all harbors combined)	CSO Studies Program			
		Philadelphia WPCPs	Philadelphia SWD[a]	Boston POTW	Boston CSO[a]
Resin pellets	728,117	12,967	1,066		456
Miscellaneous plastic pieces	16,539	17,950	118	1,331	1,074
Polystyrene pieces < 7.3 cm dia.	9,494	9,728	260		543
Plastic sheeting < 2 linear ft	6,476	60,577		6,525	161
Cigarette butts and filters	4,718	3,070	23		614
Grease balls	3,676		8		710
Polystyrene cups/bowls (pieces)	3,458		145		
Plastic food wrappers	3,410	2,612	14	783	
Foamed polystyrene spheres	3,340		462		104
Polystyrene beverage labels	2,811		33		
Cigar/cigarette wrappers/packs			17		
Miscellaneous fibrous material		22,463		4,667	
Paper towels		15,647		70,360	
Condoms (pieces)		4,408		702	
Tampons		2,642		5,588	
Polystyrene packing peanuts			114		385
Plastic cups/spoons/forks/straws			21		
Plastic cap/lid pieces					267
Polystyrene balls					126
Miscellaneous plastic items				759	
Diaper and facial wipes				1,228	
Miscellaneous paper items				1,003	

[a]Number of items per inch of rainfall.

rene pieces (<7.3 cm in diameter, which is about the size of an American baseball), cigarette butts and filters, grease balls, and foamed polystyrene spheres. The combined harbors shared only three of the most common items with the POTW and six with the combined WPCPs. It should be noted that a significant rainstorm occurred before the collection of samples from the WPCP, possibly carrying a significant amount of street litter through the CSO system (during preoverflow conditions) and into the STP. This could account not only for the higher percentage of street litter observed in the WPCP samples as opposed to that found in the POTW but also for the greater number of most common items shared with the combined harbors.

Descriptive Statistical Comparisons

Dice and Bray–Curtis similarity analyses were used to compare the CSO and Harbor Studies data for meaningful associations among sites based on the presence of debris types (Dice) and relative abundance of debris types (Bray-Curtis). Two comparisons were made, each using both analytical techniques. The first compared the average CSO and Harbor Studies data for specific harbors to look for similarities among the harbors. The second analysis compared harbor debris data with CSO debris data to evaluate similarities there. Strong similarities between CSO and harbor data in the second analysis would further confirm that CSOs are a significant source of marine debris.

The results of these statistical comparisons were inconclusive. In both cases, the Dice similarity analysis was able to cluster together sites that (1) had little debris, (2) were considered well flushed, (3) were close to the open ocean, and (4) had little recent rainfall. With a few obvious exceptions (i.e., Houston's upper, middle, and lower ship channel and Baltimore's inner and middle harbors), the clusterings were inconsistent with geographic or operational variability in the data exacerbated by the different sampling strategies used in the two programs. Sampling strategies designed to generate data specifically for use in cluster analyses of the similarities among CSO, SWD, and STP discharges and harbor debris may yield more useful results. These strategies are recommended to be incorporated in future studies.

Conclusions

As a result of comparing the two programs, this paper suggests several possible similarities and differences (both qualitatively and statistically) in the debris compositions among the five different sources examined. The results of the qualitative comparison seem to be much more conclusive than the statistical comparison in identifying CSOs and SWDs as major sources of harbor debris. However, neither comparison provides definitive results. Because of the differences in sampling methods used during each program and because of some of the problems encountered during sample collection (especially the CSO Studies Program), the magnitude of the similarities or differences among the sources is not clear; as a result, no absolute conclusions can be drawn. However, based on the apparent similarities and differences identified in the comparison, several conclusions can be inferred.

CSO and SWDs appear to be major contributors to the overall aquatic debris problem. Relative debris composition for each of the major categories for each outfall type resemble the composition patterns in harbors (EPA 1990b, 1992a, 1992b, 1992d, 1993a).

CSO and SWD discharges are similar in debris composition, and a considerable portion of the debris discharged by the two sources is composed of street litter.

Related items (sewage-, medical-, and drug-related) occur in similar percentages in the harbors and both outfalls.

Related items comprise a much greater percentage of the total debris in STPs versus the harbors and outfalls, but occur in similar percentages when comparing STPs (WPCPs and POTWs).

During storm events, POTWs and WPCPs appear to receive large amounts of street litter from CSO systems under both overflow and no-overflow conditions.

Plastic (resin) pellets are present in stormwater, sewage treatment, and CSO systems and originate directly from sources such as the plastics and related industries. Pellets can enter the aquatic environment directly from SWD and CSO outfalls.

Statistical comparison of similarities among the debris sources was ambiguous.

- When comparing all harbor studies data with that from the CSO, SWD, and STPs, Dice similarity appears to suggest two relationships: one among the Boston CSO and the three Philadelphia WPCPs, and one among those sources and several harbors. Bray–Curtis similarity suggests a relatively close relationship among the Philadelphia WPCPs and the Deer Island STP and a more loosely defined relationship among the SWD, CSO, and several harbors.

- When comparing data from four harbors to those from the CSO, SWD, and STPs, Dice similarity appears to suggest a relationship between the Deer Island STP, the Boston CSO, the Philadelphia SWD, the seven areas of the four harbors studied, and the Philadelphia STP. Bray–Curtis similarity suggests that (1) the two STPs are similar to each other, but not similar to most other sources; and (2) that the Boston CSO is marginally similar to a nine-cluster group that includes the Philadelphia SWD and eight areas of the four harbors.

The debris composition data from both the combined harbors and the targeted outfalls did not strongly resemble the composition patterns from the POTW and WPCPs. However, WPCPs (following a rain event) did contain street litter in compositions comparable to the combined harbors and the outfalls. This provides strong evidence that the material discharged by CSOs and SWDs is the same type of material entering the harbors.

24.

New York and New Jersey Beaches: "It Was a Very Good Year"

R. L. Swanson and Marci L. Bortman

Introduction

Few if any of New York and New Jersey ocean beaches were closed in the summer of 1993 because of floatable material washing ashore. The cleanliness of the beaches in the summer of 1993 probably will become the new standard by which future summers will be compared (Dave Rosenblatt, New Jersey Department of Environmental Protection and Energy, November 1993, personal communication). It was a major improvement from the summers of 1987 and 1988, when area beach closures and the perception of degraded beach and water quality were estimated to have cost the New York–New Jersey economy several billion dollars (Swanson et al. 1991). Equally important was that other measures of coastal water quality were thought to be high in 1993. Area beachgoers reported that coastal waters were clear and bright blue; ocean beach closures from high coliform counts (an indicator of raw sewage contamination) were also quite low; and measurements of summer-minimum, bottom dissolved oxygen concentrations (in the New York Bight apex) were the highest measured since 1973, further indicating low levels of sewage contamination (although no data were collected during 1986–1989).

Since 1988, significant steps have been undertaken to reduce the quantity of floatables reaching coastal waters in the New York-New Jersey metropolitan area. Included among these are monitoring and collection efforts by federal, state, and local governments. A Beach Information Network was instituted in New York State to improve communications between the scores of beach operators. Educational programs have also been developed (see Molinari, Chapter 26, this volume).

In addition to these intervention programs, weather conditions also played an important role in 1993 beach conditions. The summer was warm, 39 days with temperatures of 37.7°C (90°F) or higher at Central Park, tying the record set in 1991 (New York Times 1994). It was also dry: the frequency of combined sewer overflow (CSO) events (10 mm of rain for a period of 6.7 h) was less than normal. The winds were offshore much of the summer, transporting any floatable materials seaward.

We compared summer 1993 meteorological conditions to the summers 1987 and 1988 to estimate the likelihood that floatable materials were released to marine systems and the probability of their potential for being transported shoreward toward Long Island or New Jersey ocean beaches. We also examined the floatables intervention programs, including the U.S. Army Corps of Engineers (CoE) Drift Collection Program and the CSO retention programs in New York and New Jersey, to estimate their impacts on reducing marine debris. These analyses are beneficial for understanding the excellent conditions experienced at New York–New

Jersey ocean beaches in 1993 and why conditions might be different in the future.

Methods

Observations of wind velocities and rainfall data were compiled for this analysis. Hourly wind data were used to construct progressive vector diagrams (PVDs) and to calculate wind constancy and wind energy values for June, July, and August for a number of years with and without floatable events. Wind constancy values, c, were estimated from the relationship:

$$c = \frac{v}{u} \times 100$$

where v equals the monthly mean vector wind speed and u equals the monthly mean scalar wind speed. Energy values, E, were obtained from

$$E = \left| \frac{u}{u_{norm}} \right|^2 \times 100$$

where the u term in the numerator equals the monthly mean kinetic energy of the wind, and u_{norm} in the denominator is the average monthly mean kinetic energy over 30 years (1959–1988) (Swanson and Zimmer 1990).

Daily rainfall values were employed to construct precipitation graphs for June, July, and August. The PVDs helped to describe the wind behavior and to identify times of potential washups of floatable debris on the south coast of Long Island or the coast of New Jersey. The constancy indicated the persistency of the wind to blow from a given direction. The precipitation graphs served to identify periods of possible storm sewer and combined sewer overflow events to the New York–New Jersey Harbor Estuary (NYNJHE). The estimated time lag between an overflow event and an initial washup on ocean beaches was 3–5 days.

Wind data from John F. Kennedy International Airport were used because they are the most readily available data reflecting oceanic wind conditions. Rainfall data for Central Park in New York City were used because they represent conditions close to the major

sources of floatable wastes, namely the harbor estuary.

During June through August 1993, there were no reported incidents of ocean beach closures as a consequence of the washup of floatable debris in either New York or New Jersey. There were continuous occurrences of small numbers of sharps (e.g., syringe needles and scalpels) washing ashore on the south shore of Long Island throughout the summer and reported to the New York State Beach Information Network. A recent survey of diabetics in the New York metropolitan area suggests that there still may be as many as 9000 syringes and needles per day being disposed of in toilets (Bauer et al. 1994). HydroQual (1993) estimated that some 1850 syringes per month escape various collection systems and enter the NYNJHE.

Floatable Events and Meteorology of Summer 1993

June

The vector wind for June (Fig. 24.1) blew from a west-southwesterly direction. Sus-

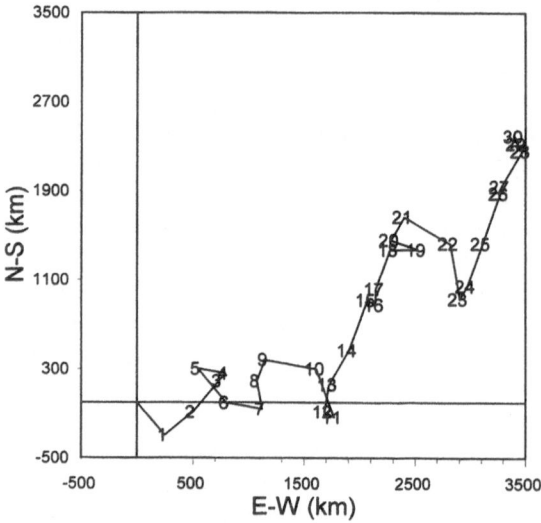

FIGURE 24.1. Progressive vector diagram, daily winds, J. F. Kennedy International Airport, June 1993.

TABLE 24.1. Monthly vector mean wind speed and direction for J. F. Kennedy International Airport, selected years for the period 1976, 1987–1993, compared with a reference period (the norm), 1959–1988.

Wind factor	Norm	Year							
		1976	1987	1988	1989	1990	1991	1992	1993
Vector wind speed, m s^{-1}									
June	1.4	2.8	1.2	1.3	0.8	2.2	0.9	1.4	1.6
July	1.7	1.9	1.1	2.4	0.9	0.8	1.0	1.2	1.0
August	1.3	1.2	0.6	2.1	0.8	0.8	1.5	0.7	1.7
Direction, °T									
June	208	200	230	280	196	217	242	235	240
July	212	240	200	210	225	206	208	234	250
August	216	250	230	210	263	189	238	238	190

From NOAA (1993a), with permission.

tained south-southwesterly winds, which were very nearly perpendicular to the general trend of the Long Island coast, blew during the periods of June 13–18 and June 24–28. In early June, the winds were westerly, and from June 21 through June 23 the winds had a strong northerly or offshore component.

Over the entire month, the vector mean wind speed was 1.6 m s^{-1} from 240°T (Table 24.1), and the scalar mean speed was 4.2 m s^{-1}. This compares with the 1959–1988 climatological norm for June of 1.4 m s^{-1} from 208°T (Table 24.1). Relative wind energy as a percent of normal was 74, and wind constancy, a measure of the persistence of the wind, was 38% for the month (Table 24.2).

Rainfall for the month, as recorded in Central Park, is shown in Fig. 24.2 The total precipitation during this month was 38 mm, which was 41% of normal monthly rainfall of

93 mm (Table 24.3). Significant rains occurred on June 5, 9, 20, and 27. Some floatable material could have been washed into the NYNJHE on these occasions. The New York City Department of Environmental Protection estimates that a rainfall intensity of 0.15 cm h^{-1} for 6.7 h will cause bypassing of 24.2 m^3 s^{-1} of stormwater and sewage (Swanson and Zimmer 1990).

July

The winds during July (Fig. 24.3) were atypical for summertime. They were mixed with a slight easterly component July 1-5. From July 8 through July 14 there was a predominant westerly component, and from July 14 through July 23 the winds were primarily out of the northwest. For the balance of the

TABLE 24.2. Monthly wind constancy and relative wind energy for J. F. Kennedy International Airport, selected years over the period 1976–1993, compared with a reference period (the norm), 1959–1988.

Wind factor	Norm	Year							
		1976	1987	1988	1989	1990	1991	1992	1993
Constancy, %									
June	34	55	26	29	19	50	23	35	38
July	41	40	29	58	21	18	28	36	23
August	33	23	13	44	19	21	37	20	46
Energy, %									
June	100	108	88	90	81	75	65	62	74
July	100	104	70	80	78	73	67	50	82
August	100	129	94	112	90	71	78	53	66

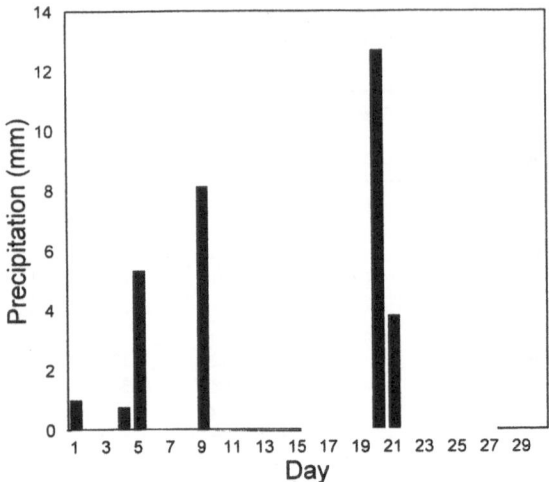

Figure 24.2. Daily rainfall (mm) at Central Park, New York City, June 1993.

month, they had a southerly component, more typical of summer conditions.

Over the entire month, the vector mean wind speed was 1.0 m s^{-1} from 250°T (Table 24.1) and the scalar mean speed was 4.2 m s^{-1}. The vector mean speed was considerably less than the 1959–1988 climatological norm for July of 1.7 m s^{-1} (Table 24.1). The monthly vector mean wind direction swung

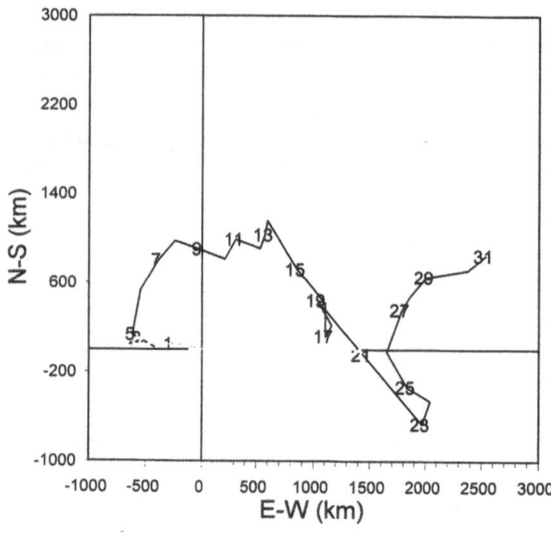

Figure 24.3. Progressive vector diagram, daily winds, J. F. Kennedy International Airport, July 1993.

around to the west compared to the norm, which is 212°T. Wind energy as a percent of normal was 82, and wind constancy was only 23% for the month (Table 24.2).

Rainfall for the month, as recorded in Central Park, is shown in Fig. 24.4. The total precipitation during this month was 43 mm, which was 39% of the normal amount for the month of July (110 mm) (Table 24.3). The few rains were scattered throughout July. Some 56% of the monthly total fell on July 2 and 3; the only other significant rains occurred on July 14, 19, and 27. These rains, particularly the ones on July 2 and 3, could have contributed floatable material to the NYNJHE.

August

The winds during August (Fig. 24.5) were primarily from the south by west although they were relatively weak compared to the norm for August. The only break in the general south by westerly flow was during the periods of August 4 through August 7 and August 18–22, when the winds were weak and variable.

Over the entire month, the vector mean wind speed was 1.7 m s^{-1} from 190°T (Table 24.1) and the scalar mean wind speed was 3.7 m s^{-1}. The vector mean wind speed was greater than the 1959–1988 climatological

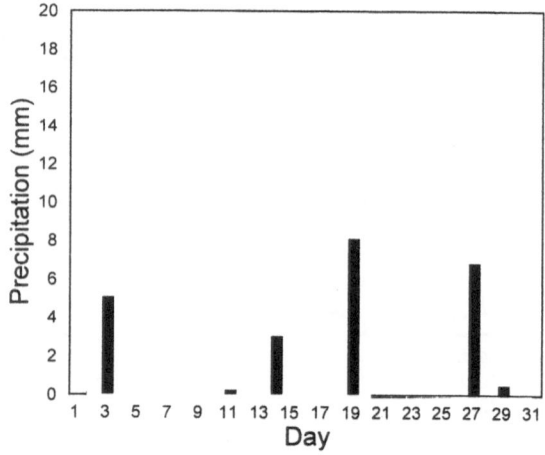

Figure 24.4. Daily rainfall (mm) at Central Park, New York City, June 1993.

TABLE 24.3. Monthly rainfall (mm) for Central Park, New York City, selected years over the period 1976–1993, compared with the norm: 1976, 1987–1993.

Month	Norm	1976	1987	1988	1989	1990	1991	1992	1993
June	93	70	100	88	223	64	106	121	38
July	110	36	105	207	130	89	116	114	43
August	102	166	125	51	110	314	181	89	137

From NOAA (1993b), with permission.

norm for August of 1.3 m s^{-1} (Table 24.1). The resultant direction, 190°T, was close to the norm, which is from 216°T. Wind energy was a percent of normal was 66, and wind constancy was 46% for the period (Table 24.2).

Rainfall for August, as recorded in Central Park, is shown in Fig. 24.6. The total precipitation during this period was 137 mm, somewhat more than the norm of 102 mm (Table 24.3). The heaviest rains occurred on August 16 and 17 (81 mm), but CSO events also occurred on August 6–7 and on August 13.

Comparison of the Summer of 1993 to the Summers of 1976, 1987, and 1988

The floatable debris problem was particularly severe on the south shore of Long Island during the summers of 1976 and 1988. Similarly, the floatable problem was significant along the coast of New Jersey during the summer of 1987. The floatable washups during these summers have been extensively documented in Swanson et al. (1978) and Swanson and Zimmer (1990). In each of these summers a clear correlation was established among the sources of floatables, meteorological conditions, and the fate or location of floatable washups. The resultant winds for each month (see Table 24.1) clearly show the winds of June 1976 and July 1988 as being the most likely to drive floatables to the southern coast of Long Island (resultant speed of 2.8 m s^{-1} and 2.4 m s^{-1} out of 200°T and 210°T, respectively). The wind constancies (Table 24.2) of 55% and 58% are much greater than the norm for the respective months (34% and 41%, respectively).

Numerous events contributed to the floatables event of 1976, including pier fires in the

TABLE 24.4. Estimated monthly total floatable debris (captured and retained) and debris remaining for transport in the New York–New Jersey Harbor Estuary (metric tons).

Combined sewer overflow (CSO) retention	
Newark Airport and Spring Creek	0.8[a]
Fresh Creek	1.2[b]
U.S. Army Corps of Engineers Drift Collection Program	9.4[c]
Total debris retained	11.4
Total floatable debris	34.9[a]
Total floatable debris remaining in harbor	23.5

[a]HydroQual (1993).
[b]Forndran and Delva (1992).
[c]Alan Dorfman, CoE, Jersey City, NJ, personal communication, April 1994.

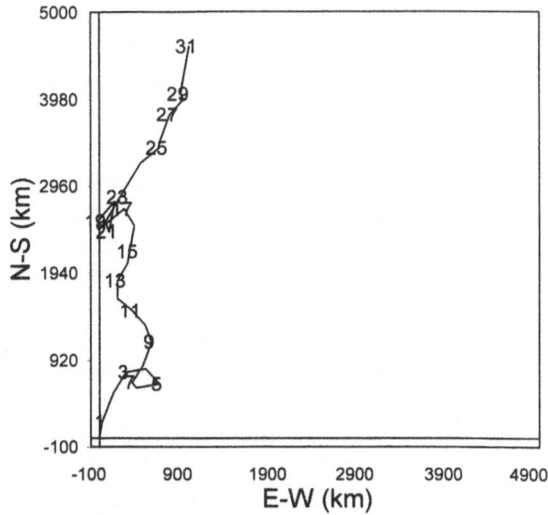

FIGURE 24.5. Progressive vector diagram, daily winds, J. F. Kennedy International Airport, August 1993.

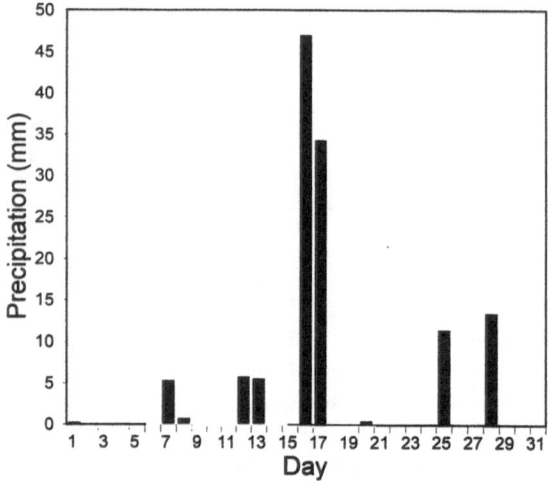

FIGURE 24.6. Daily rainfall (mm) at Central Park, New York City, August 1993.

NYNJHE and the near record high Hudson River runoff in May of that year (Swanson et al. 1978). Significant rains occurred before and during the 1988 floatable event, contributing to major CSO discharges and sewage treatment plant bypassing. The winds of August 1988 also had the potential for producing washups on the southern coast of Long Island. The magnitude of its resultant wind vector was 2.1 m s^{-1} (slightly less than July) from the same direction (210°T) with a constancy of 44%. There were no beach closures during that month because there was little rainfall and no CSO events occurred.

The washups of August 1987 on the coast of New Jersey are explicitly reflected in the weak magnitude of the resultant wind vector (0.6 m s^{-1}) and its low constancy (13%). This is an indication of variable winds counteracting the normal southerlies or southwesterlies. The washups were associated with easterlies that transported, shoreward, floatable material associated with the Hudson River plume that typically hugs the New Jersey coastline.

The magnitudes of the monthly resultant wind vectors for the summer of 1993 (Table 24.1) were slightly stronger than normal with the exception of July, which was some 47% less. Wind energy was relatively low, however, at 74%, 82%, and 66% of the norm for the respective months. The direction of the

winds for June and July were much more westerly than the norm, some 40°–50° west of the southerlies that prevailed in June 1976 and July 1988. Wind constancy for June (Table 24.2) was slightly greater than the norm, but still less than the 45%–50% typically associated with floatable events on Long Island. August winds were persistently (46%) from the south (190°T). The vector wind speed was greater than the norm, but the wind energy was only 62% of the norm.

Rainfall was only 218 mm for the three summer months of 1993, 71% of the norm. August rainfall, however, was 34% greater than the norm of 102 mm. Rainfall intensity equaled or exceeded that necessary to cause a CSO event on 5 days in August. For the entire summer, CSO events occurred on only 12 days (Beau Ranheim, New York City Department of Environmental Protection, Wards Island, NY, April 1994, personal communication). This is about 50% of that of the summer of 1989, a very wet summer. Although there were many CSO events during the summer of 1989, the beaches of Long Island and New Jersey were relatively free of floatable material, because the CSO events generally occurred when the winds did not favor shoreward transport of marine debris.

Discussion

Perhaps one of the best measures of the improvements of beach conditions is that of the increase of beach users since the summer of 1988. At that time, annual beach use, measured in user-days, was estimated to be less by about 37–127 million for New York and New Jersey ocean beaches (Swanson et al. 1991). The Jones Beach State Park (Peggy Kucija, Jones Beach State Park, Wantagh, NY, May 1994, personal communication) reported a 48% increase in user-days relative to 1988 for the months of June through August 1993 (Fig. 24.7). The increase has been nearly monotonic since 1988, indicating a gradual improvement in the public's confidence in using the beaches. There are, however, other factors to be considered in these statistics (e.g.,

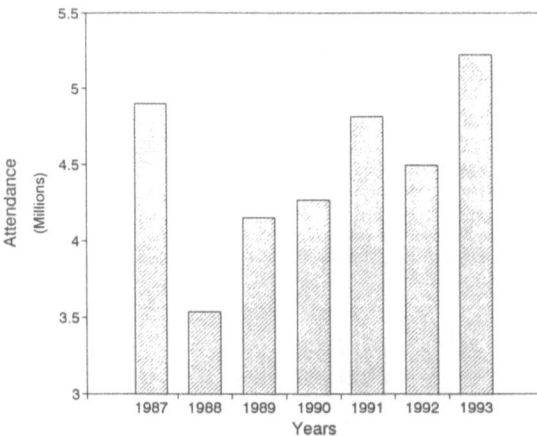

FIGURE 24.7. Attendance at Jones Beach State Park (Wantagh, NY), June, July, and August, 1987–1993.

the extremely warm summers of 1991 and 1993).

The U.S. Environmental Protection Agency (EPA 1994) reported a decrease in the number of floatable slicks observed since the agency began its floatable monitoring and surveillance program in 1989 [see Molinari, Chapter 26, this volume). In 1989, from mid-May through mid-September, some 139 slicks were observed. In 1993, during a similar period, only 35 slicks were observed. The latter figure is a clear reflection of the dry weather.

Some 280 metric tons (311 tons) of debris were removed by the CoE from the NYNJHE in June, July, and August 1993. Of this, approximately 5% (14 metric tons) was debris other than wood (Alan Dorfman, CoE, Jersey City, NJ, personal communication, April 1994). In 1989, the CoE reported 493 metric tons (544 tons) collected over the summer, beginning in mid-May (Dorfman 1989). In the latter, some 10% (49 metric tons) was estimated to be floating debris other than wood. Thus, in 1993, less debris was collected than in 1989.

There are, however, a number of prevention programs in place to reduce the amount of potentially floatable debris from reaching coastal waters. New York City is experimenting with capturing CSO for later processing during dry weather flow. HydroQual (1993) estimated that through existing and planned CSO abatement facilities in New York City, floatables from the City will be reduced about 24%. Since 1989, spillage has been controlled at Marine Transfer Stations (MTSs) by means of floating-debris curtains. Also, barges carrying garbage from the eight MTSs in New York City to the Fresh Kills landfill on Staten Island are covered with nets to prevent spillage.

A number of public awareness and public educational programs have also been implemented, including the U.S. EPA Clean Streets and Clean Beaches campaign. The National Oceanic and Atmospheric Administration Sea Grant programs also are emphasizing, through public education, the need to reduce litter, some significant fraction of which may reach storm sewers and CSOs.

One weakness in society's response to reducing the impact of marine debris in the New York–New Jersey metropolitan area is that the efforts to clean streets in New York City remain marginal. Present plans are that the budget for street cleaning in the City remain level relative to the past several years (Jerry Ring, New York City Department of Sanitation, New York, May 1994, personal communication). The City of Newark, however, has aggressively attacked the problem of littered streets, although it is not clear how effective the effort has been (Swanson et al. 1993).

Cleaner beaches, increased beach attendance, fewer observed slicks, extensive and increasing remediation efforts, and improved public education all suggest that the marine debris problem in the New York–New Jersey area may have been overcome. But, is this the case?

HydroQual (1993) estimated that over the period of a characterization and quantification study (1990 and 1991), some 2.3×10^6 floatable items (nonwood) weighing 34.9 metric tons (38.5 tons) entered the NYNJHE each month (Table 24.4). HydroQual (1993) also estimated that the CSO floatable retention facilities at Newark Airport and Spring Creek in Jamaica Bay now capture about 43,000 floatables per month of some 1.54×10^6 floatable items weighing 27.7 metric tons (30.5 tons) entering the harbor each month.

Thus, approximately 2.8%, or 0.8 metric tons (0.9 tons), of the CSO-related floatable burden is captured by these two facilities. In addition, the experimental CSO facility at Fresh Creek in Jamaica Bay is capturing about 1.2 metric tons (1.4 tons) per month. During 3 months in 1993, the CoE collected an average of 4.7 metric tons (5.2 tons) of nonwood debris per month. In 1989, a wet year, it collected about 14 metric tons (15 tons) per month.

The total amount of floatables (nonwood) now captured in the harbor is about 11.4 metric tons per month (12.6 tons) (Table 24.4), which is approximately 33% of the total floatable load by weight. Thus, during moderately wet years, perhaps 67% of the nonwood floatable load still is available for transport to coastal waters. Once in coastal waters, this load is subject to the vagaries of wind and current. The fraction that does not get transported shoreward will gradually drift seaward—some degrading, some sinking, and some drifting—well off into the Atlantic Ocean.

Summertime climatology may still be the controlling factor with regard to the amount of floatable material washing ashore on Long Island and New Jersey ocean beaches. Since July 1988 there have been few occasions when rainfall sufficient to create a major CSO event has occurred concomitantly with a persistent and energetic onshore wind to create a major floatable washup.

In 1993, the weather was particularly helpful in keeping the beaches clean. The Hudson River flow was small, rainfall was well below normal, and the winds were offshore for most of the summer—a very good year.

Conclusions

It was a very good year for New York area ocean beach activities in 1993. Now, there are many endeavors in place to inhibit floatables from exiting the NYNJHE:

CSO Abatement programs
Netting covers on garbage barges
Placing booms around marine transfer stations and the Fresh Kills Landfill
Skimming floatables in the harbor
Cleaning beaches and wetlands.

Public education programs are undoubtedly helping to some degree as well. However, even with these programs, the potential for substantial quantities of debris escaping into the marine environment is large, perhaps two-thirds the historical load, by weight.

Since 1988, summer weather conditions have not favored the washup of marine debris on either Long Island or New Jersey beaches. Time has been available to implement remediation programs. However, the job is only partially complete. The metropolitan area must do a better job of cleaning streets and highways. Streetcleaning is generally a discretionary program in municipal budgets, but when the impact of dirty streets hurts the environment and economy of communities 50–150 km downstream, somehow "discretionary" seems inappropriate. The CSO abatement program is having some success, but there are many technological and logistical impediments still to be overcome. Full implementation of the program is essential.

Indeed 1993 was a very good year at the beaches, but it is certainly no reason to be complacent with regard to the floatables problem.

Acknowledgments. The authors appreciate the help provided by Randall Young in preparing and reviewing this document. His views improved the document considerably.

25.
Sources of Plastic Pellets in the Aquatic Environment

David P. Redford, Heather K. Trulli, and Wayne R. Trulli

Introduction

The United States Environmental Protection Agency (EPA) Oceans and Coastal Protection Division of the Office of Wetlands, Oceans, and Watersheds initiated the study summarized in this paper to determine possible land-based sources of plastic pellets within the plastics industry. The study objectives were to (1) identify and locate possible sources of pellet releases into the environment, (2) evaluate the significance of each source as a pellet-release pathway, and (3) recommend mechanisms for controlling and preventing the release of pellets. The complete study, published by the EPA (1992d), represents the first comprehensive assembly of information regarding the sources of pellets in the aquatic environment and is expected to become a basic reference for the EPA and plastics industry.

Resin pellets are the raw materials from which plastic products are produced. An estimated 60 billion lb of resin are manufactured in the United States annually, most of which is pelletized. Pellets may be formed in spherical, ovoid, or cylindrical shapes, sizes ranging from 1 to 5 millimeters (mm) in diameter, and various colors (usually clear, white, or off-white). Those most commonly found in the environment are composed of polyethylene, polystyrene, and polypropylene (EPA 1990a).

Many types of resin pellets float in fresh-

water or seawater. Additives affect polymer density, thereby influencing whether a pellet will float, sink, or be suspended in water. Because salinity affects water density, a particular resin pellet could float in seawater but sink in freshwater. Hydrodynamic processes, such as turbulence and surface tension, also affect a pellet's ability to float.

Because pellets are small, buoyant, persistent, and ubiquitous in the aquatic environment, they pose a potential threat to organisms that ingest them. Documented accounts exist that describe pellet and other plastic debris ingestion by wildlife, most notably by seabirds (EPA 1990a; Ryan 1990b) and sea turtles (Balazs 1985). Although the effects of the ingested pellets are not fully known, evidence suggests that ingestion causes diminished foraging ability, lost nutrition, and absorption of plasticizers. Several authors have documented the human esthetic (Gregory 1983; Wallace 1985; Wilber 1987; Klemm and Wendt 1990) and economic (Wallace 1985) impacts of pellets in the environment.

Although plastic pellets are one of the least noticeable forms of plastic pollution, they are ubiquitous in the oceans and on beaches. They have been reported in the sediments (Carpenter et al. 1972; Morris and Hamilton 1974) and the surface waters of coastal areas and oceans (Carpenter and Smith 1972; Wong et al. 1974; Gregory 1977; Ryan 1988a) throughout the world; data are limited regarding the presence of pellets in riv-

335

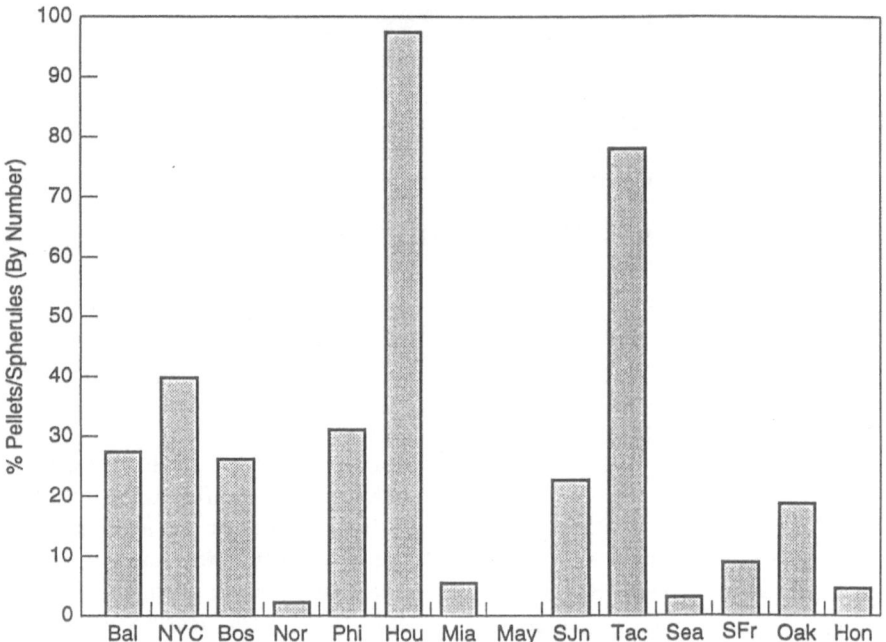

Figure 25.1. Summary of findings of the EPA Harbor Studies Program. *Bal,* Baltimore; *NYC,* New York City; *Bos,* Boston; *Nor,* Norfolk; *Phi,* Philadelphia; *Hou,* Houston; *Mia,* Miami; *May,* Mayagüez; *SJn,* San Juan; *Tac,* Tacoma; *Sea,* Seattle; *SFr,* San Francisco; *Oak,* Oakland; *Hon,* Honolulu.

ers, streams, and lakes. The widespread distribution of pellets is also demonstrated by their presence in remote areas of the world, such as beaches of the South Pacific (Gregory 1977) and Hawaii (EPA 1992b).

Since 1988, the EPA has conducted two studies, the Harbor Studies Program (EPA 1990b, 1992a; Trulli et al. 1990; Redford et al. 1992) and the Combined Sewer Overflow (CSO) Studies Program (EPA 1992c) [summarized in Chapter 23 by Trulli et al. this volume], that identified the presence of potential sources of resin pellets and other floating debris in harbors of the United States. The Harbor Studies Program and the CSO Studies Program provide direct evidence that pellets continue to enter the aquatic environment through storm sewers, CSOs, and direct spillage into waterways.

Harbor Studies Program

During the Harbor Studies Program, the EPA studied floating aquatic debris in 14 harbors along the coastal United States, Hawaii, and Puerto Rico, during a total of 21 surveys conducted between October 1988 and February 1992. The debris was collected by towing fine-mesh nets at the water surface to a maximum depth of 0.5 m. The netting mesh selected was small (0.3 mm) to facilitate collection of all types of resin pellets. More than 200 different types of manufactured debris (including resin pellets) were identified and counted. The program results were reported by the EPA (1990b, 1992a, 1992b) and summarized by Trulli et al. (1990) and Redford et al. (1992).

Plastic pellets were found in the harbors of 13 of the 14 cities surveyed (Fig. 25.1) and in 29 of the 32 sampling areas (two to four areas were sampled within each city). Plastic pellets were among the most common items found during the Harbor Studies Program, comprising approximately 94% (by number) of all debris collected. A wide variety of pellet shapes, sizes, and colors (mostly clear, white, or off-white) were collected. Visual assessments made by a polymer chemist confirmed that a variety of pure polymers and additive-containing pellets were found in the samples (EPA 1992d). Two additional pellet-related items, plastic powder and flattened pellets (pressed flat by railcar wheels), were collected in several cities.

Several pellet sources were identified based on field observations and conversations with local authorities. These sources included the following:

CSO and stormwater discharges: in the Kills area of New York Harbor, NY, the cleanliness and uniform size, shape, and color of the collected pellets indicated a possible single source, and the pellets were mixed with other debris typically discharged from storm sewers or CSOs. In Boston, MA, the majority of the pellets was collected from the Charles River on the upstream side of the locks. There is no commercial fishing on the river and no known pellet industries along the banks, which suggests that stormwater runoff or discharges are the likely sources of the pellets.

Stormwater runoff from plastics industries: in the Houston Ship Channel in Houston, TX, more than 700,000 pellets were collected during two surveys. Although pellets of many colors and shapes were collected, most of the pellets were clear, white, or off-white, and ovoid. Most of these pellets were found in Buffalo Bayou, an area inaccessible to shipping traffic and in the watershed of one of the highest concentrations of plastics industries in the United States (suggesting stormwater runoff as a possible source). Additionally, plastic powder (the raw material used to manufacture the pellets) was observed in high concentrations on the surface of the Channel.

Direct spillage into waterways: most of the pellets collected in Tacoma, WA, were found in a single sample. The pellets were of the same color (white), shape (ovoid), and size (approximately 5-mm diameter), which suggested that they were from a single or at least similar source. Local residents indicated that spillage at local loading and shipping docks occurred periodically and pellets were regularly observed on local beaches.

Commercial shipping or other international sources: during the pellet survey conducted along the Kahana Beach in Hawaii, plastic pellets were found in areas within the tidal zone farthest from the water and were most concentrated in obvious tide lines. Because of the predominant water currents in the vicinity of the Hawaiian Islands, it is improbable that pellets from local industry were deposited on Kahana Beach, and the pellets were likely transported by ocean currents from sources outside Hawaii.

Combined Sewer Overflow Studies Program

In older cities of the northeastern United States, CSO discharges of raw sewage and street litter are common during rainstorms. Studies conducted under the CSO Studies Program examined the types and amounts of floatable debris discharged from selected CSOs and storm sewers in Philadelphia, PA, and Boston, MA, and floatable debris captured by bar screens and floating in the scum of sewage treatment facilities in the two cities. Study results proved that pellets are present in municipal sewage, in CSO discharges, and in stormwater discharges. These findings support the findings of the Harbor Studies Program, which indicated that pellets enter municipal sewerage systems from land-based sources and subsequently may enter the aquatic environment through municipal discharges.

Pellet Sources

With the assistance of the Society of the Plastics Industry, Inc. (SPI), the EPA conducted a study to identify pellet-release points within the plastics industry. In 1990 and 1991, SPI arranged for each sector of the plastics industry—pellet producers, transporters and contract packagers, and processors—to be visited by the study team. Two producers, two contract packagers, and three processors voluntarily participated in the study. The representativeness of the visited companies as indicators of pellet release and containment conditions industry wide could not be determined, and the possibility exists that the visited companies represented best-case conditions. The fact that the two producers visited recognized the uniqueness of their containment systems and developed public relations materials highlighting the systems supports this possibility.

The flow of pellets through the plastics industry is summarized in Fig. 25.2. The pellet producers create the polymers and extrude the resin pellets. The pellet transporters and contract packagers are the mechanisms by which plastic pellets move from the producer to the processor. The pellet transporters convey bulk shipments of pellets from the producer to the contract packagers and processors and carry repackaged shipments from the contract packagers to the processors. Pellet processors remelt and mold the pellets into commercial and industrial plastic products.

Despite the unknown representativeness of the observed pellet-handling companies, the study team was able to identify several pellet-release points in each sector. Many of the release pathways apply to all the facilities, and other pathways are specific to one or two sectors. The release pathways can be categorized as to where deficiencies may exist: (1) communication between company and industry managers; (2) employee training; (3) containment systems and devices; (4) routine operations; (5) housekeeping practices; (6) packaging; (7) shipping practices; and (8) recycling. No attempt was made to rank the release points in order of significance or quantities released. A brief discussion of each pathway is presented next. The order of discussion does not indicate ranking, order of significance, or relative contribution of pellets released to the aquatic environment.

Communication Between and Within Sector Management

Not all company managers have recognized the pellet problem and the need to control pellet releases. Pellet spillage information, such as the condition of packages and the receipt of unsealed rail hopper cars (railcars consisting of a large container that is filled through ports at the top and emptied through valved ports at the bottom, and may be divided internally into two or more compartments) is shared among companies only occasionally. Officials of one contract packager stated that they occasionally receive feedback on package damage and pellet loss during transit, but that the majority of reported information is from international, not domestic, companies.

Employee Awareness and Training

Employees are generally unaware of the hazards posed by pellets and employee responsibility for causing and controlling releases to the environment. Officials of one processor were aware of SPI's past educational efforts, but the information was not disseminated or posted in the facility for workers to read, and workers had not been briefed or trained with regard to pellet-related concerns. These officials believed that their employees would be generally apathetic toward the pellet issue.

Although the need for employee education was met with skepticism at a few of the visited facilities, most officials were receptive to the need for increased employee awareness of the hazards of pellet releases into the environment. Officials at one of the contract packager facilities also recognized that employee awareness of the problem is directly related to the investment that management is willing to make in employee education. Some officials recommended that educational materials be made available in several languages.

Employees at any of the facilities also may lack the proper training to reduce the release of pellets into the environment. For example, a major release pathway is through package damaged caused by improper operation of forklifts, and the cause of packaging damage most frequently cited by the packagers was punctures by forklift tines.

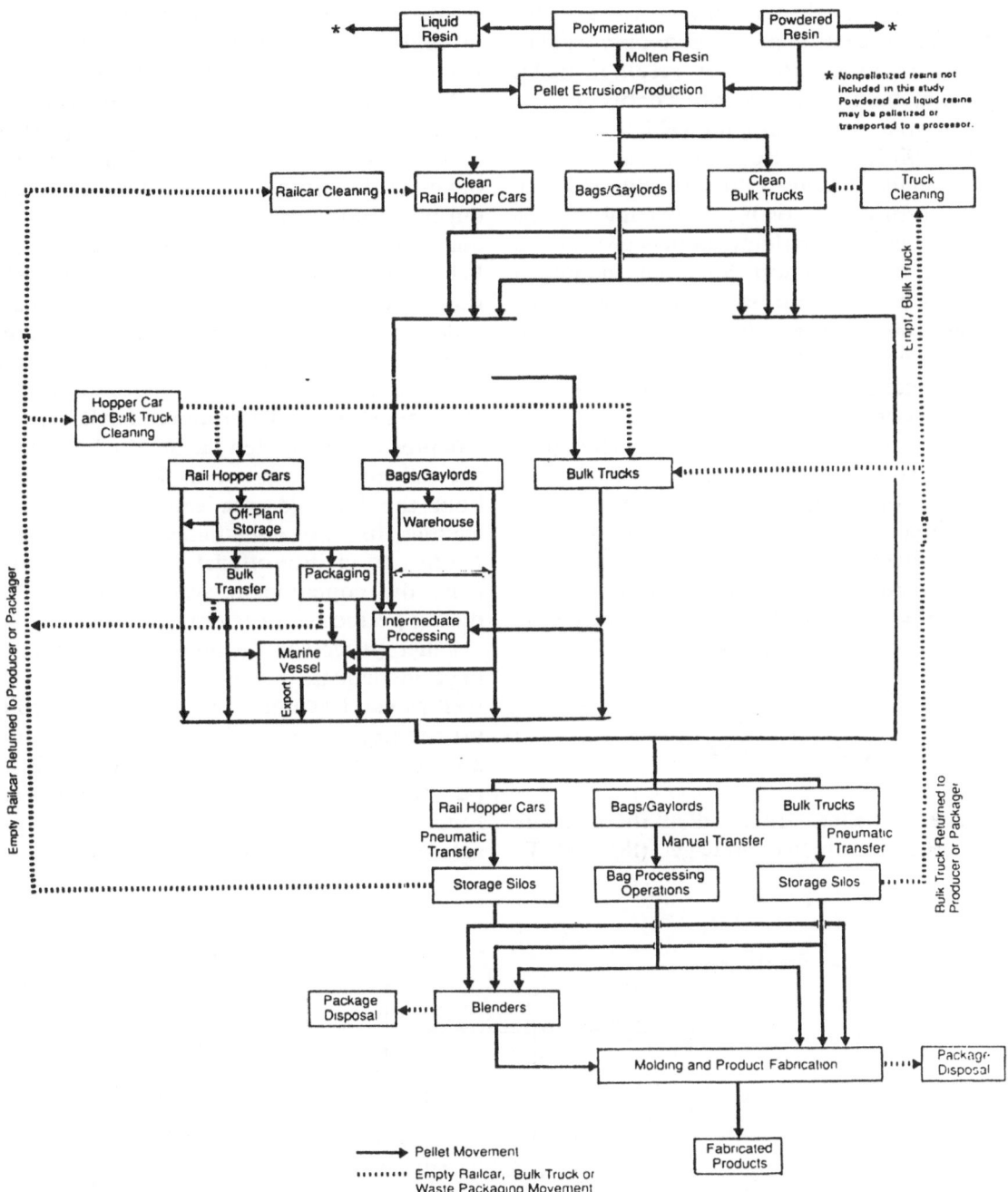

FIGURE 25.2. The flow of pellets through the plastics industry. *Solid line,* pellet movement; *dashed line,* empty railcar, bulk truck, or waste packaging movement.

Facilities

With the exception of the two producers visited during this study, companies have few or no cooling water, wastewater, or storm-water containment systems in place, such as portable screens or facility-wide containment systems. Pellets may be entrained in these waters and subsequently discharged into municipal storm and sanitary sewers or into natural drainage systems.

At contract packager facilities, the transfer of bulk shipments (e.g., in rail hopper cars) to storage or inside areas may result in pellet spills. Incomplete connections between conveying systems and shipping vehicles at packager facilities also are known to result in pellet spills. If there is no screening in place over storm drains, or if the existing screens are not cleaned routinely, pellets are carried by stormwater runoff into the municipal storm sewers or into natural drainage areas. Because six of the seven site visits were conducted during storm conditions, the transport of pellets in stormwater runoff was directly observed at six facilities.

During visits to the targeted pellet processing and packaging facilities, plastic pellets were observed in the stormwater ditches and interceptors for each type of facility. Areas where pellets are most likely to escape capture and enter the aquatic environment include parking lots, loading docks, receiving areas, and railroad sidings.

Routine Operations

Whenever pellets are handled there is the potential for pellet spillage (Fig. 25.3).

Manual pellet handling is more likely to result in spills than handling by pneumatic conveying systems (closed pipelines that move pellets by using compressed air). However, if pneumatic systems are not properly operated and maintained, pellets may leak through openings in the system and may also lead during routine maintenance, repairs, and quality control inspections of those systems. Pellets may be released also during the transfer of damaged, unrepaired packaging.

Spillage caused by carelessness in manual handling of the packaged pellets was observed in each sector of the plastics industry. During a site visit to producer facilities, pellets presumably spilled during the loading of the rail hopper cars were observed throughout the loading area and in the gutters leading to the containment system. According to facility officials, pellet overflows from the containment system are possible during periods of heavy rainfall.

Pellet processor facilities were found to have inadequate pellet-handling controls as well as careless unloading and warehousing procedures. Improperly warehoused packages (stacked haphazardly or too high or punctured while moving to storage) were found to result in pellet spills. Pellet packaging that is not carefully inspected before

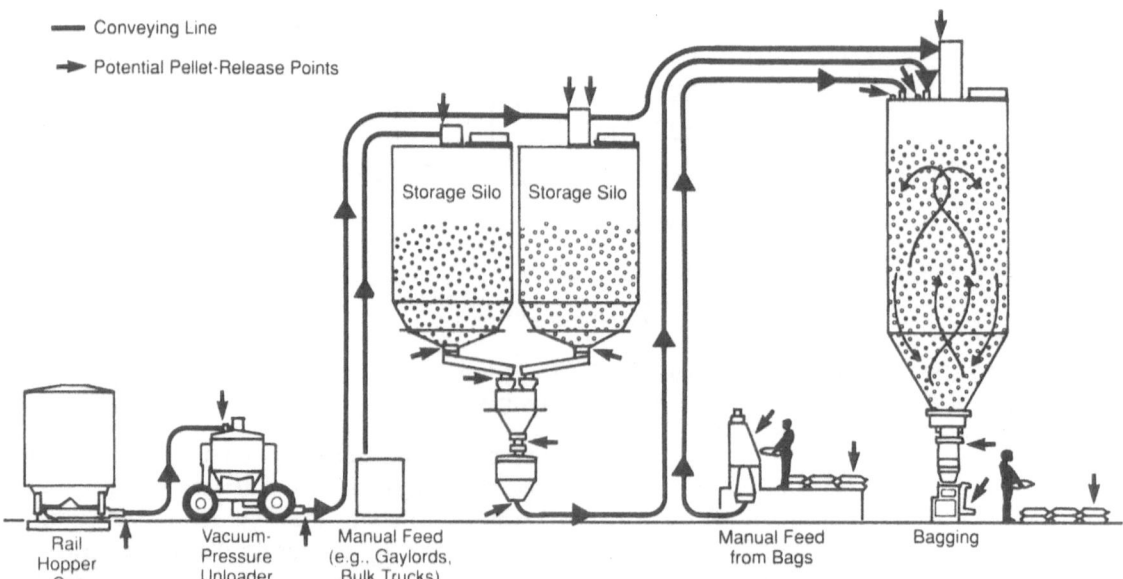

FIGURE 25.3. Potential pellet spillage points during manual and pneumatic transfer. *Solid line with arrowheads,* conveying line; *arrows,* pellet-release points.

offloading or is damaged during offloading may be spilled onto the ground and throughout the receiving area.

At the packaging facilities, plastic pellets are transferred from the rail hopper cars to the packaging areas by pneumatic conveyors. During transfer operations, pellets may spill to the ground and enter the aquatic environment despite the use of containment devices such as portable screens. Spill cleanup at these facilities was limited to only large pellet spills; stray pellets were not recovered.

Housekeeping Practices

Pellets that are not quickly recovered through good housekeeping practices may be lost to the environment. At producer facilities, rail hopper car and bulk truck cleaning operations may release pellets into the environment where they can be carried in the wastewater to the facility drainage system or washed away by stormwater runoff into areas outside of the facility. This is particularly significant at facilities lacking an adequate pellet containment system.

At processor facilities, loose pellets were observed throughout interior and exterior open areas and throughout the storage and warehouse areas. Pellets were also observed in areas where pellets are not normally handled (offices, printing shops, etc.) and in areas inaccessible to routine maintenance equipment (floor cracks, beneath molding machines, etc.). Typical housekeeping procedures are to dispose of broom-swept pellets into a trash dumpster. The dumpster contents are subsequently collected by a commercial waste hauler who disposes the pellet-containing waste into the municipal waste stream or a landfill. Once disposed, the pellets may become entrained in stormwater runoff or may be blown by wind into nearby waterways. Nearly all the facility officials believe that good housekeeping is the most effective way to control pellet loss.

Packaging

The mechanical process of packaging pellets and the use of fragile packaging material con-tribute to the release of pellets into the aquatic environment. Incompletely sealed fill valves on the packaging machines and poorly sealed packaging, including unsealed bag valves and seams, release pellets onto the floors of the facilities. At each of the two packagers, pellets were observed leaking from the valves or spouts of the bagging machines, incompletely sealed bag valves, or broken bags and were scattered on the floor beneath and in the vicinity of the machines. Pellets were also observed in the storage areas of the packagers.

The type of material used for packaging can also serve to contribute to pellet releases into the environment. Paper and cardboard are the two types of packaging material most commonly used throughout the industry. These materials are easily damaged during transport and handling, and, in fact, may be designed to be easily broken for quick loading of pellets into molding machines. Pellet spills resulting from breakage of the bags or gaylords (large cardboard boxes used for bulk shipments of pellets) seemed to be a common occurrence throughout the warehouses of the facilities visited, and loose pellets were obvious in every aisle. Damage to the packaging material may result from punctures by forklifts or improper storage (stacking too high or in an area of heavy traffic or activity). Officials at one packager believe that more pellets are lost from bag and gaylord breakage than from accidental spills or leaking bag valves.

Shipping Practices

A significant source of pellet loss to the aquatic environment is through inappropriate cargo handling and leaks from connections to rail hopper cars and bulk trucks, particularly when connecting and disconnecting the system and the valves. Pellets leak from unsealed valves and connection hoses that are not completely emptied before a rail hopper car or bulk truck is moved. Valve caps are also frequently not sealed after the hopper cars or trucks are emptied, allowing residual pellets to escape along the railroad right-of-way. Additionally, sealed or unsealed rail hopper-car ports may be opened by vandals, allowing pellets to spill along the railroad tracks.

In all sectors of the industry, most facilities receive bulk shipments of pellets in rail hopper cars from which pellets are transferred pneumatically to the storage silos. A few of the visited facilities placed screens or other containment devices beneath the valves when making and breaking pneumatic connections. When large spills occur, attempts are made to recover and recycle the pellets. However, many pellets are not captured or recovered at this point; they often become lodged between the bedding stones or scatter into adjacent areas and are subsequently carried by stormwater runoff into waterways.

Improper handling of pellet cargo at shipping docks also contributes to pellet losses. At the facilities visited, no permanent pellet containment systems were installed at the loading docks, and only one facility had placed screens over their parking lot storm drains. Pellets spilled at or near the docks could fall onto the outside pavement, be dispersed by wind or stormwater runoff, and enter the aquatic environment. At some facilities, screens similar to those placed under rail hopper cars were installed under shipping docks with concrete overhangs; however, docks lacking the overhang were not similarly equipped with screens, even within the same facility.

Recycling

Officials at one packager facility estimated that their annual pellet spillage is 10,000–30,000 lb during packaging and 2,000–20,000 lb during shipping and warehousing. All the facilities observed during the study practiced some form of housekeeping and waste pellet sweeping and collection. Some companies recycle the recovered pellets, while others do not recycle and, instead, dispose the pellets into the municipal waste stream. Disposal of waste pellets into dumpsters was noted at producer, packager, and processor facilities. Loose pellets in the municipal waste stream may be easily lost to the environment.

Employees at one processor are trained to minimize pellet spills for economic rather than environmental reasons, but collected pellets are not recycled. Instead of being reclaimed for renumeration, loose pellets are dumped into the municipal waste stream. Both the producer facilities, however, recognize both the environmental and economic advantages to recovering and recycling loose pellets. One producer estimates that between 25,000 and 60,000 lb of recyclable pellets are recovered (and recycled) each month from the rail-hopper-car cleaning area alone.

Conclusions

The seven site visits provided direct evidence of the continuing release of plastic pellets from all sectors of the plastics industry: producers, transporters and packagers, and processors. Some pellets may enter the aquatic environment directly, such as through stormwater runoff from facilities adjacent to waterways. Many pellets enter the aquatic environment after being carried into municipal drainage systems and released through CSO and stormwater discharges. The findings of the Harbor Studies Program and the CSO Studies Program also support this conclusion.

Controlling pellet releases can begin with proper training and education of plastics industry managers and employees, particularly by increasing awareness of the hazards posed by pellets and recognizing the economic benefits of controlling spills and recycling waste pellets. [In 1991, in response to this study, SPI initiated Operation Clean Sweep, a multimedia educational program designed to heighten industry awareness and promote a zero-pellet-discharge policy (Society of the Plastics Industry 1991).] Capital investments in containment systems may be necessary to control release at facilities that handle large volumes of pellets, but inexpensive control measures, such as portable screens or tarps, may be adequate for controlling releases at small-volume companies. All facilities could improve routine housekeeping measures by increasing the frequency of sweeping and including the use of vacuums to recover spilled pellets from hard-to-reach areas.

Existing U.S. regulations require controls over the release of all plastic materials, including pellets, into the aquatic environment. The recently revised Clean Water Act enables regulators to impose significant penalties on a company if pellets are present in their storm-water discharges in violation of the company's National Pollutant Discharge Elimination System (NPDES) permit. However, penalties alone cannot control the release of pellets; the penalties can only encourage companies to implement control measures. Ultimately, controlling pellet releases into the environment is the responsibility of the plastics industry, and effective controls should be continued and enhanced through voluntary industry programs.

26.

Implementation and Assessment of a Floatables Action Plan for the New York–New Jersey Harbor Complex

Paul J. Molinari

Introduction

In the summers of 1987 and 1988 significant amounts of floating debris, such as wood, paper, and medical wastes, washed up on the ocean beaches of New Jersey and the southern shore of Long Island, New York. These washups, while occurring for short periods of time, resulted in the closure of approximately 70 miles of ocean beaches in New Jersey in 1987 and approximately 70 miles of ocean beaches in New York in 1988. The beach closures, along with the public's perception of a fouled ocean, resulted in a $2 billion loss of revenue for the states of New Jersey and New York.

In November 1987, Region II of the U.S. Environmental Protection Agency (EPA) began a 3-month helicopter surveillance and a series of onsite investigations of floatables accumulations in the New York and New Jersey harbor complex (Fig. 26.1). Scientists mapped the estuaries and shorelines that were most heavily impacted and used specially marked bottles, painted floating objects, and transect lines to monitor tidal action and its effect on the movement of floatables.

This study determined that floatable pollution takes two distinct forms: dispersed quantities of free-floating garbage and wood, and floating slicks of concentrated garbage and wastewater. The dispersed floatables appeared to have a wide variety of sources, most of which contribute to the accumula-tion on a daily basis. These sources included litter from beach and pleasure vessels, debris from marine solid waste transfer operations, stormwater runoff, combined sewer over-flows (CSOs), and wood from decaying piers and vessels. Floating debris slicks appeared after rainstorms causing combined sewer overflows (CSOs) and stormwater discharges. These slicks are also formed when the high tides caused by a full or new moon resuspend the floatables deposited on the shoreline, carry them out to sea, and concentrate them. The largest observed debris slicks occurred when a resuspension and storm event occurred at the same time.

In August 1988 an interagency workgroup of local, state, and federal agencies was formed to develop an action plan to mitigate the floatables problem. During the winter, the Floatables Action Plan was adopted as a pilot project for the period of May through September 1989. The objectives of the plan were to minimize debris escaping the harbor, minimize beach closures from floating debris, maintain an effective communication net-work to coordinate slick removals, and supply timely notification of beach operators of potential debris washups.

Floatables Action Plan

The plan consisted of four key elements: surveillance, regular cleanups following moon tides and storms, nonroutine cleanups, and a communication network to facilitate the use of available resources.

The surveillance element concentrated on detecting slicks forming within the harbor where they could be collected. Surveillance activities were carried out by the EPA, New Jersey Department of Environmental Protection (NJDEP), and the U.S. Coast Guard (USCG) as follows:

1. The NJDEP helicopter provided daily surveillance of the New Jersey coast and the Lower Harbor, Monday through Sunday except for Wednesday.
2. EPA:
    a. Helicopter surveillance daily of the Harbor Complex and of the New Jersey and Long Island shores as part of the normal water quality monitoring program (Monday through Saturday).
    b. Patrol of the Harbor twice per week using the research vessel (RV) *Clean Waters*.
3. USCG:
    a. Helicopter surveillance of the New York Bight 3 days per week.
    b. Daily vessel patrols of the harbor complex and weekly vessel patrol of the New York Bight.

An integral part of the plan was the regular removal of debris from the harbor at the Narrows and the lower harbor outflow of the Arthur Kill during daylight hours on the day before, the day of, and the day after full and new moon tides, and after significant storms that caused overflow of combined wastewater. Vessels of the U.S. Army Corps of Engineers (CoE) removed debris with vessels using nets with mesh openings of less than 1.75 in. The New York City Department of Sanitation (NYCDOS) supplied a barge near one of its marine transfer stations to transport the collected debris to the Fresh Kills landfill for disposal.

An additional aspect of the plan focused on the capture of slicks spotted at other locations and times within the harbor. The CoE and fishing vessels conducted debris removal as instructed by EPA; however, collection was only possible landward of the Sandy Hook–Rockaway Point transect because of the dispersion of floatables (which made collection unlikely) and heavy seas. For slicks that were observed beyond the Sandy Hook––Rockaway Point transect, a computer model was used to predict potential impact areas [see Swanson and Bortman, Chapter 24, this section], state floatables coordinators were informed of a potential washup, and local authorities were also notified.

A communications network (Fig. 26.2) was established for reporting sightings of floatables. An EPA floatables coordinator functioned as the center of the reporting network and coordinated debris removal. All agencies involved in the operation were available 24 h a day through the use of a hotline and paging system.

Assessment of the Pilot

Despite all the rainfall the area received during the time the plan was in action, only two stretches of ocean beaches along the Long Island and New Jersey shorelines were closed during the bathing season as a result of floating debris washing ashore. On July 20, debris, including syringes, washed ashore on the beaches of Ocean City, NJ. The beaches were reopened the next day after cleanup. On August 18, after torrential rains, a full moon, and a lunar eclipse, a huge slick as wide as 100 ft in some sections, again including syringes, washed ashore on the ocean beaches of Gateway National Park, NJ, and the beaches were closed. They were reopened after 2 days and the collection of 125 syringes.

These beach closures represented a drastic reduction in the miles of ocean beaches closed during the summers of 1987 and 1988. In 1987, two floatables incidents alone closed 70 miles of ocean beaches in New Jersey for 2–3 days and in 1988, floatables incidents closed 70 miles of ocean beaches in New York. The reduction of the beach closures may be attributed partially to the Floatables Action Plan. The CoE, using their small-mesh nets, collected more than 540 tons of debris, of which more than 450 tons were collected during the regular and nonroutine cleanups

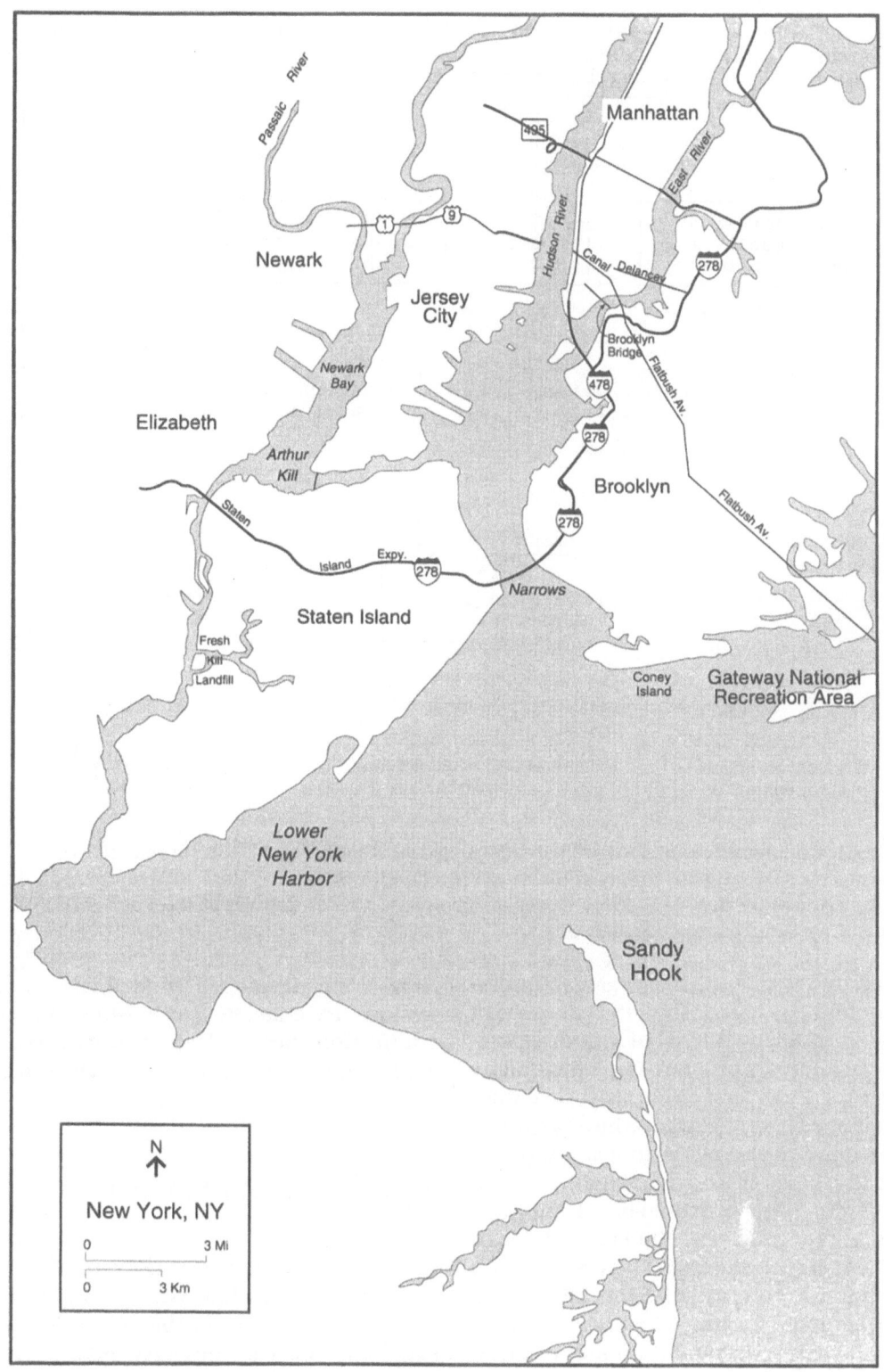

FIGURE 26.1. The New York–New Jerey Harbor complex.

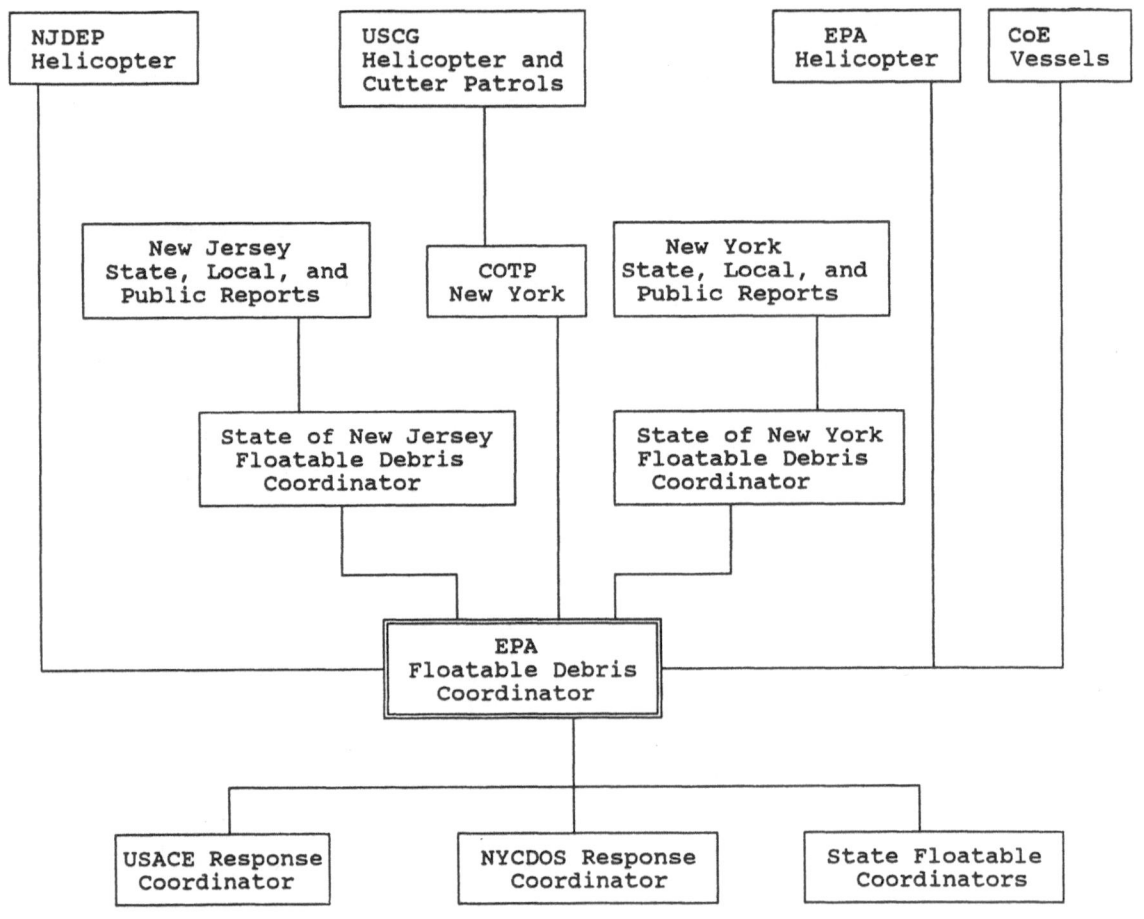

FIGURE 26.2. Communications network for reporting and responding to floatable-debris slicks. *NJ-DEP*, New Jersey Department of Environmental Protection; *USCG*, U.S. Coast Guard; *EPA*, U.S. Environmental Protection; *CoE*, U.S. Army Corps of Engineers; *COTP*, Captain of the Port, NYCDOS, New York City Department of Sanitation.

specified by the plan. Other organized operations netted 165 barrels (bbl) (55 gal each) of household trash and 30 yd^3 of wood. Operation Clean Shores, a program implemented by the State of New Jersey whereby floatables were removed before entrance into the water from 45 miles of shoreline from south of the George Washington Bridge to Highlands, NJ, from March through September 1989, was responsible for removing 3000 tons of debris from New Jersey shorelines. This program used minimum security prisoners along with the NJDEP and local municipality personnel. These activities reduced the amount of material escaping from the harbor complex.

In addition, the states of New Jersey and New York developed guidelines and held sessions to educate beach operators about notification, beach cleanup procedures, and the handling and disposal of medical wastes.

The Reasons for Success

The CoE utilization of smaller mesh nets on their drift collection vessels was a major factor in the success of the Floatables Action Plan. The specially designed nets proved to be very effective in capturing slicks without increasing the drag on the vessels. The collected material was estimated to contain 90% wood and 10% other floatable materials such as plastics, Styrofoam, paper products, tires, grasses, and reeds. This 10% other floatable materials was a significant increase from the minimal amounts collected in previous years (using nets with larger mesh openings).

The communications network that was established to coordinate the scheduling of debris collection was one of the keys to the success of the Floatables Action Plan. It greatly enhanced the efficiency of multiagency efforts to respond to floatables problems. As an example, use of helicopters as spotters allowed the extent of a slick to be quickly determined and for a vessel to be directed to the optimal collection point.

During normal business hours, communications coordinators received surveillance reports from helicopters and vessels and relayed the location and cleanup priority of slicks so that the nearest drift collection vessel could be dispatched for cleanup. Only once in more than 100 sightings was a drift collection vessel unable to locate a reported slick. During nonbusiness hours, the communications network functioned through the hotline and paging system without difficulty, with the exception of an occasional reporting of a slick too small to warrant a response.

The computer model to predict landfall of floating debris was used three times during the implementation of the plan (Table 26.1). Response time for obtaining predictive reports varied from several hours to a day, depending on the time of the request. On two occasions, the model correctly predicted the path of the slick.

Improvements to the Plan

A meeting of the agencies involved in implementing the Floatables Action Plan was held to discuss the successes and problems of the plan and to suggest improvements that could be made to the plan. Recommendations in the following areas were developed to make future cleanup operations more effective.

Floating Debris Cleanup Activities

The cleanup schedule was modified to replace the first-day-after cleanup for new and full moon tides with a second-day-after cleanup, because few slicks had been reported on the first day after and numerous slicks had been spotted on the second day after the high tides. The repositioning of the drift collection vessels from the outflow of the Arthur Kill into Raritan Bay to the Newark Bay outflow into the Arthur Kill also was recommended.

Communications Network Improvements

To ensure consistency of reporting slicks, the development of a 1-h training session for all those involved in surveillance was recommended. Expanded use of computer models and the installation of cellular phones on

TABLE 26.1. Results of computer model predictions during the implementation of the Floatables Action Plan.

Date	Description	Prediction	Observations
May 23, 1989	Slick moving out of harbor, through the Narrows	The debris would disperse and stay in harbor	The debris dispersed and no beaches were closed
August 10, 1989	A narrow slick located 3 miles east of Sandy Hook	The debris would washup on eastern shore	The slick could not be located because of poor weather, and no beaches were closed
August 17, 1989	Large slick escaped harbor through the Narrows	The debris would land fall at Sandy Hook on August 18, 1989	Gateway National Park–Sandy Hook closed beaches at 4:00 P.M. on August 18, 1989

TABLE 26.2. Amounts of floatable debris collected during the Floatables Action Plan and the New Jersey Department of Environmental Protection (NJDEP) Operation Clean Shores, 1989–1993.

Year	Tons of debris collected during the Floatables Action Plan	Tons of debris (and miles of beach covered) during the NJDEP Operation Clean Shores	Rainfall, percent above normal[a]
1989	541[a]	3000 (45)	225
1990	795[b]	4800 (48)	42
1991	701[b]	4688 (74)	28
1992	958[b]	5789 (84)	18
1993	1088[b]	5750 (67)	—

[a]May 15 to September 15 only.
[b]Year-round collection.

surveillance vessels were also suggested as areas of improvement.

Additional Recommendations

It was decided that the Floatables Action Plan should be continued throughout the year, with a less resource-intensive effort during the winter that would integrate the plan's activities into the normal winter wood collection effort. For debris slicks sighted in the ocean, a tracking system to verify movement versus model prediction would need to be implemented. The use of satellite-tracked drifters, similar to those used to track oil plumes, to reliably track the movement of debris slicks was suggested but never implemented.

Subsequent Years

In calendar year 1990, the foregoing modifications were implemented and the plan went into year-around operation. Significant amounts of debris (Table 26.2) continued to be collected during the summer and winter

months of the following years. The state of New Jersey has continued and expanded its Operation Clean Shores, removing tremendous amounts of debris from the shoreline. Beach closures were small isolated occurrences (Table 26.3).

The Floatables Action Plan has played an integral role in preventing the repeat of the large numbers of beach closures that occurred during the summers of 1987 and 1988 and in keeping the beaches clean of floating debris; however, this plan and other cleanup programs are all stopgap measures until long-term solutions can be instituted to correct the sources of the debris entering the harbor.

TABLE 26.3. Number (and miles) of beaches closed before (pre-1989) and during (1989–1993) the New Jersey Department of Environmental Protection Operation Clean Shores.

Year	Number of beaches closed	Approximate miles of beach closed
1987	Numerous	70
1988	Numerous	70
1989	2	<4
1990	1[a]	<1
1991	1	<1
1992	2	<1
1993	0	0

[a]As the result of illegal disposal of medical waste.

27.
The Control of Floating Debris in an Urban River

Emmett Durrum

Introduction

Floating debris interferes with the safe and healthy use of the Anacostia and Potomac rivers. Besides being unsightly, debris is destructive to dams, bridges, power plants, municipal water systems, recreational and commercial boats, and boating. Floating debris disrupts aquatic vegetation, which stabilizes shorelines and wetlands, aids in the removal of certain pollutants, and increases dissolved oxygen content in the water column. It interferes with the establishment of aquatic plants, which, in turn, limits spawning areas and habitat for fish and benthic organisms. Floating debris, particularly fragments of plastic and fishing line, is a hazard to wildlife both through ingestion and entanglement. This debris profoundly detracts from the natural beauty and healthy recreational enjoyment of the rivers of Washington, D.C. Finally, floating debris that is not trapped and removed in the Anacostia and Potomac rivers eventually enters and degrades the waters of the Chesapeake Bay and the open Atlantic Ocean.

The Anacostia River was the focus of a floating-debris control study initiated in August 1992. Restoration of the Anacostia River has received broad political and grassroots support since the early 1980s. This support culminated in the signing of the Anacostia Watershed Restoration Agreement of 1987. The signatories of the agreement, the District of Columbia, the state of Maryland, and Montgomery and Prince Georges counties (Maryland) have committed themselves to the following goals for the Anacostia and its tributaries:

Achievement of improved water quality and protection of aquatic life and habitat

Basinwide management of erosion, sediment, and other sources of pollution

Maintenance of the tidal portion of the Anacostia River as a navigable waterway for commercial and recreational activities

Expansion of opportunities for public recreational access

Enhancement of public interest and public participation in restoration activities

Propagation of the vision that the "entire Potomac and Anacostia River system should be a constant source of natural enjoyment, urban orientation, and visual delight".

The formulation of a floating-debris control program was a logical step in reaching the stated goals of the Anacostia Watershed Restoration Agreement by removing the most visible pollutant in the Anacostia River: floating debris.

The tidal portion of the Anacostia River is located in east Washington, D.C. The Anacostia River drains some 170 square miles of urban and suburban Montgomery and Prince Georges counties, Maryland. The Anacostia Basin is a part of the Potomac River Basin (Fig. 27.1). The tidal portion of the Anacostia

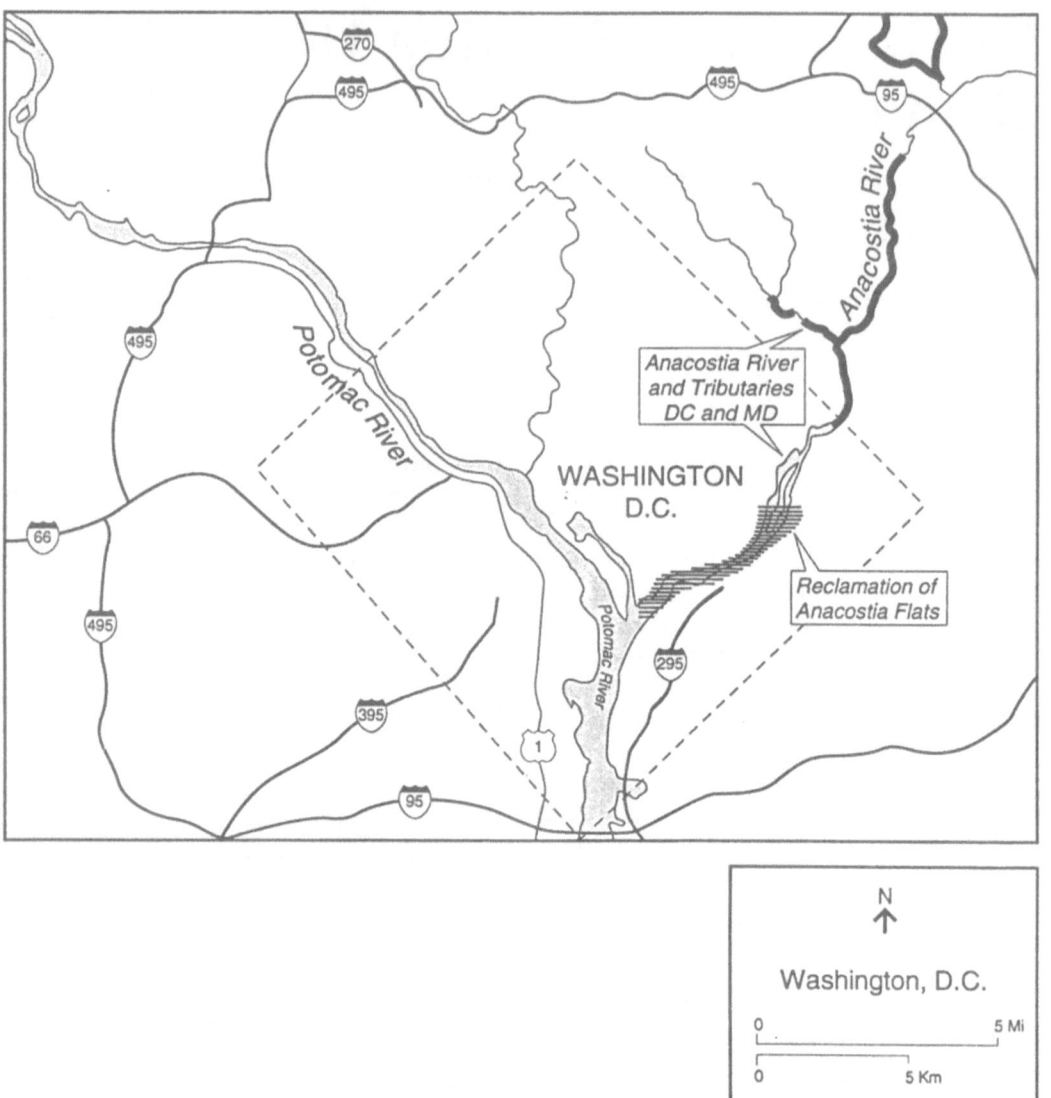

FIGURE **27.1.** Floating-debris control study area, including the tidal Anacostia and Potomac Rivers and their tributaries.

River has become a kind of trash sink for the entire Anacostia watershed. The glaring need for floating-debris control can be confirmed by observing the channel, fringe wetlands, or marinas of the Anacostia after any significant rainfall (Fig. 27.2).

The U.S. Environmental Protection Agency (EPA), after a review of a grant application from the District of Columbia Department of Public Works (DCDPW), agreed to provide $290,000 (100% federal) to do a pilot floating-debris control program on the Anacostia

River. The Director of the DCDPW authorized the program and provided administrative and operational support. The objectives of the program were as follow:

1. To determine the main sources and extent of the floating trash in the Anacostia River and Potomac river
2. To evaluate, test, and document methods and equipment used in floating-trash control appropriate for the currents, tidal influences, and hydrology of the Anacostia

Figure 27.2. Floating debris on the Anacostia River (Prince George's County, MD), 100 m north of the New York Avenue Bridge, 1992.

and Potomac river systems or similar systems
3. To establish a permanent floating-debris control program (through the experience gained during the EPA-funded pilot study) for the District of Columbia.

Methods

The study area includes the tidal Anacostia and Potomac rivers and their tributaries (see Fig. 27.1). Emphasis was placed on the Anacostia River because it is severely impacted by floating trash. The work undertaken to meet the objectives included reviews of floating-debris assessment and control measures used in other jurisdictions and evaluations of their potential for effective use in the study area; evaluations of various floating-debris collec-

tion schemes to determine their effectiveness, ease of operation, and overall operating cost; and considerations of the influence of the hydrology and geomorphology (e.g., strong tidal influences and low flow rate) of the Anacostia River system when evaluating debris control measures.

Efforts to assess the sources of floating debris included surveys of sewer outfalls and small streams that feed the Anacostia and Potomac rivers to identify major sources of debris. Containment booms were placed at suspected sources to trap and monitor debris. Also, marina operators, boaters, fishermen, and residents were encouraged to report debris sources on survey forms that were distributed by the Department of Public Works.

The operational work, necessary to meet the objectives, included the integrated use of mechanized debris removal equipment including a trashskimmer boat, an Argo amphibious vehicle, and spotter or manual collection boats. Weekly reports were prepared to identify the amount of debris removed and any operational problems encountered, and weekly meetings or telephone conferences were held to assess program effectiveness and consider new approaches to river and wetland cleanup.

Results of Objective 1 (Determine the Main Sources and Extent of the Floating Trash . . .)

Program spotter boats observed large accumulations of floating debris in virtually every part of the study area. Depending on precipitation, currents, and tidal influences, debris collected along shorelines, in mudflat and delta areas, and along fringe wetland areas. Large floating trashmats were common after any significant rainfall. Trash accumulations consisted of nearly any material that will float. Common items included glass and plastic bottles, toys, balls, motor oil containers, antifreeze containers, construction materials, all types of Styrofoam food containers, sanitary items, dead fish, and plastic bags.

Anacostia River problem areas include the shoreline near the New York Avenue Bridge, the mouth of Litter Beaver Dam Creek, the shoreline along the National Arboretum, the mouth of Hickey Run, Children's Island, Kenilworth Marsh, Kingman Lake, the PEPCO Power Plant cooling intake, the River Terrace shoreline area, Conrail Bridge, the Anacostia Park boat ramp, the Main Sewage Pumping Station area, Washington Navy Yard, Seafarers' Boat Club, Buzzards Point, and James Creek Marina. Other problem areas are the Washington Channel—particularly around the seafood market—Gangplank Marina, Washington Marina, the Fort McNair main seawall, the Watergate (near the Lincoln Memorial), and the mouth of Rock Creek.

The drainage basin of the Anacostia is approximately 170 square miles of heavily urbanized terrain. Any floatable material discarded in the Anacostia Basin can enter the tidal Anacostia. The District has approximately 145 storm sewer outfalls and 60 combined sewer outfalls (CSOs). The combined sewer system, separate storm sewer system, and Anacostia tributaries all transport floatable debris into the tidal Anacostia and Potomac after heavy rains. Debris entering the Potomac is flushed downstream by swift currents. Debris entering the Anacostia tends to stay for long periods because of low flow, weak currents, and tidal influences. Significant quantities of debris enter the Anacostia from the northeast and northwest branches of the Anacostia in Prince Georges and Montgomery counties, Maryland.

Results of Objective 2 (Evaluate, Test and Document Methods and Equipment . . .)

Floating-Debris Control Methods

The floatable-debris collection work employs a series of floating-debris booms that trap trash at stormwater outfalls, combined sewer outfalls, bridges, tributaries, and other locations where floating debris accumulates along the Anacostia and Potomac rivers. A trashskimmer boat (Fig. 27.3) is then used to collect the trash and deposit it in a holding area. The program uses an amphibious eight-wheel vehicle to collect debris from mudflats and

FIGURE 27.3. A trashskimmerboat in operation on the Anacostia River (District of Columbia) near the Conrail Bridge, 1992.

shallows. Several other work boats are used to locate work sites and maintain the collection booms.

The study area has approximately 8 square miles of water surface and 60 miles of shoreline with certain areas only accessible at low tide. To cover this large area it is necessary to trap and contain debris until it can be removed. Of the several options considered, debris booms are the most cost effective and practical containment measure.

Booms are relatively inexpensive (average $12/ft), easy to transport and deploy, and very durable. In case of storm flows one end of the boom can be secured with line designed to break and prevent destruction or loss of the boom. Booms do not cause stormwater outfalls to back up. Booms can be used to deflect or capture debris before it enters areas where it is hard to retrieve such as marinas, shallows, and mudflats. During the program, booms were deployed at outfalls, bridges, and tributaries. Trapped debris could then be collected when the trashskimmer was available and tides were favorable.

The most effective method of removing debris trapped in containment booms is with a boat specifically designed for this purpose. By lowering the conveyorized wings of the trashskimmer over the boom and operating the conveyor system, debris is drawn into the main boat conveyor for later offloading. Small quantities of debris can be removed by hand with dip nets working from utility boats. By simultaneous operation of the main trashskimmer conveyor and an offloading conveyor at the dock, debris is transferred to a dumpster for later disposal. Sorting of recyclable materials should occur at this point.

Equipment Used In the Project: Vessels

The trashskimmer used in this program is designed to collect and store solid debris floating on top of or just below the water surface. It has a waste-carrying capacity of 2400 lb or 150 ft^3. Included with the boat is an offloading conveyor system, heavy-duty transport trailer, and two gasoline motors

needed to operate the conveyor and trailer. The skimmer is designed to pick up floating objects in the upper 30 in. of water. The program also employed an amphibious vehicle capable of operating on dry land, sandbars, and mudflats, and traversing calm water of any depth. It was used to remove tires and other large embedded debris from shorelines and mudflats.

Two utility boats were used during the study. One was a 19-ft fiberglass utility boat with a center console, heavy-duty transport trailer, and a 100-HP commercial-duty outboard engine. This boat was mainly used for transporting crew and work materials, deploying containment booms, identifying and marking stormwater outfalls, and conducting inspections of all sections of the river for potential trash collection and boom deployment sites. The other boat was a 15-ft fiberglass utility boat with bench seats, heavy-duty transport trailer, and a 60-HP outboard engine. This boat was used primarily to patrol the river searching for trash concentrations, checking boom deployment, identifying and marking stormwater outfalls, and transporting supervisory personnel to locations on the river as needed.

Equipment Used In the Project: Containment Booms

More than 0.5 mile of containment boom (2700 ft) was purchased and evaluated for program use. Five different types of containment booms were tested. These booms were used to trap and contain debris for later pickup and disposal. Booms were also used to deflect debris away from sensitive areas (Fig. 27.4).

Containment booms of differing widths (10–24 in.) were ordered to determine the correct depth for the most effective control of debris. One type of boom, designed for containment of debris in flowing waters (e.g., rivers, canals, etc.), is 18 in. wide with a perforated skirt. Another boom was designed for use in shallow water, with water chambers that enable it to conform to the bottom contour during low tide conditions and float

FIGURE 27.4. A containment boom trapping debris on the Anacostia River (District of Columbia) near the Conrail Bridge, 1992.

on the surface during flood tide conditions. All booms were manufactured in 50-ft sections.

Results of Objective 3 (Establish a Permanent Floating Debris Control Program . . .)

In March 1993, the District of Columbia's Department of Public Works completed the demonstration program to assess methods of floating debris control in the Anacostia and Potomac rivers. The Demonstration Program, which ran from August 1992 until March 1993, was completely funded by EPA under Clean Water Act (CWA) section 104(b)(3). Based on preliminary results (2000

tires, 500 yd^3 floatable debris collected with the trashskimmer boat, 400 30-gal trash bags collected from the shorelines, and the obvious effectiveness of debris booms) the Director of the District's Department of Public Works requested the Water and Sewer Utility Administration (WASUA) to establish and fund a permanent floating debris control program, the Floating Debris Control Branch (FDCB), as a part of WASUA's Bureau of Sewer Services.

The Floating Debris Control Branch (FDCB)

The Floating Debris Control Branch (FDCB) was established in March 1993 as part of the Water and Sewer Utility Administration, Bureau of Sewer Services. The Floating Debris Control Branch's mission is to remove accumulated floating and other debris from the Anacostia and Potomac rivers and adjacent wetlands. The FDCB also attempts to prevent reaccumulation of debris through the widespread use of debris containment booms. In the long term, the FDCB will expand its service area, develop public information and awareness programs, and assist other area agencies with river management activities (e.g., pollution control and response, as well as fire, salvage, and navigation).

The permanent program is based on the methods that were proven effective during the demonstration phase. Six of the original staff members from the demonstration phase have been retained for the permanent program. When fully staffed, the FDCB will employ a supervisor, four boat operators, four deckhands, and an administrative assistant. All crew members will be required to be certified in boating safety and thoroughly trained in debris removal techniques. The FDCB operations site is now located near the Pennsylvania Avenue Bridge in southeastern District of Columbia. Additional offloading and disposal sites are being sought to increase FDCB efficiency.

Equipment

The FDCB program equipment includes all the equipment used during the demonstration phase: the trashskimmer boat, the amphibious craft, the 19- and 15-ft utility boats, and the 2700 ft of containment booms. Additional items include the following:

A third-generation, large trashskimmer boat with improved reliability and five times the capacity of the demonstration boat, which will increase the range and efficiency of the FDCB, was delivered in August 1994.

Two Navy surplus landing craft (LCMs) that are each 65 ft long and 20 ft wide with a capacity of 150 tons and shallow draft are being modified for use as work platforms and self-propelled barges. Cranes and davits as well as a shallow-water anchoring system are being installed on the LCMs. As modified, they will be ideal for removing embedded tires and large, cumbersome debris from mudflats. The large capacity of the LCMs will enable the crew to work more efficiently because the frequency of trips to disposal sites will be reduced. They can also be used as a motorized barge or tender in conjunction with the trashskimmer boats. The trashskimmers can offload debris directly to the LCMs, which will reduce or eliminate downtime (travel time) to disposal sites.

Another 12,000 ft of containment booms, delivered in November 1993, allow a total of 15,000 ft of containment booms to be deployed. The additional booms will allow the FDCB to service a much larger area.

Budget

For fiscal year 1994 (FY94), personnel, maintenance, and operational costs will be approximately $400,000. Additionally, a one-time capital cost of $500,000 for a 12,000-lb-capacity trashskimmer boat and 12,000 ft of debris booms will be incurred during FY94. To date, the District is the only jurisdiction in the Anacostia watershed to fund a permanent, fulltime, floating-debris abatement program for the Anacostia and Potomac rivers and their tributaries. Because much of the debris that enters the tidal Anacostia originates in Prince Georges and Montgomery counties, the District is seeking funding from the state of Maryland to help offset FDCB operations costs.

Conclusions

Control of floating debris should be a basinwide endeavor. Preventing the generation and transport of trash to receiving waters should receive the highest priority. Once debris enters a waterway, retrieval and proper disposal are the only way to prevent the inevitable degradation of marine habitat. Prevention is more cost effective than direct removal. However, so long as littering, illegal dumping, and CSO overflows are commonplace there will be a need for direct removal of floating debris.

This project (through the use of the floating-debris control methods) clearly establishes that floating debris can be successfully trapped and removed from urban waterways before it reaches the marine environment. The lessons learned during this project can be carried over to similar projects on other rivers. For example, based on demonstration project results, the amphibious craft was dropped from the permanent program because it was not durable enough for sustained use, and other amphibious craft need to be evaluated for use in the permanent program.

It is very difficult to estimate the total floating-debris load carried by the Anacostia River and therefore to determine the actual percentage of debris removed by the program. It is known, however, that since the inception of the project in August of 1992 the District's floating-debris demonstration study and the permanent Floating Debris Control Program have removed a total of 600 tons of floating trash and approximately 10,000 tires (as of August 1995), that is, 600 tons of debris that was intercepted before it could further impair the Anacostia River or degrade the lower Potomac River and ultimately the marine environment.

Another measure of program success is provided by residents living near the Anacostia and recreational boaters (there are six marinas on the tidal Anacostia) who have commented that it is markedly cleaner since the Floating Debris Control Program became fully operational. The EPA has commended the project as an excellent example of an effective environmental management program within an urban setting.

Finally, the program has been included as part of a series of field trips for international delegates to the UNEP Intergovernmental Conference on the Protection of the Marine Environment from Pollution from Land-Based Activities, held in Washington, D.C. from October 23 to November 3, 1995.

28.
Linkages Between Land-Based Sources of Pollution and Marine Debris

Michael Liffmann and Laura Boogaerts

Introduction

While acknowledging that marine pollution arises from the actions of man, research indicates that this pollution cannot be attributed solely to man's activities performed directly in the oceans. This paper identifies the primary land-based sources of marine pollution and discusses the main coastal rural and upland activities that affect the discharges of plastic waste and other floatable debris into the marine environment. Particular attention is given to the coastal zones within semienclosed areas such as the Wider Caribbean that are increasingly showing signs of pollution. The paper concludes with some ideas concerning a framework to address control and reduction of these sources.

Although experts agree that marine pollution is derived mainly from land-based sources, virtually all our attention, in regard to marine debris, has focused on ship-originated waste and its fate when cast overboard. Much of the interest can be attributed to the 1987 ratification of Annex V of the International Convention for the Prevention of Pollution from Ships (MARPOL 73/78) and the passage of national laws, such as the United States Marine Plastics Pollution Research and Control Act (MPPRCA) of 1987, which implemented Annex V. Annex V and the MPPRCA, by their content, focus on vessel-source pollution.

The two main, land-based sources of plastic and other floatable pollutants are (1) waterborne municipal and industrial wastes and (2) the terrestrial litter and refuse that are transported to sea by river systems from upland and coastal sources. Precise data concerning the primary sources of land-originated, ocean-bound waste are not available, but some estimates concerning ship-originated versus land-based marine pollution sources have been ventured. The United Nations Joint Group of Experts on the Scientific Aspects of Marine Pollution (GESAMP) has concluded that, globally, 80% of marine pollution stems from land sources (Windom 1992). Kindt (1984) noted that worldwide ". . . the amount dumped by vessels may represent only about a tenth of the pollution entering the oceans . . .". M'Gonigle (1990) speculated that 90% of all pollutants in United States coastal waters are derived from land-based sources.

Unfortunately, very few hard data have been collected to support these conclusions. There is a small body of research on sources, types, and transport of coastal and upland floatable debris and its eventual fate. But by and large, most investigators that have addressed land-based sources of marine debris have relied on speculation, considerable anecdotal information, and scant estimates.

From this research, however, it can be safely concluded that the world's large coastal population centers are the primary offenders. Increased urbanization and population growth have exacerbated the impact of waste created by the traditional coastal concentrations of vessel traffic, commercial fishing, manufacturers, and riverborne waste discharges. They have created a worldwide solid waste problem that has been generated by ". . . literally billions of consumers disposing of a myriad of plastic containers, wrappers, and other debris" (Bean 1987).

A New York Harbor floatable materials study conducted between 1989 and 1991, for instance, concluded that most of the materials washed ashore consisted of items typical of street litter: plastics (41%), polystyrene (25%), and paper (19%). The study estimated that 85% of the floatable debris entered the Harbor via combined sewer overflows and storm sewers (HydroQual 1993). Wade observed that the abundance of litter on beaches of the Kingston (Jamaica) Harbour area related to their distance downcurrent from the city of Kingston (Wade et al. 1991).

The plastics industry has been a direct source of plastic pollutants in the form of pellets, the feedstocks used by manufacturers of plastic products. The pellets enter the ocean via rivers and industrial outfalls, and from surface runoff from trucks, trains, and ships during the product's onloading, offloading, and transport phases (Pruter 1987a; Redford et al., Chapter 25, this volume). The industry is aware of this worldwide problem and is seeking to mitigate it.

While the coastal urban centers might be the most notorious contributors, there is evidence that marine debris and other forms of pollution can also be attributed to many people who do not live close to the sea or, for that matter, even visit the beaches. Worldwide, 6 of 10 people live within 60 km of the coast and the population of the coastal zone is projected to double in the next generation (Karau 1992). In the United States alone, 50% of the population resides within 50 miles of the ocean or one of the Great Lakes. Pruter (1987b) estimated that each year 9 million tons of solid waste produced in the United States are dumped at sea and an even greater amount from land reaches the ocean via rivers, drainage systems, estuaries, and other avenues.

It is also important to recognize that while the more developed nations produce far more potential pollutants in proportion to their populations than do the less developed ones, the developed nations have far more effective solid waste management systems than do the other countries. Presumably, as the less developed nations increase their levels of industrialization and well-being, an increase in total generated waste will follow. Unless the domestic solid waste systems keep pace with growth and develop the capacity to handle the increased volumes, it is reasonable to assume that there will be a greater potential contribution from land-based pollution to the sea.

Much of this pollution is likely to be in the form of plastics and other synthetic materials, although traditionally Third World domestic waste has differed in composition from that originating in developed nations. Most developing nations rely more heavily on organic components for convenience and packaging. Plastic and other synthetics have usually represented less than 4% of the domestic solid waste stream in these countries. In Honduras, plastics constitute less than 4% (by weight) of the solid waste stream and in Kenya less than 3%. This contrasts with similar figures for the United States, the Netherlands, and Japan, where the percentage of plastics in the waste stream is closer to 7%, and Sweden, where the percentage is closer to 10% (Pan American Health Organization 1982).

But the use of plastics and other synthetics in the developing world is changing, and in the past 40 years the use of such products to replace degradable items has expanded at a rapid pace. The use of these synthetics has often come to represent new lifestyles, local symbols of pride, and national progress. It is thus likely that the percentage of plastics in the waste stream of these emerging countries will increase considerably in the not too distant future.

It is also widely recognized that most developing nations, in their urgent quest to improve the social and economic status of their residents, often tend to place little emphasis on maintaining environmental quality (NATO 1986). Only those countries interested in tourism as a means for development have within the last decade recognized the linkage between sustainable between growth and environmental quality. In Curacao (the Netherland Antilles), for instance, Winkel (1982) wrote of industrial activities and an "enormous waste problem . . . that in the long run may be a drag on an important source of income, namely tourism".

For many developing nations bordering on the oceans, the coastal zone has been especially attractive for unbridled development. Lower labor and environmental costs to carry out industrial activities have attracted many firms from the developed world (NATO 1986). Tourism-related activities have increased rapidly in the last few decades and the cruise ship sector, in particular, has grown dramatically. The increasing populations, spiraling amounts of domestic and industrial waste, and the refuse generated by cruise ships and hotels have made the coastal waters and nearshore areas attractive for legal and illegal waste disposal. In many instances, these areas have become significant solid waste dumping grounds. The dumping problem has been exacerbated by the fact that plastic objects and other debris are most frequently found in the nearshore marine environment within which they are very efficiently trapped and cycled (Windom 1992).

Identification and Significance of Principal Coastal Rural and Upland Contributors

The management of solid wastes presents the single largest problem around the world relative to coastal rural and upland contributions to the overall volume of marine debris. The problem is likely to become even more acute with industrialization and population growth in the developing world. The types of waste to be disposed of, the financial constraints under which public agencies must operate, the institutional mechanisms, and the infrastructure available will all affect the

approaches taken in managing this major and expanding predicament (Khosla 1982).

Some examples of this predicament should help illustrate the nature and magnitude of the solid waste management problem in developing nations. The country-specific information that appears next was gleaned from country reports prepared in 1991 for the First Caribbean Solid Waste Management Meeting.

Uncontrolled Dumping and Solid Waste Management

The single most prevalent source of coastal rural and upland-originated marine debris is the uncontrolled dumping that remains the almost universal method of disposing of solid waste in developing nations. Those which now enjoy some level of economic prosperity have begun to use sanitary disposal in landfills (Holmes 1984). Gourlay (1992) thinks that more than 90% of domestic waste, in some cases the whole of it, in both the industrialized and developing world is taken away and dumped in landfills or added to growing mountains of refuse. The remainder is either burned or dumped at sea. Although modern containment landfills are being built to avoid groundwater and surface water contamination, heavy rains and storms have been known to cause existing landfills to overflow. Many of the older facilities are poorly located close to waterbodies that empty into the sea.

A major part of the problem is that the collection and transportation of domestic solid waste is the most difficult and expensive aspect of solid waste management. Even in developed countries, where rural areas are usually served as efficiently as most towns and cities, on average 70% of the entire solid waste-handling budget is consumed in collecting and transporting the refuse; little money is left for treatment and adequate disposal.

In developing nations, this problem is intensified because collection and transport are difficult, thus creating a situation that engenders uncontrolled dumping. Road access to outlying communities and rural locations is frequently poor or lacking. Upland and coastal remote areas are not readily accessible to collection vehicles. The Caribbean island of Dominica illustrates this situation. Refuse collection activities are only undertaken for the west coast communities. Only about 47,000 of the island's 75,000 inhabitants are served. Local authorities aspire to ". . . extend the services to the rest of the island and to encourage communities to use other disposal methods" (Williams 1991). A similar situation exists on the island of Grenada where about 45% of the island's 91,000 residents have their waste collected. The collection is "still confined to the towns and environs" (Williams 1991). Jamaica, another Caribbean example, has a population of approximately 3 million inhabitants, approximately 60% rural and 40% urban. The lack of equipment and funds limit solid waste management services primarily to urban centers and to developed rural areas.

Data from the six nations that belong to the Organization of Eastern Caribbean States (OECS) indicate that 40%–50% of all solid wastes generated in OECS cities do not reach official landfills or dumpsites. By most accounts, the solid waste collection and disposal needs of most rural areas in the member nations are rarely met, largely because of the aforementioned access problems and the limited funds of national governments. This has resulted in open dumping and uncontrolled landfilling. The closure of four coastal dumps in Puerto Rico, and the resulting consolidation of disposal and newly imposed charges for solid waste management, have resulted in widespread dumping along the island's rivers (personal communication with Ruperto Chaparro, Universidad de Puerto Rico, Mayagüez, April 1994). Observations of the types of litter found on certain St. Lucia beaches indicate that most of the plastics are from "local, land-based refuse" (Corbin and Singh 1993; Singh and Xavier, Chapter 30, this volume).

Authorities have often advocated composting organic waste, and reuse and recycling of some materials, but in many uncontrolled dumping instances the waste stream is used to reclaim and level less usable areas such as ravines, old quarries, and valleys. Many coastal communities use solid waste to help fill in and "reclaim" wetlands, notably marshes and mangroves. The 40 islands that comprise the British Virgin Islands serve as an example. Authorities stress that their most significant problem is at the four public dumpsites on the major islands. Because of the hilly terrain and high land values, it has been extremely difficult to find suitable sites for the dumps. "The best managed site can be found on Tortola. It is basically a flat, swampy area that is being filled with compacted garbage and covered with fill on a daily basis. This dumpsite is located very close to a number of aquifers. On the island of Anegada, the garbage is dumped on the surface and burned within a fenced area. On the island of Virgin Gorda the major problem is the unavailability of fill material. On that island the garbage is burnt and pushed off regularly" (Williams 1991).

Even in the rare instances when the systems and budgets are adequate for collection, safe disposal of collected wastes often remains a problem. Sanitary landfills have become the norm in only a handful of locations. More controlled sanitary landfills are desperately needed to bring some semblance of order and hygiene to the dumps and to help reduce the risk of pollutants, including plastics and other floatables, from entering the ocean.

In Antigua, an island with 80,000 residents, sanitary landfilling is not an established method of garbage disposal. Open dumping or crude tipping is still the means for final disposal of solid waste. In Jamaica, disposal also leaves much to be desired. There are currently 22 disposal sites, none of which would qualify as a conventional landfilling operation. Each is located in an environmentally vulnerable situation. Burning and burial (not incineration and sanitary landfilling) are the methods used (Williams 1991). Some of the more densely populated parts of the world have, out of sheer necessity, adopted controversial incineration systems. For instance, in Gibraltar, reportedly the fourth most densely populated place on earth, the incinerator was out of commission for 6 weeks. During this period, the authorities sent the daily refuse for 30,000 residents down a chute into the Mediterranean Sea (Gourlay 1992).

Litter, Storm Drains, and Sewage Systems

Storm drains and sewage systems also carry litter and other debris to the sea. Coombe writes that approximately 75% of the "dangerous human refuse" that moves into California's bays comes from individuals who live 30 miles inland. The process is quite simple, he observes. Storm drains carry water off the streets as quickly as possible. This water makes its way into the bays completely untreated and includes "just about everything thrown on our streets" (Coombe 1989). The aforementioned New York Harbor report notes that ". . . upland runoff is carried directly across land surfaces to area streams or drainage ditches and subsequently transported into New York Harbor" by way of numerous tributaries.

Sewage is perceived by many as the most significant pollution problem. The discharge of excessive amounts of human wastes is perhaps the most significant pollution problem facing the nearshore: inadequate or nonexistent facilities for the treatment of sewage before discharge, and sewage inputs that often include plastic debris. Although this problem primarily affects the world's major population centers, it also poses a major problem for smaller population hubs in the more remote areas of many countries. In Indonesia, very little wastewater that is discharged into watercourses is treated. Many of the outlying areas are served by systems consisting of concrete-lined open channels that clog with street refuse and debris. The rivers of the Indian subcontinent are notorious for the quantity of sewage and other debris they receive. In the early 1980s, the Ganges River was described as "gradually turning from river to drain, where raw sewage, garbage and muck pour continuously

into the river from 361 outfalls" (Gourlay 1992).

Most of the rural communities in the Caribbean region are on the ocean's edge, and within them are some of the most difficult solid waste and sewage disposal problems. Vast quantities of debris are contained in the poorly treated sewage that is disposed of through outfalls to the ocean or into the rivers that ultimately empty into the Caribbean. The United Nations Environmental Programme (UNEP) reported in 1984 that fewer than 10% of the sewage systems around the Caribbean had treatment facilities and that very little debris was filtered out.

The United Nations Environmental Programme has noted that littering of recreational beaches is a common occurrence, and not just by careless beachgoers. Frequently, after heavy rains, waste that has been deposited by residents in or near streams and in low-lying upland and coastal areas is carried by runoff out to sea. Storm tides and waves redeposit the trash on beaches and in marshes and mangroves (Henneman 1988). This was corroborated by Wade in his recent studies of Jamaican beaches (Wade 1991).

Even where public works have improved and sanitary landfills are in place, litter remains a major problem. Where storm drains have replaced earthen ditches, they often become choked with the garbage. In areas where sewerage is provided, there are stories of manhole covers being removed to ease garbage dumping. The sewers then become blocked. In other areas where special dumps or public receptacles are provided, the garbage is heaped outside (Pan American Health Organization 1992).

Nearly 90% of St. Kitts' 43,000 residents have a waste collection and disposal service. Local authorities estimate that approximately 75% of the refuse generated in the areas being served is collected. "The remaining 25 percent is disposed of by unapproved methods, namely indiscriminate dumping and backyard burning" (Williams 1991).

In the city of Georgetown, Guyana (population 260,000), the collection system ran smoothly until the early 1970s, when Greater Georgetown was added to the city without an accompanying budget or staff increase. Where refuse had been collected as often as three times per week, it has now dropped to once per month in several service areas. The outcome has been the establishment of illegal minidumps and dumping into "narrow and shallow drains." These practices have severely stressed the already inadequate drainage systems, have increased the incidence of localized flooding, and in general have resulted in major public health hazards. Public littering of the streets and parapets is very common.

Current and Needed Efforts to Control and Reduce Coastal and Upland Discharges

A variety of strategies are being employed to address all aspects of land-based marine pollution, including plastics and other man-made debris. They include strategies based on legislation, local and regional infrastructure development, and international agreements.

Institutional Considerations

In most developed nations, institutional factors rather than technical problems are the target of strategies. Coastal pollution by litter arises as a consequence of deficient storm drain and sewage disposal infrastructure, inadequate solid waste management regulations, or the lax enforcement of existing regulations. For example, in the United States plastics as nonhazardous waste are and will likely remain principally a state and local responsibility as assigned by the Resource Conservation and Recovery Act of 1985 (a national law). Thus, U.S. strategies aimed at reducing the amount of marine plastic pollution from such sources will necessarily be directed by state and local governments. State-level initiatives might, for instance, address increased recycling efforts and the enactment

of laws requiring degradable containers and beverage container return deposit laws. Nevertheless, the law provides for a vigorous federal government role of technical and financial assistance to help the states reduce wastes, promote recycling, and better manage wastes (Bean 1987; Nollkaemper, Chapter 22, this volume).

Many U.S. cities, including Baltimore, Glen Cove (New York), Washington, D.C., Atlanta, and New York City are also involved in programs to combat marine plastic pollution through efforts to abate floatables (Molinari, Chapter 26, this volume; Durrum, Chapter 27, this volume). The 1987 and 1988 floatable incidents in New York and New Jersey drew considerable attention, and New York City sponsored studies to address potential methods for abatement (HydroQual 1993). Many of the proposed methods apply where the sources of debris are combined sewer overflows, raw sewage discharges, barging of refuse, marine transfer stations, landfills, and recreational boating (Table 28.1).

It is also most apparent that many developing nations must enact national legislation to adequately address environmental matters (Brathaug and Barnett 1994). Many states of the Wider Caribbean Region, for instance, lack laws to deal with environmental pollution. Where laws exist, they are woefully outdated. The laws do not reflect current issues and do not encompass the control of polluting substances (Wade, Chapter 15, this volume).

For instance, legislation and guidance specifying management measures and practices for sources of both point and non-point pollution in coastal waters are needed. Point source is defined in the U.S. Clean Water Act as any "discernible, continued, and discrete conveyance," including pipes, outfalls, etc., from which pollutants are or may be discharged (EPA 1993b). Because the sources of non-point pollution are not readily discernible, the range of these pollutants is of even greater concern. Plastics and related debris are transported from non-point sources to surface water by a variety of means, including runoff.

TABLE 28.1. Floatable material control techniques.

Solid waste handling	Wastewater systems	Shoreline sources
Landfills	Water pollution control plants	Piers and waterfront structures
Good housekeeping	Bar screens	Demolition
Booms	Clarifier skimming	Booming
Water skimming	Combined storm and sewer overflows	Abandoned vessels
Marine transfer stations	Source reduction	Removal
Good housekeeping	Public education	Improper disposal
Booms	Packaging minimization	Public education
Manual dipnetting	Recycling	Interdiction
Barges	Product reformulation	Oil terminals
Cover nets	Streetcleaning	Booming
Containerization	Catch basin cleaning	Marinas
	End-of-pipe netting	Public education
	Swirl concentrators	Shoreline cleanup
	Booming containment	
	Containment tanks	

From Hydro Qual (1993).

Countries must move to specify economically feasible management objectives that will control the addition of plastic and other pollutants to coastal waters. Appropriate and sound practices that can help implement the specific management objectives also need to be discussed and tested. Without national laws, no enforcement is possible. MARPOL, for one, cannot be implemented by a state until the state has laws in place to enforce the international convention. To date, MARPOL (Annex V) has been ratified by 17 of the 35 Wider Caribbean states and territories (Barnett, Chapter 14, this volume, Table 14.2).

In other nations like Mexico, India, and Colombia, the influence of regulations has been limited by the resources of the enforcement agencies, inefficient legal processes, and the low levels of fines (Eskeland and Jimenez 1991). Incentive systems must also be put in place to encourage the proper disposal at reasonable costs (Brathaug and Barnett 1994).

Infrastructure Development

Even if the regulatory systems were adequate, they would serve little purpose for controlling solid waste dumping and littering. In most developing nations, the infrastructure needed to receive and dispose of these materials is grossly inadequate or not in place. It simply has not been an urgent issue. Public authorities have frequently needed considerable persuasion to believe that investments in landfill, plant, equipment, and manpower are really priorities.

Many Caribbean nations have, however, come to appreciate the potential impact of waste handling and disposal on coastal resources. The waste management project for the OECS nations serves as an example of such recognition and demonstrates how international bodies are seeking to provide financial and technical assistance to help address the issue (Barnett, Chapter 14, this volume). The World Bank and the Global Environment Facility (GEF) have agreed to fund waste reception and disposal facility projects in the OECS nations and the Wider Caribbean. The project's initial impetus was provided by the growing problems associated with cruise ship-generated waste, the designation of the Region as a MARPOL Annex V Special Area, and the urgent need to invest in port reception facilities. But the offloading of the solid waste generated onboard the cruise ships merely transferred additional problems to the island nations and compounded the islands' longer term solid waste management problems. These infrastructure investments are intended to relieve such problems. Existing dumps will be either closed or converted to sanitary landfills; new sanitary landfills will be built; and recycling and composting will receive a great deal of attention.

Infrastructure financing, construction, and technical assistance, however, are only the beginning. The infusion of tremendous amounts of external aid will not effectively and ultimately resolve the solid waste and marine debris problems of the developing nations without strong governmental and civic support and ongoing public education efforts. There must be a commitment to carry on once the aid ceases.

Global and Regional Strategies

The need to control the land-based sources of marine pollution is not a new issue. Most nations of the world are aware that the problem of waste disposal pollution at sea is beyond unilateral action. Great attention has been given during the past three decades to devising mechanisms to protect the coastal waters of one country from pollutants released by neighboring nations. Bilateral, and to a considerably lesser extent, regional agreements to protect regional seas have met with varying degrees of success (Nollkaemper, Chapter 22, this volume). For the most part, countries have handled land-based, marine pollution on a national basis.

There is a clear consensus that land-based sources of marine pollution are not adequately addressed and that further serious degradation of the marine environment will

occur without concerted new actions. Most international marine pollution legal and policy experts advocate a worldwide, strategic approach. Nollkaemper (1992) observed that regional approaches can only provide solutions if they are complemented by "an adequate global strategy". Kindt (1984) noted that although there is an international consensus that certain waste materials cause damage to the environment and public health, no common approach or agreement on a system of regulations has emerged. Karau (1992), in concurring with this assertion, indicated that most efforts have provided helpful information on various approaches for pollution control yet have not addressed the strategic aspects. There is a need, he states, for a global framework or mechanism for reviewing and harmonizing the effectiveness of these various national and regional efforts.

At present only two global conventions exist to combat marine pollution: MARPOL 73/78 and the London Dumping Convention (LDC). MARPOL addresses the problems of shipborne pollution and the LDC, those of direct disposal (Nauke and Holland 1992). Both are relatively small sources of pollution. Nauke and Holland concluded that additional international agreements will be needed to address land-based aspects of marine pollution and that such developments should draw on the experiences gained from the two global conventions.

The legal authorities that apply or could be made to apply to the problem of solid waste and debris pollution in the marine environment are currently inadequate to redress this problem in a comprehensive fashion.

29.

Upland Sources of Marine Debris on the Shorelines of Puerto Rico

Ruperto Chaparro and Javier Vélez

Introduction

The territory of Puerto Rico contains one large island and several small islands including Vieques, Culebra, and Mona. Puerto Rico is bordered on the north by the Atlantic Ocean, on the east by the Virgin Passage, on the south by the Caribbean Sea, and on the west by the Mona Passage. The principal island of Puerto Rico is one of the largest of the West Indies. Puerto Rico's greatest east-to-west distance is about 180 km (110 mi), and its extreme north-to-south distance is about 65 km (40 mi). The highest point is 1338 m (4389 ft) at Cerro La Punta in the central town of Jayuya.

The large, mountainous island of Puerto Rico is located directly in the path of the trade winds. These account for its tropical rain forest and tropical wet and dry climate. The central, east-to-west mountain range extends almost the entire length of the island. The average elevation of these mountains, which include the Cordillera Central and the Sierra de Luquillo, is about 915 m (about 3000 ft). Although hills adjacent to the mountains cover most of Puerto Rico, in the north lies a coastal plain approximately 19 km ($\approx$ 12 mi) wide and a narrower coastal plain approximately 13 km ($\approx$ 8 mi) wide that extends along the southern coast. For most of its length, the mountain system is nearer to the southern than the northern coast, and the slopes are generally steeper on the southern side. At the eastern end of the island, however, the mountains curve toward the northern corner.

Average annual precipitation ranges from 30 in. (along the western parts of its offshore islands and on the southwest coast) to about 200 in. in the northeastern area of the island (Gómez-Gómez and Heisel 1980). The average annual rainfall in Puerto Rico is about 70 in. ($\approx$ 11,600 million gal per day). Of this amount, 65% ($\approx$ 7,540 million gal per day) is lost through transpiration and evaporation. Nearly 32% ($\approx$ 3,700 million gal per day) of the island's annual precipitation becomes surface water runoff.

Puerto Rico has had abundant surface- and groundwater resources throughout the history of its development. However these resources are unevenly distributed. The northern part of the island has a far more abundant water supply than the southern area. The rainy season usually lasts from August to November, and the dry season is usually from January to April (Torres and Rodríguez 1987). Traditionally, towns were established on the coastal plains and near rivers, and their economic activities were controlled by the availability of nearby water resources. Projects aimed at developing water resources included reservoirs and complex systems of tunnels, pipelines, and canals. The development of large water supply projects began early in this century and contributed to the extensive population growth on the island.

According to the 1990 census, Puerto Rico (including Vieques and Culebra) had 3,522,037 inhabitants, an increase of about 10.2% from 1980. The average population density in 1990 was 387 persons per square kilometer (1,002 persons per square mile), which is a much higher density than for any U.S. state except New Jersey and Rhode Island. In 1990 approximately 71% of the island's inhabitants lived in areas defined as urban and the rest lived in rural areas. In the early 1980s, nearly 90% of the adult population was literate compared with 67% in 1940.

Surface water is the major source of freshwater throughout Puerto Rico. More than 100 streams and rivers flow to the ocean. For the purposes of this study, the island of Puerto Rico was divided into four major areas. The northern coastal area extends from the municipality of Isabela to the municipality of Luquillo; the eastern coastal area extends from the municipality of Fajardo to the municipality of Maunabo; the southern coastal area extends from the municipality of Patillas to the municipality of Lajas; and the western coastal area extends from the municipality of Cabo Rojo to the municipality of Aguadilla. These areas are further clarified by the U.S. Geological Survey's 1987 National Water Summary:

The principal streams of the northern coastal area are the Río Grande of Loíza, Río de la Plata, and Río Grande of Arecibo. Río Grande of Loíza is the principal source of water for metropolitan San Juan. Río de la Plata and Río Grande of Arecibo are regulated for water supply and power generation. In the eastern coastal area, the Río Blanco is the principal source of water for public supply. Río Grande of Patillas, Río Toa Vaca, Río Jacaguas, Río Loco, Río Yauco, Río Coamo and Río Guamaní are the principal sources for water supply and irrigation in the southern coastal areas. Río Grande of Añasco and Río Guanajibo satisfy the water supply demands for most of the western coastal area (Torres and Rodríguez 1987).

The Marine Debris Problem in Puerto Rico

Puerto Rico's shorelines and coastal areas were once famous for their white sands, crystalline waters, and abundant species of plants and animals, but these natural resources are now under considerable pressure. Island shorelines and coastal areas support coastal economies, provide recreational opportunities, and serve as nursery areas for marine plants and animals. At the same time, these areas are used as dumpsites for trash and other wastes. In addition to being unpleasant to look at, in the marine environment this debris has serious impacts on wildlife and the economy.

Puerto Rico is a third-world country that, because of its political relationship with the United States, receives U.S. federal funds for social programs, infrastructure development, and technology transfer. However, the attitude of most Puerto Ricans toward their natural resources has not advanced at the same pace as Puerto Rico's economic and industrial development. It is not uncommon to see people dropping or tossing litter out of their cars into the streets, rivers, beaches, sinkholes, illegal dumps, and reservoirs. These practices are detrimental to the environment because of the persistence of synthetic materials like plastics, Styrofoam, and glass that accumulate in bodies of water. In addition to these irresponsible attitudes toward the environment, garbage collection services do not reach all residents, and poorly designed, managed, and operated landfills are often located close to or in sensitive areas.

It has been surmised that vessels are the largest producers of marine debris, accounting for nearly 60% of all marine-generated waste (Eastern Research Group 1988). Since the 1987 ratification of Annex V of the International Convention for the Prevention of Pollution from Ships (MARPOL 73/78), much of the recent attention concerning marine debris and its sources has focused on vessel-generated waste and its fate when thrown overboard. Additionally, Puerto Rico's relationship with the United States means that the programs, regulations, and initiatives developed for the U.S. mainland are commonly implemented in Puerto Rico without taking into consideration cultural, geographic, environmental, and economic differences. Researchers, politicians, and managers of the resources in Puerto Rico agreed with early U.S. views concerning the sources of marine debris. This view identified vessels as generators of most of the marine debris. While this may have been the case for waters and shorelines of the United States, it is certainly not the case for Puerto Rico.

Marine Debris Sources

Before the Sea Grant Program of the University of Puerto Rico (SGPUPR) developed its "Adopt a Beach" educational campaign in 1986, there were no data available regarding types, amounts, distribution, and sources of marine debris on Puerto Rico. Some of the first data gathered during beach cleanups identified cruise ships as a point source of Puerto Rico's marine debris (e.g., shampoo bottles, plastic cups, napkins). At the same time, other vessel sources were also identified as significant contributors to the marine debris problem (e.g., commercial fishing and recreational vessels). These data helped to reaffirm the theory accepted in the United States: vessels were responsible for most marine debris.

However, while vessel sources of marine debris were initially evident in Puerto Rican waters and shorelines, data gathered during 6 years of marine debris research and beach cleanups by the SGPUPR, and coordinated with the Center for Marine Conservation (CMC), suggest that the marine debris problem in Puerto Rico goes beyond vessel-source debris (Coe et al., Chapter 3, this volume).

Evaluation of the data collected during beach cleanups shows that land-based sources are a much more significant factor than are vessel sources. Litter discarded more than 20–30 miles inland becomes marine debris when it is carried by streams, rivers, or sewers into the ocean or reservoirs. This is more noticeable after heavy rains when large amounts of waste that has been dumped by residents in or near rivers and streams is carried to the sea and eventually makes its way back to Puerto Rico's beaches. One of the indicators of debris discharged into the ocean by rivers and streams are logs and pieces of wood ("ballaos"), which are brought back to the coast by wave action and ocean currents. These "ballaos" travel down the river with rubber tires, refrigerators, washing machines, plastic gallon jugs, trash bags, plastic bags, furniture, and other debris. Similar amounts of debris and "ballaos" have been recorded by volunteers during cleanups of artificial lakes and reservoirs. These bodies of water cannot receive debris from ocean sources; however, during the cleanup of the Dos Bocas reservoir three complete bedroom sets were collected together with 26 tons of garbage. This garbage included refrigerators, washing machines, water heaters, plastic gallon jugs, motor oil containers, glass, and Styrofoam.

According to the Coastal Zone Management Program of Puerto Rico, all 3,500,000 residents of the island live within 18 miles of the coast and are considered coastal residents. Puerto Ricans produce 57,600,000 lb of solid waste every year—more than 4 lb per person each day—which is one of the highest per capita rates among industrialized nations. A large part of this litter ends up on the coasts and beaches via rivers and streams and via direct littering by marine recreationists, coastal residents, and through other avenues.

Irresponsible (or at least uneducated) attitudes toward the environment have a significant influence on the sources of debris entering the coastal areas. Many contributors to the marine debris problem in Puerto Rico are rural and urban residents who never visit the beaches or recognize that they are part of the problem. However, the data demonstrate that beachgoers and marine recreationists are also significant sources of marine debris in Puerto Rico. Taken together, these land-based sources of marine debris pose a continuing threat to tourism, marine recreation industries, and wildlife.

Data collected by the volunteers show that each beach can be classified according to the sources and types of debris found. In Puerto Rico, there are several major land sources of marine debris: rural-upland discharges transported by rivers and streams, coastal discharges by local communities, industrial wastes, construction wastes, recreational wastes, medical wastes, and agricultural discharges. While some beaches are affected by a combination of land sources, others beaches are only affected by a single source.

Examples of beaches affected by one land source of solid wastes include el Combate and Boquerón in the municipality of Cabo Rojo, Isla Verde in the municipality of Carolina, and Las Playuelas in the municipality of Aguadilla. Recreationists and beachgoers are the main land source of debris affecting these beaches. Marine debris recorded by volunteers includes plastic (bags, beverage bottles, cigarette filters, cups, utensils, diapers, straws, six-pack holders), Styrofoam (meat trays, pieces, plates, cups), glass (beverage bottles and pieces), and metal (beverage cans). The beach of Ballenas in the municipality of Guánica and the beach of the municipality of Mayagüez are examples of beaches that receive several land sources of debris. Debris found in Ballenas is composed of plastics discharged by rural and urban residents into the Río Loco and Río Yauco Rivers, as well as litter from recreationists and beachgoers. At this beach, marine debris recorded by volunteers includes plastic (bleach containers, bags, gallon milk jugs, cigarette butts, straws, gallon water jugs, toys, caps, and lids), glass (beverage bottles and pieces), and metal (beverage cans). The source

of the debris recorded at the beach of Mayagüez can be traced to the Río Yaguez, Río Grande of Añasco, and Río Guanajibo rivers. Debris at this beach also comes from illegal dumping (by local residents). Marine debris recorded by volunteers at Mayagüez includes plastic (bottles, toys, cables, bags, syringes, other medical wastes, automobile and motorcycle parts, and furniture), Styrofoam (meat trays, packaging material), construction wastes, glass (beverage bottles, television screens, light bulbs, and various other pieces), rubber (tires), metal (refrigerators, water heaters, washing machines, and automobile and motorcycle parts), and paper (bags, cardboard, and various other pieces).

Data gathered by volunteers [and coordinated with the Center for Maritime Conservation (CMC)] during the past 6 years of beach cleanups suggests that more than 80% of the marine debris found in the coastal waters and on the shorelines of Puerto Rico is derived from land-based sources. A better understanding of the sources, amounts, distribution, fate, and effects of debris (plastics and other man-made floatables) in the marine environment is needed if the problem is to be solved.

30.

Land-Based Sources of Marine Debris and Contamination of the Coastal Areas of the Caribbean Islands of St. Lucia, Dominica, and the British Virgin Islands

Joth G. Singh and Boniface Xavier

Introduction

In recent years, attention to marine debris and its associated problems has been increasing in the Caribbean region. This debris consists mostly of persistent material such as plastics, metal, glass, and rubber. The sources of this litter include maritime activities associated with shipping and fisheries, beach recreation activities, and land-based refuse. Marine debris poses serious threats to marine wildlife (through entanglement and ingestion), vessels (e.g., entanglement in propellers), and the aesthetic qualities of marine areas. Also, debris on shorelines is a potentially serious problem for Caribbean island states because it threatens the safety of beach users and negatively impacts the economy—tourism is a major source of revenue in many Caribbean island states. While regional experts increasingly identify the problem of marine debris as one of primary concern, many Caribbean states and territories have ratified conventions (Barnett, Chapter 14, this volume, Table 14.2) and initiated surveys and monitoring programs that address marine debris in the Wider Caribbean (Coe et al., Chapter 3, this volume).

Land-based sources (and the effects) of marine debris in the British Virgin Islands, Dominica, and St. Lucia are reviewed in this paper, because of the variation in their methods of solid waste disposal and solid waste management practices. This paper also provides preliminary information on the geographic distribution of marine debris in the Eastern Caribbean, as the British Virgin Islands, Dominica, and St. Lucia are well spaced in the Lesser Antilles and each generated comparable data sets on marine debris.

Geography

The British Virgin Islands, Dominica, and St. Lucia are three of an archipelagic clustering of oceanic islands in the Caribbean Sea known collectively as the Lesser Antilles. This biogeographic grouping sweeps in a graceful curve from Puerto Rico (in the north) to Trinidad (in the south) and is notable among tourists and scholars for its cultural, environmental, and geomorphological diversity.

The British Virgin Islands (B.V.I.), the northernmost of the three countries discussed in this paper, consist of a grouping of approximately 36 islands with a total land area of 153 km² and a combined population of about 17,000. All these islands, with the exception of Anegada, are hilly with steep slopes. Anegada, in contrast, is very flat with virtually no topographical features. Tortola, the largest and most developed island, has a pronounced ridge running the length of the island. A number of natural drainage channels or ghuts run down the ridges. These channels are generally dry and run only after heavy rainfall. The economy of this British territory is supported primarily by tourism and its internationally popular commercial center in Road Town, Tortola. The B.V.I. are a popular stopover for cruise ships operating in the region and are also a sailing haven of international repute. The population is mostly concentrated in the major towns of these islands, but the islands of Tortola and Virgin Gorda have many small villages scattered along their coastlines and interior.

The island of Dominica is the largest of the English-speaking Eastern Caribbean Islands (i.e., the Lesser Antilles). Dominica is also the most mountainous island in the region with an area of 752 km^2 and a population of about 82,000. There are numerous villages along the coastline of Dominica, but the majority of the population is concentrated in the city of Roseau and the town of Portsmouth. The island is covered by lush green rain forests and has 365 rivers. These rivers cascade along mountain slopes and through valleys, eventually winding their way (through many Dominican communities) to coastal waters. Agriculture is the mainstay of the economy, and like other Windward Islands Dominica is an exporter of bananas to Europe. During the past few years attempts have been made to lay the foundation for a tourist industry with strong emphasis on eco-tourism. Currently, Dominica is a common port of call for cruise ships plying the Caribbean. Solid waste is collected from these ships, for a fee, and disposed of at the landfill.

St. Lucia is a volcanic island with a rugged, interior topography and majestic beaches along its coast. The island has an area of 620 km^2 and a total population of about 151,000. The population is concentrated in the northeastern portion of the country, with the Castries and Gros Islet areas containing about half the population. Like Dominica, St. Lucia has several small villages along its east and west coasts. Agriculture and tourism are the mainstay of the economy, and St. Lucia is also a popular stopover island for cruise ships in the Caribbean Sea.

The CEHI and IOCARIBE Marine Debris Study

During 1991, the problems associated with increasing amounts of marine debris (and the lack of reliable data) moved the Intergovernmental Oceanographic Commission Sub-Commission for the Caribbean (IOCARIBE) to initiate a monitoring program for the region. Consequently, the Caribbean Environmental Health Institute (CEHI) was awarded a contract that provided funding, by IOCARIBE, for the generation of marine debris data in St. Lucia, Dominica, and the British Virgin Islands. The monitoring program was conducted with assistance from local agencies of all three governments.

The site selection and monitoring protocol of the CEHI monitoring program were determined by the recommendations of IOCARIBE (IOC 1990). The two types of beaches were selected for the program. The beach selection was based on (1) high recreational usage and (b) isolation. The rationale for these two criteria was that recreational beaches are of economic importance and that isolated beaches indicate contamination from maritime sources rather than from localized land-based activities.

The beaches selected for the survey were two recreational beaches on the west coast and one nonrecreational beach on the east coast of St. Lucia, three recreational beaches on Dominica's west coast, and three recreational beaches in the British Virgin Islands (Great Camanoe Island, Prickly Pear Island, and Anegada Island). Four fixed transects, 5 m wide and evenly spaced along the beach front, were sampled. The transect extended from the low water line to the vegetation line. Samples were collected monthly between January 1991 and December 1992. All visible, persistent litter items larger than 1–2 cm were collected, cleaned, and sorted. These were tallied in one of nine categories: plastics, Styrofoam, fishing gear, glass, metal, rubber, driftwood, cloth, and other. Results were recorded as numbers and weights of items per meter of beach front.

The results obtained for these beaches are represented in Fig. 30.1. Overall, in terms of

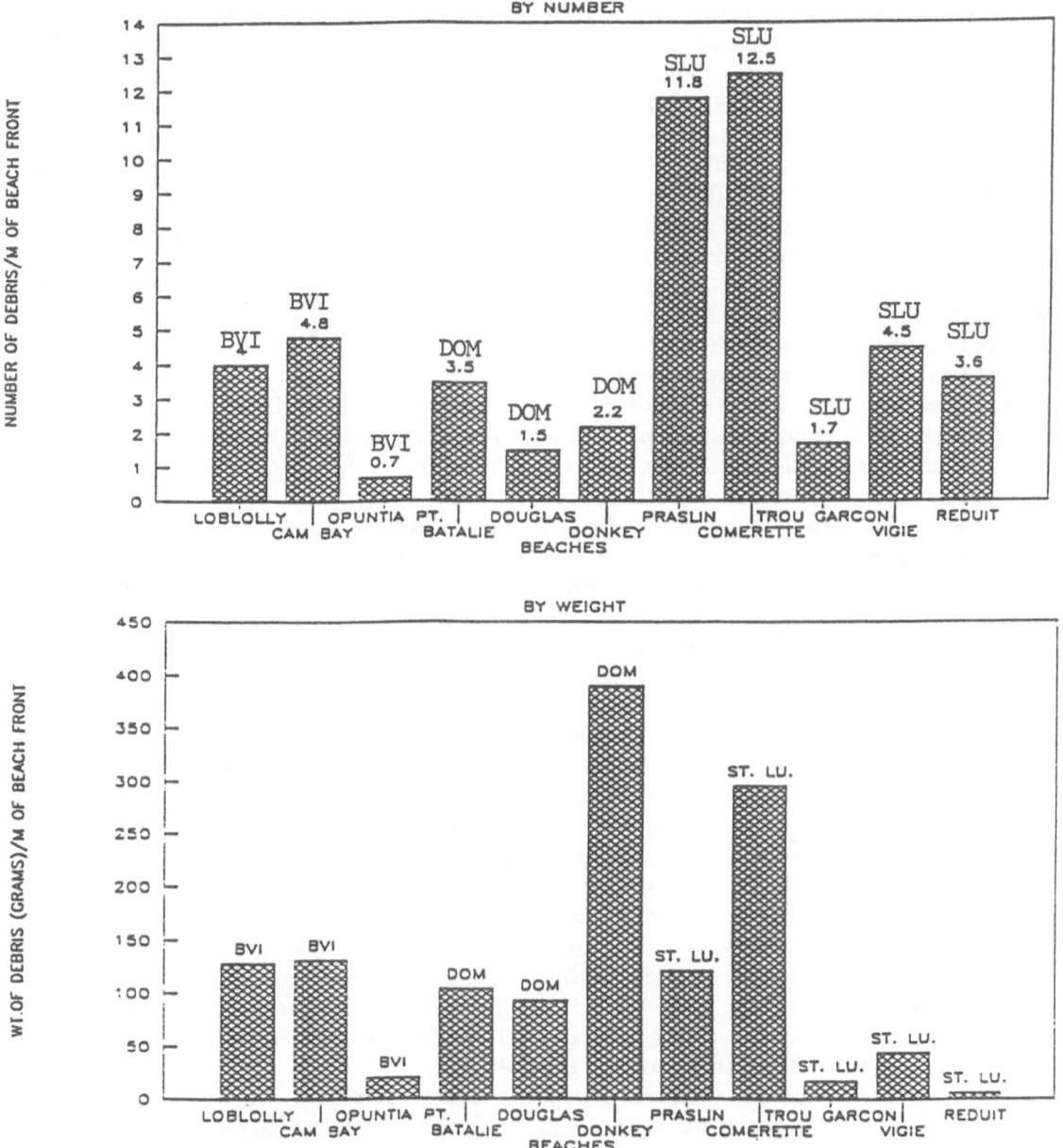

FIGURE **30.1.** A Caribbean Environmental Health Institute (CEHI) monitoring program results (by number) from St. Lucia, Dominica, and the British Virgin Islands, 1991–1992. B CEHI monitoring program results (by weight) from St. Lucia, Dominica, and the British Virgin Islands, 1991–1992.

number of items (Fig. 30.1A), the two east coast beaches of St. Lucia had the highest levels of debris, a monthly average of 12.5 items per meter for Anse Comerette and 11.7 for Praslin. Opuntia Point Beach, in the British Virgin Islands, had the lowest level with a monthly average of 0.7 items per meter. Debris levels on the other beaches ranged from 1.4 to 4.5 items per meter. When compared by weight, Donkey Beach in Dominica had the highest levels followed by Anse Comerette with monthly average of 309.3 and 294.9 grams per meter (g/m), respectively. For the remaining beaches in the survey, the highest levels (by weight) ranged from 92.7 to 131.4 g/m at Cam Bay, Loblolly,

Batalie, Dongles, and Praslin. The lowest levels (by weight) were found at Trou Gascon, Vigie, Reduit, and Opuntia Point, ranging from 6 to 43.9 g/m (Fig. 30.1B).

The IOCARIBE data, from isolated beaches in Puerto Rico, reflect an average abundance of 3.9 items/m and 945 g/m (IOC 1991). The only isolated beach studied as part of the CEHI survey was at Anse Comerette where levels of 13 items/m and of 295 g/m were recorded. Additionally, IOCARIBE's preliminary data from Puerto Rico show that recreational beaches had 50% more debris than isolated beaches. This was not found in the CEHI survey of St. Lucia, suggesting differences in solid waste disposal practices by beach users (at the survey sites) or differences in frequency of cleaning (of the surveyed recreational beaches).

The results of the CEHI survey are consistent with other worldwide surveys in which plastics have been identified as the major category of marine debris reported by weight or number or both (Caulton and Mocogni 1987; Shiber 1987; Vauk and Schrey 1987; IOC/FAO/UNEP 1989).

In Dominica, at Donkey Beach, a large quantity of driftwood accounts for the high weight with a low count. This is unusual when compared to other beach surveys. A number of coastal communities in this island practice fishing, and reports indicate that a large number of abandoned wooden fishing vessels wash ashore.

Observations of the type of litter collected during this survey indicate that most of the plastic is from local land-based refuse. Exceptions are on the east coast beaches in St. Lucia where labels on glass and plastic bottles suggest an origin as far north as the U.S. Virgin Islands. These beaches, which are difficult to access, also show a substantially higher percentage of plastics compared to recreational west coast beaches where debris items are more varied.

The findings of the CEHI survey must be viewed in the context in which data were generated. This survey was structured deliberately to avoid beaches that would have been directly impacted by wastes generated from land-based activities. The study was

essentially aimed at assessing the impact on beaches of ocean-transported debris from maritime operations; however, the data pointed to significant input from land-based activities. The extent of the impacts of debris generated from land was not determined by this study.

CEHI's St. Lucia Beach Debris Survey

Also during 1991, CEHI undertook a rapid survey of selected beaches in St. Lucia to assess debris levels on beaches impacted by various sources. This work was done independently from the CEHI-IOCARIBE Survey (previously reported in this paper). Eleven beaches were selected for this study; the locations of these sites are shown in Fig. 30.2. The sites included some of the major recreational beaches used by tourists and local residents. The sites were also adjacent to towns and fishing villages on both the Atlantic and Caribbean coasts. The beach survey data were collected by walking the entire length of the beach and recording the number and categories of marine debris observed. Where the level of marine debris was too high to permit a realistic survey of the entire beach, a minimum representative distance of 10 m was chosen. The results of the island-wide survey are summarized in Table 30.1 and presented graphically in Fig. 30.3.

Levels of debris found on Tapion Beach were exceedingly high (10,600 items/100 m). This area is not used as a recreational beach, and the levels of debris are very much atypical. Much of the waste deposited in the Castries Harbour via the Castries River (and to a lesser extent from direct dumping) accumulates in the Tapion Beach area because of local ocean currents and wind patterns. This result confirms the effect of solid waste disposal practices in the city and on surrounding coastal environments.

Beaches situated on St. Lucia's northwest coast are used mainly for recreational activities. These beaches include Choc, La Toc,

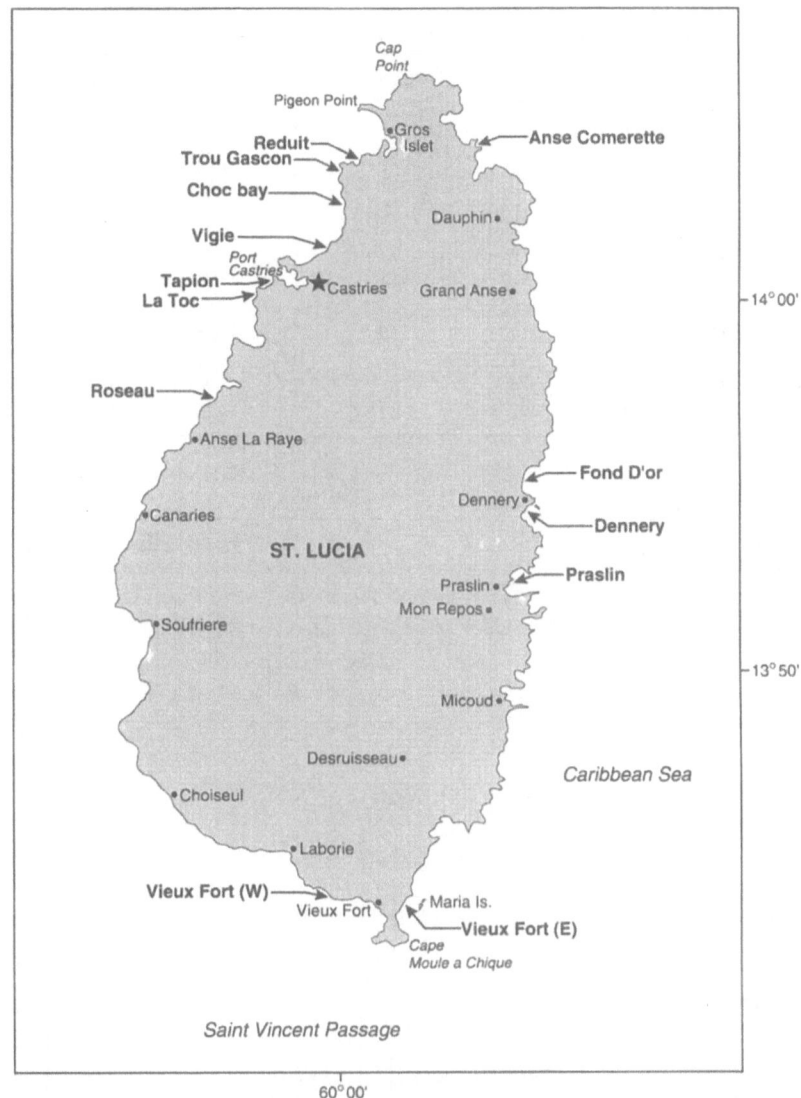

FIGURE 30.2. CEHI St. Lucia beach debris survey sites, 1991.

Pigeon Point, Roseau, Reduit, and Vigie. The average level of marine debris in this area is low (≈ 14 items/100 m) with the major source of pollution being beachgoers (local and tourists). Regular beach cleanups, the presence of disposal containers, and minimal input from the Caribbean Sea are all reflected in the low levels of debris found on St. Lucia's northwest coast.

The only area studied on St. Lucia's northeastern coast was Cas-En-Bas. This beach is not situated close to any major urban settlement or recreational attraction. The levels of marine debris are somewhat higher (47 items/100 m) than those obtained on the northwestern coast. Indiscriminate dumping by beachgoers coupled with the lack of cleanup contribute to the problem. The input from the sea tends to be more significant in this area because the area is exposed to the Atlantic Ocean and the prevailing winds generate onshore movement.

The beaches with the highest levels (>100 items/100 m) of marine debris represent areas on the east and southeast coastlines. A number of factors contribute to this:

The marine input on beaches situated on the east coast is greater, as evidenced at Cas-En-Bas.

TABLE 30.1. Quantities of marine debris per 100 m of shorefront on St. Lucia.

Location	Plastics	Glass	Styrofoam	Rubber	Metal	Paper	Wood	Cloth	Other	Total
Cas En Bas	36	1	3	1	2	1	2	1	0	47
Choc	5	1	1	3	3	3	<1	1	0	17
Dennery	64	3	4	0	5	10	3	1	0	90
La Toc	1	0	<1	0	0	0	<1	0	0	2
Pigeon Point	10	2	1	<1	3	6	<1	<1	0	22
Praslin	88	2	2	0	1	5	10	2	0	110
Reduit	14	<1	<1	<1	1	6	1	2	0	25
Roseau	8	1	<1	<1	1	3	0	1	0	14
Tapion	8,850	250	175	150	225	100	750	100	0	10,600
Vieux Fort (W)	40	4	1	2	18	2	4	292	4	367
Vieux Fort (E)	91	4	4	0	8	5	10	1	3	126
Vigie	8	1	2	<1	3	3	1	1	0	19

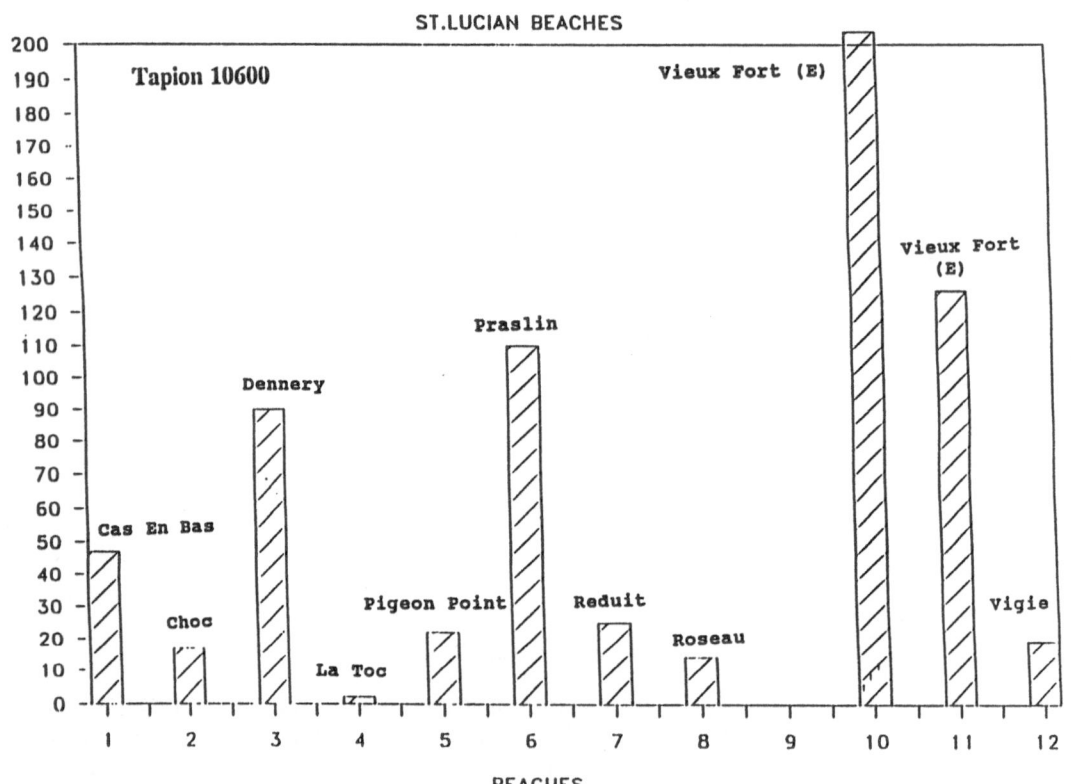

FIGURE 30.3. Results (by number) of CEHI St. Lucia beach debris survey, 1991.

There is less control of dumping, a lack of adequate waste disposal facilities, and little or no beach cleanup.

There is greater impact from rivers that flow through a number of rural settlements and deposit debris either directly on beaches or into the sea where ocean currents and winds can return the debris to the shore.

The most common litter on all the beaches, with the exception of West Vieux Fort, was plastics. Bags and cups were the plastic items most frequently observed. In addition, significant amounts of paper and metal were commonly recorded. Certain areas reflected localized disposal practices, such as at West Vieux Fort with a level of cloth of almost

80%, and at Choc with a 20% level of rubber (mainly in the form of automobile tires). The composition of the observed debris suggests a combination of input from both marine- and land-based sources that varied from beach to beach.

Solid Waste Disposal and Land-Based Sources of Debris

The data from the CEHI-IOCARIBE and the CEHI survey indicate large-scale contributions to beach debris from land-based sources, pointing to the need for investigation of solid waste disposal practices in the islands studied. Identification of the specific land-based sources of marine debris would confirm the role of land-based activities in the marine debris problem.

Dominica

In November of 1989, the Government of Dominica introduced a new method of waste storage and collection using refuse skips [dumpsters]. This system incorporated an element of community participation in that the residents carry their refuse to the storage bins or skips. Immediately after the new system of collection and storage was established, the recipient communities accepted the system and abandoned their commonly used open dumps. Cleanup campaigns helped to eliminate about 90% of the open dumps on the west coast of the island. The collected wastes are deposited in a sanitary landfill created for that purpose. This system serves the western coast of the island from Scotts Head to Portsmouth, encompassing about 45% of the population. This service is not available in the eastern and southern parts of the island, and open dumping continues.

There were 214 refuse skips distributed in the collection area. By the middle of 1992, a repair program was undertaken to combat the rapid deterioration of the skips. However, because of lack of funds repairs could not keep pace with the deterioration. The subsequent shortage of skips for refuse storage caused communities to revert to dumping along the coastline, rivers, and hillsides.

Other sources of marine debris on Dominica are from the construction industry, small industries, and the agricultural sector. Cement bags, plastics, steel, and lumber are disposed among rubble and excavation material. These materials are used for reclaiming land along the boarders of highways cut off by deluge during the rainy season. Waste generated by small skilled industries such as furniture shops, auto-body repair, and mechanic shops are dumped on the outskirts of communities where the trades are undertaken. Derelict vehicles, tires, and abandoned household appliances are also dumped into ravines and on the beaches. Some of these are also taken to the landfills, but the country does not possess the capability to effectively shred, crush, or reduce these bulky materials.

The banana industry continues to be the biggest marine debris source in the agricultural sector because of increasing use of sleeving material and boxes for packaging bananas. Poor handling and disposal practices result in these materials being washed away from slopes into fast-flowing rivers that later enter into the sea. Also washed away with these materials are agriculture-based wastes contributed by the coconut and bayoil industries. Along with these synthetic materials, large amounts of vegetation are also washed into the marine environment. Solid waste from commercial activities, especially from grocery shops and self-service outlets in rural areas, is often dumped on the coasts and over cliffs. These wastes are composed of plastic, cardboard, cans, and bottles.

Other important sources of debris originate from storm drains in urban and rural areas. Although an antilitter act exists in Dominica, lack of enforcement results in widespread littering, and large volumes of material are washed into the sea: Beaches strewn with debris after heavy downpours are common sights. However, new approaches are presently under consideration to improve the solid waste management program in Dominica [e.g., the Organization of Eastern Caribbean States' (OECS) Solid-Waste Management Project] and to expand collection services throughout the island (Barnett, Chapter 14, this volume).

St. Lucia

Solid waste management is a comparatively high priority for the government of St. Lucia. Waste disposal is a major problem, requiring urgent attention and improvement. Solid waste collection services are provided to the majority of people living in urban areas, but there is no collection system in the rural areas. At present, people living outside principal district towns are advised to bury non-combustible refuse and burn the combustible portion (e.g., paper, plastic, and garden waste).

The two major dumpsites in St. Lucia are located in Ciceron and Vieux Fort, respectively (see Fig. 30.2). The dumpsites are used for disposal of waste collected from the urban areas in the vicinity of the dumpsites, including the towns of Gros-Islet, Castries, Vieux-Fort, and Soufriere.

Several villages in the rural areas have more crude means of disposing of their wastes. In the village of Anse La Raye, garbage is disposed at a dump on the bank of an estuary. There is no cover, and the garbage is regularly burned. At Canaries, crude dumping takes place on the Canaries River bank and on the coast. Until 1993, crude dumping was also practiced at village dumpsites on the coast at Micoud and Dennery. A recommendation was made in September 1992 by Scheu (1992a) that the Micoud and Dennery dumping area be closed and the waste be transported to the dump at Vieux-Fort in the south. This recommendation was implemented in 1993.

In areas where collection services exist, but are not efficiently operated (e.g., Castries), solid waste dumping in rivers, ghuts, and drains is widely practiced. This leads to severe contamination of the coastal areas when these dumped wastes are washed down during heavy rains. The severe contamination of Castries Harbour is caused by these practices, which result in garbage being brought down by the Castries River.

It is estimated that approximately 60% of the waste generated in St. Lucia is collected. About 85% of the total waste collected in St. Lucia is generated in Castries, Gros Islet, and Vieux. In addition to domestic waste from residential areas, other sources such as waste from commercial enterprises, industries, institutions, streetsweeping and drain cleaning activities, agricultural waste, and garden and bulky waste must also be considered if the land-based sources of marine debris in St. Lucia are to be controlled. The problem is the same in Dominica.

British Virgin Islands (Tortola)

In past decades, solid waste disposal in Tortola was used to reclaim wetlands, in particular shallow mangrove ponds in the coastal region. Virtually all larger mangrove areas in the vicinity of Road Town are already destroyed, namely the Duffs Bottom area (landfill until 1986), Pockwood Pond (totally destroyed by an oil spill), and the Flamingo Pond at Coxhealth (landfill since mid-1986). Following a proposal made by Dillon (1988), the government agreed to build a municipal waste incinerator. Reasons for the choice of incineration technology included these:

Reduction of the amount of waste to save landfill capacity (most of the island is already developed and almost the entire island is private property)
Reduction in the cost of cover material
Avoidance of open burning at the landfill.

The incinerator has been constructed at Pockwood with the Coxhealth landfill site used for the disposal of residues from the incinerator. Concern over the land-based sources of marine debris in Tortola is expected to be dramatically reduced, because it is believed that the main source of contamination of the coastal area was from the Coxhealth dump, which is approximately 50 m from the coastline. Collection services are reliable and regular, and the responsible Department's budget seems to be sufficient. Disposal sites in Tortola will handle primarily residues from the incinerator.

British Virgin Islands (Virgin Gorda)

Refuse collected in Virgin Gorda is disposed at a site in the Cooper Mine area that has operated for more than 25 years. This site is located very close to the coastline, and debris contamination of the coast waters is quite common. Disposal standards at this site are poor because both supervision and equipment are inadequate. Open burning is practiced, and waste is never covered. The site at Cooper Mine is being replaced by a new site at Valley Hill. It is expected that this change will significantly reduce the risk of contamination of the coastal waters because the new site is much further inland. However, it has been suggested by Scheu (1993) that the Cooper Mine site, if adequately maintained, could have less overall impact on the environment.

British Virgin Islands (Anegada)

Anegada is a very flat limestone (coral) island with maximum elevation of some 20 ft. Most of its approximately 160 inhabitants live in Settlement, a small town close to the south coast. A private contractor is responsible for domestic refuse collection, which is carried out once weekly. An estimated amount of less than 1 ton weekly is collected and disposed to a refuse tip [dumpsite] at Nutmeg Point. It is believed that the contamination of coastal waters from the Nutmeg Point dump is negligible.

In the other islands of the British Virgin Islands domestic garbage is transported to the larger islands for disposal. However, the temptation to dump garbage is always present and ocean dumping of garbage does occur. Dumping of garbage outside designated areas in B.V.I. occurs on a very limited scale. Dumping occurs in Hansome Bay and in Gum Creek in Virgin Gorda. The responsible authority (Solid Waste Management Department) is cognizant of these occurrences, and cleanup measures are undertaken weekly. Dumping in ghuts and in coastal waters also occurs in Tortola, and cleanup measures are implemented as necessary.

Conclusion and Recommendations

Solid waste disposal is a serious environmental issue in all countries in the Lesser Antilles. Official and unofficial means of disposing of refuse continue to have undesirable impacts on the natural environment of these island systems. Present waste disposal sites, some formally operated and others informally tolerated by governments, are of concern for several reasons. In addition to the direct health impacts, there are negative impacts on the tourism and investment sectors of the economy from aesthetic concerns (e.g., litter, overflowing garbage collection containers, disagreeable odors, unsightly dumps, and contamination of the coastal areas). Many disposal sites are located in wetlands close to coastlines, where they destroy productive plant communities, displace wildlife, and reduce marine water quality via surface runoff and toxic leachates. Away from the urban population centers, the number of officially designated areas for solid waste disposal is inadequate to meet rural demands, with disposal taking place at ad hoc "unofficial" dumps or in watercourses, beaches, and the sea.

St. Lucia and Dominica share similar problems in disposing of solid waste. Systems for collection are neither maintained nor operated efficiently. Budgetary restrictions and ineffective management of the systems in place are the overriding reasons for this situation. Responsibility for solid waste collection and disposal resides with the Ministries of Health in the respective countries, and this sometimes lead to diffused focus on this issue. In direct contrast, the British Virgin

Islands has a Solid Waste Department dealing specifically with waste collection and disposal, and this department operates very efficiently. An adequate budget for the execution of the Department's responsibilities is the most critical factor in effective control of the land-based source of marine debris.

Scheu proposed the sanitary landfill technique as the most economical method of solid waste disposal for St. Lucia and Dominica (1992b). The incineration of municipal refuse before disposal of the residues to landfill sites is obsolete, because refuse in St. Lucia and Dominica is unsuitable for self-sustaining combustion. Composting of vegetable waste may prove feasible to reduce the amount of waste before disposal at landfill sites. Regarding recycling of valuables, Scheu recommended an evaluation of the potential markets in the Eastern Caribbean region, which should include identification of market prices for glass, metals, paper, and plastics as well as a review of the experiences of neighboring countries in recycling endeavors.

The review of existing disposal practices in small towns clearly shows that continuing the operations of small disposal sites is likely to lead to a large number of crude dumping areas in the country, each contributing to coastal pollution. It is, therefore, recommended that the number of landfill sites be kept to a minimum. This can be achieved by introducing common disposal from several communities at centralized, well-managed sanitary landfill sites.

Public Education Campaigns are essential in alleviating the solid waste problem in these countries. This is particularly important because crude dumping has been practiced for generations and may even be regarded as a cultural feature in some small rural villages. Programs on waste reduction, composting, and reuse are essential, and governments in the region need to encourage implementation of such programs through general education and technical training. Markets for recyclable products must be developed and accepted in the Caribbean region before a successful program in this area can be mounted.

Finally, monitoring programs for assessing marine debris need to incorporate approaches for determining both land-based and maritime contamination sources. This information is critical to the formulation of effective mitigative measures to reduce and control the problem of marine debris.

31.
Strategies to Reduce, Control, and Minimize Land-Source Marine Debris

Michael Liffmann, Bob Howard, Kathy O'Hara, and James M. Coe

Introduction

While acknowledging that marine debris arises from the actions of man, it has become increasingly clear that it cannot be attributed solely to man's activities in the oceans. In this paper, we (1) identify the primary land-based sources of marine debris; (2) discuss the main coastal and upland activities that affect the discharges of plastic waste and other floatable debris into the marine environment; and (3) conclude with some ideas concerning strategies and activities to reduce, control, and minimize these sources.

The management of solid wastes presents a growing problem around the world, directly affecting the volume of marine debris. The problem is likely to become even more acute with industrialization and population growth, particularly in coastal areas. The quantity and the constituents of solid wastes vary throughout the world, because both are determined by social customs and living standards. The specific nature of the problems varies from country to country and from cities and towns to rural communities.

It is difficult to trace common debris items to a single source. Items found along a beach could be the remains of a nearby resident's picnic, trash from a careless boater far up a nearby river or out at sea, or from litter washed down a city's storm sewer. Precise data concerning the primary sources of land-originated, ocean-bound wastes are not available, but some estimates have been ventured. The United Nations Joint Group of Experts on the Scientific Aspects of Marine Pollution (GESAMP) has concluded that, globally, 80% of marine pollution stems from land sources and that waste materials reach the oceans via "direct discharges, runoff and to a lesser extent, the atmosphere."

Although large coastal population centers are the primary land-based offenders, there is evidence that marine debris and other forms of pollution can be attributed to residents and recreationists who may not live close to the sea or, for that matter, even visit the beaches. Worldwide, 6 of 10 people live within 60 km of the coast and the population of the coastal zone is projected to double in the next generation (Karau 1992). In the United States, 50% of the population resides within 50 miles of the ocean or one of the Great Lakes.

It is difficult to enforce laws that prevent an individual from leaving trash where it can be washed into nearby storm sewers or from carelessly tossing an article onto the street or into a nearby waterway. Efforts have been made by governments to control urban sources of debris through educational programs to reduce generation, as well as through control technologies to collect debris before and after it reaches the waterways. However, preventing the discharge of persistent waste in urban, rural, and recreational areas has not been considered a high priority by many governments and societies. Much remains to be done to effectively control these sources of debris.

In considering the possible solutions to the land-based sources of marine debris, two points should be kept in mind. First, the relative contributions of urban and coastal rural and upland areas differ by region, depending on the degree of industrialization of countries, their population densities, and the extent of maritime and other offshore activities. Furthermore, the single biggest difference between rural and urban sources, particularly in developing regions, is access to solid waste management systems. While urban districts might be poorly served, their infrastructure, coverage, and services are typically far better than those of rural areas.

Second, it should be recognized that policies addressing specific environmental quality problems, such as solid waste and marine debris, differ between developed and developing areas. There are many impediments to action. Even when straightforward ways of tackling marine litter and other environmental problems exist, governments have often found it difficult to translate them into effective policy. The reasons for the gap between intentions and performance include political pressures; formidable financial, institutional, and infrastructure obstacles; an absence of data and knowledge; and limited awareness and participation of local people in finding solutions.

Sources of Marine Debris from Land

The three main land-based sources of plastic and other floatable debris are (1) waterborne municipal and industrial solid wastes, (2) sewage, and (3) the terrestrial litter refuse that are transported to sea by natural and man-made drainage systems from upland and coastal urban and rural sources.

Even in developed countries where rural areas are usually served as efficiently as most towns and cities, the collection and transportation of domestic solid waste is the most difficult and expensive aspect of solid waste management. On average, 70% of the entire solid waste-handling budget is consumed in collecting and transporting, leaving little money for treatment and adequate disposal. In developing nations, this problem is intensified because collection and transport are difficult. Access to outlying communities and rural locations frequently is poor or nonexistent. Upland and coastal remote areas are not readily accessible by collection vehicles. People in these areas often resort to composting of organic waste and to reuse of some materials, but most of the waste is discarded in open dumps or is used to reclaim and level less usable areas like ravines, old quarries, and valleys. Many coastal communities use solid waste to help fill in and "reclaim" wetlands, notably marshes, salt ponds, and mangroves.

Even in the rare instances when the systems and budgets are adequate for collection, safe disposal of collected wastes often remains a problem. Sanitary landfills have become the norm in only a handful of locations. More controlled sanitary landfills are desperately needed to bring some semblance of order and hygiene to the dumps and to help reduce the risk of pollutants, including plastics and other floatables, entering the ocean.

Open dumping and uncontrolled landfilling remain the almost universal methods for disposing of solid waste in developing nations, although those countries that enjoy some level of economic prosperity have begun to use sanitary disposal in landfills (Holmes 1984). Gourlay (1992) note that more than 90% of domestic waste, in some cases all of it, in both the industrialized and the developing world is taken away and dumped in landfills or added to growing mountains of refuse. The remainder is either burned or dumped at sea. Although modern containment landfills are being built to avoid groundwater and surface water contamination, heavy rains and storms have been known to cause existing landfills to overflow, and many of the older facilities are located close to water bodies that empty into the sea.

Storm drain and sewage systems also carry trash to the sea. Generally, rainwater washes pollutants, dirt, land debris, and refuse from streets, parks, hillsides, and gullies into natural as well as man-made drainage systems. Although many people do not realize it, these drainage systems are direct links to rivers and oceans that are sometimes miles away. In coastal cities, virtually everything found in the streets and gutters can be swept into the ocean.

Inadequate (or nonexistent) facilities for the treatment of sewage before discharge allow the release of large amounts of plastic debris into aquatic systems. Although this problem primarily affects the world's major population centers, it also poses problems for smaller population hubs in the more remote areas of many countries. In some urban areas, stormwater systems are connected directly to municipal sanitary sewers. As a result, heavy rainfall can cause the capacity of these systems to be exceeded, discharging nondegradable wastes along with untreated sewage. Typical indicators of debris from these combined sewer overflows (CSOs) are tampon applicators and latex condoms. The progressive degradation of rivers, estuaries, and coastal waters by sewage and floatable debris is a massive worldwide public and environmental health problem.

Even where public works have improved and sanitary landfills are in place, residential and recreational litter remains a major problem. Because of their sheer abundance, recreational users have the potential to be one of the largest sources of marine debris in many areas. This is particularly problematic because recreational users concentrate their activities in coastal and upland areas that are most vulnerable to the negative impacts of marine debris. Currently there are no reliable quantitative estimates of the amount of debris generated by campers, boaters, divers, fishers, or beachgoers, but the amounts are widespread and substantial.

Debris generated during recreational use of marine and upland areas may be grouped under two general types: (1) packaging from food and domestic goods and (2) items associated with the specific activities conducted by recreational users. The former includes items such as plastics (i.e., cups, plates, and eating utensils), beverage cans and bottles, bags and wrappers, beverage ringcarriers, straws, Styrofoam coolers, popsicle sticks, sunscreen and insect-repellent containers, and diapers (nappies). The most common debris item found in the International Coastal Cleanups arises from littering: the cigarette filter. These synthetic fiber filters are left behind by beachgoers, washed off coastal and upland streets, and flicked into waterways, worldwide.

Key Players in the Solution

Any strategy to reduce and control the land-based sources of marine debris must incorporate the efforts of a wide variety of governments, organizations, institutions, and businesses. The following is a preliminary list of the types of entities that should play important roles in the solutions to land-based marine debris.

International organizations (e.g., International Maritime Organization [IMO], United Nations Environment Programme [UNEP], World Health Organization [WHO], World Bank, and regional development banks)
National governments and agencies
Local and state governments (e.g., municipal water, solid waste and sewage treatment authorities)
Urban planning and zoning authorities
Solid waste management industry including waste management companies, recyclers, and industry trade associations
Tourist boards and associations
Outdoor recreational industries
Manufacturers, vendors and suppliers that provide goods to outdoor recreation industries
Hotel and resort operators
Press and media
Celebrities
Civic and religious organizations
Nongovernmental groups
Consumer organizations
Libraries

These players can work independently and in various combinations to reduce the generation and improve the disposal of potential marine debris.

Strategies for Land-Source Marine Debris

There is a clear consensus that land-based sources of marine debris are not adequately addressed and that further serious degradation of the marine environment will occur without concerted new actions. Most international legal and policy experts on marine debris advocate a worldwide strategic approach to legitimize and guide actions to control pollution at all levels.

The problems caused by marine debris from land sources have stimulated actions at all levels. There are international agreements, state and local legislation, and regulations that address solid waste and marine debris (e.g., litter and floatables). In general, international agreements address regulatory, enforcement, and research policy, and are considered good-faith understandings between countries. Policies identified in these agreements are expected to be implemented through the domestic environmental policies and statutes of the participating countries. Hence, national interests provide the real authority and power to enact the recommendations outlined in international agreements.

National policies and statutes provide an important impetus for action at other levels and promote consistency among a nation's various governmental activities. Actions at the state and local level are probably the most important and effective means of reducing marine debris: local stormwater management and sewage treatment programs can be significant in reducing these sources of debris; local recycling, reuse and waste reduction efforts, streetsweeping, and catch basins for land-generated wastes have resulted in the reduction of urban-source marine debris; beach and shoreline cleanups and skimmer vessels have effectively controlled adverse impacts from marine debris after it reaches surface waters. Some communities have used minimum-security prisoners to help in beach and shoreline cleanups. Volunteer organizations have been an important force for education, cleanup, and political support at the international, national, and local levels.

Within this framework of international to local efforts, the key interdependent strategies for affecting long-term reduction in the land-based sources of marine debris include the following:

Adopting and enforcing legal and administrative measures at all levels

Strengthening education and awareness efforts

Strengthening institutions for the development of a solid waste management infrastructure

Developing and implementing appropriate technologies

Building political commitment

Strengthening assessment and monitoring.

These strategies are applicable at all organizational levels, from households to cities to the United Nations. They require mutually supportive actions at various levels: international (global, regional, bilateral), national, and local, as well as supportive technical and financial assistance.

The will to address the marine debris issue is a precondition for the successful implementation of many of these strategies. In developing countries in particular, political commitment must occur at the national and local levels if investments in infrastructure are to be realized. Governments must be made aware of the adverse economic implications of unabated marine debris. This need becomes particularly critical given population increases, industrialization trends, tourism development schemes, and increasing dependency on the coastal areas. Global consensus on principles and strategies that apply to land-based marine debris, as well as financial and technical assistance, can help build the necessary national commitments.

The Objectives

The obvious worldwide scope of the marine debris problem and the major contribution from land-based sources certified in this chapter justify a broad strategy for control. Each element of the strategy must be associated with one or more equally broad objective(s) to give overall direction to the efforts expended at each level. For example, if the strategy is to strengthen education and awareness, the objective might be to inform government officials about the problems of and solutions to the land-based sources of marine debris. This objective can then be interpreted for different governments and levels of government so that specific actions suitable for advancing the objective can be selected and refined.

With the understanding that each objective is to be interpreted at any level at which it may be applicable, the objectives of a broad, land-based-source control program for the next several decades should include the following:

1. Implement the Global Programme of Action to Protect the Marine Environment from Land-Based Activities' (GPA) targets for solid waste disposal, litter control, and sewage treatment worldwide (Nollkaemper, Chapter 22, this volume)
2. Advance technologies for source reduction, recycling and degradable materials
3. Establish continuous, regional and national education programs for governments and the public on the need for marine debris control
4. Create and fund institutions to build, and maintain, solid waste infrastructure in developing countries
5. Advance technologies for solid waste management and disposal in developing countries
6. Promulgate and enforce uniform solid waste, litter, and floatables control and containment laws and policies
7. Establish regional, marine debris monitoring networks.

Activities to Achieve the Objectives

This section lists examples of actions that could assist in reaching the foregoing objec-

tives at a number of levels, thereby reducing the production of marine debris from land-based sources. There are five points in the general life cycle of land-source marine debris where interventions may be taken. Those points are source generation (including product demand and behavior), on-land management, transport, deposition, and impacts. The marine debris reduction actions that can be taken at these points include demand reduction, source reduction, improving management of potential debris material, cleaning up waterways (including beaches and ocean bottoms), and remedial actions such as beach closures, animal rescues, and vessel towing and repair. The tools available to reduce marine debris at these opportunity points can include improved technology, organizational and operational changes, education, regulation and enforcement, and economic adjustments (Laska, Chapter 13, this volume). The following recommended actions fall within this general framework.

1. Implement the GPA Targets for Solid Waste Disposal, Litter Control, and Sewage Treatment Worldwide.
    a. International organizations (e.g., UNEP, IMO, WHO, IOC, [Intergovernmental Oceanographic Commission], PAHO [Pan American Health Organization]), and international funding agencies (e.g., World Bank, the Global Environment Facility [GEF], regional development banks, and the European Union) should be encouraged to consider the GPA's measures in policy negotiations, investment plans, national environmental action plans, and coastal zone management plans.
    b. National governments should be encouraged to support, through legislation, policy, and resource commitment, the broad achievement of the objectives of the GPA.
    c. National, state, and local governments should immediately develop long-range solid waste management plans that lead to full collection and disposal services.
2. Advance Technologies for Source Reduction, Recycling, and Degradable Materials

a. Source reduction can be promoted through efforts to support the use of reduced or alternate packaging and product materials and by promoting recycling. Efforts should be targeted toward materials, packaging, and products that are most typically found as marine debris. The following were identified as potential source-reduction actions:

(1) Reduce packaging for all items.

(2) Increase efforts to develop and transfer technology to produce alternative products and packaging less likely to become marine debris, utilize raw materials having less impact on the marine environment, and increase recycling and reuse.

(3) Promote the use of standardized or alternative materials that are easier to recycle and have less potential impact.

(4) Encourage the development and marketing of degradable and compostable alternative materials for the manufacture of common litter and marine debris items.

(5) Implement programs to encourage collection and recycling of particularly problematic debris (e.g., monofilament fishing line and fishing nets).

3. Establish Continuous Regional and National Education Programs for Governments and the Public on the Need for Marine Debris Control

a. Education must be a significant component of an effective marine debris control program. The residential, rural, commercial, industrial, and governmental communities must understand the importance of debris control, the impacts of debris in the marine environment, and the actions that can be taken to reduce debris production. The following activities will help reduce the production of marine debris from land-based sources:

(1) Promote programs to educate the public and industry about the sources and environmental, economic, and other impacts of marine debris as well as the available control measures.

(2) Promote awareness of the role waste minimization, recycling, and reuse can have in reducing the extent of marine debris.

(3) Inform the public and local governments about the role stormwater management has in reducing urban contributions to marine debris and the need for debris control strategies.

(4) Promote programs for storm drain stenciling to increase awareness of the connection of storm sewers to the oceans.

(5) Inform the public about volunteer opportunities to clean up land, beach, waterway, and ocean-bottom debris.

(6) Promote activities that inform the public of the types of products and packaging that increase (or minimize) the potential for creating marine debris.

(7) Develop and promote programs in elementary schools regarding marine debris and actions that can be taken by students to address this problem.

(8) Expand the International Coastal Cleanup Campaign to inland waterways to raise awareness and knowledge of land-based debris sources.

(9) Create a partnership with the manufacturers of products used by beachgoers that develops a widespread public awareness campaign on marine debris problems (e.g., imprinting antilitter messages on product containers, in-store displays, print advertisements, and toll-free phone numbers).

(10) Create an international marine debris educational materials resource guide (in a variety of languages) that would be updated annually and maintained by the international coastal cleanup coordinator network.

(11) Create an international postage stamp with a marine debris theme.

(12) Hotel and resort area businesses should provide information (e.g., tent cards, brochures, and litter bags) on marine debris to their guests.

(13) Insert literature in products commonly purchased by recreational user groups that will enable individuals to appeal to appropriate agencies for the provision of adequate waste reception facilities at appropriate sites.

(14) Use population education techniques that are culturally sensitive.

(15) Encourage local initiatives that promote a sense of "ownership" of natural resources.

(16) Recognize and award communities—particularly children, nongovernment, and other community-based organizations—that prevent and reduce marine debris.

(17) At the regional level, replicate the IOC/UNEP CEPPOL education and outreach model, which emerged from the Marine Debris and Waste Management Action Plan for the Caribbean (Barnett, Chapter 14, this volume).

4. Create and Fund Institutions to Build, and Maintain, Solid Waste Infrastructure In Developing Countries

a. Encourage "earmarking" of source-specific funds (tariffs and taxes) to support solid waste management programs.

b. Encourage the establishment of national solid waste authorities, or appropriate local or regional entities, to be charged with solid waste collection transport and disposal.

c. Promote the development of local, stormwater, and solid waste utility districts to generate revenue and implement controls related to urban marine debris.

d. Encourage financing the proper maintenance and operations of solid waste management systems through creation of an international revolving fund under the GPA.

e. Encourage the planning, financing, location, construction, and equipping of solid waste management systems by drawing on the experience and assistance of international organizations and developed countries.

f. Encourage comprehensive waste management and reduction through establishment of national and local authorities specifically for solid waste management.

g. Develop partnerships with local agencies, private sector, and other organizations (e.g., NGOs, CBOs, and educational institutions) to help design, locate, construct, operate, and maintain waste reduction, collection, and disposal systems.

5. Advance Technologies for Solid Waste Management and Disposal in Developing Countries.

a. Management and disposal of debris before it becomes waterborne and dispersed provides significant efficiencies in manpower and greater debris reduction effectiveness. The following actions can reduce the production of marine debris:

(1) Strengthen solid waste and sewage management through technical assistance and training provided by developed countries and international organizations.

(2) Adapt sanitary landfill technology and practices to meet the wide variety of solid waste disposal challenges of developing countries.

(3) Investigate the feasibility of broader utilization of small-scale, solid waste-handling and disposal equipment such as compactors, incinerators, and composters.

(4) Construct and improve maintenance of catch basins, settling ponds, and other control methods for sewage overflows, storm drains, and CSOs.

(5) Use silt fences and improve trash management at construction, commercial, and industrial sites.

(6) Increase efficiency, effectiveness,

and use of streetsweeping in targeted urban and recreational areas to reduce contributions to storm sewers and area surface waters.

(7) Include floatables controls in coastal cities' stormwater control and sewage treatment plants.

(8) Ensure adequate waste reception facilities in a variety of port and coastal recreational sites and ensure that the waste in these facilities is collected as part of a total solid waste management plan.

b. Locating and subsequently removing debris from surface waters, shorelines, and ocean bottoms is the last opportunity to intervene in the life cycle of marine debris. However, these activities can be costly and resource intensive. Locating and removing debris at the closest point to the source is cost efficient and reduces impacts. The following actions will reduce land-source marine debris:

(1) Develop and transfer improved technologies to remove and handle debris in surface waters, on shorelines, and underwater.

(2) Promote the use of debris booms, boat skimmers, and other concentration and removal technologies as part of a debris management and removal program.

(3) Enlist the assistance of area residents, marine users, local civic organizations, area business, industry, and others in cleanup projects.

(4) Develop a communications network to locate, track, and assess urban floatables events.

6. Promulgate and Enforce Uniform Solid Waste, Litter, and Floatables Control and Containment Laws and Policies

a. No marine debris control program can be successful without an effective enforcement component. In any population there are individuals, institutions, and industries that will not conform to established norms and are not affected by peer pressure to change practices that produce marine debris. Enforcement is necessary to get compliance from these parties and to establish a situation of fairness for those complying with established norms. The following actions should be considered to improve enforcement related to the land sources of marine debris:

(1) Designate a responsible coordinating agency within governments.

(2) International institutions should assist in the development of national laws, regulations and enforcement systems.

(3) Increase priorities and resources dedicated by national, state, and local governments to address marine debris problems through enforcement of existing statutes and regulations.

(4) Encourage citizens to report illegal dumping and to aid in enforcement.

(5) Publicize violations of marine debris laws and the amounts of assessed fines.

(6) Develop more effective methods for surveillance and identification of debris discharges.

(7) Support programs to cross-train local, state, and federal officers to enforce marine debris-related statutes.

(8) Evaluate the effectiveness of existing international agreements and national, state, and local statutes and regulations addressing land sources of marine debris.

(9) Develop national, state, and local stormwater control strategies that include floatable control requirements and enforcement authorities.

b. To remedy compliance problems, states should strongly consider applying economic instruments, such as fees or taxes, to ensure that consumers and manufacturers take into account true costs of their waste disposal decisions.

(1) Create financial or regulatory incentives for recycling and reuse.

(2) Consider taxing the producers,

handlers, sellers, and consumers of high-potential marine debris items based on the cost of marine debris cleanup efforts.

(3) Develop financial incentives for producers, handlers, sellers, and consumers to convert to nondebris-producing materials, processes, products, and packaging.

(4) Increase local, state, and national priorities and resources for collection, disposal, and enforcement actions for materials likely to become marine debris.

(5) Establish programs to recognize and reward individuals, institutions, and industries who reduce sources of debris.

(6) Consider the use of court-ordered community service and low security risk prisoners to collect and dispose of debris.

7. Establish Regional Marine Debris Monitoring Networks

a. Adequate monitoring is necessary for targeting and effectively preventing marine debris. Monitoring should yield data on the land sources of marine debris and its effects. Source monitoring for specific waste types (e.g., medical waste) should be considered. All surveys and monitoring programs should be carefully designed to ensure the statistical validity of their results. As part of the monitoring effort, the following survey activities should be undertaken and coordinated at national and regional levels.

(1) Solid waste surveys, including identification of disposal methods and compositions of the waste stream.

(2) Photographic surveys of beaches.

(3) Rapid assessment surveys, such as for roadsides, streams, lakes, and beaches.

(4) Surveys at specific sites to establish transport via rivers, the types and quantity of the materials, and the mobility of the waste.

(5) Conduct periodic surveillance activities to identify surface water, shore-

line, and underwater areas where debris collects.

(6) Divers should be enlisted in monitoring programs for underwater debris, both for qualitative studies as well as topics of special interest such as ghost-fishing and benthic species impacts.

b. International monitoring programs should be developed, strengthened, or continued, where appropriate, to assess the effects of debris on regional ecosystems and to identify and transboundary effects of marine debris.

Conclusion

Appreciation of the true scale of challenges presented by the land-based sources of marine debris is just beginning. At current levels of understanding, it is difficult to elicit consensus on the utility of all the solution strategies presented in this paper. Many other priorities such as public safety, public health, economics, and energy conservation, for example, may play significant roles in determining which activities are feasible to reduce marine debris. Nevertheless, the strategies outlined in this paper can provide a starting point for considering actions that can and should be taken to address the land-source marine debris problem.

Reduction of the land sources of marine debris is a socioeconomically complex and potentially resource-intensive undertaking. To effectively address these problems will require support from citizens, industries, and governments. All three sectors must commit to changing their behavior with regard to waste management practices. All parties must consider the potential for waste production in lifestyle, purchasing, and resource commitment decisions. Additional attention will have to be given to enforcement and measurement to assure equity and verify the effects of programs and actions. And, finally, there also will have to be improvements in technology for every part of the process.

General awareness of the challenges posed

by the land-based sources of marine debris is rising, but much remains to be done. Public, industry, and government priorities do not yet incorporate this issue and its important connection with the solutions to long-term public and environmental health problems. The land-source marine debris problem is a symptom of the continuing lag between the rising consumption associated with economic development and the waste disposal capacity that society is willing to support. A pertinent question at this juncture is "How much waste are we willing to live with?" Clearly, if the answer is to be "less than we live with now," greater effort must be taken to draw immediate attention to solving this global, land-based pollution problem.

Literature Cited

Afelin, C., and B. Puleloa. 1982. Marine turtles. *Scientific Event Alert Network (SEAN) Bulletin* 7(1):13.

Ainley, D. G., W. R. Fraser, and L. B. Spear. 1990a. The incidence of plastic in the diets of Antarctic seabirds. In Proceedings of the Second International Conference on Marine Debris, 1989, ed. R. S. Shomura and M. L. Godfrey. U.S. Department of Commerce. NOAA-TM-NMFS-SWFSC-154:682–691.

Ainley, D. G., L. B. Spear, and C. A. Ribic. 1990b. The incidence of plastic in the diets of pelagic sedabirds in the eastern equatorial Pacific region. In Proceedings of the Second International Conference on Marine Debris, 1989, ed. R. S. Shomura and M. L. Godfrey. U.S. Department of Commerce. NOAA-TM-NMFS-SWFSC-154:653–664.

Ajzen, I., and M. Fishbein. 1980. *Understanding Attitudes and Predicting Social Behavior.* Englwood Cliffs, NJ: Prentice-Hall, Inc.

Akerlof, G. 1980. A theory of social custom of which unemployment may be one consequence. *Quarterly Journal of Economics* (June):749–775.

Akerlof, G. 1982. Labor contracts as partial gift exchange. *Quarterly Journal of Economics* 97:543–569.

Akerlof, G. 1983. Loyalty filters. *American Economic Review* 73(1):54–63.

Akers, R. L. 1985. *Deviant Behavior: A Social Learning Approach.* 3d ed. Belmont, CA: Wadsworth.

Akers, R. L., M. D. Krohn, L. Lonza-Kaduce, and M. Radosevitch. 1979. Social learning and deviant behavior: A specific test of a general theory. *American Sociological Review* 44:635–655.

Alaska Sea Grant. 1988. Oceans of plastic: A workshop on fisheries-generated marine debris and derelict fishing gear. Portland, OR. Alaska Sea Grant College Program, University of Alaska, Fairbanks.

Aldershoff, W. G. 1982. Gezondheidsaspecten van PVC voor levens-middelen. *Plastica* 35:58–60.

Allingham, M., and A. Sadnmo. 1972. Income tax evasion: A theoretical analysis. *Journal of Public Economics* (September).

Alverson, D. L., and J. A. June, eds. 1988. In Proceedings of the North Pacific Rim Fishermen's Conference on Marine Debris, 1987. Seattle. Natural Resources Consultants. 460.

Andrady, A. L. 1990. Environmental degradation of plastics under land and marine exposure conditions. In Proceedings of the Second International Conference on Marine Debris, 1989, ed. R. S. Shomura and M. L. Godfrey. U.S. Department of Commerce. NOAA-TM-NMFS-SWFSC-154:848–869.

Anonymous. 1975. Plastic cups found in fish. *Marine Pollution Bulletin* 6:148.

Anonymous. 1981. Galapagos tainted by plastic pollution. *Geo* 3:137.

Anonymous. 1988. Plastic may have caused beached whales death. *Sport Fishing Institute Bulletin* no. 395.

Anonymous. 1992. Report on assessment and avoidance of incidental mortality in the Convention Area 1991/1992. Hobart, TAS Australia. Committee for the Conservation of Antarctic Marine Living Resources. CCAMLR-XI/BG/7.

Anonymous. 1994. *Fishing News International* (February).

Arbuthnot, J., R. Tedeschi, M. Wayner, J. Turner, S. Kressel, and R. Rush. 1977. The induction of sustained recycling behavior through foot-in-the-door technique. *Journal of Environmental Systems* 6(4):355–368.

Armstrong, W. P. 1994. Floaters. *Sea Frontiers* 40:24–30.

Arnold, G., T. Clayton, J. Craig, G. Lewis, and M. Cornelius. 1994. Litter associated with storm water discharges in Auckland City. In Poster Abstracts and Manuscripts from the Third International Conference on Marine Debris, 1994, ed. J. C. Clary. U.S. Department of Commerce. NOAA-TM-NMFS-AFSC-51.

Arnould, J. P. Y. 1993. Beach debris survey—Main Bay, Bird Island, South Georgia 1990/91. Hobart, TAS Australia. Committee for the Conservation of Antarctic Marine Living Resources. CCAMLR-XII/BG/3:5.

Aronfreed, J. 1968. *Conduct and Conscience*. New York: Academic Press.

Aronfreed, J. 1969. The problem of imitation. In *Advances in Child Development and Behavior*, ed. L. P. Lipsutt and H. W. Reese. Vol. 4. New York: Academic Press.

Arrow, K. R., R. Solow, P. R. Portney, E. E. Leamer, R. Radner, and H. Schuman. 1993. Report of the NOAA panel on contingent valuation. *Federal Register* 58, no. 20 (15 January):4601.

Ashmole, N. P. 1971. Seabird ecology and the marine environment. In *Avian Biology*, ed. D. S. Farner and J. R. King, 223–286. New York: Academic Press.

Augerot, X. 1988. Background paper for a national workshop on fisheries-generated marine debris and incentive-based regulatory systems. In Oceans of Plastic. Alaska Sea Grant Report no. 88-7.

Azzarello, M. Y., and E. S. Van Vleet. 1987. Marine birds and plastic pollution. *Marine Ecology Progress Series* 37:295–305.

Baba, N., M. Kiyota, and K. Yoshida. 1990. Distribution of marine debris and Northern fur seals in the eastern Bering Sea. In Proceedings of the Second International Conference on Marine Debris, 1989, ed. R. S. Shomura and M. L. Godfrey. U.S. Department of Commerce. NOAA-TM-NMFS-SWFSC-154:419–430.

Balazs, G. H. 1978. A hawksbill turtle in Kanehoe Bay, Oahu. *Élepaio* 38(11):128–129.

Balazs, G. H. 1980. Synopsis of biological data on the green turtle in the Hawaiian Islands. U.S. Department of Commerce. NOAA-TM-NMFS-SWFC-7:141.

Balazs, G. H. 1983. Sea turtles and their traditional usage in Tokelau. *Atoll Research Bulletin* 279:1–29.

Balazs, G. H. 1985. Impact of ocean debris on marine turtles: Entanglement and ingestion. In Proceedings of the Workshop on the Fate and Impact of Marine Debris, 1984, ed. R. S. Shomura and H. O. Yoshida. U.S. Department of Commerce. NOAA-TM-NMFS-SWFSC-54:387–429.

Baltz, D. M., and G. V. Morejohn. 1976. Evidence from seabirds and plastic particle pollution off central California. *Western Birds* 7:111–112.

Bandura, A. 1969. Social-learning theory of identificatory processes. In *Handbook of Socialization Theory and Research*, ed. D. A. Goslin. Chicago: Rand McNally.

Barros, N. B., D. K. Odell, and G. W. Patton. 1990. Ingestion of plastic debris by stranded marine mammals from Florida. In Proceedings of the Second International Conference on Marine Debris, 1989, ed. R. S. Shomura and M. L. Godfrey. U.S. Department of Commerce. NOAA-TM-NMFS-SWFSC-154:746.

Battelle Memorial Institute (BMI). 1990. The study of floatable debris in U.S. waters (Harbor Studies Program), November 1988 through February 1989. U.S. Environmental Protection Agency Office of Water. Final report EPA 503/4-90-003.

Battelle Ocean Sciences (BOS). 1992. The study of floatable debris in U.S. waters (Harbor Studies Program), March 1989 through April 1991. U.S. Environmental Protection Agency Office of Water. Final report EPA 842 R-92-001.

Bauer, H., K. Mitchell, M. Bortman, and R. L. Swanson. 1994. Analysis and interpretation of a needle/syringe disposal survey and the feasibility of a needle/syringe return program. In Reduction in Beach Debris Through Public Education, a report prepared for the U.S. EPA. New York. U.S. EPA Region 2. Grant no. 431-4395A.

Baumol, W., and W. Oates. 1979. *Economics, Environmental Policy, and the Quality of Life*. Englewood Cliffs, NJ: Prentice-Hall, Inc.

Bean, M. J. 1987. Legal strategies for reducing persistent plastics in the marine environment. *Marin Pollution Bulletin* 18:357–360.

Beck, C. A., and N. B. Barros. 1991. The impact of debris on the Florida manatee. *Marine Pollution Bulletin* 22(10):508–510.

Becker, G. 1968. Crime and punishment: An economic approach. *Journal of Political Economy* 76(2):169–217.

Beggs, S., S. Cardell, and J. A. Hausman. 1981. Assessing the potential demand for electric cars. *Journal of Econometrics* 16:1–19.

Bell, F. W., and V. R. Leeworthy. 1990. Recreational demand by tourists for saltwater beach days. *Journal of Environmental Economics and Management* 18:189–205.

Below, T. H. 1979. First reports of pellet ejection in 11 species. *Wilson Bulletin* 91:626–628.

Bengtson, J. L., C. W. Fowler, H. Kajimura, R. L. Merrick, K. Yoshida, and S. Nomura. 1988. Fur seal entanglement studies: Juvenile males and newly-weaned pups, St. Paul Island, Alaska. In Fur Seal Investigations 1985, ed. P. Kozloff and H. Kajimura. U.S. Department of Commerce. NOAA-TM-NMFS-F/NWC-146:34–57.

Bengtson, J. L., B. S. Stewart, L. M. Ferm, and R. L. DeLong. 1989. The influence of entanglement in marine debris on the diving behavior of subadult male northern fur seals. In Fur Seal Investigations 1986, ed. H. Kajimura. NOAA-TM-NMFS-F/NWC-174:48–56.

Bentham, J. 1789. *An Introduction to the Principles of Morals and Legislation,* ed. W. Harrison. Oxford: Basil Blackwell.

Benton, T. 1991. Oceans of garbage. *Nature* 352:113.

Benton, T. 1995. From castaways to throwaways: Marine litter in the Pitcairn Islands. *Biological Journal of the Linnean Society* 56:415–422.

Berger, J. D., and C. E. Armistead. 1987. Discarded net material in Alaskan waters, 1982–84. U.S. Department of Commerce. NOAA-TM-NMFS F/NWC-110:66.

Berry, A. R. 1994. Effects of macro-rubbish on mangrove swamps, *Avicennia marina* var. *resinifera.* Unpublished report, School of Biological Sciences, University of Auckland.

Bertelsen, E., and H. Ussing. 1936. Marine tropical animals carried to the Copenhagen Sydhavn on a ship from the Bermudas. Videnskabelige Meddelelser. Køhbenhavn. *Dansk Naturfredning* 100:237–245.

Bingel, F., D. Avsar, and M. Ünsal. 1987. A note on plastic materials in trawl catches in the Northeastern Mediterranean. *Meeresforschung* 31:227–233.

Bingel, F. 1989. Plastic in the Mediterranean Sea. IOC/UNESCO:1–65.

Bjorndal, K. A., and A. B. Bolton. 1994. Effects of marine debris on sea turtles. In Poster Abstracts and Manuscripts from the Third International Conference on Marine Debris, 1994, ed. J. C. Clary. U.S. Department of Commerce. NOAA-TM-NMFS-AFSC-51.

Bjorndal, K. A., A. B. Bolton, and C. J. Lagueux. 1994. Ingestion of marine debris by juvenile sea turtles in coastal Florida habitats. *Marine Pollution Bulletin* 28(3):154–158.

Blasi, A. 1980. Bridging moral cognition and moral action. *Psychological Bulletin* 88:1–45.

Blue Plan. 1987. Mediterranean basin environmental data (natural environment and resources). Mediterranean Blue Plan Regional Activity Center, Valbonne, France.

Bockstael, N. W., K. E. McConnell, and I. E. Strand. 1988. Benefits from improvements in Chesapeake Bay water quality. Report to the U.S. EPA. Washington, D.C. CR-811043-01-0.

Boesch, D. F. 1977. Application of numerical classification in ecological investigations of water pollution. Ecological Research Series, EPA-600/3-77-033.

Bolten, A. B., and K. A. Bjorndal. 1991. Effect of marine debris on juvenile, pelagic sea turtles. Interim Project Report to the National Marine Fisheries Service Marine Entanglement Research Program. Seattle. Northwest and Alaska Fisheries Science Center. 53.

Bonner, W. N., and T. S. McCann. 1982. Neck collars on fur seals, *Arctocephalus gazella,* at South Georgia. *British Antarctic Survey Bulletin* 57:73–77.

Bourne, W. R. P. 1976. Seabirds and pollution. In *Marine Pollution, ed. R. Johnston, 403–502. New York: Academic Press.*

Bourne, W. R. P. 1977. Nylon netting as a hazard to birds. *Marine Pollution Bulletin* 8(4):75–76.

Bourne, W. R. P., and G. C. Clark. 1984. The occurrence of birds and garbage at the Humbolt Front off Valparaiso, Chile. *Marine Pollution Bulletin* 15:343–344.

Bourne, W. R. P., and M. J. Imber. 1982. Plastic pellets collected by a prion on Gough Island, central South Atlantic Ocean. *Marine Pollution Bulletin* 13(1):20–21.

Brathaug, H., and F. G. Barnett. 1994. MARPOL Annex V and special areas status report. Nassau. Third Caribbean Marine Debris Workshop.

Braune, B. M., and D. E. Gaskin. 1982. Feeding ecology of nonbreeding populations of lairds off Deer Island, New Brunwick. *Auk* 99(1):67–76.

Breen, P. A. 1987. Mortality of Dungeness crabs caused by lost traps in the Fraser River estuary, British Columbia. *North American Journal of Fisheries Management* 7:429–435.

Breen, P. A. 1990. A review of ghost fishing by traps and gillnets. In Proceedings of the Second International Conference on Marine Debris, 1989, ed. R. S. Shomura and M. L. Godfrey. U.S. Department of Commerce. NOAA-TM-NMFS-SWFSC-154:571–599.

Broadrick, T. 1982. Marine turtles. *Scientic Event Alert Network (SEAN) Bulletin* 7(6):18.

Brongersma, L. D. 1968. Notes upon some turtles from the Canary Islands and from Madeira. *Proceedings of the Koninklijke Nederlandse Akademie van Wetenschappen Series C Biological and Medical Sciences* 71:128–136.

Brongersma, L. D. 1972. European Atlantic turtles. *Zoologische Verhandelingen (Leiden)*

121:1–318.

Brothers, G. 1989. Plastic debris summary of lost fishing gear retrieval projects. Information Paper prepared for the Workshop on Plastic Debris and Los and Abandoned Fishing Gear in the Aquatic Environment, 1989. Ottawa, Ontario. Canadian Department of Fisheries and Oceans, Oceanography and Contaminants Branch. 6.

Brothers, G. 1992. Lost or abandoned fishing gear in the Newfoundlands aquatic environment. Paper presented to the C-Merits Symposium "Marine Stewardship in the Northwest Atlantic." St. Johns, Newfoundland, Canada. Department of Fisheries and Oceans.

Brown, C. H., and W. M. Brown. 1982. Status of sea turtles in the Southeastern Pacific: Emphasis on Peru. In *Biology and Conservation of Sea Turtles,* ed. K. A. Bjorndal, 235–240. Washington, D.C.: Smithsonian Institution Press.

Bruner, R. G. 1990. Plastic industry and marine debris: Solutions through education. In Proceedings of the Second International Conference on Marine Debris, 1989, ed. R. S. Shomura and M. L. Godfrey. U.S. Department of Commerce. NOAA-TM-NMFS-SWFSC-154.

Bryan, W. B. 1971. Coral Sea drift pumice on Eua Island, Tonga, in 1969. *Geological Society of America Bulletin* 82:2799–2812.

Butler, J. J., B. F. Morris, J. Cadwallader, and A. W. Stoner. 1983. Studies of *Sargassum* and the *Sargassum* community. Bermuda Biological Station. Special Publication no. 10.

Buxton, R. 1990. Plastic debris and lost and abandoned fishing gear in the aquatic environment. A workshop paper. Halifax, Nova Scotia. Department of Fisheries and Oceans.

Calder, D. R. 1986. Hydroida (hydroid polyps). In *Marine Fauna and Flora of Bermuda,* ed. W. Sterrer, 128–140. New York: John Wiley & Sons.

Caldwell, M. C., D. K. Caldwell, and J. B. Siebenaler. 1965. Observations on captive and wild Atlantic bottlenose dolphins, *Tursiops truncatus,* in the Northeastern Gulf of Mexico. *Los Angeles County Museum Contributions in Science* 91:1–10.

Calkins, D. G. 1985. Steller sea lion entanglement in marine debris. In Proceedings of the Workshop on the Fate and Impact of Marine Debris, 1984, ed. R. S. Shomura and H. O. Yoshida. U.S. Department of Commerce. NOAA-TM-NMFS-SWFSC-54:308–314.

Cameron, T. A. 1988. A new paradigm for valuing non-market goods using referendum data. *Journal of Environmental Economics and Management* 15:355–379.

Cannon, G. A., R. K. Reed, and P. E. Pullen. 1985. Comparison of El Niño events off the Pacific Northwest. In *El Niño North—Niño Effects in the Eastern Subarctic Pacific Ocean,* ed. W. S. Wooster and D. L. Fluharty, 75–84. Seattle: Washington Sea Grant Program at the University of Washington.

Cantin, J., J. Eyraud, and C. Fenton. 1990. Quantitative estimates of garbage generation and disposal in the U.S. Maritime Sectors before and after MARPOL Annex V. In Proceedings of the Second International Conference on Marine Debris, 1989, ed. R. S. Shomura and M. L. Godfrey. U.S. Department of Commerce. NOAA-TM-NMFS-SWFSC-154.

Carlton, J. T. 1987. Patterns of transoceanic marine biological invasions in the Pacific Ocean. *Bulletin of Marine Science* 41:452–465.

Carlton, J. T., and J. B. Geller. 1993. Ecological roulette: The global transport of nonindigenous marine organisms. *Science* 261:78–82.

Carpenter, E. J., and K. L. Smith. 1972. Plastics on the Sargasso Sea surface. *Science* 175:1240–1241.

Carpenter, E. J., S. J. Anderson, G. R. Harvey, H. P. Miklas, and B. B. Peck. 1972. Polystyrene spherules in coastal waters. *Science* 178:749–750.

Carr, A. 1986. Rips, FADS, and little loggerheads. *BioScience* 36:92–100.

Carr, A. 1987. Impact of nondegradable marine debris on the ecology and survival outlook of sea turtles. *Marine Pollution Bulletin* 18(6B):352–356.

Carr, A., and S. Stancyk. 1975. Observations on the ecology and survival outlook of the hawksbill turtle. *Biological Conservation* 8:161–172.

Carr, H. A. 1986. Observation on the occurrence of impacts of ghost gillnets on Jeffrey's Ledge. In Program and Abstracts Sixth International Ocean Disposal Symposium, 1986. Pacific Grove, CA. 134–135.

Carr, H. A. 1988. Long term assessment of a derelict gill net found in the Gulf of Maine. Oceans 88 Proceedings. 984–986.

Carr, H. A., and R. A. Cooper. 1987. Manned submersible and ROV assessment of ghost gillnets in the Gulf of Maine. Oceans 87 Proceedings. Sandwich, MA. Massachusetts Division of Marine Fisheries.

Carr, H. A., E. A. Amaral, A. W. Hulbert, and R. A. Cooper. 1985. Underwater survey of simulated lost demersal and lost commercial gill nets off New England. In Proceedings of the Workshop on the Fate and Impact of Marine Debris, 1984, ed. R. S. Shomura and H. O. Yoshida. U.S.

Department of Commerce. NOAA-TM-NMFS-SWFSC-54:438–447.

Carr, H. A., A. J. Blott, and P. G. Caruso. 1992. A study of ghost gillnets in the inshore waters of southern New England. In Proceedings of the 1992 Marine Technology Society Meeting, October 19–21, 1992. Washington, D.C. pp. 361–366.

Carson, R. T., R. C. Mitchell, W. M. Hanemann, R. J. Kopp, S. Presser, and P. A. Ruud. 1992. A contingent valuation study of los passive use values resulting from the Exxon Valdez oil spill. Report to the Attorney General of the State of Alaska. La Jolla, CA: Natural Resource Damage Assessment, Inc.

Carson, R. T., W. M. Hanemann, R. J. Kopp, J. A. Krosnick, R. C. Mitchell, S. Presser, P. A. Ruud, and V. K. Smith. 1994. Prospective interim lost use value due to DDT and PCB contamination in the Southern California Bight. Report to National Oceanic and Atmospheric Administration. La Jolla, CA. Natural Resource Damage Assessment, Inc.

Cartagena Convention. 1983. Convention for the Protection and Development of the Marine Environment of the Wider Caribbean Region, March 24, 1983. Cartagena de Indias, TIAS.

Cary, J. L., J. E. Robinson, and K. A. Grey. 1987. Collected technical reports on the Marmion Marine Park, Perth, Western Australia. Perth, Western Australia. Environmental Protection Authority. Technical Series 19:201–208.

Caulton, E., and M. Mocogni. 1987. Preliminary studies of man-made litter in the Firth of Forth, Scotland. *Marine Pollution Bulletin* 18:446–450.

Cavaliere, L. A. 1989. Trash management at sea. In Dealing with Garbage Under MARPOL Annex V: Examples of Compliance Approaches Used by the Shipping Industry. Alexandria, VA. A.T. Kearney, Inc. 14–17.

Cawthorn, M. W. 1985. Entanglement in, and ingestion of, plastic litter by marine mammals, sharks, and turtles in New Zealand waters. In Proceedings of the Workshop on the Fate and Impact of Marine Debris, 1984, ed. R. S. Shomura and H. O. Yoshida. U.S. Department of Commerce. NOAA-TM-NMFS-SWFSC-54:336–343.

CCAMLR (Commission for the Conservation of Antarctic Marine Living Resources). 1993. Guidelines for conducting surveys of beached marine debris. Hobart, TAS Australia. CCAMLR-XII/BG/5.

Centaur Associates, Inc. 1986. Issue report and work plan for the development of a marine debris education program for the Northwestern Atlantic and Gulf of Mexico. Report to the National Marine Fisheries Service. Seattle. U.S. Department of Commerce, Northwest and Alaska Fisheries Center. 62.

Chalmers, M. O., J. P. Coxall, R. I. Lewis Smith, and J. R. Shears. 1991. Beach litter surveys Signy Islands, South Orkney Islands 1990/91. Hobart, TAS Australia. Commission for the Conservation of Antarctic Marine Living Resources. CCAMLR-XI/BG/14:10.

Clarke, K. R., and M. Ainsworth. 1993. A method linking multivariate community structure to environmental variables. *Marine Ecology Progress Series* 92:205–219.

Cole, C. A., J. P. Kumer, D. A. Manski, and D. V. Richards. 1990. Annual report of national park marine debris monitoring program, 1989. U.S. Department of Interior. National Park Service Technical Report NPS/NRWV/NRTR-90/04.

Cole, C. A., W. P. Gregg, D. V. Richards, and D. A. Manski. 1992. Annual report of national park marine debris monitoring program, 1991, with summary data from 1988 to 1991. U.S. Department of Interior. National Park Service Technical Report NPS/NRWV/NRTR-92/10.

Cole, C. A., W. P. Gregg, and D. A. Manski. 1995. Annual report of the national park marine debris monitoring program, 1992. Marine debris surveys with summary of data from 1988–1992. U.S. Department of Interior. National Park Service Report, Washington, D.C. 56p.

Colton, J. B., Jr., F. D. Knapp, and B. R. Burns. 1974. Plastic particles in surface waters of the Northwestern Atlantic. *Science* 185:491–497.

Conant, S. 1984. Man-made debris and marine wildlife in the Northwestern Hawaiian Islands. *'Elepaio* 44(9):87–88.

Connors, P. G., and K. G. Smith. 1982. Oceanic plastic particle pollution: Suspected effect on fat deposition in red phalaropes. *Marine Pollution Bulletin* 13(1):18–20.

Coombe, T. 1989. Beach sweeping and ocean keeping. *Sea Frontiers* (March-April):31–36.

Corbin, C. J., and J. G. Singh. 1993. Marine debris contamination of beaches in St. Lucia and Dominica. *Marine Pollution Bulletin* 26(6):325–328.

Crichton, M. 1988. *Travels.* New York: Ballantine.

Croxall, J. P., S. Rodwell, and I. L. Boyd. 1990. Entanglement in man-made debris of Antarctic fur seals at Bird Island, South Georgia. *Marine Mammal Science* 6(3):221–233.

Cruise Industry News. 1993/94. Nissen Lie Communications, Inc., New York.

Cummings, R. G., and G. W. Harrison. 1994. Was

the Ohio Court well informed in its assessment of the accuracy of the contingent valuation method? *Natural Resources Journal* 34(winter):1–36.

Dahlberg, M. L., and R. H. Day. 1985. Observations of man-made objects on the surface of the North Pacific Ocean. In Proceedings of the Workshop on the Fate and Impact of Marine Debris, 1984, ed. R. S. Shomura and H. O. Yoshida. U.S. Department of Commerce. NOAA-TM-NMFS-SWFC-54:198–212.

Day, R. H. 1980. The occurrence and characteristics of plastic pollution in Alaska's marine birds. Master's thesis, University of Alaska, Fairbanks.

Day, R. H. 1988. Quantitative distribution and characteristics of neustonic plastic in the North Pacific. Final Report to the Auke Ba Laboratory. Juneau, AK. National Marine Fisheries Service. 73.

Day, R. H., and D. G. Shaw. 1987. Patterns in the abundance of pelagic plastic and tar in the North Pacific Ocean, 1976–1985. *Marine Pollution Bulletin* 19(6B):311–316.

Day, R. H., D. G. Shaw, and S. E. Ignell. 1990a. The quantitative distribution and characteristics of marine debris in the North Pacific Ocean, 1984–88. In Proceedings of the Second International Conference on Marine Debris, 1989, ed. R. S. Shomura and M. L. Godfrey. U.S. Department of Commerce. NOAA-TM-NMFS-SWFSC-154:182–211.

Day, R. H., D. G. Shaw, and S. E. Ignell. 1990b. The quantitative distribution of characteristics of neuston plastic in the North Pacific Ocean, 1985–88. In Proceedings of the Second International Conference on Marine Debris, 1989, ed. R. S. Shomura and M. L. Godfrey. U.S. Department of Commerce. NOAA-TM-NMFS-SWFSC-154:247–266.

Day, R. H., D. H. S. Wehle, and F. C. Coleman. 1985. Ingestion of plastic pollutants by marine birds. In Proceedings of the Workshop on the Fate and Impact of Marine Debris, 1984, ed. R. S. Shomura and H. O. Yoshida. U.S. Department of Commerce. NOAA-TM-NMFS-SWFSC-54:344–386.

Debenham, P. 1990. Education and awareness: Keys to solving the marine debris problem. In Proceedings of the Second International Conference on Marine Debris, 1989, ed. R. S. Shomura and M. L. Godfrey. U.S. Department of Commerce. NOAA-TM-NMFS-SWFSC-154:1100–1114.

Debenham, P., and L. K. Younger. 1991. Cleaning North America's beaches—1900 beach cleanup results. Washington, D.C.: Center for Marine

Conservation.

DeGange, A. R., and T. C. Newby. 1980. Mortality of seabirds and fish in a lost salmon driftnet. *Marine Pollution Bulletin* 11:322–323.

DeLong, R. L., P. Dawson, and P. J. Gearin. 1988. Incidence and impact of entanglement in netting debris on northern fur seal pups and adult females, St. Paul Island, Alaska. In Fur Seal Investigations 1985, ed. P. Kozloff and H. Kajimura. U.S. Department of Commerce. NOAA-TM-NMFS-F/NWC-146:58–68.

Department of Fisheries and Oceans. 1993. St. Lawrence action plan, characterisation of plastic debris from the St. Lawrence Estuary. Final Report.

Devitt, M., Q. Fong, and J. G. Sutinen. 1994. Incentives vs. suasion: A review of the evidence. National Sea Grant Office. Final Report R/SP-04.

Diamond, P. A., and J. A. Hausman. 1994. Contingent valuation: Is some number better than no number. *Journal of Economic Perspectives* 8:45–64.

Dillion Consulting Engineers. 1988. Incinerator and sewage treatment study. British Virgin Islands.

Dixon, T. R., and A. J. Cooke. 1977. Discarded containers on a Kent beach. *Marine Pollution Bulletin* 8:105–109.

Dixon, T. R., and T. J. Dixon. 1980. Marine litter research program stage 2. Marine litter surveillance at two sites on the western Cherbourg Peninsula and West Jutland shores of the English Channel and southern North Sea. Keep Britain Tidy Group, Brighton.

Dixon, T. R., and T. J. Dixon. 1983. Marine litter research program stage 5. Marine litter surveillance of the North Atlantic Ocean shores of Portugal and Western Isles of Scotland. Keep Britain Tidy Group, Buchinghamshire College of Higher Education, Brighton.

Dorfman, A. 1989. Floatables and medical type wastes on region's beaches, 1989 vs. 1988. In Proceedings of the Workshop at the Marine Sciences Research Center, The University at Stony Brook, NY.

Duguy, R. 1983. La tortues luth (*Dermochelys coriacea*) sur les coates de France. *Annales de la Societé Sciences Naturelles de la Charente-Maritime (Supplment)* 38.

Duguy, R., and M. Duron. 1981. Observations de tortues luth sur les cotes de France en 1980. *Annales de la Societe Sciences Naturelles de la Charente-Maritime* 6(8):819–825.

Duguy, R., and M. Duron. 1982. Observations de tortues luth sur les cotes de France en 1981. *Annales de la Societe Sciences Naturelles de la*

Charente-Maritime 6(9):1015–1020.

Duguy, R., M. Duron, and C. Alzieu. 1980. Observations de tortues luth (*Dermochelys coriacea* L.) dans les pertuis charentais en 1979. *Annales de la Societe Sciences Naturelles de la Charente-Maritime* 6(7):681–691.

Duron, M., and P. Duron. 1980. Des tortues luth dans le pertuis charentais. *Courrier de la Nature* 69:37–41.

Eastern Research Group. 1988. An economic evaluation and environmental assessment of regulations implementing Annex V. Marpol 73/78.

Economic Intelligence Unit. 1989. *International Tourism Report*. London: Economist Publications Ltd.

Economist. 1990. (July 21):73.

Edwards, D., J. Pound, L. Arnold, G. Arngold, and M. Lapwood. 1992. A survey of beach litter in Marmion Marine Park, July 1992. Unpublished report, Friends of Marmion Marine Park. Perth, Western Australia.

EPA (U.S. Environmental Protection Agency). 1990a. Methods to manage and control plastic wastes. Report to the Congress. Washington, D.C. EPA/530-SW-89-051.

EPA (U.S. Environmental Protection Agency). 1990b. The study of floatable debris in U.S. waters (Harbor Studies Program), November 1988 through February 1989. Washington, D.C. EPA 503/4-90-003.

EPA (U.S. Environmental Protection Agency). 1992a. The study of floatable debris in U.S. waters (Harbor Studies Program), March 1989 through April 1991. Washington, D.C. EPA 842-R-92-001:242.

EPA (U.S. Environmental Protection Agency). 1992b. Harbor studies program survey at Honolulu, HI, and Kahana Bay Beach observations, February 1992. Final report to the Environmental Protection Agency. Duxbury, MA. Battelle Ocean Sciences, Contract no. 68-C8-0105.

EPA (U.S. Environmental Protection Agency). 1992c. Pilot study to characterize floatable debris discharged from combined sewer overflows and storm drains. Duxbury, MA. Battelle Ocean Sciences, Contract no. 68-C8-0105.

EPA (U.S. Environmental Protection Agency). 1992d. Plastic pellets in the aquatic environment: Sources and recommendations. Washington, D.C. EPA 842-B-92-010.

EPA (U.S. Environmental Protection Agency). 1993a. Pilot study to characterize floatable debris discharged from combined sewer overflows and stormwater discharges, September 1989 through May 1990. Final report to the Environmental Protection Agency. Duxbury, MA. Bat-

telle Ocean Sciences, Contract no. 68-C8-0105.

EPA (U.S. Environmental Protection Agency). 1993b. Guidance specifying management measures for sources of nonpoint pollution in coastal waters. Washington, D.C. EPA Office of Water.

EPA (U.S. Environmental Protection Agency). 1994. New York Bight water quality, summers of 1992 and 1993. New York. Environmental Services Division, Region 2.

Eskeland, G. S., and E. Jimenez. 1991. Curbing pollution in developing countries. *Finance and Development* (March).

Etzioni, A. 1988. *The Moral Dimension*. London: Free Press. 314.

Fahy, E. 1993. Inventory of enmeshing gears in European waters. Dublin. Commission of the European Communities, Directorate-General for Fisheries.

Fahy, E., S. O'Donoghue, and M. O'Driscoll. 1992. Estimating effort by the Irish gillnet fishery. Dublin. Dept. of Marine. Fishery Leaflet 152.

Federal Republic of Germany. 1985. Verschmutzung der Nordsee Durch ol und schiffsmull. A report of Umweltbundesmamt (October).

Fedoryako, B. I. 1988. Accumulation of fish near stationary buyos in the ocean. *Oceanology* 28:521–523.

Feldcamp, S. D. 1985. The effects of net entanglement on the drag and power output of a California sea lion, *Zalophus californianus*. *Fisheries Bulletin* 83(4):692–695.

Feldcamp, S. D., D. P. Costa, and G. K. DeKrey. 1988. Energetic and behavioral effects of net entanglement on juvenile northern fur seals, *Callorhinus ursinus*. *Fisheries Bulletin* 87(1):86–94.

Fine, M. L. 1970. Faunal variation on pelagic *Sargassum*. *Marine Biology* 7:112–122.

Fiorentini, L., and K. Hansen. 1994. Characteristics of high strength knotted and knotless netting in comparison to traditional ones. Montpellier, France. ICES Fishing Technology and Fish behaviour Working Group.

Fishbein, M., and I. Ajzen. 1975. *Belief, Attituäe, Intention, and Behavior: An Introduction to Theory and Research*. Reading, MA: Addison-Wesley.

Fletcher, E. 1982. Marine turtles. *Scientific Event Alert Network (SEAN) Bulletin* 7(5):15.

Forndran, A., and E. Delva. 1992. Flow balancing method for CSO abatement. *Clearwaters* (fall):7–11.

Forsyth, M. L. 1993. Dune litter pollution, its ecological impacts with special reference to meiofauna. Unpublished honours thesis, University of Port Elizabeth. 26.

Foster, B. A. n.d. Barnacles indicated direction of drift of wrecked trimaran. Unpublished manuscript.

Fowler, C. W. 1982. Interactions of northern fur seals and commercial fisheries. In Transactions of the 47th North American Wildlife and Natural Resources Conference. Washington, D.C. Wildlife Management Institue. 278–292.

Fowler, C. W. 1985. An evaluation of the role of entanglement in the population dynamics of northern fur seals on the Pribilof Islands. In Proceedings of the Workshop on the Fate and Impact of Marine Debris, 1984, ed. R. S. Shomura and H. O. Yoshida. U.S. Department of Commerce. NOAA-TM-NMFS-SWFSC-54:291–307.

Fowler, C. W. 1987. Marine debris and northern fur seals: A case study. *Marine Pollution Bulletin* 18(6B):326–335.

Fowler, C. W. 1988. A review of seal and sea lion entanglement in marine debris. In Proceedings of the North Pacific Rim Fishermen's Conference on Marine Debris, 1987, ed. D. L. Alverson and J. A. June, 16–63. Seattle. Natural Resources Consultant.

Fowler, C. W., J. Scordino, T. R. Merrell, and P. Kozloff. 1985. Entanglement of fur seals from the Pribilof Islands. In Fur Seal Investigations 1982, ed. P. Kozloff. U.S. Department of Commerce. NOAA-TM-NMFS F/NWC-71:22–23.

Fowler, C. W., R. Merrick, and J. D. Baker. 1990. Studies of the population level effects of entanglement on northern fur seals. In Proceedings of the Second international Conference on Marine Debris, 1989, ed. R. S. Shomura and M. L. Godfrey. U.S. Department of Commerce. NOAA-TM-NMFS-SWFSC-154:453–474.

Fowler, C. W., R. Ream, B. Robson, and M. Kiyota. 1992. Entanglement studies, St. Paul Island, 1991 juvenile male northern fur seals. Seattle. U.S. Department of Commerce, Alaska Fisheries Science Center. AFSC Processed Report 92-07:45.

Fowler, C. W., J. Baker, R. Ream, B. Robson, and M. Kiyota. 1993. Entanglement studies, St. Paul Island, 1992 juvenile male northern fur seals. Seattle. U.S. Department of Commerce, Alaska Fisheries Science Center. AFSC Processed Report 93-03:42.

Frank, R. H. 1985. *Choosing the Right Pond.* New York: Oxford University Press.

Frank, R. H. 1987. If Homo Economicus could choose his own utility function, would he want one with a conscience? *American Economic Review* 77(4):593–604.

Frank, R. H. 1988. *Passions Within Reason.* New York: W. W. Norton.

Franklin Associates, Ltd. 1994. Characterization of municipal solid waste in the United States, 1994 update. Prepared for the U.S. Environmental Protection Agency. Washington, D.C. Report no. EPA 530-S-94-042.

Freeman, A. M. 1993a. The economics of valuing marine recreation: A review of empirical evidence. Economics working paper, Bowdoin College. Brunswick, ME. 93–102.

Freeman, A. M. 1993b. *The Measurement of Environmental and Resource Values: Theory and Methods.* Washington, D.C.: Resources For The Future.

French, D. P., and M. Reed. 1990. Potential impact of entanglement in marine debris on the population dynamics of northern fur seal, *Callorhinus ursinus.* In Proceedings of the Second International Conference on Marine Debris, 1989, ed. R. S. Shomura and M. L. Godfrey. U.S. Department of Commerce. NOAA-TM-NMFS-SWFSC-154:431–474.

Friday, S. 1990. Marine debris: North Carolina's solutions through education. In Proceedings of the Second International Conference on Marine Debris, 1989, ed. R. S. Shomura and M. L. Godfrey. U.S. Department of Commerce. NOAA-TM-NMFS-SWFSC-154.

Friedrich, H. 1969. *Marine Biology.* Seattle: University of Washington Press. 474.

Fritts, T. H. 1982. Plastic bags in the intestinal tracts of leatherback marine turtles (*Dermochelys coriacea*). *Herpetological Review* 13(3):72–73.

Fry, D. M., S. I. Fefer, and L. Sileo. 1987. Ingestion of plastic debris by Laysan albatrosses and wedge-tailed shearwaters in the Hawaiian Islands. *Marine Pollution Bulletin* 18(6B):339–343.

Furness, B. L. 1983. Plastic particles in three procellariiform seabirds from the Benguela Current, South Africa. *Marine Pollution Bulletin* 14(8):307–308.

Furness, R. W. 1985a. Ingestion of plastic particles by seabirds at Gough Island, South Atlantic Ocean. *Environmental Pollution Series A Ecological and Biological* 38:261–272.

Furness, R. W. 1985b. Pastic particle pollution: Accumulation by procellariiform seabirds at Scottish colonies. *Marine Pollution Bulletin* 16(3):103–106.

Gabrielides, G. P., A. Golik, L. Loizides, M. G. Marino, F. Bingel, and M. V. Torregrossa. 1991. Man-made garbage pollution on the Mediterranean coastline. *Marine Environmental Research* 23:437–441.

Galil, B., A. Golik, and M. Turkay. 1995. Litter at the bottom of the sea: A sea bed survey in the eastern Mediterranean. *Marine Pollution Bulletin* 30:22–24.

Gauch, H. G., Jr. 1982. *Multivariate Analysis in Community Ecology.* Cambridge: Cambridge University Press.

Geerken, M., and W. Gove. 1975. Deterrence: Some theoretical considerations. *Law and Society Review* (spring):498–513.

Geoguide. 1994. *National Geographic,* February.

Gerrodette, T. 1985. Towards a population dynamics of marine debris. In Proceedings of the Workshop on the Fate and Impact of Marine Debris, 1984, ed. R. S. Shomura and H. O. Yoshida. U.S. Department of Commerce. NOAA-TM-NMFS-SWFSC-54:508–518.

GESAMP (Group of Experts on the Scientific Aspects of Marine Pollution). 1991. *The State of the Marine Environment.* London: Blackwell Scientific Publications. 146.

Gómez-Gómez, F., and J. E. Heisel. 1980. Summary appraisals of the nation's ground water resources—Caribbean Region: U.S. Geological Survey. Professional Paper 813-U.

Golik, A., and Y. Gertner. 1990. Solid waste on the Israeli coast: composition, sources, and management. In Proceedings of the Second International Conference on Marine Debris, 1989, ed. R. S. Shomura and M. L. Godfrey. NOAA-TM-NMFS-SWFSC-154. U.S. Department of Commerce, Washington, DC, pp. 369–378.

Golik, A., and Y. Gertner. 1992. Litter on the Israeli coastline. *Marine Environmental Research* 33:1–15.

Gochfeld, M. 1973. Effect of artifact pollution on the viability of seabird colonies on Long Island, New York. *Environmental Pollution* 4:1–6.

Goodall, R. N. P. 1990a. Surveys of marine debris on the coasts of Argentina and Uruguay. Unpublished report, U.S. Marine Mammal Commission. MM 4465864-1.

Goodall, R. N. P. 1990b. Persistent marine debris on selected beaches in Tierra del Fuego, Argentina. Unpublished report, U.S. Marine Mammal Commission. MM 4465864-1.

Goodall, R. N. P. 1990c. Beach debris at Deception Island. Unpublished report, U.S. Marine Mammal Commission.

Gots, B., K. Ronald, and J. Dougan. 1992. The newsletter of the league for the conservation of the monk seal. Ontario, Canada. Institute for Environmental Policy and Stewardship at the Arboretum, University of Guelph. 62.

Gourlay, K. A. 1992. *World of Waste: Dilemmas of Industrial Development.* London: Zed Books Ltd.

Grace, R. V. 1994. Oceanic debris observations in the Indian Ocean whale sanctuary and Eastern Mediterranean Sea. Report to International Whaling Commission Scientific Committee SC/46/0 26.

Gramentz, D. 1988. Involvement of loggerhead turtle with the plastic, metal, and hydrocarbon pollution in the central Mediterranean. *Marine Pollution Bulletin* 19(1):11–13.

Greenpeace. 1991. Antarctic Expedition Report, Greenpeace 1990/91.

Gregory, M. R. 1977. Plastic pellets on New Zealand beaches. *Marine Pollution Bulletin* 8:82–84.

Gregory, M. R. 1978. Accumulation and distribution of virgin plastic granules on New Zealand beaches. *New Zealand Journal of Marine and Freshwater Research* 12:399–414.

Gregory, M. R. 1983. Virgin plastic granules on some beaches of Eastern Canada and Bermuda. *Marine Environmental Research* 10:73–92.

Gregory, M. R. 1987. Plastics and other seaborne litter on the shores of New Zealand's subantarctic Islands. *New Zealand Antarctic Record* 7(3):32–47.

Gregory, M. R. 1990a. Environmental and pollution aspects. In *Antarctic Sector of the Pacific,* ed. G. P. Glasby. Amsterdam: Elsevier.

Gregory, M. R. 1990b. Plastics: Accumulation, distribution and environmental effects of mseo-, macro-, and megalitter in surface waters and on shores of the Southwest Pacific. In Proceedings of the Second International Conference on Marine Debris, 1989, ed. R. S. Shomura and M. L. Godfrey. U.S. Department of Commerce. NOAA-TM-NMFS-SWFSC-154:55–84.

Gregory, M. R. 1991. The hazards of persistent marine pollution: Drift plastics and conservation islands. *Journal of the Royal Society of New Zealand* 21:83–100.

Gregory, M. R. 1993. Plastics and the coast: The Waikato region. In Proceedings of Region Resource Futures Conference, 1991, ed. W. Boyce, N. Erickson, and N. Hingley, 300–307. Centre for Environmental and Resource Studies, University of Waikato.

Gregory, M. R., R. M. Kirk, and M. C. G. Mabin. 1984. Pelagic tar, oil, plastics and other litter in the surface waters of the New Zealand sector of the Southern Ocean, and on Ross Dependency shores. *New Zealand Antarctic Record* 6(1):12–28.

Guillet, J. E. 1974. Plastics, energy, and ecology— a harmonious triad. *Plastics Engineering* 30:48–56.

Hanemann, W. M. 1984. Welfare evaluations in contingent valuation experiments with discrete responses. *American Journal of Agriculture Economics* 66:332–341.

Hanemann, W. M. 1994. Valuing the environment through contingent valuation. *Journal of Economic Perspectives* 8:19–44.

Hardin, G. 1968. The tragedy of the commons. *Science* 162(3859):1243–1248.

Hare, M. P., and J. G. Mead. 1987. Handbook for determination of adverse human-marine mammal interactions from necropsies. Seattle. U.S. Department of Commerce, Northwest and Alaska Fisheries Center. NWAFC Processed Report 87-06:35.

Hargreaves, N. B., and E. W. Carter. 1989. Summary of lost or discarded driftnets reported on the coast of British Columbia in 1989. Nanamio, British Columbia. Department of Fisheries and Oceans, Pacific Biological Station.

Harms, J. 1990. Marine plastic litter as an artificial hard bottom fouling ground. *Helgolaender Meeresuntersuchungen* 44:503–506.

Harper, P. C., and J. A. Fowler. 1987. Plastic pellets in New Zealand storm-killed prions (*Pachyptilla* spp.), 1958–1977. *Notornis* 34:65–70.

Harrison, G. S., T. S. Hida, and M. P. Seki. 1983. Hawaiian seabird feeding ecology. *Wildlife Monographs* 85:1–71.

Harrision, P. 1983. *Seabirds: An Identification Guide.* Boston, MA: Houghton Mifflin Company. 448.

Hartog, J. C. den. 1980. Notes on the food of sea turtles: *Eretmochelys imbricata* (Linnaeus). *Netherlands Journal of Zoology* 30(4):595–610.

Hartog, J. C. den, and M. M. Van Nierop. 1984. A study of the gut contents of six leatherback turtles *Dermochelys coriacea* (Linnaeus) (Reptilla: Testudines: Dermochelyidae) from British waters and from the Netherlands. *Zoologische Verhandelingen (Leiden)* 209:1–36.

Harvell, C. D. 1984. Predator-induced defense in a marine bryozoan. *Science* 224:1357–1359.

Hausman, D. M., and M. S. McPherson. 1993. Taking ethics seriously: Economics and contemporary moral philsophy. *Journal of Economic Literature* XXXI (2):671–731.

Hayward, B. W. 1984. Rubbish trends—beach litter surveys at Kawerua, 1974–1982. *Tane* 30:209–217.

Heatwole, H., and R. Levins. 1972. Biogeography of the Puerto Rican Bank: Flotsam transport of terrestrial animals. *Ecology* 53:112–117.

Heckman, J. 1979. Sample bias as a specification error. *Econometrica* 47:153–162.

Heineke, J. M., ed. 1978. *Economic Models of Criminal Behavior.* Amsterdam: North-Holland; New York: Elsevier North-Holland.

Heislers, D. 1994. Eastern Victorian coastal trek: An environmental snapshot. Mt. Waverly, Victoria, Australia: Victorian Coastal Trekkers Inc.

Henderson, J. R. 1984. Encounters of Hawaiian monk seals with fishing gear at Lisianski Island, 1982. *Marine Fisheries Review* 46(3):59–61.

Henderson, J. R. 1985. A review of Hawaiian monk seal entanglements in marine debris. In Proceedings of the Workshop on the Fate and Impact of Marine Debris, 1984, ed. R. S. Shomura and H. O. Yoshida. U.S. Department of Commerce. NOAA-TM-NMFS-SWFSC-54:326–335.

Henderson, J. R. 1988. Marine debris in Hawaii. In Proceedings of the North Pacific Rim Fishermen's Conference on Marine Debris, 1987, ed. D. L. Alverson and J. A. June. Seattle. Natural Resources Consultants.

Henderson, J. R. 1990. Recent entanglements of Hawaiian monk seals in marine debris. In Proceedings of the Second International Conference on Marine Debris, 1989, ed. R. S. Shomura and M. L. Godfrey. U.S. Department of Commerce. NOAA-TM-NMFS-SWFSC-154:540–553.

Heneman, B. 1988. Persistent marine debris in the North Sea, Northwest Atlantic Ocean, wider Caribbean area, and the West Coast of Baja California. A Report to the Marine Mammal Commission and the National Ocean Pollution Program Office. Washington, D.C.

Heneman, B., and J. M. Coe. 1989. Persistent marine debris in the Wider Caribbean Area. Report submitted to the Intergovernmental Oceanographic Commission, Sub-Commission for the Caribbean and Adjacent Regions (June).

Hentschel, E. 1922. Über den Bewuch auf den treibenden Tang der Sargassosee. *Jahr buch der Hamburg der Wissenschaften Anstalt* 38:1–26.

Heyerdahl, T. 1971. *The 'Ra' Expeditions.* London: George Allen and Unwin Ltd. 344.

Heyning, J. E., and T. D. Lewis. 1990. Entanglements of baleen whales in fishing gear off Southern California. In 40th Report of the International Whaling Commission. Cambridge, England. 427–431.

High, W. L. 1976. Escape of Dungeness crabs from pots. *Marine Fisheries Review* 38(4):19–23.

High, W. L. 1985. Some consequences of lost fishing gear. In Proceedings of the Workshop on the Fate and Impact of Marine Debris, 1984, ed. R. S. Shomura and H. O. Yoshida. U.S. Department of Commerce. NOAA-TM-NMFS-SWFSC-54:430–437.

High, W. L., and D. D. Worlund. 1979. Escape of king crab, *Paralithodes camtschatica,* from derelict pots. NOAA Technical Report NMFS SSRF-734. Washington, D.C. GPO.

Hildebrand, H. H. 1980. Report on the incidental capture, harassment, and mortality of sea turtles in Texas. La Jolla, CA. U.S. Department of Commerce, Southwest Fisheries Center, National Marine Fishers Service. Contract no. NA80-GG-A-00160:33.

Hirth, H. F. 1971. Synopsis of biological data on the green turtles, *Chelonia mydas* (Linnaeus) 1758. *FAO Fisheries Biology Synopsis* 85(1):1–8:19.

Hodge, K., J. Glen, and D. Lewis. 1993. *1992 National Coastal Cleanup Report,* ed. R. Bierce and K. O'Hara. Washington, D.C.: Center for Marine Conservation.

Hoffman, M. 1977. Moral internalization: Current theory and research. In *Advances in Experimental Social Psychology,* ed. L. Berkowitz. Vol. 10. New York: Academic Press.

Hollin, D., and M. Liffman. 1991. Use of MARPOL Annex V reception facilities and disposal systems at selected Gulf of Mexico ports, private terminals and recreational boating facilities. A report to the Texas General Land Office. Austin, TX.

Hollin, D., and M. Liffman. 1993. Survey of Gulf of Mexico operations and recreational interests: Monitoring of MARPOL Annex V compliance trends. Final report to the U.S. Environmental Protection Agency. Region 6.

Hollström, A. 1975. Plastic films on the bottom of the Skagerrak. *Nature* 255:622–623.

Holmes, J. R., ed. 1984. In *Managing Solid Waste in Developing Countries.* Chichester: John Wiley & Sons.

Horsman, P. V. 1982. The amount of garbage pollution from merchant ships. *Marine Pollution Bulletin* 13:167–169.

Hoss, D. E., and L. R. Settle. 1990. Ingestion of plastics by teleost fishes. In Proceedings of the Second International Conference on Marine Debris, 1989, ed. R. S. Shomura and M. L. Godfrey. U.S. Department of Commerce. NOAA-TM-NMFS-SWFSC-154:693–709.

Hughes, G. R. 1970. Further studies on marine turtles in Tongaland, III. *Lammergeyer* 12:7–25.

Hughes, G. R. 1974a. The sea turtles of Southeast Africa. I. Status, morphology and distribution. *Oceanographic Research Institute (Durban) Investigational Report* 35:144.

Hughes, G. R. 1974b. The sea turtles of Southeast Africa. II. The biology of the Tongaland loggerhead turtles *Caretta caretta* L. with comments on the leatherback turtle *Dermochelys coriacea* L. and the green turtle *Chelonia mydas* L. in the study region. *Oceanographic Research Institute (Durban) Investigational Report* 36:96.

Humpback Whale Recovery Team. 1991. Final recovery plan for the humpback whale *Megaptera novaeangliae.* Silver Spring, MD. National Marine Fisheries Service. 105.

Humphreys, J. 1990. Marine debris demonstration and education project at Squalicum Harbor, Bellingham, Washington, USA. In Proceedings of the Second International Conference on Marine Debris, 1989, ed. R. S. Shomura and M. L. Godfrey. U.S. Department of Commerce. NOAA-TM-NMFS-SWFSC-154.

Hunter, J. R. 1968. Fishes beneath flotsam. *Sea Frontiers* 14:280–288.

Hunter, J. R., and C. T. Mitchell. 1966. Association of fishes with flotsam in the offshore waters of central America. *Fishery Bulletin* 66:13–27.

Hutchins, L. W. 1941. The growth and distribution of salt-water Bryozoa in relation to salinity. Unpublished Ph.D. dissertation, Yale University, New Haven, Connecticut.

HydroQual, Inc. 1993. City-wide floatables study. New York. New York City Department of Environmental Protection.

Hyman, L. H. 1959. *The Invertebrates.* Vol. 5. New York: McGraw-Hill. 783.

IMO (International Maritime Organization). 1988. Guidelines for the Implementation of Annex V of MARPOL 73/78. London: IMO.

IMO (International Maritime Organization). 1992. MARPOL 73/78 Conslidated Edition, 1991. IMO, London.

IMO (International Maritime Organization). 1993. IMO News no. 3. IMO, London.

IMO (International Maritime Organization). 1995. Report of the 37th meeting of the Marine Environment Protection Committee of the International Maritime Organization. IMO, London.

INPFC (International North Pacific Fisheries Commission). 1986. Canadian, United States, and Japanese statements on agenda item 12. In Proceedings of the 33rd Annual Meeting of INPFC.

IOC (Intergovernmental Oceanographic Commission) 1990. Guidelines for monitoring persistent synthetic materials. Kingston, Jamaica.

IOC (Intergovernmental Oceanographic Commission). 1991. Preliminary report on the results of the Caribbean marine debris monitoring pilot project. IOCARIBE, Cartagena, Colombia.

IOC (Intergovernmental Oceanographic Commission). 1994. Marine debris: Solid waste management action plan for the Wider Caribbean. IOC technical Series 41:20, UNESCO.

IOC/FAO/UNEP (Intergovernmental Oceanographic Commission, the Food and Agriculture Organization of the United Nations, and the United Nations Environment Programme). 1989. Report of the IOC/FAO/UNEP review meeting of the persistent synthetic materials pilot survey. Athens, Greece. Unit for the Mediterranean Action Plan.

ITF (Interagency Task Force). 1988. Report of the interagency task force on persistent marine debris. Washington, D.C. U.S. Department of Commerce, National Oceanic and Atmospheric Administration.

Jameson, G. L. 1986. Trial systematic salvage of beach-cast sea otter, *Enhydra lutris,* carcasses in the central and southern portion of the sea otter range in California. Report to the U.S. Marine Mammal Commission. Springfield, VA. National Technical Information Service PB87-108288:60.

Jarman, C. 1988. Background paper: A review of the legal structure enabling federal or state enactment of the various types of incentive systems. In Oceans of Plastic. Alaska Sea Grant Report no. 88-7.

Jenkin, M. 1990. Plastic a problem for Tasmania's wildlife. Hobart, TAS, Australia. The Southern Star Newspaper (19 November).

Johnson, S. W. 1989. Deposition, fate, and characteristics of derelict trawl web on an Alaskan beach. *Marine Pollution Bulletin* 20:164–168.

Johnson, S. W. 1990a. Distribution, abundance, and source of entanglement debris and other plastics on Alaskan beaches, 1982–88. In Proceedings of the Second International Conference on Marine Debris, 1989, ed. R. S. Shomura and M. L. Godfrey. U.S. Department of Commerce. NOAA-TM-NMFS-SWFSC-154:331–348.

Johnson, S. W. 1990b. Entanglement debris on Alaska beaches, 1989. Juneau, AK. Auke Bay Laboratory NWAFC Processed Report 90-10:16.

Johnson, S. W. 1993. Entanglement and other plastic debris on Alaskan beaches, 1990–92. Juneau, AK. Auke Bay Laboratory AFSC Processed Report 93-02:12.

Johnson, S. W. 1994. Deposition of trawl web on an Alaska beach after implementation of MARPOL Annex V Legislation. *Marine Pollution Bulletin 28:477–481.*

Jokiel, P. L. 1989. Rafting of reef corals and other organisms at Kwajalen Atoll. *Marine Biology* 101:483–493.

Jones, L. J., and R. C. Ferrero. 1985. Observations of net debris and associated entanglements in the North Pacific Ocean and Bering Sea, 1978–1984. In Proceedings of the Workshop on the Fate and Impact of Marine Debris, 1984, ed. R. S. Shomura and H. O. Yoshida. U.S. Department of Commerce. NOAA-TM-NMFS-SWFSC-54:183–196.

Kasperson, R. E., and K. D. Pijawka. 1985. Societal response to hazards and major hazard events: Comparing natural and technological hazards. *Public Administration Review* 45:7–17.

Karau, J. 1992. The control of land-based sources of marine pollution: Recent international initiatives and prospects. *Marine Pollution Bulletin* 25:80–81.

Kartar, S., R. Milne, and M. Sainsbury. 1973. Polystyrene waste in the Severn Etuary. *Marine Pollution Bulletin* 4:144.

Kartar, S., F. Abou-Seedo, and M. Sainsbury. 1976. Polystyrene spherules in the Severn Estuary: A progress report. *Marine Pollution Bulletin* 7:52.

Kasteleine, R. A., and M. S. S. Lavaleije. 1992. Foreign bodies in the stomach of a female harbor porpoise (*Phocoena phocoena*) from the North Sea. *Aquatic Mammals* 18(2):40–46.

Kelman, H. 1961. Three processes of social influence. *Public Opinion Quarterly* 25:57–69.

Khosla, A. 1982. Environmentally sound management of solid wastes. In Solid Waste Disposal and Utilization in Developing Countries, ed. R. M. Schelhaas. Amsterdam. Department of Agricultural Research.

Kimker, A. 1992. Tanner crab survival in closed pots. Anchorage. Alaska Department of Fish and Game. Regional Information Report no. 2A92-21.

Kindt, J. W. 1984. Solid wastes and marine pollution. *Catholic University Law Review* 34:37–100.

Kinoshita, R. K., A. Greig, J. D. Hastie, and J. M. Terry. 1993. Economic status of the groundfish fisheries off Alaska, 1993 (preliminary). Seattle. U.S. Department of Commerce, National Marine Fisheries Service.

Kishino, H., H. Kato, F. Kasamatsu, and Y. Fujise. 1989. Statistical method for the estimation fo age composition and biological parameters of the population. Scientific Sub-Committee of the 21st International Whaling Commission. IWC/SC/41/SHM13.

Klemm, B., and D. Wendt. 1990. Beach confetti. *Sea Frontiers* (3-4):28–29.

Kohlberg, L. 1969. Stage and sequence: The cognitive-development approach to socialization. In *Handbook of Socialization Theory and Research,* ed. D. A. Goslin. New York: Rand McNally.

Kohlberg, L. 1976. Moral stages and moralization. In *Moral Development and Behavior,* ed. T. Likona. New York: Holt, Rinehart and Winston.

Kohlberg, L. 1981. *Essays on Moral Development.* Vol. 1, *The Philosophy of Moral Development: Moral Stages and the Idea of Justice.* San Francisco: Harper & Row.

Kohlberg, L. 1984. *Essays on Moral Development.* Vol. 2. San Francisco: Harper & Row.

Kraus, S. D. 1990. Rates and potential causes of mortality in North Atlantic right whales (*Eubalaena glacialis*). *Marine Mammal Science* 6(4):278–291.

Kreps, D. M. 1990. *A Course in Microeconomic Theory.* Princeton, NJ: Princeton University Press.

Krouse, J. S. 1994. U.S. commercial and recreational American lobster, *Homarus americanus,* landings and fishing effort by state, 1970–1992. Report to the New England Regional Fishery Management Council, Gloucester, Massachusetts.

Kruse, G. H., and A. Kimker. 1993. Degradable escape mechanisms for gear: A summary report to the Alaska Board of Fisheries. Juneau. Alaska Department of Fish and Game. Regional Information Report no. 5J93-01.

Kubota, T. 1990. Synthetic materials found in the stomachs of longnose lancetfish from Suruga Bay, central Japan. In Proceedings of the Second International Conference on Marine Debris, 1989, ed. R. S. Shomura and M. L. Godfrey. U.S. Department of Commerce. NOAA-TM-NMFS-SWFSC-154:710–717.

Kuperan, K., and J. G. Sutinen. 1994. An econometric analysis of compliance and enforcement in Malaysian fisheries. Unpublished working paper, Department of Natural Resource Economics, Universiti Pertanian Malaysia, Serdang.

Kuzen, A. E. 1990. A study of the effects of commercial fishing debris on *Callorhinus ursinus* from breeding islands in the Western Pacific. In Proceedings of the Second International Conference on Marine Debris, 1989, ed. R. S. Shomura and M. L. Godfrey. U.S. Department of Commerce. NOAA-TM-NMFS-SWFSC-154.

Laist, D. W. 1987. Overview of the biological effects of lost and discarded plastic debris in the marine environment. *Marine Pollution Bulletin* 18(6B):319–326.

Land, L. S. 1979. Carbonate mud: Production by epibiont growth on *Thalassia testudinum. Journal of Sedimentary Petrology* 40:1361–1363.

Laska, S. 1990. Designing effective educational programs: The attitudinal basis of marine litter-ing. In Proceedings of the Second International Conference on Marine Debris, 1989, ed. R. S. Shomura and M. L. Godfrey. U.S. Department of Commerce.NOAA-TM-NMFS-SWFSC-154:1179–1190.

Lee, D. S., and W. M. Palmer. 1981. Records of leatherback turtles, *Dermochelys coriacea* (Linnaeus), and other marine turtles in North Carolina waters. *Brimleyana* 5:95–106.

Levine, F. J., and J. L. Tapp. 1977. The dialectic of legal socialization in community and school. In *Law, Justice and the Individual in Society: Psychological and Legal Issues,* ed. J. L. Tapp and F. J. Levine. New York: Holt, Rinehart and Winston.

Lewis, G., T. Clayton, G. Arnold, and J. Craig. 1994. Patterns of occurrence of marine debris: Hauraki Gulf, Auckland, New Zealand. Report by Island Care New Zealand Trust.

Lough, R. G., P. C. Valentine, D. C. Potter, P. J. Auditore, G. R. Boltz, J. D. Neilson, and R. I. Perry. 1989. Ecology and distribution of juvenile cod and haddock in relation to sediment type and bottom currents on Eastern Georges Bank. *Marine Ecology Progress Series* 56:1–12.

Loughlin, T., P. J. Gearin, R. L. DeLong, and R. L. Merrick. 1986. Assessment of net entanglement on northern sea lions in the Aleutian Islands, 25 June–15 July 1985. Seattle. U.S. Department of Commerce, Northwest and Alaska Fisheries Center. NWAFC Processed Report 86-02:50.

Low, L., R. E. Nelson, Jr., and R. E. Narita. 1985. Net loss from trawl fisheries off Alaska. In Proceedings of the Workshop on the Fate and Impact of Marine Debris, 1984, ed. R. S. Shomura and H. O. Yoshida. U.S. Department of Commerce. NOAA-TM-NMFS-SWFSC-54:130–153.

Lowry, L. 1993. Foods and feeding ecology. In *The Bowhead Whale,* ed. J. J. Burns, J. J. Montague, and C. J. Cowles, 201–238, no. 2. Lawrence, KS: The Society for Marine Mammology.

Lucas, Z. 1992. Monitoring persistent litter in the marine environment on Sable Island, Nova Scotia. *Marine Pollution Bulletin* 24(4):192–199.

Ludwig, J. A., and J. F. Reynolds. 1988. *Statistical Ecology: A Primer on Methods and Computing.* New York: John Wiley & Sons.

Lutgens, F. K., and E. J. Tarbuck. 1992. *The Atmosphere: An Introduction to Meteorology.* 5th ed. Englewood Cliffs, NJ: Prentice Hall. 430.

Lutjeharms, J. R. E., L. V. Shannon, and L. J. Beekman. 1988. On the surface drift of the Southern Ocean. *Journal of Marine Research* 46:267–279.

Lutz, P. L. 1990. Studies on the ingestion of plastics and latex by sea turtles. In Proceedings of the Second International Conference on Marine Debris, 1989, ed. R. S. Shomura and M. L. Godfrey. U.S. Department of Commerce. NOAA-TM-NMFS-SWFSC-154:179–735.

Macfie, C. 1989. DOC as a catalyst for beach clean-ups and public education. In Proceedings of National Workshop on Marine Debris in New Zealand's Coastal Waters. Wellington, New Zealand. Department of Conservation. 11–12.

Mann, J., R. A. Smolker, and B. B. Smuts. 1995. Response to calf entanglement in free-ranging bottlenose dolphins. *Marine Mammal Science* 11(1):100–106.

Manooch, C. S., III. 1973. Food habits of yearling and adult striped bass, *Marone saxatalis* (Walbaum), from Albemarle Sound, North Carolina. *Chesapeake Science* 14:73–86.

Manooch, C. S., III, and W. T. Hogarth. 1983. Stomach contents and giant trematodes from wahoo, *Acanthocybium solanderi,* collected along the South Atlantic and Gulf coasts of the United States. *Bulletin of Marine Science* 33:227–223.

Manooch, C. S., III, and D. L. Mason. 1983. Comparative food studies of yellowfin tuna, *Thunnus albacares,* and blackfin tuna, *Thunnus atlanticus* (Pisces: Scombridae) from the Southeastern and Gulf Coasts of the United States. *Brimleyana* 9:33–52.

Manooch, C. S., III, D. L. Mason, and R. S. Nelson. 1984. Food and gastrointestinal parasites of dolphin, *Coryphaena hippurus,* collected along the Southeastern and Gulf Coasts of the United States. *Bulletin of the Japanese Society of Scientific Fisheries* 50:1511–1525.

Manooch, C. S., III, D. L. Mason, and R. S. Nelson. 1985. Foods and little tuny, *Euthynnus alletteratus,* collected along the Southeastern and Gulf Coasts of the United States. *Bulletin of the Japanese Society of Scientific Fisheries* 51:1207–1218.

Mansbridge, J., ed. 1990. *Beyond Self-Interest.* Chicago: University of Chicago Press; London: University of Chicago Press. 402.

Manski, D. A., W. P. Gregg, C. A. Cole, and D. V. Richards. 1991. Annual report of the National park marine debris monitoring progra, 1990. U.S. Department of Interior. National Park Service Technical Report NPS/NRWV/NRTR-91/07.

Manville, A. M. 1990. A survey of plastics on Western Aleutian Island beaches and related wildlife entanglement. In Proceedings of the Second International Conference on Marine Debris, 1989, ed. R. S. Shomura and M. L. Godfrey. U.S. Department of Commerce. NOAA-TM-NMFS-SWFSC-154:349–362.

Marcus, R. 1941. Sôbre os Briozoa do Basil. II. Boletin da Faculdade de Filosofia, Ciências e Letras, Universidad de São Paulo, Vol. 25. *Zoologica* 6:57–106.

Marine Mammal Commission. 1993. Annual report to Congress 1992. Washington, D.C. 226.

Marino, M. G., P. Aranzadi, and J. Sobrino. 1989. Litter on the beaches and littoral waters of the Spanish Mediterranean coast. Progress report submitted to the UNEP/IOC/FAO. Haifa, Israel, pp. 1–8.

Marsden, J. E. 1992. Standard procedures for monitoring and sampling zebra mussels. *Illinois Natural History Survey Biological Notes* 138:1–40.

Martin, A. R., and M. R. Clarke. 1986. The diet of sperm whales (*Physeter macrocephalus*) captured between Iceland and Greenland. *Journal of the Marine Biological Association of the United Kingdom* 66:779–790.

Mate, B. R. 1985. Incidents of marine mammal encounters with debris in active fishing gear. In Proceedings of the Workshop on the Fate and Impact of Marine Debris, 1984, ed. R. S. Shomura and H. O. Yoshida. U.S. Department of Commerce. NOAA-TM-NMFS-SWFSC-54:453–457.

Matson, S. 1989. Summary of DOC clean-ups and publicity work to date. In Proceedings of National Workshop on Marine Debris in New Zealand's Coastal Waters. Wellington, New Zealand. Department of Conservation. 13.

Matthews, W. 1975. Marine litter: Assessing potential ocean pollutants. National Academy of Sciences, National Research Council, Washington, DC, pp. 405–438.

Maturo, F. J. 1991. Self-fertilization in gymnolaemate Bryozoa. *Bulletin de la Societe des Sciences Naturelles de l'Ouest de la France* 1:572.

McCarron, D., and T. Hoops. 1991. Massachusetts lobster fishery statistics. Massachusetts Division of Marine Fisheries Technical Series 26.

McCarron, D., and T. Hoops. 1991. Massachusetts lobster fishery statistics. Massachusetts Division of Marine Fisheries Technical Series 27.

McConnell, K. E. 1977. Congestion and willingness to pay: A study of beach use. *Land Economics* 53:185–195.

McConnell, K. E. 1990. Models for referendum data: The structure of discrete choice models for contingent valuation. *Journal of Environmental Economics and Management* (January):19–34.

McCoy, F. W. 1988. Floating megalitter in the eastern Mediterranean. *Marine Pollution Bulletin* 19:25–28.

McFadden, D. 1974. Conditional logit analysis of

qualitative choice behavior. In *Frontiers in Econometrics,* ed. P. Zarembka, 105–142. New York: Academic Press.

Merrell, T. R., Jr. 1980. Accumulation of plastic litter on beaches of Amchitka Island, Alaska. *Marine Environmental Research* 3:171–184.

Merrell, T. R., Jr. 1984. A decade of change in nets and plastic litter from fisheries off Alaska. *Marine Pollution Bulletin* 15:378–384.

Merrell, T. R., Jr. 1985. Fish nets and other plastic litter on Alaska beaches. In Proceedings of the Workshop on the Fate and Impact of Marine Debris, 1984, ed. R. S. Shomura and H. O. Yoshida. U.S. Department of Commerce. NOAA-TM-NMFS-SWFSC-54:160–182.

Meylan, A. B. 1978. The behavioral ecology of the West Caribbean green turtle (*Chelonia mydas*) in the internesting habitat. Master's thesis, University of Florida, Gainesville. 131.

Meylan, A. B. 1983. Marine turtles of the Leeward Islands, Lesser Antilles. *Atoll Research Bulletin* 278:1–24.

M'Gonigle, M. R. 1990. Developing sustainability and the emerging norms of international environmental law: The case of land-based marine pollution control. *Canadian International Yearbook.* 169–224.

Miller, J. E. 1993. Marine debris investigation, Padre Island National Seashore. Corpus Christi, TX. National Park Service Report.

Miller, J. E., S. W. Baker, and D. L. Echols. 1995. Marine debris point source investigation, 1994–1995. U.S. Department of Interior. Corpus Christi, TX: Padre Island National Seashore.

Mio, S., and S. Takehama. 1988. Estimation of distribution of marine debris based on the 1986 sighting survey. In Proceedings of the North Pacific Rim Fishermen's Conference on Marine Debris, 1987, ed. D. L. Alverson and J. A. June, 64–94. Seattle. Natural Resources Consultants.

Mio, S., S. Takehama, and S. Matsumura. 1989. Distribution and density of floating objects in the North Pacific based on 1987 sighting survey. In Proceedings of the Second International Conference on Marine Debris, 1989, ed. R. S. Shomura and M. L. Godfrey. U.S. Department of Commerce. NOAA-TM-NMFS-SWFSC-154.

Mischel, W., and H. N. Mischel. 1976. A cognitive social-learning approach to morality and self-regulation. In *Moral Development and Behavior,* ed. T. Lickona. New York: Holt, Rinehart and Winston.

Mitchell, R. C., and R. T. Carson. 1989. *Using Surveys to Value Public Goods: The Contingent Valuation Method.* Washington: Resources For The Future.

Montevecchi, W. A. 1991. Incidence and types of plastic in gannets' nests in the Northwest Atlantic. *Canadian Journal of Zoology* 69:295–297.

Mooney, J., and J. Naughton. 1981. Marine turtles. *Scientific Event Alert Network (SEAN) Bulletin* 6(6):10.

Morell, J. M. 1992. Marine debris pilot monitoring project for the greater Caribbean. Mayaguez. Department of Marine Science at the University of Puerto Rico. UNESCO Contract no. SC/FIT/218.112.0:20.

Morin, M. P. 1987. Laysan finches drown as a result of marine debris. *'Elepaio* 47(11):107–108.

Morris, R. J. 1980. Floating plastic debris in the Mediterranean. *Marine Pollution Bulletin* 11:125.

Morris, A. W., and E. I. Hamilton. 1974. Polystyrene spherules in the Bristol Channel. *Marine Pollution Bulletin* 5:26–27.

Morris, R. J. 1980. Plastic debris in the surface waters of the South Atlantic. *Marine Pollution Bulletin* 11(6):164–166.

Moser, M. L., and D. S. Lee. 1992. A fourteen-year survey of plastic ingestion by Western North Atlantic seabirds. *Colonial Waterbirds* 15(1):83–94.

Mudar, M. J. 1991. Reducing plastic contamination of the marine environment under MARPOL Annex V: a model for recreational harbors and ports. Ph.D. dissertation, Rensselaer Polytechnic Institute, Troy, NY.

Muir, W. D., J. T. Durkin, T. C. Coley, and G. T. McCabe, Jr. 1984. Escape of captured Dungeness crabs from commercial crab pots in the Columbia River estuary. *North American Journal of Fisheries Management* 4:552–555.

Muller, E. 1979. *Aggressive Political Participation.* Princeton, NJ: Princeton University Press.

Müller-Karger, F. E., C. R. McClain, T. R. Fisher, W. E. Esaias, and R. Varela. 1989. Pigment distribution in the Caribbean Sea: Observations from space. *Progress in Oceanography* 23:23–64.

Nash, A. D. 1992. Impacts of marine debris on subsistence fishermen: An exploratory study. *Marine Pollution Bulletin* 24:150–156.

Nasu, K., and Y. Shimadzu. 1969. A method of estimating whale population by sighting observation. Scientific Sub-Committee of the 21th International Whaling Commission. IWC/SC/2122.

National Marine Fisheries Service. 1987. Fisheries of the United States. Current Fishery Statistics 8700.

National Marine Fisheries Service. 1991. Recovery plan for the humpback whale (*Megaptera novaeangliae*). Silver Spring, MD. Humpback Whale Recovery Team. 105.

National Marine Fisheries Service. 1993. Conservation plan for the northern fur seal *Callorhinus ursinus*. Seattle. National Marine Mammal Laboratory and the Alaska Fisheries Science Center; Silver Spring, MD. Office of Protected Resources. 80.

NATO Advanced Study Institute. 1986. *Strategies and Advanced Techniques for Marine Pollution Studies: Mediterranean Sea,* ed. C. S. Giam. Berlin: Springer-Verlag.

Nauke, M., and G. L. Holland. 1992. The role and development of global marine conventions, two case histories. *Marine Pollution Bulletin* 25:74–79.

New York Times. 1994. New York's weather for 1993 (9 January):L31.

NOAA (National Oceanic and Atmospheric Administration). 1993a. Local climatological data, 1976, 1987–1993, monthly summary at John F. Kennedy International Airport for June, July, and August. Asheville, NC. National Climatic Data Center.

NOAA (National Oceanic and Atmospheric Administration). 1993b. Local climatological data, 1976, 1987–1993, monthly summary at New York City Central Park Observatory for June, July, and August. Asheville, NC. National Climatic Data Center.

Nollkaemper, A. 1992. Marine pollution from land-based sources: Towards a global approach. *Marine Pollution Bulletin* 24:8–12.

O'Callaghan, P. 1993. Sources of coastal shoreline litter near three Australian cities. Queenscliffe, Victoria. Victorian Institute of Marine Science.

Ogi, H. 1990. Ingestion of plastic particles by sooty and short-tailed shearwaters in the North Pacific. In Proceedings of the Second International Conference on Marine Debris, 1989, ed. R. S. Shomura and M. L. Godfrey. U.S. Department of Commerce. NOAA-TM-NMFS-SWFSC-154:635–652.

O'Hara, K. J. 1990. National marine debris data base: Findings on beach debris reported by citizens. In Proceedings of the Second International Conference on Marine Debris, 1989, ed. R. S. Shomura and M. L. Godfrey. U.S. Department of Commerce. NOAA-TM-NMFS-SWFSC-154:379–391.

O'Hara, K. J., and L. K. Younger. 1990. Cleaning North America's beaches—1989 beach cleanup results. Washington, D.C. Center for Marine Conservation.

Onions, C., and G. Rees. 1992. An assessment of the environmental impacts of carriers discarded in the marine environment and the benefits derived from those fabricated from a photodegradable plastic giving enhanced degradability. Wigan, Great Britain. The Tidy Britain Group. 19.

Pan American Health Organization. 1982. Health conditions in the Americas, 1977–1980. Washington, D.C. World Health Organization.

Pan American Health Organization. 1992. Reunion centroamericana sobre aseo Urbano, Agosto. Washington, D.C. World Health Organization.

Parker, P. A. 1990. Clearing the oceans of the plastics threat. *Sea Frontiers* 36:18–27.

Parslow, J. L. F., and D. J. Jefferies. 1972. Elastic thread pollution of puffins. *Marine Pollution Bulletin* 3:43–45.

Paul, J. M., A. J. Paul, and A. Kimker. 1993. Starvation resistance in Alaskan crabs: Interim report. Anchorage. Alaska Department of Fish and Game. Regional Information Report no. 2A93-03.

Pemberton, D., N. P. Brothers, and R. Kirkwood. 1992. Entanglement of Australian fur seals in man-made debris in Tasmanian waters. *Wildlife Research* 19:151–159.

Pestana, H. 1985. Carbonate sediment production by *Sargassum* epibionts. *Journal of Sedimentary Petrology* 55:184–186.

Pettit, T. N., G. S. Grant, and G. C. Whittow. 1981. Ingestion of plastics by Laysan albatross. *Auk* 98:839–841.

Philo, L. M., J. C. George, and T. F. Albert. 1992. Rope entanglement of bowhead whales (*Balaena mysticetus*). *Marine Mammal Science* 8(3):306–311.

Plotkin, P., and A. F. Amos. 1990. Effects of anthropogenic debris on sea turtles. In Proceedings of the Second International Conference on Marine Debris, 1989, ed. R. S. Shomura and M. L. Godfrey. U.S. Department of Commerce. NOAA-TM-NMFS-SWFSC-154:736–743.

Podolsky, R. H., and S. W. Kress. 1989. Plastic debris incorporated into double-crested cormorant nests in the Gulf of Maine. *Journal of Field Ornithology* 60:248–250.

Portney, P. R. 1994. The contingent valuation debate: Why economists should care. *Journal of Economic Perspectives* 8:3–18.

Pruter, A. T. 1987a. Sources, quantities and distribution of persistent plastics in the marine environment. *Marine Pollution Bulletin* 18(6B):305–310.

Pruter, A. T. 1987b. Plastics in the marine environment. *Fisheries* 12:16–17.

Pyle, D. J. 1983. *The Economics of Crime and Law Enforcement.* New York: St. Martin's Press.

Rabalais, S. C., and N. N. Rabalais. 1980. The occurrence of sea turtles on the South Texas Coast. *Contributions in Marine Science* 23:123–129.

Ramirez, G. D. 1984. Capture of southern sea lions (*Otaria flavescens*) in the wild by drug immobilization. Abstract from the First Conference on the Work of Experts on Marine Mammals of South America, 1984. Buenos Aires, Argentina.

Ramirez, G. D. 1986. Rescue of entangled South American sea lions (*Otaria flavescens*). A Report for the Center for Environmental Education. Washington, D.C. 16.

Read, G. B., and D. P. Gordon. 1991. Adventive occurrence of the fouling serpulid *Ficopomatus enigmaticus* (Polychaeta) in New Zealand. *New Zealand Journal of Marine and Freshwater Research* 25:269–273.

Redford, D. P., W. R. Trulli, and H. K. Trulli. 1992. Composition of floating debris in harbours of the United States. *Chemistry and Ecology* 7:75–92.

Reed, D. K., and J. D. Schumacher. 1985. On the general circulation in the subarctic Pacific. In Proceedings of the Workshop on the Fate and Impact of Marine Debris, 1984, ed. R. S. Schomura and H. O. Yoshida. U.S. Department of Commerce. NOAA-TM-NMFS-SWFSC-54:483–496.

Reed. S. 1981. *Wreck of Kerguelen* 28:239–240.

Reid, K. 1993. Beach debris survey—Main Bay, Bird Island, South Georgia, 1991–1992. Hobart, TAS Australia. Commission for the Conservation of Antarctic Marine Living Resources. CCAMLR-XII/BG/4:6.

Ribic, C. A. 1991. Design of shoreline surveys for aquatic litter pollution. Washington, D.C. U.S. Environmental Protection Agency, Office of Marine and Estuarine Protection. Final report NTIS 911790511/AS:18.

Ribic, C. A., T. R. Dixon, and I. Vining. 1992. Marine debris survey manual. U.S. Department of Commerce. NOAA Technical Report NMFS 108. Washington, D.C.: GPO.

Rice, D. W. 1977. A list of the marine mammals of the world. NOAA Technical Report NMFS SSRF-711. Washington, D.C.: GPO.

Robards, M. D. 1993. Plastic ingestion by North Pacific seabirds. U.S. Department of Commerce. NOAA-43ABNF203014.

Robards, M. D., J. F. Piatt, and K. D. Wohl. 1995. Increasing frequency of plastic particle ingestion by seabirds in the subarctic North Pacific. *Marine Pollution Bulletin* 30:151–157.

Rodwell, S. 1990. Beach debris survey—Main Bay, Bird Island, South Georgia, 1989–1990. Hobart, TAS Australia. Commission for the Conservation of Antarctic Marine Living Resources. CCAMLR-IX/BG/4:19.

Rundgren, D. C. 1992. Aspects of pollution of False Bay, South Africa. Unpublished Master's thesis, University of Cape Town.

Russell, M. C. 1896. Current papers, no. 2. *Journal of the Royal Society of New South Wales* 30:202–210.

Ryan, P. G. 1985. Plastic pollution at sea and in seabirds off Southern Africa. In Proceedings of the Workshop on the Fate and Impact of Marine Debris, 1984, ed. R. S. Shomura and H. O. Yoshida. U.S. Department of Commerce. NOAA-TM-NMFS-SWFSC-54:523.

Ryan, P. G. 1986. The incidence and effects of ingested plastic on seabirds. Master's thesis, University of Cape Town.

Ryan, P. G. 1987a. The origin and fate of artefacts stranded on islands in the African sector of the Southern Ocean. *Environmental Conservation* 14(4):341–346.

Ryan, P. G. 1987b. The incidence and characteristics of plastic particles ingested by seabirds. *Marine Environmental Research* 23:175–206.

Ryan, P. G. 1987c. The effects of ingested plastic on seabirds: Correlations between plastic load and body condition. *Environmental Pollution* 46:119–125.

Ryan, P. G. 1988a. The characteristics and distribution of plastic particles at the sea-surface off the Southwestern Cape Province, South Africa. *Marine Environmental Research* 25:249–273.

Ryan, P. G. 1988b. Effects of plastic ingestion on seabird feeding: Evidence from chickens. *Marine Pollution Bulletin* 19:125–128.

Ryan, P. G. 1990a. The marine plastic debris problem off Southern Africa: Types of debris, their environmental effects, and control measures. In Proceedings of the Second International Conference on Marine Debris, 1989, ed. R. S. Shomura and M. L. Godfrey. U.S. Department of Commerce. NOAA-TM-NMFS-SWFSC-154.

Ryan, P. G. 1990b. The effects of ingested plastic and other marine debris on seabirds. In Proceedings of the Second International Conference on Marine Debris, 1989, ed. R. S. Shomura and M. L. Godfrey. U.S. Department of Commerce. NOAA-TM-NMFS-SWFSC-154.

Ryan, P. G. 1990c. The intraspecific variation in plastic ingestion by seabirds and the flux of plastic through seabird populations. *Condor* 90:446–452.

Ryan, P. G. 1991. The impact of the commercial lobster fishery on seabirds at the Tristan da Cunha Islands, South Atlantic. Ocean. *Biological Conservation* 57:339–350.

Ryan, P. G. 1995. Updating marine debris research in South Africa. *Marine Debris Worldwide* 1:6–7.

Ryan, P. G., and M. W. Fraser. 1988. The use of great skua pellets as indicators of plastic pollution in seabirds. *Emu* 88:16–19.

Ryan, P. G., and C. L. Maloney. 1990. Plastic and other artefacts on South African beaches: Temporal trends in abundance and composition. *South African Journal of Science* 86:450–452.

Ryan, P. G., and C. L. Moloney. 1993. Marine litter keeps increasing. *Nature* 361:23.

Ryan, P. G., and B. P. Watkins. 1988. Accumulation of stranded plastic objects and other artefacts at Inaccessible Island, central South Atlantic Ocean. *South African Journal of Antarctic Research* 18:11–13.

Ryan, P. G., A. D. Connell, and B. D. Gardner. 1988. Plastic ingestion and PCBs in seabirds: there is a relationship? *Marine Pollution Bulletin* 19:174–176.

Sadove, S. S., and S. J. Morreale. 1990. Marine mammal and sea turtle encounters with marine debris in the New York Bight and the Northeast Atlantic. In Proceedings of the Second International Conference on Marine Debris, 1989, ed. R. S. Shomura and M. L. Godfrey. U.S. Department of Commerce. NOAA-TM-NMFS-SWFSC-154:562–570.

Salvador, A. 1978. Materiales para una Herpetofauna Balearica. 5. Las salamanquesas y tortugas del archipelago de Caberera. Donana. *Acta Vertebratica* 5:5–17.

Saydam, C., I. Salihoglu, M. Sakarya, and A. Yilmaz. 1985. Dissolved and dispersed petroleum hydrocarbon, suspended sediment, plastic, pelagic tar, and other litter in the Northeastern Mediterranean. In CIESM/PNUE COIU VII journees d'etudes sur les pollutions marine en Mediterranee. Workshop on Pollution of the Mediterranean, 1984. CIESM, Monaco, 509–518.

Scarsbrooke, J. R., G. A. McFarlane, and W. Shaw. 1988. Effectiveness of experimental escape mechanisms in sablefish traps. *North American Journal of Fisheries Management* 8:158–161.

SCCAMLR (Scientific Committee for the Conservation of Antarctic Marine Living Resources). 1992. Report of the Eleventh Meeting of the Scientific Committee, 1992. Hobart, TAS Australia.

Schelling, T. 1984. Self-command in practice, in policy, and in a theory of rational choice. *American Economic Review* 74(2):1–11.

Scheu, M. 1992a. Solid waste management in St. Lucia. St. Lucia, West Indies. Caribbean Environmental Health Institute, Institutional Strengthening Project.

Scheu, M. 1992b. Solid waste disposal in Dominica. St. Lucia, West Indies. Caribbean Environmental Health Institute, Institutional Strengthening Project.

Scheu, M. 1993. Municipal solid waste disposal in British Virgin Islands. Caribbean Environmental Health Improvement Project.

Schoelkopf, R. 1981. Marine turtles. *Scientific Event Alert Network (SEAN) Bulletin* 6(10):17.

Schoener, A., and G. T. Rowe. 1970. Pelagic *Sargassum* and its presence among the deep sea benthos. *Deep-Sea Research* 17:923–925.

Schrey, E., and G. J. M. Vauk. 1987. Records of entangled gannets (*Sula bassana*) at Helgoland, German Bight. *Marine Pollution Bulletin* 18(6B):350–352.

Schwartz, M. L., A. A. Hohn, H. J. Bernard, S. J. Chivers, and K. M. Peltier. 1992. Stomach contents of beach-cast cetaceans collected along the San Diego County Coast of California, 1972–1991. U.S. Department of Commerce, Southwest Fisheries Science Center. Administrative Report LJ-92-18.

Scordino, J. 1985. Studies of fur seal entanglement 1981–1984, St. Paul Island, Alaska. In Proceedings of the Workshop on the Fate and Impact of Marine Debris, 1984, ed. R. S. Shomura and H. O. Yoshida. U.S. Department of Commerce. NOAA-TM-NMFS-SWFSC-54:278–290.

Scordino, J., H. Kajimura, N. Baba, and A. Furuta. 1988. Fur seal entanglement studies in 1984, St. Paul Island, Alaska. In Fur Seal Investigations 1985, ed. P. Kozloff and H. Kajimura. NOAA-TM-NMFS-F/NWC-146:70–91.

Scott, G. 1975. The growth of plastics packaging litter. *International Journal of Environmental Studies* 7:131–132.

Scott, G. 1990. The philosophy and practice of degradable plastics. In Proceedings of the Second International Conference on Marine Debris, 1989, ed. R. S. Shomura and M. L. Godfrey. U.S. Department of Commerce. NOAA-TM-NMFS-SWFSC-154:827–847.

Seber, G. A. F. 1982. *The Estimation of Animal Abundance and Related Parameters.* 2d ed. London: Griffin.

Shaughnessy, P. D. 1980. Entanglement of cape fur seals with man-made objects. *Marine Pollution Bulletin* 11(11):332–336.

Shaw, D. G. 1977. Pelagic tar and plastic in the Gulf of Alaska and Bering Sea: 1975. *Science of the Total Environment* 8:13–20.

Shaw, D. G., and R. H. Day. 1994. Colour- an form-dependent loss of plastic and micro-debris from the North Pacific Ocean. *Marine Pollution Bulletin* 28:39–43.

Shaw, D. G., and G. A. Mapes. 1979. Surface circulation and the distribution of pelagic tar and plastic. *Marine Pollution Bulletin* 10:160–162.

Shears, J., R. I. Lewis-Smith, and M. Chalmers. 1992. Beach litter survey Signy Island, South Orkneys, 1991–1992. Hobart, TAS Australia. Commission for the Conservation of Antarctic Marine Living Resources. CCAMLR-XI/BG/14:10.

Sheavly, S. B. 1995. *1994 International Coastal Cleanup Results,* ed. R. Bierce and K. O'Hara. Washington, D.C.: Center for Marine Conservation.

Sheldon, W. W., and R. L. Dow. 1975. Trap contribution to losses in the American lobster fishery. *National Marine Fisheries Service Fisheries Bulletin* 73:449–451.

Shiber, J. G. 1979. Plastic pellets on the coast of Lebanon. *Marine Pollution Bulletin* 10:28–30.

Shiber, J. G. 1982. Plastic pellets on Spain's Costa del Sol beaches. *Marine Pollution Bulletin* 13:409–412.

Shiber, J. G. 1987. Plastic pellets and tar on Spain's Mediteranean beaches. *Marine Pollution Bulletin* 18:84–86.

Shiber, J. G., and J. M. Barrales-Rienda. 1991. Plastic pellets, tar, and megalitter on Beirut beaches, 1977–1988. *Environmental Pollution* 71:17–30.

Shomura, R. S., and M. L. Godfrey, eds. 1990. Proceedings of the Second International Conference on Marine Debris, 1989. U.S. Department of Commerce. NOAA-TM-NMFS-SWFSC-154.

Shomura, R. S., and H. O. Yoshida, eds. 1985. Proceedings of the Workshop on the Fate and Impact of Marine Debris, 1984. U.S. Department of Commerce. NOAA-TM-NMFS-SWFSC-54.

Shoop, C. R., and C. A. Ruckdeschel. 1989. Analysis of sea turtles gut contents for nonfood components. Final report to the U.S. Department of Commerce. Solicitation no. 52-EANF-7-00067.

Sibley, T. H., and R. M. Strickland. 1989. Potential effects of marine debris on benthic communities. Seattle. University of Washington. School of Fisheries Report FRI-UW-8915:32.

Sileo, L., P. R. Sievert, and M. D. Samuel. 1990a. Causes of mortality of albatross chicks at Midway Atoll. *Journal of Wildlife Diseases* 26(3):329–338.

Sileo, L., P. R. Sievert, M. D. Samuel, and S. I. Fefer. 1990b. Prevalence and characteristics of plastic ingested by Hawaiian seabirds. In Proceedings of the Second International Conference on Marine Debris, 1989, ed. R. S. Shomura and M. L. Godfrey. U.S. Department of Commerce. NOAA-TM-NMFS-SWFSC-154:665–681.

Siung-Chang, A., and A. Deane. 1984. A survey of beach litter in the northwest peninsula of Trinidad, W.I. Association of Island Marine Laboratories of the Caribbean (18th annual meeting, August), University of the West Indies, St. Augustine, Trinidad.

Slater, J. 1990. Can you teach an old sea-dog new tricks? A case study in educating industry. In Our Common Future: Pathways for Environmental Education, ed. S. Blight, R. Sautter, J. Sibly, and R. Smith, 235–242. Proceedings of the Australian Association for Environmental Education International Conference, 1990. University of Adelaide, South Australia.

Slater, J. 1991. Flotsam and jetsam. In Marine Debris Bulletin 1. Beach Survey Results, January 1990 to June 1991. Hobart, TAS Australia. Department of Parks, Wildlife and Heritage.

Slater, J. 1992. The incidence of marine debris in the southwest of the World Heritage area. *The Tasmanian Naturalist* 111:32–35.

Slater, J. 1994. Plastic ingestion by seabirds in Tasmania, Australia. In Poster Abstracts and Manuscripts from the Third International Conference on Marine Debris, 1994, ed. J. C. Clary. U.S. Department of Commerce. NOAA-TM-NMFS-AFSC-51.

Slip, D. J. 1990. Fishing debris brings a new problem to the Southern Ocean. Australian National Antarctic Research Expedition. ANARE News (June). 9.

Slip, D. J., and H. R. Burton. 1990. The composition and origin of marine debris stranded on the shores of subantarctic Macquarie Island. In Proceedings of the Second International Conference on Marine Debris, 1989, ed. R. S. Shomura and M. L. Godfrey. U.S. Department of Commerce. NOAA-TM-NMFS-SWFSC-154:403–416.

Slip, D. J., and H. R. Burton. 1991. Accumulation of fishing debris, plastic litter, and other artefacts on Heard and Macquarie Islands in the Southern Ocean. *Environmental Conservation* 18:249–254.

Smith, A. 1759. *The Theory of Moral Sentiments.* London: A Millar.

Smith, A. M., and C. S. Nelson. 1994. Calcification rates of rapidly colonizing bryozoans in Hauraki Gulf, Northern New Zealand. *New Zealand Journal of Marine and Freshwater Research*

28:227–234.

Smith, C. R. 1992. Whale falls: Chemosynthesis on the deep seafloor. *Oceanus* 35:74–78.

Smith, J. M. B. 1990. Drift disseminules on Fijian beaches. *New Zealand Journal of Botany* 28:13–20.

Smith, J. M. B. 1992. Patterns of disseminule dispersal by drift in the Southern Coral Sea. *New Zealand Journal of Botany* 30:57–67.

Smith, J. M. B., P. Rudall, and P. L. Keage. 1989. Driftwood on Heart Island. *Polar Record* 25:223–228.

Smith, N. P. 1981. An investigation of seasonal upwelling along the Atlantic Coast of Florida. In *Ecohydrodynamics,* ed. J. C. J. Nihoul, 79–98. Amsterdam: Elsevier.

Smith, P., and J. Tooker. 1990. Marine debris on New Zealand coastal beaches. Auckland. Greenpeace.

Smith, V. K. 1993. NonMarket valuation of environmental resources: An interpretive appraisal. *Land Economics* 69:1–26.

Smolowitz, R. J. 1978. Trap design and ghost fishing: An overview. *Marine Fisheries Review* 40(5-6).

Snedecor, G. W., and W. G. Cochran. 1967. *Statistical Methods,* 6th ed. Ames, IA: Iowa State University Press.

Society of the Plastics Industry, Inc. 1986. *Facts and Figures of the U.S. Plastics Industry.* New York.

Society of the Plastics Industry, Inc. 1991. Operation clean sweep: A manual on preventing pellet loss. Washington, DC.

Southward, A. J. 1986. Class Cirripedia (barnacles). In *Marine Fauna and Flora of Bermuda,* ed. W. Sterrer. New York: John Wiley & Sons. 299–305.

Spear, L., D. G. Ainley, and C. A. Ribic. 1995. Incidence of plastic in seabirds from the tropical Pacific, 1984–91: Relation with distribution of species, sex, age, season, year, and body weight. *Marine Environmental Research* 40:123–146.

Stevens, L. M. 1992. Marine plastic debris: Fouling and degradation. Master's thesis, University of Auckland.

Stewart, B. S., and P. K. Yochem. 1987. Entanglement of pinnipeds in synthetic debris and fishing net and line fragments at San Nicholas and San Miguel Islands, California, 1978–1986. *Marine Pollution Bulletin* 18(6B):336–339.

Stewart, B. S., and P. K. Yochem. 1990. Pinniped entanglement in synthetic materials in the Southern California Bight. In Proceedings of the Second International Conference on Marine Debris, 1989, ed. R. S. Shomura and M. L. Godfrey.

U.S. Department of Commerce. NOAA-TM-NMFS-SWFSC-154:554–561.

Stewart, B. S., J. Bengston, and N. Baba. 1989. Northern fur seals tagged and observed during entanglement studies, St. Paul Island, Alaska. In Fur Seal Investigations 1986, ed. H. Kajimura. NOAA-TM-NMFS F/NWC-174:61–62.

Summers, R., and A. Heston. 1991. The Penn World Table. *Quarterly Journal of Economics* 327–368.

Sutinen, J. G. 1988. Remarks to the oceans of plastic workshop: Economic evaluation of incentives. In Oceans of Plastic. Alaska Sea Grant Report no. 88-7.

Sutinen, J. G., A. Rieser, and J. Gauvin. 1990. Measuring and explaining noncompliance in federally mananaged fisheries. *Ocean Development and International Law* 21:335–372.

Swanson, R. L., and R. L. Zimmer. 1990. Meteorological conditions leading to the 1987 and 1988 washups of floatable wastes on New York and New Jersey beaches and comparison of these conditions with the historical record. *Estuarine, Coastal and Shelf Science* 30:59–78.

Swanson, R. L., T. M. Bell, J. Kahn, and J. Olha. 1991. Use impairments and ecosystem impacts of the New York Bight. *Chemistry and Ecology* 5:99–127.

Swanson, R. L., H. M. Stanford, and J. S. O'Connor. 1978. June 1976 pollution of Long Island ocean beaches. *Journal of Environmental Engineering Division ASCE* 104(EE6):1067–1085.

Swanson, R. L., V. Breslin, S. Reaven, S. Ross, R. Young, and R. Becker. 1993. An assessment of impacts associated with implementation of the Suffolk County plastics law. Stony Brook, NY. Waste Management Institute, Marine Science Research Center, The University at Stony Brook.

Swartzman, G. L., C. A. Ribic, and C. P. Huang. 1990. Simulating the role of entanglement in northern fur seal, *Callorhinus ursinus,* population dynamics. In Proceedings of the Second International Conference on Marine Debris, 1989, ed. R. S. Shomura and M. L. Godfrey. U.S. Department of Commerce. NOAA-TM-NMFS-SWFSC-154:513–530.

Takatani, S., T. Sagi, and M. Imai. 1986. Distributions of floating tars and petroleum hydrocarbons at the surface in the Western North Pacific. *Oceans Magazine* 36:33–42.

Tapp, J. L., and L. Kohlberg. 1977. Developing senses of law and legal justice. In *Law, Justice and the Individual in Society: Psychological and Legal Issues,* ed. J. L. Tapp and F. J. Levine. New York: Holt, Rinehart and Winston.

Tarpley, R. J. 1990. Plastic ingestion in a pygmy

sperm whale, *Kogia breviceps*. In Proceedings of the Second International Conference on Marine Debris, 1989, ed. R. S. Shomura and M. L. Godfrey. U.S. Department of Commerce. NOAA-TM-NMFS-SWFSC-154:745.

Teas, W. G., and W. N. Witzell. 1994. Effects of anthropogenic debris on marine turtles in the Western North Atlantic. In Poster Abstracts and Manuscripts from the Third International Conference on Marine Debris, 1994, ed. J. C. Clary. U.S. Department of Commerce. NOAA-TM-NMFS-ASFC-51.

Thaler, R. 1991. *Quasi-Rational Economics.* New York: Russell Sage Foundation. 367.

Torres, D., and A. Aguayo. 1993. Impacto antrópico en Cabo Shirreff, Isla Livingston, Antártica. *Cient INACH* 43:93–108.

Torres, D., and M. Gajardo. 1985. Informacion preliminar sabre desechos plásticos hallados en Cabo Shirreff, Isla Livingston Shetland der Sur. *Boletin Anatartico Chileno* 5:12–13.

Torres, H., and T. Rodríguez. 1987. National water summary 1987—water supply and use: State summaries. U.S. Geological Survey. Professional Paper 2350.

Transport Canada. 1990. Directory of reception facilities for marine waste in Canada. Canadian Coast Guard.

Trulli, W. R., H. K. Trulli, and D. P. Redford. 1990. Characterization of marine debris in selected harbors of the United States. In Proceedings of the Second International Conference on Marine Debris, 1989, ed. R. S. Shomura and M. L. Godfrey. U.S. Department of Commerce. NOAA-TM-NMFS-SWFSC-154:309–324.

Turk, S. M. 1982. Influx of warm-water drift animals into Bristol and English Channels, summer 1981. *Journal of the Marine Biological Association of the United Kingdom* 62:487–489.

Tyler, T. 1990. *Why People Obey the Law.* New Haven: Yale University Press; London: Yale University Press. 173.

Uchida, I. 1990. On the synthetic materials found in the digestive systems of, and discharged by, sea turtles collected in waters adjacent to Japan. In Proceedings of the Second International Conference on Marine Debris, 1989, ed. R. S. Shomura and M. L. Godfrey. U.S. Department of Commerce. NOAA-TM-NMFS-SWFSC-154:744.

Uchida, R. N. 1985. The types and amounts of fish net deployed in the North Pacific. In Proceedings of the Workshop on the Fate and Impact of Marine Debris, 1984, ed. R. S. Shomura and H. O. Yoshida. U.S. Department of Commerce. NOAA-TM-NMFS-SWFSC-54:37–108.

UNEP (United Nations Environment Programme). 1984. The state of marine pollution in the Wider Caribbean Region. UNEP Regional Seas Reports and Studies 36:45.

UNEP (United Nations Environment Programme). 1992. Environmental quality criteria for coastal zones in the Wider Caribbean Region. Caribbean Environment Programme Technical Report 14.

UNEP (United Nations Environment Programme). 1994. Regional overview of land-based sources of pollution in the Wider Caribbean Region (revised). Cartagena Workshop Group 14/4.

UNEP/IOC/FAO (United Nations Environment Programme, Intergovernmental Oceanographic Commission, and the Food and Agriculture Organization of the United Nations). 1991. Assessment of the state of pollution of the Mediterranean Sea by persistent synthetic material which may float, sink or remain in suspension. Mediterranean Action Plan Technical Report 56:1–103.

United States of America. 1991. Report on assessment and avoidance of incidental mortality in the Convention area 1990/91. Hobart, TAS Australia. Report of the U.S. Delegation to the Commission for the Conservation of Antarctic Marine Living Resources. CCAMLR-X/BG/7:5.

U.S. Fish and Wildlife Service. 1980. California brown pelican recovery plan. Portland, OR. U.S. Fish and Wildlife Service Regional Office.

van Dolah, R. F., V. G. Burrell Jr., and S. B. West. 1980. The distribution of pelagic tar and plastics in the South Atlantic Bight. *Marine Pollution Bulletin* 11:352–356.

Van Franeker, J. A. 1983. Plastics—en bedreiging voor zeevogels. *Nieuwsbrief NSO* 4:41–61.

Van Franeker, J. A. 1985. Plastic ingestion in the North Atlantic fulmar. *Marine Pollution Bulletin* 16(9):367–369.

Van Franeker, J. A., and P. J. Bell. 1988. Plastic ingestion by petrels breeding in Antarctica. *Marine Pollution Bulletin* 19(12):672–674.

Van Nierop, M. M., and J. C. den Hartog. 1984. A study on the gut contents of five juvenile loggerhead turtles, *Caretta caretta* (Linnaeus) (Reptilia, Cheloniidae), from the southeastern part of the North Atlantic Ocean, with emphasis on coelenterata identification. *Zoologische Mededelingen (Leiden)* 59(4):35–54.

Vauk, G. J. M., and E. Schrey. 1987. Litter pollution from ships in the German Bight. *Marine Pollution Bulletin* 18(6B):316–319.

Vauk-Hentzelt, K. 1982. Misbildungen, verletzungen, und krankheiten auf Helgoland erlegter silbermowen (Deformities, injuries, and illnesses of silver gulls that died in Helgoland).

Niedersachsischer Jager 15:700–702.

Vogel, J. 1974. Taxation and public opinion in Sweden: An interpretation of recent survey data. *National Tax Journal* (December):499–513.

Wace, N. M. 1990. Flotsam and jetsam in the Southern Ocean: Monitoring ocean litter on Australia coasts. *Centre for Resource and Environmental Studies Bulletin* 13(2):1–2.

Wace, N. M. 1991. Garbage in the oceans. *Bogong* 12(1):15–18.

Wace, N. M. 1994. Beachcombing for ocean litter. *Australian Natural History* 24:46–52.

Wade, B. A., B. Morrison, and M. A. J. Jones. 1991. A study of beach litter in Jamaica. *Caribbean Journal of Science* 27:190–197.

Walker, W. A., and J. M. Coe. 1990. Survey of marine debris ingestion by odontocete cetaceans. In Proceedings of the Second International Conference on Marine Debris, 1989, ed. R. S. Shomura and M. L. Godfrey. U.S. Department of Commerce. NOAA-TM-NMFS-SWFSC-154:747–774.

Wallace, B. 1990. Shipping industry marine education plan. In Proceedings of the Second International Conference on Marine Debris, 1989, ed. R. S. Shomura and M. L. Godfrey. U.S. Department of Commerce. NOAA-TM-NMFS-SWFSC-154.

Wallace, N. 1985. Debris entanglement in the marine environment: A review. In Proceedings of the Workshop on the Fate and Impact of Marine Debris, 1984, ed. R. S. Shomura and H. O. Yoshida. U.S. Department of Commerce. NOAA-TM-NMFS-SWFSC-54:259–277.

Walters, L. J. 1992. Post-settlement success of the arborescent bryozoan *Bugula meritina* (L.): The importance of structural complexity. *Journal of Experimental Marine Biology and Ecology* 164:55–71.

Ward, N. 1994. Hitchhikers on the larval highway. *Ocean Explorer* 3:6–7.

West, C. J. 1981. The significance of small plastic boats as seed dispersal agents. *Tane* 27:175.

Westat Inc. 1980. Individual Income Tax Compliance Factors Study, Qualitative Research. U.S. Internal Revenue Service.

Wheeler, W. M. 1916. Ants carried in a floating log from the Brazilian mainland to San Sebastian Island. *Psyche* (Cambridge) 23:180–183.

White, H. 1982. Maximum likelihood estimation of misspecified models. *Econometrica* 50:1–76.

Wilber, R. J. 1987. Plastic in the North Atlantic. *Oceanus* 30(3):61–68.

Wiley, D. N., R. A. Asmutis, T. D. Pitchford, and D. P. Gannon. 1995. Stranding and mortality of humpback whales, *Megaptera novaeangliae,* in the mid-Atlantic and Southeast United States. *Fisheries Bulletin* 93(1):196–205.

Williams, A. T., S. L. Simmons, and A. Fricker. 1993. Off-shore sinks of marine litter: A new problem. *Marine Pollution Bulletin* 26:404–405.

Williams, R. A. 1991. Report on First Caribbean Solid Waste Management Meeting. Washington, D.C. American Health Organization.

Windom, H. L. 1992. Contamination of the marine environment from land-based sources. *Marine Pollution Bulletin* 25:32–36.

Winkel, I. R. 1982. Waste disposal problems in the Netherland Antilles. In Solid Waste Disposal and Utilization in Developing Countries, ed. R. M. Schelhaas. Amsterdam. Department of Agricultural Research.

Winston, J. E. 1982a. Marine bryozoans (Ectoprocta) of the Indian River area (Florida). *Bulletin of the American Museum of Natural History* 173:99–176.

Winston, J. E. 1982b. Drift plastic: An expanding niche for a marine invertebrate? *Marine Pollution Bulletin* 73:348–357.

Witte, A., and D. Woodbury. 1985. The effect of tax laws and tax administration on tax compliance: The case of the U.S. individual income tax. *National Tax Journal* (March):1–13.

Wolf, R. 1982. Marine turtles. *Scientific Event Alert Network (SEAN) Bulletin* 7(5):15.

Wong, C. S., D. R. Green, and W. J. Cretney. 1974. Quantitative tar and plastic waste distribution in the Pacific Ocean. *Nature* 247:30–32.

Wong, C. S., D. R. Green, and W. J. Cretney. 1976. Distribution and source of tar on the Pacific Ocean. *Marine Pollution Bulletin* 7:102–106.

Woods Hole Oceanographic Institution. 1952. Marine fouling and its prevention. Annapolis, MD. United States Naval Institute. 388.

World Bank. 1991. IMO Port Reception and Disposal Facilities for Garbage in the Wider Caribbean. Washington, D.C. (October).

Worth, D. F., and M. L. Hollinger. 1977. Nearshore marine ecology at Hutchinson Island, Florida, 1971–1974. Florida Department of Natural Resources. Florida Marine Research Publication no. 23:25–85.

WRI (World Resources Institute). 1992. *World Resources (1992–1993).* New York: University Press.

Wrong, D. 1980. *Power: Its Forms, Bases and Uses.* New York: Harper & Row.

Yohisda, K., and N. Baba. 1985. The problem of fur seal entanglement in marine debris. In Proceedings of the Workshop on the Fate and Impact of Marine Debris, 1984, ed. R. S. Shomura and H. O. Yoshida. U.S. Department of Commerce. NOAA-TM-NMFS-SWFSC-54:448–452.

Yoshida, K., N. Baba, M. Nakajima, Y. Fujimaki, S. Furuta, S. Nomura, and K. Takahashi. 1985. Fur seal entanglement survey report test study at a breeding facility, 1983. Translated by W. C. Atlemson. Provided to the North Pacific Fur Seal Commission. Fur Seal Resource Section Far Seas Fisheries Research Laboratory-Izu-Mito Sea Paradise. 20.

Young, C. 1993. Plastic peril. *Western Fisheries* (spring):45–49.

Younger, L. K., and K. L. Hodge. 1992. *1991 International Coastal Cleanup Results,* ed. R. Bierce and K. O'Hara. Washington, D.C.: Center for Marine Conservation.

Zhang, X., V. K. Smith, and R. B. Palmquist. 1992a. Marine debris focus group I: Summary report I. Raleigh, NC. Resource and Environmental Economics Program, North Carolina State University. Unpublished paper.

Zhang, X., V. K. Smith, and R. B. Palmquist. 1992b. Marine debris focus group II: Summary report I. Raleigh, NC. Resource and Environmental Economics Program, North Carolina State University. Unpublished paper.

Zhang, X., V. K. Smith, and R. B. Palmquist. 1993. Marine debris focus group III: Summary report I. Raleigh, NC. Resource and Environmental Economics Program, North Carolina State University. Unpublished paper.

Index